T0224884

Algebra

From *rings* to *modules* to *groups* to *fields*, this undergraduate introduction to abstract algebra follows an unconventional path. The text emphasizes a modern perspective on the subject, with gentle mentions of the unifying categorical principles underlying the various constructions and the role of universal properties. A key feature is the treatment of *modules*, including a proof of the classification theorem for finitely generated modules over Euclidean domains. Noetherian modules and some of the language of exact complexes are introduced. In addition, standard topics—such as the Chinese Remainder Theorem, the Gauss Lemma, the Sylow Theorems, simplicity of alternating groups, standard results on field extensions, and the Fundamental Theorem of Galois Theory—are all treated in detail. Students will appreciate the text's conversational style, 400+ exercises, appendix with complete solutions to around 150 of the main text problems, and appendix with general background on basic logic and naïve set theory.

Paolo Aluffi is Professor of Mathematics at Florida State University. Aluffi earned a Ph.D. from Brown University with a dissertation on the enumerative geometry of plane cubic curves, under the supervision of William Fulton. His research interests are in algebraic geometry, particularly intersection theory and its application to the theory of singularities and connections with theoretical physics. He has authored about 70 research publications and given lectures on his work in 15 countries. Beside *Notes from the Underground*, he has published a graduate-level textbook in algebra (*Algebra: Chapter 0*, AMS) and a mathematics book for the general public, in Italian (*Fare matematica*, Aracne Editrice).

CAMBRIDGE MATHEMATICAL TEXTBOOKS

Cambridge Mathematical Textbooks is a program of undergraduate and beginning graduate-level textbooks for core courses, new courses, and interdisciplinary courses in pure and applied mathematics. These texts provide motivation with plenty of exercises of varying difficulty, interesting examples, modern applications, and unique approaches to the material.

A complete list of books in the series can be found at www.cambridge.org/mathematics
Recent titles include the following:

Algebra

PAOLO ALUFFI

Florida State University

CAMBRIDGE
UNIVERSITY PRESS

University Printing House, Cambridge CB2 8BS, United Kingdom

One Liberty Plaza, 20th Floor, New York, NY 10006, USA

477 Williamstown Road, Port Melbourne, VIC 3207, Australia

314–321, 3rd Floor, Plot 3, Splendor Forum, Jasola District Centre, New Delhi – 110025, India

79 Anson Road, #06–04/06, Singapore 079906

Cambridge University Press is part of the University of Cambridge.

It furthers the University's mission by disseminating knowledge in the pursuit of education, learning, and research at the highest international levels of excellence.

www.cambridge.org
Information on this title: www.cambridge.org/9781108958233
DOI: 10.1017/9781108955911

First published 2021

A catalogue record for this publication is available from the British Library.

ISBN 978-1-108-95823-3 Paperback

Contents

Introduction

This is an introductory textbook on abstract algebra. A glance at the table of contents will immediately reveal its basic organization and salient features: it belongs to the 'rings-first' camp, and places an unusual emphasis on *modules,* which (to my puzzlement) are often essentially omitted in textbooks at this level. A few section titles will also reveal passing nods to the notion of category, which is not developed or used in the main body of the text, but is also not intentionally hidden from sight.

Why 'rings first'? Why not 'groups'? This textbook is meant as a first approach to the subject of algebra, for an audience whose background does not include previous exposure to the subject, or even very extensive exposure to abstract mathematics. It is my belief that such an audience will find *rings* an easier concept to absorb than *groups.* The main reason is that rings are defined by a rich pool of axioms with which readers are already essentially familiar from elementary algebra; the axioms defining a group are fewer, and they require a higher level of abstraction to be appreciated. While $\mathbb{Z}$ is a fine example of a group, in order to view it as a group rather than as a ring, the reader needs to forget the existence of one operation. This is in itself an exercise in abstraction, and it seems best to *not* subject a naïve audience to it. I believe that the natural port of entry into algebra is the reader's familiarity with $\mathbb{Z}$, and this familiarity leads naturally to the notion of *ring.* Natural examples leading to group theory could be the symmetric or the dihedral groups; but these are not nearly as familiar (if at all) to a naïve audience, so again it seems best to wait until the audience has bought into the whole concept of 'abstract mathematics' before presenting them.

Thus, my choice to wait until later chapters before introducing groups is essentially dictated by the wish to provide a *gentle* introduction to the subject, where the transition to the needed level of abstraction can happen gradually. The treatment in the first several sections of this book is intentionally very elementary, with detailed explanations of comparatively simple material. A reader with no previous exposure to this material should be able to read the first two chapters without excessive difficulty, and in the process acquire the familiarity needed to approach later chapters. (Rings are introduced in Chapter 3.) The writing becomes more demanding as the material is developed, as is necessary—after all, the intended readership is expected to reach, by the end of the book, a good comfort level with rather sophisticated material, such as the basics of Galois theory. It is my hope that the reader will emerge from the book well equipped to approach graduate-level algebra. In fact, this is what many students in my undergraduate algebra courses have done.

The book could be used for a group-first approach, provided that the audience is 'mature' enough to cope right away with the higher level of abstraction of groups, and that the instructor is willing to occasionally interpolate references to previous chapters. In fact, the reliance of the chapters on groups in Part III on material developed in previous chapters is minimal, and mostly contextual (with the exception of a treatment of cyclic groups, which is carried out in Part II). However, the writing in Part III is naturally terser and puts more demands on the reader; so, again, this is not recommended for a naïve audience. Part IV deals with fields, and here the audience is certainly expected to be comfortable handling abstract concepts, perhaps at the level of a Master's student. I would definitely not recommend a 'fields-first' approach in an introduction to algebra.

More than 50 years ago, Atiyah and Macdonald wrote in their *Introduction to Commutative Algebra*: "...following the modern trend, we put more emphasis on modules and localization." Apparently, what was the 'modern trend' in 1970 has not really percolated yet to the standard teaching of an introductory course in algebra in 2020, since it seems that most standard textbooks at this level omit the subject of modules altogether. In this book, *modules* are presented as equal partners with other standard topics—groups, rings, fields. After all, most readers will have had some exposure to linear algebra, and therefore will be familiar to some extent with vector spaces; and they will emerge from studying rings with a good understanding of ideals and quotient rings. These are all good examples of modules, and the opportunity to make the reader familiar with this more encompassing notion should not be missed.

The opportunity to emphasize modules was in itself sufficient motivation to write this book. I believe that readers will be more likely to encounter modules, complexes, exact sequences, introduced here in §9.2, in their future studies than many other seemingly more standard traditional topics covered in introductory algebra books (and also covered in this text, of course).

Having made this choice, it is natural to view abelian groups as particular cases of modules, and the classification theorem for finitely generated abelian groups is given as a particular case of the analogous results for finitely generated modules over Euclidean domains (proved in full). Groups are presented as generalizations of abelian groups; this goes counter to the prevailing habit of presenting abelian groups as particular cases of groups. After treating groups, the text covers the standard basic material on field extensions, including a proof of the Fundamental Theorem of Galois Theory.

On the whole, the progression rings–modules–abelian groups–groups–fields simply seems to me the most natural in a first approach to algebra directed at an audience without previous exposure to the subject. While it is possible to follow a different progression (e.g., start with groups and/or avoid modules) I would have to demand more of my audience in order to carry that out successfully.

I could defend the same conclusion concerning categories. My graduate(-leaning) algebra textbook[1] has somehow acquired the reputation of 'using categories'. It would be more accurate to state that I did not go out of my way to *avoid* using the basic language

---

[1] *Algebra: Chapter 0*. Graduate Studies in Mathematics, No. 104. American Mathematical Society, Providence, RI.

of categories in that textbook. For this textbook, aimed at a more naïve audience, I have resolved to steer away from direct use of the language. I have even refrained from giving the definition of a category! On the other hand, the material itself practically demands a few mentions of the concept—again, if one does not go out of one's way to avoid such mentions. It is possible, but *harder,* to do without them. It seemed particularly inevitable to point out that many of the constructions examined in the book satisfy universal properties, and in the exercises the reader is encouraged to flesh out this observation for some standard examples, such as kernels or (co)products. My hope is that the reader will emerge from reading these notes with a natural predisposition for absorbing the more sophisticated language of categories whenever the opportunity arises.

The main prerequisite for this text is a general level of familiarity with the basic language of naive set theory. For the convenience of the reader, a quick summary of the relevant concepts and notions is included in the Appendix A, with the emphasis on equivalence relations and partitions. I describe in some detail a decomposition of set-functions that provides a template for the various 'isomorphism theorems' encountered in the book. However, this appendix is absolutely not required in order to read the book; it is only provided for the convenience of the reader who may be somewhat rusty on standard set-theoretic operations and logic, and who may benefit from being reminded of the various typical 'methods of proof'. There are no references to the appendix from the main body of the text, other than a suggestion to review (if necessary) the all-important notions of equivalence relations and partitions. As stressed above, the text begins very gently, and I believe that the motivated reader with a minimum of background can understand it without difficulty.

It is also advisable that readers have been exposed to a little linear algebra, particularly to benefit the most from examples involving, e.g., matrices. This is not strictly necessary; linear algebra is not used in any substantial way in the development of the material. The linear algebra needed in Part II is developed from scratch. In fact, one subproduct of covering modules in some detail in this text should be that readers will be better equipped to understand linear algebra more thoroughly in their future encounters with that subject.

Appendix B includes extensive solutions to about one-third of the exercises listed at the end of each chapter; these solved problems are marked with the symbol ▹. These are all (and only) the problems quoted directly from the text. In the text, the reader may be asked to provide details for parts of a proof, or construct an example, or verify a claim made in a remark, and so on. Readers should take the time to perform these activities on their own, and the appendix will give an opportunity to compare their work with my own 'solution'. I believe this can be very useful, particularly to readers who may not have easy access to an alternative source to discuss the material developed in this book. In any case, the appendix provides a sizeable amount of material that complements the main text, and which instructors may choose to cover or not cover, or assign as extra reading (*after* students have attempted to work it out on their own, of course!).

How to use this book? I cover the material in this book in the order it is written, in a two-semester sequence in abstract algebra at the undergraduate level at Florida State University. Instructors who, like me, have the luxury of being able to spend two semesters on this material are advised to do the same. One semester will suffice for the first part, on rings; and some time will be left to begin discussing modules. The second semester will complete modules and cover groups and fields.

Of course, other ways to navigate the content of this book are possible. By design, the material on groups has no hard references back to Chapter 6, 7, or 9. Therefore, if only one semester is available, one could plan on covering the core material on rings (Chapters 1–5), modules (Chapter 8), and groups (Chapter 11). It will be necessary to also borrow some material from Chapter 10, such as the definition and basic features of cyclic groups, as these are referenced within Chapter 11. This roadmap should leave enough time to cover parts of Chapter 12 (more advanced material on groups) or Chapter 13 (basic material on field extensions), at the discretion of the instructor.

The pedagogical advantages of the rings-first approach have been championed in Hungerford's excellent undergraduate-level algebra text, which I have successfully used several times for my courses. Readers who are familiar with Hungerford's book will detect an imprint of it in this book, particularly in the first several sections and in the choice of many exercises.

Thanks are due to the students taking my courses, for feedback as I was finding this particular way of telling this particular story. I also thank Ettore Aldrovandi for spotting a number of large and small errors in an earlier version of these notes, and several anonymous reviewers for very constructive comments. I thank Sarah Lewis for the excellent copyediting work, and Kaitlin Leach and Amy Mower at Cambridge University Press for expertly guiding the book from acquisition through production. Lastly, thanks are due to the University of Toronto and to Caltech, whose hospitality was instrumental in bringing this text into the present form.

Before we Begin

Dear reader: This book introduces you to the subject of *abstract algebra,* and specifically the study of certain structures—rings, modules, groups, and fields—which are the basic pillars of abstract algebra. While learning this material, you are not the end-user of a set of tools (as a calculus student may be); rather, you are taking a guided tour through the factory that produces those tools, and you are expected to learn how the tools themselves are put together. Your attention should be directed towards the deep reasons that make things work the way they do. These reasons are encapsulated in the various statements we will encounter, and these statements will deal with the precise definitions we will introduce. Your focus will be on *understanding* these definitions, these statements, and why the statements are true.

Unavoidably, in the main text I must assume that you have reached a certain level of 'mathematical maturity' and can read and write proofs, follow and construct logical arguments, and are familiar with the basic language of naive set theory. If you do not have this background, or if it dates too far back, you will want to spend some time perusing Appendix A, which attempts to cover these preliminary notions in a concise fashion. In any case, I have written the first several sections of the main text with particular abundance of details and explanations; a minimum level of familiarity with the way the notation is used should suffice in order to understand the material in these early sections, and the practice acquired in the process should be helpful in reading the later chapters, where the writing is necessarily more terse and your task is altogether more demanding.

The book includes a large number of exercises. They were also chosen so as to be more straightforward in earlier sections and gradually increase in complexity as the text progresses. Many exercises (marked with the symbol ▷ in the lists given at the end of each chapter) are quoted from within the text: these may ask you to complete a proof, or to fill in the details of an example, or to verify a claim made in a remark, and so on. While reading a mathematical text, it is very common to have to stop and carry out such verifications; indeed, this is an integral part of the process of understanding mathematics. My hope is that these exercises will help you gain this healthy habit. Appendix B includes the solutions to *all* these exercises. Comparing your work to my solution will hopefully provide you with useful feedback. Of course you shouldn't look up the solution to an exercise until you have given it your best shot; and you are warmly encouraged to try and solve *every* exercise in the book, whether a solution is provided or not.

Three indices are provided: the first two list definitions and theorems, and the third one is a standard subject index. The *definitions* are probably the most important component in a subject like algebra: they trace the path along which the material is developed, and most of the theorems one proves at this level have the purpose of expressing the precise way in which these definitions are related to one another. You should expect to look up the exact meaning of specific terms introduced in this book very often. The corresponding index will help you do this quickly when the need arises.

'Algebra' as we understand it currently, and as I will attempt to present it in these notes, was developed over the course of many decades if not centuries, and reached a very mature form long ago. It deals with certain types of objects (rings, groups, ...), that are introduced by choosing suitable collections of axioms to describe their general features precisely. These axioms are often chosen to model features of 'concrete' examples, such as the conventional algebra of integers; and indeed, this text begins with a discussion of the algebra of ordinary integers from a viewpoint which naturally leads to the definition of the notion of *ring*. The other algebraic structures treated in this text arise just as naturally. These structures have been extremely successful in the development of higher-level concepts and important branches of modern mathematics. Indeed, algebra is the language in which entire modern fields of mathematics are 'spoken': for example, algebraic geometry and algebraic number theory; but also seemingly distant fields like particle physics rely on sophisticated algebraic notions. Making any sense of such fields will require the background you are beginning to acquire by reading this book.

The book consists of four 'parts': on *rings,* on *modules,* on *groups,* and on *fields.* These are listed roughly in order of increasing abstraction. By the end of the text, we will gain an appreciation of *Galois theory,* which will answer very sophisticated questions about polynomials with integer coefficients. Such questions could be posed (and indeed were posed) before any of the more abstract scaffolding was put into place; but answering such questions really only becomes possible by using the abstract tools we will develop.

Before beginning in earnest, I will mention a concept that is even more basic and abstract than the main structures we are going to study. In fact, this concept gives a unifying view of such structures, and you have likely (and unwittingly) gathered an intuitive feeling for it in earlier studies. At some level, you are familiar with *sets:* they are a formalization of the notion of 'collections' of arbitrary things (called 'elements'); and we use *functions* to transfer information from a set to another set. (If you are not familiar with these notions, spend some time with Appendix A in this book!) You are also (likely) familiar with *vector spaces:* vector spaces are certain objects you study in linear algebra, and in that case also there is a way to 'transfer information' from one object to another object. Specifically, *linear maps* act between vector spaces, and you can describe them efficiently by means of matrices. Depending on your mathematical history to date, you may have run across the same template elsewhere. For example, if you have seen a little topology, then you have encountered *topological spaces,* and ways to go from one topological space to another one, that is, *continuous functions.*

After running into several such examples, you get used to the idea that studying a specific subject amounts to understanding the *objects* you are dealing with (for example, sets, vector spaces, topological spaces, and so on), and the type of functions that may be used to move from one object to another object of the same kind. In general, these 'functions' are called *morphisms*. (For example, morphisms are ordinary functions for sets, linear maps for vector spaces, continuous functions for topological spaces, and so on.) A 'category' is an abstract entity that captures this general template: categories consist of *objects*, and of *morphisms* between objects. There are a few axioms spelling out general requirements on objects and morphisms, which we do not need to list now but which would look familiar based on our experience with sets and functions, and with the other examples mentioned above. (For example, every object is required to have an 'identity morphism', just as every set has an 'identity function' and every vector space has an 'identity linear map', corresponding to the identity matrix.)

Large areas of mathematics can be understood, at least as a first approximation, as the study of certain categories. This unifying principle underlies *everything we will cover in these notes*. There is a category of *rings*, a category of *groups*, and several other categories of interest in the field of algebra. These categories have several common features. For example, we will be able to define 'quotients' of rings and 'quotients' of groups. These constructions all have a common thread, and this will become evident when we go through them. The reason is that they are all instances of general constructions that could be defined simultaneously for certain types of categories.

Studying algebra at this introductory level means to a large extent understanding how these constructions are performed in several key instances. We will not really use the language of categories in these notes, but it will secretly underlie most of what we will do. Some latent awareness of this fact may be beneficial to you, as it is to me. In any case, you will learn much more about this point of view when you move on to texts covering this material at a more advanced level.

Exercises

0.1 Read (or at least skim through) the Wikipedia page on 'category theory'.

Part I

Rings

1 The Integers

1.1 The Well-Ordering Principle and Induction

Our general goal is the generalization of common algebraic structures; we will try to capture their essence by determining axioms which are responsible for their properties. We start with the set of integers $\mathbb{Z}$, considered along with the two operations of addition (+) and multiplication ($\cdot$). We will spend some time trying to understand how $\mathbb{Z}$ is put together with respect to these two operations, and we will identify several key properties. We will then take a selection of those properties, the so-called *ring axioms,* and eventually aim at studying *all* structures that are defined by requiring a set R along with two operations (which will be called + and $\cdot$ even if they may have nothing to do with the conventional + and $\cdot$) to satisfy the ring axioms. These structures will be called *rings:* from this perspective, $\mathbb{Z}$ is a particular example of a ring. Other examples you are (hopefully) familiar with are $\mathbb{Q}$ ('rational' numbers), $\mathbb{R}$ ('real' numbers), $\mathbb{C}$ ('complex' numbers); but many more exist, and most of them have *nothing* to do with numbers. We will study 'all' of them at once in the sense that we will determine several features that every such structure (as opposed to specific examples like $\mathbb{Z}$ or $\mathbb{Q}$) must have. We will (implicitly) define a *category* of rings by specifying certain types of functions (which we will call 'homomorphisms') between rings.

In any case, we begin with $\mathbb{Z}$. We will start by recalling several known facts about $\mathbb{Z}$, from a perspective that will perhaps be a little more modern and rigorous than what may be seen in high-school math or Calculus. Everything we will see should sound familiar, but the viewpoint may seem unusual at first—the goal will be to single out the key facts that are responsible for 'the way $\mathbb{Z}$ works'. These facts will be useful in studying other examples, particularly coming from 'modular arithmetic', and the study of these examples will guide us in choosing the axioms that we will use in order to define what a ring is.

We do not really start with a blank slate: I will assume familiarity with the basic, elementary-school properties of addition and multiplication between integers. I will also assume familiarity with the notion of 'ordering' in $\mathbb{Z}$: if a and b are integers, we write $a \leq b$ to mean that a is 'less than or equal to' b, in the ordinary sense. This ordering behaves in predictable ways with respect to the operations: for example, if $a \leq b$ and $c \geq 0$, then $ac \leq bc$; and similar statements you already know.

We can take these basic properties for granted, but it is helpful to introduce some terminology. We will begin by focusing on the fact that 'division' behaves rather peculiarly in $\mathbb{Z}$. For example, we can divide 18 by 3, getting a quotient of 6, which is an element of $\mathbb{Z}$. We can also divide 1 by 2, but the quotient is *not* an element of $\mathbb{Z}$: there is no integer c such that $1 = 2c$. We need some terminology to be able to deal with this distinction.

DEFINITION 1.1   Let $a, b \in \mathbb{Z}$. We say that 'b divides a', or 'b is a divisor of a', or 'a is a multiple of b', and write $b \mid a$, if there is an *integer* c such that $a = bc$.

Thus, 3 'divides' 18 since $18 = 3 \cdot 6$ and 6 is an integer; while 2 does *not* divide 1, because $1 = 2 \cdot \frac{1}{2}$ but $\frac{1}{2}$ is not an integer. The divisors of 12 are ± 1, ± 2, ± 3, ± 4, ± 6, and ± 12. Every integer divides 0, since $0 = b \cdot c$ for some integer c; as it happens, $c = 0$ works. On the other hand, the only integer that 0 divides is 0 itself.

With this understood, we can already record a useful (if completely elementary) fact.

LEMMA 1.2   *If $b \mid a$ and $a \neq 0$, then $|b| \leq |a|$.*

Proof   Indeed, by definition of divisibility we have $a = bc$ for some integer c; in particular, both b and c are nonzero since $a \neq 0$. Since $c \in \mathbb{Z}$ and $c \neq 0$, we have $|c| \geq 1$. And then

$$|a| = |b| \cdot |c| \geq |b| \cdot 1 = |b|$$

as claimed.   □

'Divisibility' defines an order relation on the set of nonnegative integers (Exercise 1.2). What Lemma 1.2 says is that, to some extent, this 'new' order relation is compatible with the ordinary one $\leq$.

What if b does *not* divide a? We can still divide b into a (at least if $b \neq 0$), but we have to pay a price: we get a *remainder* as we do so. Even if this fact is familiar to you, we will look into why it works the way it does, since it is something special about $\mathbb{Z}$—we have no need for this complication when we divide *rational* numbers, for example. So there must be some special property of $\mathbb{Z}$ which is responsible for this fact.

This property is in fact a property of the ordering $\leq$ we reviewed a moment ago. It will be responsible for several subtle features of $\mathbb{Z}$; in fact, besides the basic high-school properties of addition and multiplication, it is essentially the *one* property of $\mathbb{Z}$ that makes it work the way it does. So we focus on it for a moment, before returning to the issue of divisibility.

As recalled above, $\mathbb{Z}$ comes endowed with an order relation $\leq$. In fact, this is what we call a 'total ordering'. By this terminology we mean that given any two integers a and b, one of three things must be true: $a < b$, or $a = b$, or $a > b$. The same can be said of other sets of numbers, such as the set of *rational* numbers $\mathbb{Q}$ and the set of *real* numbers $\mathbb{R}$; but there is something about the ordering relation on $\mathbb{Z}$ that makes it very special. Terminology: $\mathbb{Z}^{\geq 0}$ stands for the set of *nonnegative* integers:

$$\mathbb{Z}^{\geq 0} = \{a \in \mathbb{Z} \mid a \geq 0\} = \{0, 1, 2, 3, \dots\}.$$

(Similarly $\mathbb{Z}^{>0}$ stands for the set of positive integers, $\mathbb{Q}^{\leq 0}$ is the set of nonpositive rational numbers, and so on.) Another name for $\mathbb{Z}^{\geq 0}$ is $\mathbb{N}$, the set of 'natural numbers'. The fact we need has everything to do with $\mathbb{Z}^{\geq 0}$.

FACT (Well-ordering principle) *Every nonempty set of nonnegative integers contains a least element.*

We can summarize this fact by saying that $\mathbb{Z}^{\geq 0}$ is 'well-ordered' by the relation $\leq$. A 'well-ordering' on a set S is simply an order relation such that every nonempty subset of S has a minimum.

The well-ordering principle should sound reasonable if not outright obvious: if we have many bags of potatoes, there will be one (or more) bags with the least number of potatoes. But whether this is obvious or not is a matter of perspective—if we were to attempt to *define* $\mathbb{Z}$ rigorously, the well-ordering principle for nonnegative integers would be one of the axioms we would explicitly adopt; there is (to my knowledge) no direct way to derive it from more basic properties of the set of integers.

Also note that $\leq$ is *not* a well-ordering on the set $\mathbb{Q}^{\geq 0}$ of nonnegative rationals, or on the set $\mathbb{R}^{\geq 0}$ of nonnegative reals. For example, the set of positive rationals is a nonempty subset of $\mathbb{Q}^{\geq 0}$, but it does not have a 'least' element. (If $q > 0$ were such an element, then $\frac{q}{2}$ would be even smaller and still rational and positive, giving a contradiction.) So the well-ordering principle is really a rather special feature of $\mathbb{Z}^{\geq 0}$. We will derive several key properties of $\mathbb{Z}$ from it, granting (as we already did above) simple facts about how the ordering $\leq$ behaves with respect to the operations $+, \cdot$ on $\mathbb{Z}$.

Even before we see the well-ordering principle in action in 'algebra' proper, it is useful to observe that it already plays a role in a specific logical tool with which you are already familiar: the process of *induction* depends on it. Every proof by induction can be converted into an argument appealing to the well-ordering principle. (In fact, my preference is often to do so.) Why?

As you likely know, induction works as follows. We want to prove a certain property $P(n)$ for all integers $n \geq 0$. Suppose we manage to prove that

(i) $P(0)$ holds: the property is true for $n = 0$; and

(ii) the implication $P(n) \implies P(n + 1)$ holds for all $n \geq 0$.

Then induction tells us that indeed, our property $P(n)$ holds for all $n \geq 0$. This smacks of magic, particularly the first few times you see it in action, but it is in a sense 'intuitively clear': the 'seed' $P(0)$ holds because you have proved it by hand in (i); and then $P(1)$ holds since $P(0)$ holds *and* you have proved in (ii) that $P(0) \implies P(1)$; and then $P(2)$ holds since $P(1)$ holds *and* you have proved that $P(1) \implies P(2)$; and then $P(3)$ holds since $P(2)$ holds *and* you have proved that $P(2) \implies P(3)$; and so on forever.

The problem with this argument is that 'so on forever' is not mathematics. There is an alternative argument which *is* rigorous, *once you grant the truth of the well-ordering principle*. Here it is.

Induction from the well-ordering principle   Assume we have established (i) and (ii). We have to prove that $P(n)$ holds for all $n \geq 0$.

Let $F \subseteq \mathbb{Z}^{\geq 0}$ be the set of nonnegative integers n such that $P(n)$ does *not* hold; then we have to prove that $F = \emptyset$. We will prove that this is necessarily the case by showing that $F \neq \emptyset$ leads to a contradiction.

Assume then that our set F *is a nonempty set of nonnegative integers.* By the well-ordering principle, F has a least element $\ell \in \mathbb{Z}^{\geq 0}$: that is, $P(\ell)$ *does not hold,* and ℓ is the least nonnegative integer with this property.

By (i) we know that $P(0)$ holds, and therefore $\ell > 0$. Then $n = \ell - 1$ is a nonnegative integer, and $n < \ell$, therefore $P(n)$ holds, since ℓ is the *least* nonnegative integer for which the property does not hold. By (ii), $P(n) \implies P(n+1)$; so $P(n+1)$ holds. But $n + 1 = \ell$, so this shows that $P(\ell)$ *does* hold.

We have reached a contradiction: $P(\ell)$ would both hold and not hold. Therefore, the assumption that $F \neq \emptyset$ leads to a contradiction, and we must conclude that $F = \emptyset$, as we needed.   □

Several proofs in what follows could be handled by induction or interchangeably by an appeal to the well-ordering principle. Which to use is essentially a matter of taste. I will often insist on using the well-ordering principle, to stress that we are really using a specific feature of $\mathbb{Z}$. Also, I often find it somewhat easier to write a lean, rigorous argument by directly invoking the well-ordering principle rather than induction.

1.2   'Division with Remainder' in $\mathbb{Z}$

The first substantial application of the well-ordering principle is the fact that in $\mathbb{Z}$ we can perform a 'division with remainder', as mentioned above. For example, $13 = 4 \cdot 3 + 1$: 13 divided by 4 gives 3 with a remainder of 1. The official statement is the following.

THEOREM 1.3 (Division with remainder)   *Let a, b be integers, with $b \neq 0$. Then there exist a unique 'quotient' $q \in \mathbb{Z}$ and a unique 'remainder' $r \in \mathbb{Z}$ such that*

$$a = bq + r \qquad \text{with } 0 \leq r < |b|.$$

Remark 1.4   The 'uniqueness' part is important. It is clear that we can write a division-with-remainder in many ways: $13 = 4 \cdot 4 - 3 = 4 \cdot 3 + 1 = 4 \cdot 2 + 5 = \cdots$; but the theorem claims that *one and only one* of these ways will satisfy the condition that the remainder is ≥ 0 and < 4.   ⌋

Proof of the theorem   We can assume that $b > 0$. Indeed, if $b < 0$, we can apply the statement to $-b > 0$ and then flip the sign of q after the fact. Switching the sign of b does not change $|b|$, so the condition on r is unchanged.

Assume then that $b > 0$, and consider all integer linear combinations of a and b of the form $a - bx$ with $x \in \mathbb{Z}$. Note that there are nonnegative integers of this type: if a is itself nonnegative, then $a - b \cdot 0 = a$ is nonnegative; and if a is negative, then $a - ba = a(1 - b)$ is nonnegative because $1 - b \le 0$. Therefore, the set

$$S = \{a - bx \text{ with } x \in \mathbb{Z}, \text{ such that } a - bx \ge 0\}$$

is a nonempty set of nonnegative integers. By the well-ordering principle, it contains a least element r: that is, there is some $x = q$ such that $r = a - bq \ge 0$ is smaller than any other nonnegative number of the form $a - bx$. I claim that these q and r are the unique numbers whose existence is claimed in the statement.

Indeed: by construction we have that $a = bq + r$ and that $r \ge 0$; we have to verify that (1) $r < b$ (note that $|b| = b$ since we are assuming $b > 0$) and (2) the numbers q and r are unique with these properties.

For (1), argue by contradiction. If we had $r \ge b$, then we would have $r - b \ge 0$; and

$$0 \le r - b = (a - bq) - b = a - b(q + 1).$$

This would show that $r - b$ is an element of S, since it is a nonnegative integer linear combination of a and b. But that is a contradiction, since $r - b < r$ while r was chosen to be the least element of S. Since $r \ge b$ leads to a contradiction, we can conclude that $r < b$ as we claimed.

For (2), assume that q', r' are also integers satisfying the requirement spelled out in the theorem: that is, $a = bq' + r'$, and $0 \le r' < b$. We will show that necessarily $q' = q$ and $r' = r$ (this is what we mean by stating that q and r are 'unique').

Since $a = bq + r = bq' + r'$, we have

$$b(q - q') = r' - r. \tag{1.1}$$

Now, since both r and r' are in the interval $[0, b - 1]$, their difference cannot exceed $b - 1$. In other words, $|r' - r| \le b - 1$, and in particular $|r' - r| < |b|$. But (1.1) shows that b is a divisor of $r' - r$; if $r' - r \ne 0$, then by Lemma 1.2 we would necessarily have $|b| \le |r' - r|$, contradicting the fact that $|b| > |r' - r|$ we observed a moment ago. The only possibility then is that $r' - r = 0$. This shows that $r' = r$, and further $b(q - q') = 0$. Since $b \ne 0$, this implies that $q - q' = 0$, and it follows that $q' = q$. We are done. □

1.3 Greatest Common Divisors

The next item in our agenda is an important notion, with which again you are likely to be familiar, but possibly in a slightly different form.

DEFINITION 1.5 Let $a, b \in \mathbb{Z}$. We say that a nonnegative integer d is the 'greatest common divisor' of a and b, denoted $\gcd(a, b)$ or simply (a, b), if

- $d \mid a$ and $d \mid b$; and
- if $c \mid a$ and $c \mid b$, then $c \mid d$.

If at least one of a and b is not 0, then $d = \gcd(a, b) \ne 0$. Indeed, d must divide both a and b according to the first requirement in the definition (this makes d a 'common divisor'), so d cannot be 0 unless both a and b are 0—keep in mind that 0 only divides 0. The second requirement in the definition says that if c is also a common divisor, then it must divide d. By Lemma 1.2, this implies that $|c| \le |d|$ if $d \ne 0$, and hence $c \le d$ since d is nonnegative. Therefore d is the 'greatest' (as in 'largest') common divisor of a and b *if a and b are not both 0.*

It is clear that the gcd *exists:* if you have two integers a and b, not both equal to 0, then you can simply list the divisors of a, lists the divisors of b, and then $\gcd(a, b)$ is simply the largest integer in the intersection of these two sets.

Example 1.6 To find that $\gcd(30, -42) = 6$, we can list all divisors of 30:

$$-30, -15, -10, -6, -5, -3, -2, -1, 1, 2, 3, 5, \underline{\mathbf{6}}, 10, 15, 30$$

and all divisors of -42:

$$-42, -21, -14, -7, -6, -3, -2, -1, 1, 2, 3, \underline{\mathbf{6}}, 7, 14, 21, 42$$

and then note that 6 is the largest integer that occurs in both lists.

Beware, however, that this method to find a gcd is extremely inefficient: finding the divisors of an integer is a time-consuming business, and if the integer is largish (say, a few thousand digits) then I suspect that its factorization may take current algorithms longer than the age of the universe to be carried out on the fastest imaginable digital computers. Can you think of a *faster* way to find the gcd of two integers? We will soon encounter a very efficient alternative to the 'inspection' method given above (Theorem 1.14).

Remark 1.7   You may wonder why I have not replaced the second requirement with something like

- if $c \mid a$ and $c \mid b$, then $c \le d$,

which would seem to justify the terminology 'greatest common divisor' in a more direct way. (This may be the definition you have run into in previous encounters with this notion.) The main reason is that when we recast this notion for more general 'rings', the relation of divisibility will be available, while the (perhaps) simpler relation $\le$ will *not*. Also, Definition 1.5 makes good sense for all possible integers a, b, including the case $a = b = 0$: in this case, since *every* number divides 0, the second requirement says that *every integer should divide* $\gcd(0, 0)$. Since 0 is the only number divisible by every integer, this tells us that $\gcd(0, 0) = 0$ according to Definition 1.5. Even in this case, $\gcd(a, b)$ is 'greatest'; but with respect to the divisibility relation from Exercise 1.2 rather than the ordinary order relation $\le$.

The following fact is another easy consequence of the well-ordering principle.

THEOREM 1.8   *Let a, b be integers. Then the greatest common divisor $d = \gcd(a, b)$ is an integer linear combination of a and b. That is, there exist integers m and n such that $d = ma + nb$.*

> *In fact, if a and b are not both 0, then* gcd(a, b) *is the smallest positive linear combination of a and b.*

For example, I pointed out above that $(30, -42) = 6$. The theorem then tells me that there must be integers m and n such that $30m - 42n = 6$. Does this seem 'obvious' to you? It does not look obvious to me, but a couple of attempts quickly yield $m = 3$, $n = 2$, which give $3 \cdot 30 - 2 \cdot 42 = 90 - 84 = 6$. (By the end of the section we will see a 'systematic' way to find such integers, cf. Remark 1.16.)

Notice that it is clear that the integers m and n are *not* going to be unique, since if $d = ma + nb$, then also $d = (m - b)a + (n + a)b$, $d = (m + b)a + (n - a)b$, and so on. For instance,

$$(3 + 42) \cdot 30 - (2 + 30) \cdot 42 = 45 \cdot 30 - 32 \cdot 42 = 1350 - 1344 = 6,$$

therefore $m = 45$, $n = 32$ also work in our example. Doing this type of experimentation is excellent practice, but it is kind of clear that mindless arithmetic will not prove the theorem in general. The missing ingredient is the well-ordering principle, and the beginning of the proof will remind you of the beginning of the proof of Theorem 1.3.

Proof of Theorem 1.8   If $a = b = 0$, then gcd(a, b) = gcd$(0, 0)$ = 0, which is a linear combination of a and b (because $0 = 1 \cdot 0 + 1 \cdot 0$). Therefore, we may assume that a and b are not both 0. Consider all linear combinations $ma + nb$ of a and b. I claim that some of them are positive. Indeed, take $m = a$, $n = b$; then $ma + nb = a^2 + b^2$, and this number is positive since a and b are not both 0.

Therefore, the set

$$S = \{ma + nb \mid m \in \mathbb{Z}, n \in \mathbb{Z}, \text{and } ma + nb > 0\}$$

is nonempty. This is the standard setting to apply the well-ordering principle: since S is a nonempty set of nonnegative integers, it must have a least element. Let d be this element, and let m and n be integers such that $d = ma + nb$. That is, d is the smallest positive linear combination of a and b. We are going to verify that d is the gcd of a and b, and this will prove the theorem.

Since $d \in S$, then $d > 0$: so d is nonnegative and not equal to 0. In order to prove that d is the gcd of a and b, we have to prove that

(i) $d \mid a$ and $d \mid b$; and
(ii) if $c \mid a$ and $c \mid b$, then $c \mid d$.

If $a = 0$, $d \mid a$ is automatic. If $a \neq 0$, we can use division with remainder: by Theorem 1.3, we know that there exist integers q and r such that $a = dq + r$, with $0 \leq r < d$. What can we say about r? Note that

$$r = a - dq = a - (ma + nb)q = a(1 - m) + b(-nq):$$

this shows that r is a linear combination of a and b. Can it be an element of S? No! Because $r < d$, and d has been chosen to be the smallest element of S. But then r cannot be positive, since r is a linear combination of a and b, and S contains all *positive* linear combinations of a and b. Since $r \geq 0$ and r is not positive, it follows that $r = 0$. This proves that $a = dq$, showing that $d \mid a$.

By essentially the same argument, we can deduce that $d \mid b$. (The roles of a and b are intechangeable.) This takes care of (i).

We still have to prove (ii). Suppose we have a common divisor c of a and b: $c \mid a$ and $c \mid b$. Then we have $a = uc$, $b = vc$ for some integers u and v. This gives

$$d = ma + nb = m\,(uc) + n\,(vc) = (mu + nv)\,c$$

and proves that $c \mid d$, as we needed.   □

Theorem 1.8 has nice, if somewhat technical, applications. Here is one.

COROLLARY 1.9   *Let a, b be integers. Then* $\gcd(a, b) = 1$ *if and only if 1 may be expressed as a linear combination of a and b.*

Proof   If $\gcd(a, b) = 1$, then the number 1 may be expressed as a linear combination of a and b by Theorem 1.8. On the other hand, if 1 may be expressed as a linear combination of a and b, then 1 is necessarily the *smallest* positive linear combination of a and b, because 1 is the smallest positive integer. It follows that $\gcd(a, b) = 1$, again as a consequence of Theorem 1.8.   □

DEFINITION 1.10   We say that a and b are *relatively prime* if $\gcd(a, b) = 1$.

For example, 3 and 7 are relatively prime. It follows that *every* integer is a linear combination of 3 and 7: indeed, 1 is a linear combination, and a multiple of a linear combination is a linear combination. For example, if you give me a random integer, say 173238384731, then I can do a small computation and tell you that

$$173238384731 = 866191923655 \cdot 3 + (-346476769462) \cdot 7\,.$$

(Exercise: How did I do that?)

And here is another application of Theorem 1.8. This will become handy in an important proof later on.

COROLLARY 1.11   *Let a, b, c be integers. If $a \mid bc$ and $\gcd(a, b) = 1$, then $a \mid c$.*

Proof   By Theorem 1.8, $1 = ma + nb$ for some integers m and n. Multiply by c:

$$c = 1 \cdot c = (ma + nb)\,c = mac + nbc = (mc)\,a + n\,(bc)\,.$$

Now the hypothesis tells us that $a \mid bc$, so $bc = ra$ for some integer r. Then

$$c = (mc)\,a + n\,(ra) = (mc + nr)\,a\,,$$

and this shows that $a \mid c$ as needed.   □

Remark 1.12 The hypothesis that $\gcd(a, b) = 1$ is necessary. For example, we have that $10 \mid (6 \cdot 5)$, and yet 10 divides neither 6 nor 5. This is not contradicting the statement, since $\gcd(10, 6) = 2 \neq 1$. ⌟

Next we come back to the question of *how* we could compute the gcd of two numbers more efficiently than just applying the definition (as we did at the beginning of this section). Given two integers a and b, we may assume that $b \neq 0$ (if $b = 0$, then $\gcd(a, b) = a$, so there is nothing to 'compute'). We can then apply division with remainder repeatedly, like this:

$$a = b q_0 + r_0 \qquad 0 \leq r_1 < b$$
$$b = r_0 q_1 + r_1 \qquad 0 \leq r_1 < r_0$$
$$r_0 = r_1 q_2 + r_2 \qquad 0 \leq r_2 < r_1$$
$$r_1 = r_2 q_3 + r_3 \qquad 0 \leq r_3 < r_2$$
$$\cdots \qquad\qquad \cdots$$

We can keep going so long as the remainder of the last division is not 0. Since at each step the remainder decreases and remains nonnegative, at some point it *must* become 0:

$$\cdots \qquad\qquad \cdots$$
$$r_{n-2} = r_{n-1} q_n + r_n \qquad 0 \leq r_n < r_{n-1}$$
$$r_{n-1} = r_n q_{n+1} + 0$$

Example 1.13 Let's run this procedure with $a = 30$, $b = -42$ (cf. Example 1.6):

$$30 = -42 \cdot 0 + 30$$
$$-42 = 30 \cdot (-2) + 18$$
$$30 = 18 \cdot 1 + 12$$
$$18 = 12 \cdot 1 + \underline{6}$$
$$12 = 6 \cdot 2 + 0.$$

The last nonzero remainder is 6, and we had observed that $\gcd(30, -42) = 6$. This is not a coincidence!

THEOREM 1.14 (Euclidean algorithm) *Let a, b be integers, with $b \neq 0$. Then, with notation as above, $\gcd(a, b)$ equals the last nonzero remainder r_n.*
  More explicitly: let $r_{-2} = a$ and $r_{-1} = b$; for $i \geq 0$, let r_i be the remainder of the division of r_{i-2} by r_{i-1}. Then there is an integer n such that $r_n \neq 0$ and $r_{n+1} = 0$, and $\gcd(a, b) = r_n$.

Applying the Euclidean algorithm to two integers only requires being able to carry out the division with remainder, which is very fast. No factorization is required!

Theorem 1.14 is perhaps easier to prove than it may seem at first. In fact, it will be clear once we understand the following statement.

LEMMA 1.15   *Let a, b, q, r be integers, with $b \neq 0$, and assume that $a = bq + r$. Then $\gcd(a, b) = \gcd(b, r)$.*

Proof   It is enough to prove that the pairs a and b, b and r have the same common divisors: then the largest among the common divisors will be the gcd of both (a, b) and (b, r), proving the statement.

Let c be a common divisor of a and b: $c \mid a$ and $c \mid b$. Then $a = cs$, $b = ct$ for some integers s and t. It follows that

$$r = a - bq = cs - ctq = c(s - tq),$$

and this shows that $c \mid r$. Since we already have that $c \mid b$, we see that c is a common divisor of b and r.

One applies the same technique to show that if c is a common divisor of b and r, then it is a common divisor of a and b. The reader should work this out.   □

Why does Lemma 1.15 imply Theorem 1.14? In the situation described above, the lemma tells us that

$$\gcd(a, b) = \gcd(b, r_0) = \gcd(r_0, r_1) = \cdots = \gcd(r_{n-1}, r_n) = \gcd(r_n, 0) = r_n,$$

and this is precisely what Theorem 1.14 claims.

Remark 1.16   If necessary, one can chase the Euclidean algorithm backwards and find integers m, n such that $\gcd(a, b) = ma + nb$. Rather than giving a formal description of this process, let's see how it works in Example 1.13. The fourth (i.e., second-to-last) line in the 'algorithm' tells us

$$\gcd(30, -42) = 6 = 18 - 12 \cdot 1 = 18 - 12.$$

According to the third line, $12 = 30 - 18$, so this says

$$\gcd(30, -42) = 18 - (30 - 18) = 2 \cdot 18 - 30.$$

According to the second line, $18 = -42 - 30 \cdot (-2)$, therefore

$$\gcd(30, -42) = 2 \cdot (-42 - 30 \cdot (-2)) - 30 = 3 \cdot 30 + 2 \cdot (-42),$$

and we see that $m = 3$, $n = 2$ work.   ⌐

1.4   The Fundamental Theorem of Arithmetic

Everything we have done so far assumed familiarity with the ordinary operations of addition and multiplication in $\mathbb{Z}$. You may feel that addition is a simpler business than multiplication, and there are technical reasons why this is indeed the case. Everything about addition boils down to the number 1: every positive integer can be obtained by adding 1 to itself a number of times; and if you allow subtractions, then you can 'generate' every integer from 1. The number 1 is the one brick you need to build the whole of $\mathbb{Z}$, if you can only use addition. (We will see in due time (§10.2) that this makes the *group* $(\mathbb{Z}, +)$ 'cyclic'.) From this point of view, multiplication is clearly a different story: we cannot obtain every integer by multiplying a fixed number by itself. Even if you allow division, *one* number won't do. As it happens, however, again we can build the whole of $\mathbb{Z}$ with multiplication, but we need many more bricks—an infinite number of them.

DEFINITION 1.17 An integer p is 'irreducible' if $p \neq \pm 1$ and the only divisors of p are $\pm 1, \pm p$.
    An integer $\neq 0, \neq \pm 1$ is 'reducible' or 'composite' if it is not irreducible.

(Note that 0 and ± 1 are neither irreducible nor composite according to this definition.)

For example, 10 is not irreducible (thus, it is a composite integer): indeed, 2 is a divisor of 10 that does not equal $\pm 1, \pm 10$. The first few irreducible positive integers are

$$2, 3, 5, 7, 11, 13, 17, 19, 23, 29, 31, 37, 41, 43, \ldots .$$

One straightforward way to generate the list of irreducible positive integers is the *Sieve of Eratosthenes:* list all integers ≥ 2, then cross out all multiples of 2 bigger than 2, then all multiples of 3 bigger than 3, and so on: cross out all multiples of the integers that have not been crossed out yet. You will get

$$2 \quad 3 \quad \cancel{4} \quad 5 \quad \cancel{6} \quad 7 \quad \cancel{8} \quad \cancel{9} \quad \cancel{10} \quad 11 \quad \cancel{12} \quad 13 \quad \cancel{14} \quad \cancel{15} \quad \cancel{16} \quad 17 \quad \cancel{18} \quad 19 \quad \cdots .$$

It should be clear that the integers that are *not* crossed out are all and only the positive irreducible integers.

We will see that every integer can be written as a product of irreducible integers, and further that this can be done in an essentially unique way. Before we get there, we have to look into this definition a little more carefully. Here is a consequence which relates Definition 1.17 with the material covered in §1.3.

LEMMA 1.18 *Assume that p is an irreducible integer and that b is not a multiple of p. Then b and p are relatively prime, that is, $\gcd(p, b) = 1$.*

Proof If c is a common divisor of p and b, then in particular $c = \pm 1$ or $c = \pm p$, since $c \mid p$ and p is irreducible. Since c is also a divisor of b, and p is not a divisor of b by hypothesis, we deduce that necessarily $c = \pm 1$. This is true of *all* common divisors of p and b, so it is in particular true for $\gcd(p, b)$. Since $\gcd(p, b) \geq 0$, the only possibility is $\gcd(p, b) = 1$. $\qquad\square$

This is already interesting. If p is irreducible, and b is not a multiple of p, we now know that there exist integers m and n such that $mp + nb = 1$. Was this obvious to you? It was not to me, but it is now since we have proved Lemma 1.18 and established Corollary 1.9 earlier.

Maybe you were familiar with the notion defined in Definition 1.17, but you may have expected me to call these integers *prime* rather than *irreducible*. For *prime* integers, I will adopt the following different definition.

DEFINITION 1.19   An integer p is 'prime' if $p \neq \pm 1$ and whenever p divides the product bc of two integers b, c, then p divides b or p divides c.

For example, 10 is not prime: indeed, 10 divides $30 = 6 \cdot 5$, even if 10 divides neither 6 nor 5. On the other hand, 0 *is* prime according to Definition 1.19: prove this as an exercise.

Caveat. This is an unusual choice: most references do *not* include 0 as a prime integer, and this inclusion may even be viewed as distasteful. I have made this nonstandard choice to align with some terminology involving *ideals,* which we will encounter in §6.1; and my convention will allow me to occasionally state sharper statements (cf. Examples 3.30 and 6.6). ⌙

Remark 1.20   (i) Reminder: in mathematics, 'or' is not exclusive. The definition says '...p divides b or p divides c': this allows for the possibility that p may divide *both* b and c.

(ii) It follows from the definition that if p is prime, and p divides a product of several integers a_i: $p \mid a_1 \cdots a_s$, then p must divide (at least) one of the integers a_i. This probably seems fairly clear, but would you be able to prove it 'formally'? (Exercise 1.17.) ⌙

THEOREM 1.21   *Let $p \in \mathbb{Z}$, $p \neq 0$. Then p is prime if and only if it is irreducible.*

Proof   If $p = \pm 1$, then p is neither irreducible nor prime, so the statement is true in this case. We can therefore assume that p is an integer other than 0 or ± 1, and we have to prove two things: (i) if p is prime, then it is irreducible; and (ii) if p is irreducible, then it is prime.

(i) Assume p is prime, and let d be a divisor of p. We have to show that $d = \pm 1$ or $d = \pm p$. Since $d \mid p$, there exists an integer q such that $p = dq$. But then p divides dq (since $dq = 1 \cdot p$ in this case), and since p is prime we can deduce that $p \mid d$ or $p \mid q$. If $p \mid d$, then we have that both $d \mid p$ and $p \mid d$; by Lemma 1.2, in this case we have that $|d| \leq |p|$ or $|p| \leq |d|$. Then we can conclude $|d| = |p|$, that is, $d = \pm p$. On the other hand, if $p \mid q$, then there exists an integer e such that $q = ep$. In this case, $p = dq = dep$, and hence $p(de - 1) = 0$. Since $p \neq 0$, we deduce $de - 1 = 0$, that is, $de = 1$. Then d divides 1, and this implies $d = \pm 1$. Summarizing, we have shown that if $d \mid p$, then $d = \pm p$ or $d = \pm 1$, and this shows that p is irreducible, as we needed.

(ii) For the converse, we assume that p is irreducible; so its divisors are precisely ± 1 and $\pm p$. In order to prove that p is prime, we assume that $p \mid bc$ and have to show that $p \mid b$ or $p \mid c$. If p happens to divide b, we have nothing to prove, so we may assume that p does *not* divide b, and we have to prove that in this case p necessarily divides c. By Lemma 1.18, we have that $\gcd(p, b) = 1$. But then p must indeed divide c, by Corollary 1.11. (I told you this corollary would come in handy!) □

You may really wonder why I am making a distinction between 'irreducible' and 'prime', considering that these two notions are almost identical, as we have just proved. We could adopt the convention (as many do) of simply avoiding 0 when dealing with *prime* integers, and then 'prime' and 'irreducible' would be synonyms. I prefer otherwise: it does not seem necessary to treat 0 as a 'special' number in this context. More importantly, the reason to avoid identifying the notions of 'irreducible' and 'prime' is that when we view these notions from a more general perspective later on (§6.2), we will discover that they are indeed substantially different. There are *rings* in which the analogues of Definitions 1.17 and 1.19 are really non-equivalent notions, even for nonzero elements. As it happens, nonzero prime elements are always irreducible in 'integral domains', but the converse will not be true in general. Somehow, the first part of the proof of Theorem 1.21 will generalize broadly, but the second part will not, unless the ring is very special. The ring $\mathbb{Z}$ *is* very special!

The next theorem is also an expression of how special $\mathbb{Z}$ is. It is so important that it is called the *fundamental* theorem of arithmetic. It is the formalization of the claim I made earlier, to the effect that the multiplicative structure of $\mathbb{Z}$ is built on infinitely many bricks. The bricks are precisely the prime/irreducible integers.

> THEOREM 1.22 (Fundamental Theorem of Arithmetic) *Every integer $n \neq 0, \neq \pm 1$ is a product of finitely many irreducible integers: $\forall n \in \mathbb{Z}$, $n \neq 0$, $n \neq \pm 1$, there exist irreducible integers $q_1, \ldots, q_r$ such that*
>
> $$n = q_1 \cdots q_r. \tag{1.2}$$
>
> *Further, this factorization is unique in the sense that if*
>
> $$n = q_1 \cdots q_r = p_1 \cdots p_s,$$
>
> *with all q_i, p_j irreducible, then necessarily $s = r$ and after reordering the factors we have $p_1 = \pm q_1$, $p_2 = \pm q_2$, ..., $p_r = \pm q_r$.*

Remark 1.23   As mentioned in this statement, an expression such as (1.2) is called a 'factorization', and specifically a *factorization into irreducibles,* or *prime factorization.* The statement does not exclude that $r = 1$: this happens precisely when $n = q_1$ is itself irreducible. We could in fact even agree that the 'empty product' equals 1 (it is not unreasonable to do so), and then we would have that even $n = 1$ has an irreducible factorization: the factorization consisting of no factors(!), corresponding to $r = 0$.   ⌐

Incidentally, note that the statement of this theorem would be quite a bit more awkward if we had not ruled out ± 1 in our definition of 'irreducible'. If we allowed 1 as an 'irreducible factor', then we could include any number of such factors in any factorization, and the uniqueness part of the statement would simply not be true.

Proof of Theorem 1.22   It is enough to prove the statement for positive integers n, since we can incorporate signs in the factors. Let's first deal with the *existence* of factorizations for integers $n > 1$, and then with the *uniqueness* of the factorization.

Both existence and uniqueness are consequences of the well-ordering principle. For existence, let

$$S = \{n \in \mathbb{Z}, n > 1 \mid n \text{ is } not \text{ a product of finitely many irreducibles}\}.$$

We have to prove that S is empty: this will show that *every* integer $n > 1$ does have a factorization into irreducibles.

By contradiction, let's assume that S is *not* empty. In this case S is a nonempty set of nonnegative integers, therefore S has a least element by the well-ordering principle. Let this integer be n.

Since $n \in S$, n is not a product of irreducibles and in particular n is not irreducible itself. Then n has divisors other than ± 1, $\pm n$: that is, $n = ab$ for some integers a, b different from ± 1, $\pm n$. We may assume both a and b are positive, since n is positive. Since $a \mid n$ and $b \mid n$, and neither is 1, we have that $1 < a < n$ and $1 < b < n$. (We are again using Lemma 1.2!) But n is the smallest integer in S, so $a \notin S$ and $b \notin S$. This tells us that both a and b do have factorizations into irreducibles:

$$a = q_1' \cdots q_u', \quad b = q_1'' \cdots q_v''$$

for irreducible integers q_i', q_j''. We have reached a contradiction, because this would imply that

$$n = q_1' \cdots q_u' \cdot q_1'' \cdots q_v'',$$

saying that n *does* have a factorization into irreducibles—this would mean that $n \notin S$, while n was an element of S to begin with.

Thus, our assumption that S be nonempty leads to a contradiction. It follows that S *is* empty, and this proves the first part of the theorem: every integer $\neq 0, \pm 1$ has factorizations.

Now we have to deal with uniqueness. If n and the factors q_i are all *positive*, then 'uniqueness' just means that the factorization (1.2) is unique up to the order of the factors. This is what we will show; the general case follows easily from this.

Well, assume that that is not necessarily the case, and let T be the set of nonnegative integers n for which the factorization is *not* unique: so we are assuming that T is not empty, and we would like to derive a contradiction from this assumption.

Hopefully you see where I am going with this. By definition, T is a set of nonnegative integers. If T is nonempty, then *it has a minimum n by the well-ordering principle:* that is, the factorization of n is not unique, but the factorization of any integer smaller than n *is* unique. Write two distinct factorizations for n:

$$n = q_1 \cdots q_r = p_1 \cdots p_s, \tag{1.3}$$

with all q_i, p_j irreducible and positive. Both r and s are ≥ 1, since n is not equal to 1. We are assuming that these two factorizations are *not* the same up to reordering, and we aim at reaching a contradiction.

Look at (1.3): it tells us in particular that p_1 divides the product $q_1 \cdots q_r$. Since p_1 is irreducible, it is prime by Theorem 1.21, and therefore (cf. Remark 1.20) it must divide one of the integers q_i. After rearranging the factors q_i, we may assume that p_1 divides q_1.

Next, q_1 is itself irreducible, so its divisors are ± 1, $\pm q_1$. Since p_1 is irreducible, it is not equal to ± 1; and since we have established that p_1 is a divisor of q_1, necessarily $p_1 = \pm q_1$. Both q_1 and p_1 are positive, so we can conclude that $p_1 = q_1$.

At this point we can rewrite (1.3) as

$$n = q_1 \cdot q_2 \cdots q_r = q_1 \cdot p_2 \cdots p_s. \tag{1.4}$$

By cancellation ($q_1 \neq 0$) it follows that

$$q_2 \cdots q_r = p_2 \cdots p_s.$$

But this number equals the integer n/q_1, and in particular it is less than n. Since n was the least positive integer for which the factorization was not unique, the factorization *is* unique for n/q_1. This implies that $r - 1 = s - 1$ and $q_i = p_i$ for $i \geq 2$ up to reordering. We had already established that $q_1 = p_1$, so it follows that $r = s$ and $q_i = p_i$ for $i \geq 1$, up to reordering. This says that the two factorizations presented in (1.3) *do* coincide, contradicting our assumption.

Summarizing, the assumption that $T \neq \emptyset$ leads to a contradiction. Therefore necessarily $T = \emptyset$, that is, the factorization is unique for all $n > 1$, and we are done.   □

As I mentioned, the fact that the Fundamental Theorem of Arithmetic is true tells us that $\mathbb{Z}$ is rather special. When we learn a bit of the language of rings, we will recognize that this theorem tells us that $\mathbb{Z}$ is a 'UFD' ('unique factorization domain'). This means that $\mathbb{Z}$ is a domain (a rather nice type of ring) in which every element admits a factorization into irreducibles, and further that this factorization is unique in a suitable sense, as spelled out in Theorem 1.22. We will glance at more general UFDs in §6.4.

If you have paid close attention to the argument, you may have noticed that we did not need to use Theorem 1.21 in the first part of the proof. In $\mathbb{Z}$, the *existence* of factorizations only depends on the notion of irreducibility, while the *uniqueness* requires knowing that irreducible integers are prime. This is a facet of the more general theory of factorization in rings.

Remark 1.24   I used the well-ordering principle in the argument for uniqueness. This is one example in which I feel that this leads to a leaner argument than induction. You are welcome to try to rewrite the same argument using induction; but if you do, do it 'formally'. It is tempting to write (1.3), deduce (1.4) as I did, then cancel q_1 and simply say that 'by the same argument, up to a reordering we have $q_2 = p_2$, then $q_3 = p_3$, and so on'. This sounds reasonable, but it leaves to the reader the task of formalizing the argument: 'so on' is not mathematics.

Induction, maybe in the form of the well-ordering principle, is the way to formalize such arguments. You should develop the habit of noticing when induction is running under the hood, even if the words 'induction' or 'well-ordering principle' are not mentioned explicitly for the sake of leaner prose.   ⌟

As a consequence of Theorem 1.22, every positive integer $\neq 1$ determines a well-defined set of positive irreducibles, possibly appearing with multiplicity: for example, the irreducible factorization $12 = 2 \cdot 2 \cdot 3$ tells us that 12 determines the 'multiset' consisting of 2, 2, 3. By the uniqueness part, the product of *no* other collection of irreducibles equals 12. Dealing with negative integers is no harder: we could, for example, insist that exactly one of the irreducible factors is taken with a negative sign.

Putting it yet another way, we can write any positive integer n as follows:

$$n = 2^{v_2} 3^{v_3} 5^{v_5} 7^{v_7} 11^{v_{11}} \cdots \qquad (1.5)$$

where the 'infinite product' on the right consists of powers of all the irreducible elements, and only finitely many of the exponents v_i are not equal to 0. For example, 12 is obtained by letting $v_2 = 2$, $v_3 = 1$, and setting all other exponents v_i to 0. We can even do this when n equals 1: this case corresponds to choosing all the exponents to be 0. The content of the Fundamental Theorem of Arithmetic is that *every* positive integer n can be expressed in this way, and that the ordered list of exponents $(v_2, v_3, v_5, \ldots)$ is determined *uniquely* by n.

As an application, we get a different point of view on divisibility and on greatest common divisors. First, if n is given as in (1.5) and another positive integer c is given by

$$c = 2^{\gamma_2} 3^{\gamma_3} 5^{\gamma_5} 7^{\gamma_7} 11^{\gamma_{11}} \cdots ,$$

then $c \mid n$ if and only if $\gamma_i \leq v_i$ for all i. (Make sure you understand why! Exercise 1.21.) This gives us a way to write down all divisors of a given integer, with the following useful consequence.

PROPOSITION 1.25   *Let a, b be nonzero integers, and write*

$$a = \pm 2^{\alpha_2} 3^{\alpha_3} 5^{\alpha_5} 7^{\alpha_7} 11^{\alpha_{11}} \cdots ,$$
$$b = \pm 2^{\beta_2} 3^{\beta_3} 5^{\beta_5} 7^{\beta_7} 11^{\beta_{11}} \cdots ,$$

as above. Then the gcd of a and b is the positive integer

$$d = 2^{\delta_2} 3^{\delta_3} 5^{\delta_5} 7^{\delta_7} 11^{\delta_{11}} \cdots ,$$

where $\delta_i = \min(\alpha_i, \beta_i)$ for all i.

Proof   If $c = 2^{\gamma_2} 3^{\gamma_3} 5^{\gamma_5} 7^{\gamma_7} \cdots$ is a common divisor of a and b, as observed above, we must have $\gamma_i \leq \alpha_i$ and $\gamma_i \leq \beta_i$ for all i.

Since $\delta_i = \min(\alpha_i, \beta_i)$ is indeed $\leq \alpha_i$ and $\leq \beta_i$, we see that $d = 2^{\delta_2} 3^{\delta_3} 5^{\delta_5} 7^{\delta_7} \cdots$ is a common divisor of a and b, giving requirement (i) in Definition 1.5. Further, we note that if γ_i is less than both α_i and β_i, then $\gamma_i \leq \min(\alpha_i, \beta_i) = \delta_i$. This implies that every common divisor of a and b divides d, and shows that d also satisfies requirement (ii) in Definition 1.5.   □

Essentially the same argument will show that the *least common multiple* (lcm) is similarly obtained by taking exponents $\max(\alpha_i, \beta_i)$. (The 'least common multiple' of a and b is what it says it should be: a number ℓ that is a multiple of a and a multiple of b, and such that every common multiple of a and b is a multiple of ℓ.)

Example 1.26 Going back once more to Example 1.6:

$$30 = 2 \cdot 3 \cdot 5 = 2^1 \cdot 3^1 \cdot 5^1 \cdot 7^0 \cdot 11^0 \cdot 13^0 \cdots$$
$$-42 = -2 \cdot 3 \cdot 7 = -2^1 \cdot 3^1 \cdot 5^0 \cdot 7^1 \cdot 11^0 \cdot 13^0 \cdots,$$

and therefore

$$\gcd(30, -42) = 2^1 \cdot 3^1 \cdot 5^0 \cdot 7^0 \cdot 11^0 \cdot 13^0 \cdots = 2 \cdot 3 = 6.$$

We also have

$$\mathrm{lcm}(30, -42) = 2^1 \cdot 3^1 \cdot 5^1 \cdot 7^1 \cdot 11^0 \cdot 13^0 \cdots = 2 \cdot 3 \cdot 5 \cdot 7 = 210.$$

Of course this is again *not* a fast way to compute gcds; the Euclidean algorithm (cf. Theorem 1.14) is much faster. In practice, finding the irreducible factorization of an integer is computationally expensive. The best algorithms known for factorizations would run on *quantum computers,* and practical quantum computers are not around yet (but that may change soon).

In any case, we can note the following simple consequence of these considerations.

COROLLARY 1.27 *Two nonzero integers a, b are relatively prime if and only if they have no common irreducible factor.*

Proof Write $a = \pm 2^{\alpha_2} 3^{\alpha_3} 5^{\alpha_5} \cdots$, $b = \pm 2^{\beta_2} 3^{\beta_3} 5^{\beta_5} \cdots$ as above.

By Proposition 1.25, a and b are relatively prime, that is, $\gcd(a, b) = 1$, precisely when $\min(\alpha_i, \beta_i) = 0$ for all i, that is, if and only if $\alpha_i = 0$ or $\beta_i = 0$ for all i. This means precisely that a and b have no common irreducible factor. □

Prime numbers are still very mysterious. A few things can be said in general:

- There are infinitely many of them. Find out why as an exercise (Exercise 1.26 to be precise); you will again be in the company of Euclid, who is credited with the first proof of this fact, and you will use the Fundamental Theorem in the process.
- If $\pi(n)$ denotes the number of positive primes less than or equal to n (so $\pi(1) = 0$, $\pi(2) = 1$, $\pi(3) = 2$, $\pi(4) = 2$, $\pi(5) = 3$, etc.), then as $n \to \infty$, $\pi(n)$ grows at the same rate as $n/\ln n$. More precisely, $\lim_{n \to \infty} \frac{\pi(n)}{n/\ln n} = 1$. This is the so-called *Prime Number Theorem.* Do *not* try this as an exercise, it is too hard. I first saw a proof of this fact when I took a graduate course on complex analysis.
- The *Riemann hypothesis* is one of the most important open problems in mathematics, and it amounts to a sophisticated conjecture on the distribution of primes. If you solve this one, you will get a million dollars from the Clay Mathematics Institute. Google 'Clay Riemann' if you want to know more.

The Sieve of Eratosthenes is clever, but of essentially no practical utility. If you need to test whether a decent-sized integer—say, with a hundred digits—is prime, then you need to be cleverer. In a little while (Theorem 2.18) we will prove something quite amazing: if p is prime, and a is not a multiple of p, then $a^{p-1} - 1$ *is* a multiple of p. Is this obvious to you? It is extremely non-obvious to me, even though its proof is rather easy. (Never mind that: even if a proof is easy to understand, *finding* that proof may be quite hard.) This fact can be used to test if a number is 'likely' to be prime: the remainder of the division of $a^{p-1} - 1$ by p can be computed quickly, even for large integers; if it turns out that $a^{p-1} - 1$ is a multiple of p for many numbers a, then one can be quite confident that p is prime. This is not as fuzzy as I am making it sound: this type of primality testing is implemented in computer systems, and used routinely (by your laptop, for example) as a tool in the encryption of information sent over the Internet. We will see how this works soon enough (§2.5).

Problems marked with the symbol ▷ are particularly important for later developments of the material, or are quoted from the text.

Exercises

1.1   Find an example illustrating why the hypothesis that $a \neq 0$ is necessary in the statement of Lemma 1.2.

1.2   ▷ Prove that divisibility is an order relation on the set $\mathbb{Z}^{\geq 0}$ of nonnegative integers. That is, prove that for all positive integers a, b, c:

(i) $a \mid a$;

(ii)   if $a \mid b$ and $b \mid c$, then $a \mid c$;

(iii) if $a \mid b$ and $b \mid a$, then $a = b$.

Is this order relation a *total* ordering? Does it have a 'maximum'? (That is, is there an integer m such that $\forall a \in \mathbb{Z}^{\geq 0}, a \mid m$?)

1.3   Show that $\mathbb{Z}$ is not well-ordered by $\leq$.

1.4   Prove that for all $a \in \mathbb{Z}$, $\gcd(a, 0) = |a|$.

1.5   Prove that every odd integer is either of the form $4k + 1$ or of the form $4k + 3$ for some integer k.

1.6   Prove that the cube of an integer a has to be exactly one of these forms: $9k$ or $9k + 1$ or $9k + 8$ for some integer k.

1.7   Prove that there is no integer a such that the last digit of a^2 is 2.

1.8   ▷ Let n be a positive integer, and let a and b be integers. Prove that a and b have the same remainder when divided by n if and only if $a - b = nk$ for some integer k.

The following problems have to do with greatest common divisors. First, work them out by only using the definitions and facts covered in §1.3.

1.9   Let $n \in \mathbb{Z}$. What are the possible values of $\gcd(n, n + 2)$? Of $\gcd(n, n + 6)$?

1.10   ▷ Suppose $\gcd(a, b) = 1$. Prove that if $a \mid c$ and $b \mid c$, then $ab \mid c$.

1.11 Prove that if $a \mid (b + c)$ and $\gcd(b, c) = 1$, then $\gcd(a, b) = 1$.

1.12 Prove that if $\gcd(a, c) = 1$ and $\gcd(b, c) = 1$, then $\gcd(ab, c) = 1$.

1.13 Prove that if $\gcd(a, b) = 1$, then $\gcd(a^r, b^s) = 1$ for all positive integers r, s.

1.14 Prove that if $c > 0$, then $\gcd(ca, cb) = c \cdot \gcd(a, b)$.

1.15 Prove that if $c \mid ab$ and $\gcd(a, c) = d$, then $c \mid db$.

1.16 Show that in order to verify that an integer $n > 1$ is irreducible, it suffices to verify that no integer a with $1 < a \le \sqrt{n}$ is a divisor of n.

1.17 ▷ Prove that if p is prime, and p divides a product of integers $a_1 \cdots a_s$, then p divides one of the integers a_i. (I agree that this fact looks reasonable, but can you prove it 'formally'? Try induction, or another use of the well-ordering principle.)

1.18 Let $p \ne 0, \pm1$ be an integer. Prove that p is prime if and only if for all $a \in \mathbb{Z}$, either $p \mid a$ or $\gcd(a, p) = 1$.

1.19 Prove that if p is prime, $a, n \in \mathbb{Z}$, with $n \ge 0$, and $p \mid a^n$, then $p^n \mid a^n$.

1.20 Let $q > 3$ be an irreducible integer. Prove that $q^2 + 2$ is composite. (*Hint:* Consider the remainder of q after division by 3.)

1.21 ▷ Let

$$n = 2^{\nu_2} 3^{\nu_3} 5^{\nu_5} 7^{\nu_7} 11^{\nu_{11}} \cdots$$

and

$$c = 2^{\gamma_2} 3^{\gamma_3} 5^{\gamma_5} 7^{\gamma_7} 11^{\gamma_{11}} \cdots$$

be positive integers (cf. §1.4). Prove that $c \mid n$ if and only if $\gamma_i \le \nu_i$ for all i. (*Note:* If you think this is obvious, you may be missing the point. The *uniqueness* of irreducible factorizations is crucial here.)

1.22 Work out again several of the problems on gcds listed above; this time, use the description of the gcd in terms of factorizations, obtained in Proposition 1.25. (You will find that your proofs are often (even) more straightforward.)

1.23 ▷ Let a, b be positive integers. Prove that $\gcd(a, b) \operatorname{lcm}(a, b) = ab$.

1.24 Let $a, b \in \mathbb{Z}$, and assume that $\gcd(a, b) = 1$ and that ab is a perfect square (that is, $ab = n^2$ for some integer n). Prove that a and b are perfect squares.

1.25 ▷ Let $p > 0$ be an irreducible integer. Prove that $\sqrt{p}$ is irrational. (*Hint:* If not, then $\sqrt{p} = \frac{a}{b}$ for some integers a and b. This implies $a^2 = pb^2$. What does the Fundamental Theorem tell you in this case?)

1.26 ▷ Prove that there are infinitely many primes, as follows. By contradiction, assume that there are only finitely many primes $p_1, \ldots, p_k$, and consider the number $n = p_1 \cdots p_k + 1$. What does the Fundamental Theorem tell you about n? Why does this lead to a contradiction?

2 Modular Arithmetic

2.1 Equivalence Relations and Quotients

Now that we have a clearer picture of how $\mathbb{Z}$ is put together, we can move on to examine other structures which share many features with it, and that in fact are derived from it. These will also be examples of *rings;* they will help us decide which properties we want to abstract into the official definition of *ring*.

These new examples will be constructed by endowing with operations (which will still be denoted + and ·, as in the case of integers) sets obtained as 'quotients of $\mathbb{Z}$ by certain equivalence relations'.

REMINDER ON EQUIVALENCE RELATIONS. The notions of 'equivalence relation' and 'partition' are going to be very important for the material that we need to cover, and I must recommend that you review them before moving ahead. I will summarize the story very briefly in the next few paragraphs, and this will likely suffice if you have studied these notions carefully in the not-too-distant past. Otherwise, I hope that a perusal of §A.8 in Appendix A may be helpful at this time.

- A relation $\sim$ on a set A is an 'equivalence' relation if it is reflexive, symmetric, and transitive. These properties generalize properties that hold for the $=$ relation; equality may be viewed as the prototypical (albeit trivial) example of an equivalence relation.

- If we have an equivalence relation $\sim$ on a set A, then we can construct a new set, often denoted by $A/\sim$: this set is called the 'partition' determined by $\sim$ on A, or the 'quotient' of A 'modulo' $\sim$. The elements of the set $A/\sim$ are the 'equivalence classes' determined by $\sim$. The equivalence class of $a \in A$ is the set $[a]_\sim$ consisting of all elements of A which are in relation with a: that is,

$$[a]_\sim = \{b \in A \mid a \sim b\}.$$

One verifies that the equivalence classes are disjoint subsets of A whose union is the whole of A.

- The function $\pi\colon A \to (A/\sim)$ defined by setting $\pi(a) = [a]_\sim$ is surjective, and $\pi(a) = \pi(b) \iff a \sim b$.

- This last property completely describes the quotient $A/\sim$, in the sense that if we have a surjective function $f\colon A \to B$ such that $f(a) = f(b) \iff a \sim b$, then there automatically is a one-to-one correspondence between $A/\sim$ and B. In fact, in this case there is an 'induced function' $\tilde{f}: (A/\sim) \to B$, defined by setting $\tilde{f}([a]_\sim) = f(a)$, which turns out to be injective and surjective.

If this last point sounds a little mysterious, don't worry too much: we will work things out from scratch anyway. After the fact, you will probably agree that this general observation considerably clarifies what is going on. (If you care, it is an instance of a 'universal property', something having to do with category theory; cf. Exercise 0.1. We will run into several 'universal properties' along the way.) But everything else should already look quite familiar to you. If not, please review §A.8 before moving on, or chase these definitions in your favorite text dealing with them.

2.2   Congruence mod n

Many different equivalence relations may be defined on $\mathbb{Z}$. Here are the ones we want to study more carefully. In the following definition, $n\mathbb{Z}$ denotes the set of multiples of n: $n\mathbb{Z} = \{nk \mid k \in \mathbb{Z}\} = \{\cdots, -2n, -n, 0, n, 2n, \cdots\}$.

DEFINITION 2.1   Let $n \geq 0$ be an integer, and let $a, b \in \mathbb{Z}$. We say that 'a is congruent to b modulo n', denoted $a \equiv b \bmod n$, if $b - a \in n\mathbb{Z}$.

In other words, two integers a and b are congruent modulo n if n divides the difference $b - a$. In symbols:

$$a \equiv b \bmod n \iff n \mid (b - a).$$

Remark 2.2   We could define congruence modulo negative integers as well; but the relations corresponding to n and $-n$ coincide, because $n\mathbb{Z} = (-n)\mathbb{Z}$, so we may as well take $n \geq 0$. Also note that 'congruence modulo 0' is simply the relation $=$. Indeed, $0\mathbb{Z} = \{0\}$, so the requirement that $b - a \in 0\mathbb{Z}$ simply means that $a = b$.   ⌐

It is an easy exercise to verify that, for n positive, $n \mid (b - a)$ if and only if a and b have the same remainder after division by n: in fact, this is Exercise 1.8, which you have hopefully worked out already. This fact makes it particularly easy to verify that the relation introduced in Definition 2.1 is an *equivalence* relation. Do this in your head, now.[1] It is in any case essentially immediate to verify this fact using Definition 2.1 directly, for all $n \geq 0$; for example, here is how you would prove *transitivity*. Suppose $a \equiv b \bmod n$; then by definition there exists an integer k such that $b - a = nk$. Suppose $b \equiv c \bmod n$; then by definition there exists an integer ℓ such that $c - b = n\ell$. But then

$$c - a = (c - b) + (b - a) = nk + n\ell = n(k + \ell),$$

and this shows that $c - a \in n\mathbb{Z}$, proving that $a \equiv c \bmod n$.

Now that we have one equivalence relation in $\mathbb{Z}$ for every $n \geq 0$, we can consider

---

[1] So, you will be thinking: the relation is reflexive because a has the same remainder as a; it is symmetric because if a has the same remainder as b, then b has the same remainder as a; and it is transitive because if a has the same remainder as b, and b has the same remainder as c, then a has the same remainder as c. Or some such rigmarole.

the corresponding equivalence classes. For any nonnegative integer n and any integer a, we denote by $[a]_n$ the equivalence class of a with respect to the relation of congruence modulo n. Explicitly:

$$[a]_n = \{b \in \mathbb{Z} \,|\, a \equiv b \bmod n\} = \{b \in \mathbb{Z} \,|\, n \text{ divides } (b - a)\}$$
$$= \{b = a + nk \,|\, k \in \mathbb{Z}\} \,.$$

We call $[a]_n$ the 'congruence class' of a mod n, and we say that a is a 'representative' of $[a]_n$. An alternative notation for the same set is $a + n\mathbb{Z}$, which is a shorthand for the last description of $[a]_n$ given above.

From the general facts about equivalence relations and partitions recalled in §2.1, we know that two equivalence classes $[a]_n$, $[b]_n$ for a fixed n are either *the same* or *disjoint*. In fact, we know that

$$[a]_n = [b]_n \iff a \equiv b \bmod n \,, \tag{2.1}$$

so whatever information is carried by the congruence relation, it is also packaged in its equivalence classes (that is, in the congruence classes). We do not need to do any work to verify this, since it is something that is true for *every* equivalence relation. Please see §A.8 if necessary. It would of course be quite easy to verify this fact directly from Definition 2.1 and from the definition of congruence class.

Example 2.3   For $n = 2$ there are *two* congruence classes:

$$[0]_2 = \{\cdots, -6, -4, -2, 0, 2, 4, 6, \cdots\} = \{0 + 2k \,|\, k \in \mathbb{Z}\} \,,$$
$$[1]_2 = \{\cdots, -5, -3, -1, 1, 3, 5, 7, \cdots\} = \{1 + 2k \,|\, k \in \mathbb{Z}\} \,.$$

As it happens, we have a name for these sets: $[0]_2$ is the set of *even* integers, while $[1]_2$ is the set of *odd* integers. Using the alternative notation proposed above, we could call the set of odd integers $1 + 2\mathbb{Z}$: after all, an odd integer is simply '1 plus an even integer'.

Example 2.4   For $n = 3$, there are *three* congruence classes:

$$[0]_3 = \{\cdots, -9, -6, -3, 0, 3, 6, 9, \cdots\} = \{0 + 3k \,|\, k \in \mathbb{Z}\} \,,$$
$$[1]_3 = \{\cdots, -8, -5, -2, 1, 4, 7, 10, \cdots\} = \{1 + 3k \,|\, k \in \mathbb{Z}\} \,,$$
$$[2]_3 = \{\cdots, -7, -4, -1, 2, 5, 8, 11, \cdots\} = \{2 + 3k \,|\, k \in \mathbb{Z}\} \,.$$

As far as I know, we have no special names for these sets; the first one is of course the set of all multiples of 3 ('threeven' numbers?).

Congruence classes for a fixed n form a *partition* of $\mathbb{Z}$, and this is apparent in Examples 2.3 and 2.4: that is, in each case we have a selection of subsets of $\mathbb{Z}$ with the property that every element of $\mathbb{Z}$ appears in exactly one subset. If you feel a bit foggy about equivalence classes and such, please spend a moment staring at these examples. The numbers -4 and 5 are not congruent mod 2, and indeed they are in different congruence classes mod 2 (in Example 2.3); they are congruent mod 3, and indeed they are in the same congruence class mod 3 (in Example 2.4). As (2.1) teaches us, $[-4]_2 \neq [5]_2$ while $[-4]_3 = [5]_3 = [2]_3$. These equalities are *really* equalities: $[-4]_3$ and $[2]_3$ are quite simply the same subset of $\mathbb{Z}$; this subset just happens to have several different names.

Where are we? As mentioned in my quick reminder in §2.1, now that we have an equivalence relation on a set, we can consider the corresponding partition as a set in its own right. Its elements are the equivalence classes—in our case, the subsets $[a]_n$ of $\mathbb{Z}$. This is a powerful idea: we consider a whole subset of $\mathbb{Z}$ as a single element of a new set.

DEFINITION 2.5   For any integer $n \geq 0$ we denote by $\mathbb{Z}/n\mathbb{Z}$ the set of congruence classes mod n in $\mathbb{Z}$. This set is called the 'quotient of $\mathbb{Z}$ modulo $n\mathbb{Z}$', or simply '$\mathbb{Z}$ mod n'.

Other popular notations are $\mathbb{Z}/\equiv_n$, $\mathbb{Z}/(n)$, $\mathbb{Z}/n$, or $\mathbb{Z}_n$; this latter is used particularly frequently, and that is a pity because this notation means something else in a different context ($\mathbb{Z}_p$ is the set of 'p-adic numbers', which are something else entirely). I will just use the notation $\mathbb{Z}/n\mathbb{Z}$.

Summarizing, the elements of $\mathbb{Z}/n\mathbb{Z}$ are of the form $[a]_n$, for some integer a. *Please get used to the fact that $[a]_n$ is *not* itself an integer.* That is, resist with all your might the temptation to think of $\mathbb{Z}/n\mathbb{Z}$ as a subset of $\mathbb{Z}$. The sets $\mathbb{Z}$ and $\mathbb{Z}/n\mathbb{Z}$ are closely related, but in other ways. We will come back to this (Example 4.19).

Each *element* of $\mathbb{Z}/n\mathbb{Z}$ may be viewed as a subset of $\mathbb{Z}$; but the *set* $\mathbb{Z}/n\mathbb{Z}$ of such elements should not be viewed as a subset of $\mathbb{Z}$. The element $[2]_3$ of the set $\mathbb{Z}/3\mathbb{Z}$ is not a fancy new name for the number 2: indeed, the number 2 is not the same as the number 5, and yet $[2]_3$ and $[5]_3$ really *are* the same element of $\mathbb{Z}/3\mathbb{Z}$, because they are the same subset of $\mathbb{Z}$. (Look again at Example 2.4.)

Thinking of an entity such as $[5]_3$ *both* as a subset of $\mathbb{Z}$ and as an element of the set $\mathbb{Z}/3\mathbb{Z}$ requires a certain amount of mathematical sophistication; you should have gotten used to this strange idea in your first encounter with equivalence relations. In any case, we now have a brand new set $\mathbb{Z}/n\mathbb{Z}$ for every nonnegative integer n, and we want to study it.

The first question we can ask is 'how big' $\mathbb{Z}/n\mathbb{Z}$ is. For example, we saw in Examples 2.3 and 2.4 that $\mathbb{Z}/2\mathbb{Z}$ consists of *two* elements, which we could denote by $[0]_2$ and $[1]_2$, and that $\mathbb{Z}/3\mathbb{Z}$ consists of *three* elements, which we could denote by $[0]_3, [1]_3$, and $[2]_3$ (or also by $[-6]_3, [7]_3$, and $[5]_3$ if we felt like it: this is the same list). Let's summarize in a statement something that should be clear at this point. Recall from §1.2 that $\mathbb{Z}$ is endowed with a *division with remainder:* if $n > 0$, we can divide any integer a by n, with a remainder r such that $0 \leq r < n$.

LEMMA 2.6   *Let n be a positive integer.*

- *Let $a \in \mathbb{Z}$, and let r be the remainder after division of a by n. Then $[a]_n = [r]_n$.*
- *The classes $[0]_n, [1]_n, \ldots, [n-1]_n$ are all distinct.*

Proof   First point: If $a = nq + r$, then $a - r = nq \in n\mathbb{Z}$, therefore $a \equiv r \bmod n$, and this implies $[a]_n = [r]_n$ by definition of congruence classes.

Second point: Let's verify that if $0 \leq r_1 < n$ and $0 \leq r_2 < n$, and $[r_1]_n = [r_2]_n$, then $r_1 = r_2$; this is a spelled-out version of the statement. If $0 \leq r_1 < n$ and $0 \leq r_2 < n$, then $0 \leq |r_2 - r_1| < n$. If now $[r_1]_n = [r_2]_n$, then $n \mid (r_2 - r_1)$, and hence $n \mid |r_2 - r_1|$. If $|r_2 - r_1| \neq 0$, this implies $n \leq |r_2 - r_1| < n$, a contradiction. (I am using the helpful Lemma 1.2 again.) This contradiction shows that necessarily $|r_2 - r_1| = 0$, that is, $r_1 = r_2$ as needed. $\qquad\square$

The following statement is then an immediate consequence.

THEOREM 2.7   *For all positive integers n, the set $\mathbb{Z}/n\mathbb{Z}$ consists of exactly n elements:* $[0]_n, [1]_n, \ldots, [n-1]_n$.

Proof   By the first point in the lemma every class agrees with one of these, and by the second point these elements of $\mathbb{Z}/n\mathbb{Z}$ are all different from each other. $\qquad\square$

At the risk of sounding like a broken record, I will stress once more that the choice $[0]_n, [1]_n, \ldots, [n-1]_n$, while reasonable looking, is only one of infinitely many viable choices. We can label the elements of $\mathbb{Z}/5\mathbb{Z}$ by $[160]_5, [-324]_5, [17]_5, [48]_5$, and $[-4329349871]_5$ if we feel like it; despite appearances, this is *precisely* the same list as the more conventional-looking $[0]_5, [1]_5, [2]_5, [3]_5$, and $[4]_5$. The conventional choice has the psychological advantage of being memorable, but that is all there is to it.

As an extreme example, there is *only one* element in $\mathbb{Z}/1\mathbb{Z}$, which we could denote $[a]_1$ for any integer a whatsoever. There is absolutely no way to distinguish $[0]_1$ from $[1]_1$ or $[-234938512983712983192389834594859485743498579438573974 98]_1$.

Remark 2.8   If $n = 0$, we can still make sense of the quotient, i.e., of $\mathbb{Z}/0\mathbb{Z}$. In this case the congruence relation is just equality (Remark 2.2), so the equivalence class $[a]_0$ of an integer a consists of the single element a: $[a]_0 = \{a\}$. Therefore, the set $\mathbb{Z}/0\mathbb{Z}$ is just a copy of $\mathbb{Z}$ itself. In this sense $\mathbb{Z}$ may be viewed as one of the structures $\mathbb{Z}/n\mathbb{Z}$, that is, the special case $n = 0$. For $n \neq 0$, the set $\mathbb{Z}/n\mathbb{Z}$ is *finite*, as we proved in Theorem 2.7. ⌐

2.3   Algebra in $\mathbb{Z}/n\mathbb{Z}$

This is a text on *algebra*, so we should try to do some algebra on the new sets $\mathbb{Z}/n\mathbb{Z}$. We can do this by virtue of simple compatibilities of the congruence relation with respect to the ordinary operations $+, \cdot$ in $\mathbb{Z}$.

LEMMA 2.9   *Let $n \geq 0$ be an integer, and let a, a', b, b' be integers.*
    If $a \equiv a' \bmod n$ and $b \equiv b' \bmod n$, then

(i) $a + b \equiv a' + b' \bmod n$,
(ii) $a \cdot b \equiv a' \cdot b' \bmod n$.

Proof   Since $a \equiv a' \bmod n$ and $b \equiv b' \bmod n$, there exist integers k and ℓ such that

$a' - a = nk$, $b' - b = n\ell$. Then

$$(a' + b') - (a + b) = a' - a + b' - b = nk + n\ell = n(k + \ell).$$

This shows that $n \mid (a' + b') - (a + b)$, that is, $a + b \equiv a' + b' \bmod n$, and proves (i). With the same notation,

$$a'b' - ab = (a + nk)(b + n\ell) - ab = ab + na\ell + nbk + n^2k\ell - ab = n(a\ell + bk + nk\ell),$$

showing that $n \mid a'b' - ab$, that is, $a \cdot b \equiv a' \cdot b' \bmod n$, giving (ii).   □

Armed with Lemma 2.9, we can make two important definitions: we define a notion of 'addition' and of 'multiplication' on the set $\mathbb{Z}/n\mathbb{Z}$, for every n.

DEFINITION 2.10   Let $n \geq 0$ be an integer, and let $[a]_n$, $[b]_n$ be elements of $\mathbb{Z}/n\mathbb{Z}$. We define

$$[a]_n + [b]_n := [a + b]_n,$$
$$[a]_n \cdot [b]_n := [a \cdot b]_n.$$

(*Note:* The symbols $+$, $\cdot$ on the right-hand side are the familiar addition and multiplication in $\mathbb{Z}$; the symbols $+$, $\cdot$ on the left-hand side are the new operations we are defining in $\mathbb{Z}/n\mathbb{Z}$.)

This may look perfectly reasonable, but there is an important subtlety, which I feel you should ponder carefully—at least the first time you run into it. Take $n = 2$ to deal with a small example. Then according to Definition 2.10 we have (for instance)

$$[0]_2 + [1]_2 = [1]_2.$$

The elements $[0]_2$, $[1]_2$ are the two subsets in the partition determined by the congruence relation mod 2: as we saw in Example 2.3,

$$[0]_2 = \{\cdots, -6, -4, -2, 0, 2, 4, 6, \cdots\} = \{0 + 2k \mid k \in \mathbb{Z}\},$$
$$[1]_2 = \{\cdots, -5, -3, -1, 1, 3, 5, 7, \cdots\} = \{1 + 2k \mid k \in \mathbb{Z}\}.$$

Could I make a definition in the same style, using some *other* partition of $\mathbb{Z}$ into two subsets? Let's try!

Example 2.11   Take some other equivalence relation $\sim$ splitting $\mathbb{Z}$ into two equivalence classes, for example

$$[0]_\sim = \mathbb{Z}^{\leq 0} = \{\cdots, -4, -3, -2, -1, 0\},$$
$$[1]_\sim = \mathbb{Z}^{>0} = \{1, 2, 3, 4, 5, \cdots\}.$$

Can I still define an addition for these classes, using the same idea as in Definition 2.10? For example, I would have $[0]_\sim + [1]_\sim = [0 + 1]_\sim = [1]_\sim$, just as before. This seems harmless—and yet, it does not work! The subset $[0]_\sim$ is *the same as* the subset $[-4]_\sim$: these are just different names for the first equivalence class. Let's call this subset A for a moment. Similarly, the subset $[1]_\sim$ is the same as the subset $[2]_\sim$: this is the second equivalence class. Let's call it B, and note that $A \neq B$. But then, what should $A + B$ be? According to the analogue of Definition 2.10, we would have

$$A + B = [0]_\sim + [1]_\sim = [0 + 1]_\sim = [1]_\sim = B,$$

and at the same time

$$A + B = [-4]_\sim + [2]_\sim = [-4 + 2]_\sim = [-2]_\sim = [0]_\sim = A.$$

This is nonsense, since $A \neq B$.

What do we learn from this example? We would say that the operation we just tried to define on our equivalence classes $[a]_\sim$ is *not well-defined*. It is not a viable definition, because the outcome of the operation depends on how we choose to write our elements. Whether you decide to call the first class $[0]_\sim$ or $[-2]_\sim$ cannot affect the result of an operation depending on this class: the operation should depend on the *class* itself, not on your mood of the moment on how you decide to call the class.

In other words, a viable definition should not depend on the representatives you choose in order to apply it. Example 2.11 shows a case in which the definition *does* depend on the representatives. As a result, we throw it away.

Now we see in what sense Definition 2.10 is subtle. I claim that the problem I just detected with the equivalence relation in Example 2.11 does *not* affect Definition 2.10: the proposed operations $+, \cdot$ in $\mathbb{Z}/n\mathbb{Z}$ *are* 'well-defined'.

Is this obvious? Nearly so, but only because we wisely took care of some preparatory work in Lemma 2.9. I will formalize the statement as follows.

CLAIM 2.12 *The operations defined in Definition 2.10 do not depend on the choice of the representatives. That is: if $n \geq 0$ is an integer, and $[a]_n = [a']_n$, $[b]_n = [b']_n$, then*

(i) $[a + b]_n = [a' + b']_n$,
(ii) $[a \cdot b]_n = [a' \cdot b']_n$.

Proof   This is no more and no less than a restatement of Lemma 2.9.   □

The preceding discussion may look exceedingly long-winded, but the issue of whether a notion is 'well-defined' is very important. It is easy to become a little sloppy on this point, which can lead very quickly to nonsensical statements. What we hopefully understand now is that the relations of 'congruence mod n' are very special, because they are compatible with the ordinary operations among integers in such a way that they allow us to define analogous operations among the corresponding congruence classes. As we have seen, one should not expect this from an arbitrarily chosen equivalence relation. This is the reason why we focus on the particular equivalence relations introduced in Definition 2.1.

2.4   Properties of the Operations +, · on $\mathbb{Z}/n\mathbb{Z}$

Let's see where we stand. In Chapter 1 we looked rather carefully at the 'structure' consisting of $\mathbb{Z}$ along with the ordinary operations + and ·. In §2.2 we defined new structures $\mathbb{Z}/n\mathbb{Z}$, one for every nonnegative integer n, and in §2.3 we endowed these structures with operations + and ·. By the way, we refer to these algebraic objects by still calling them $\mathbb{Z}$, $\mathbb{Z}/n\mathbb{Z}$, but we really intend these sets *along with* the natural operations +, · we have defined. If we need to emphasize the operations, we may use the notation $(\mathbb{Z}/n\mathbb{Z}, +, \cdot)$.

We have observed that $\mathbb{Z}/0\mathbb{Z}$ is 'a copy of $\mathbb{Z}$' (Remark 2.8), so we concentrate on $\mathbb{Z}/n\mathbb{Z}$ for $n > 0$. These gadgets are all different, and all different from $\mathbb{Z}$. The set of integers $\mathbb{Z}$ is infinite, while we have seen that the set $\mathbb{Z}/n\mathbb{Z}$ consists of exactly n elements (which we may denote $[0]_n, \ldots, [n-1]_n$; this is part of the content of Theorem 2.7). It is sometimes useful to draw 'tables' for the operations on these sets; for example, here is how + and · work on $\mathbb{Z}/4\mathbb{Z}$:

+	0	1	2	3
0	0	1	2	3
1	1	2	3	0
2	2	3	0	1
3	3	0	1	2

·	0	1	2	3
0	0	0	0	0
1	0	1	2	3
2	0	2	0	2
3	0	3	2	1

I am being lazy typesetting these tables, so I am writing a for what really is[2] $[a]_4$. For example, the entry in the rightmost column of the last line of the second table is 1, signifying that '$3 \cdot 3 = 1$'. You are supposed to interpret this as

$$[3]_4 \cdot [3]_4 = [1]_4 ,$$

which of course is true: $3 \cdot 3 = 9$, therefore $[3]_4 \cdot [3]_4 = [9]_4 = [1]_4$ by our definition of product of congruence classes (Definition 2.10).

It is important to note that there are properties of the operations in these different sets which seem qualitatively different. For example: we know that, in $\mathbb{Z}$,

$$ab = 0 \implies a = 0 \text{ or } b = 0. \tag{2.2}$$

This property is also true in, for example, $\mathbb{Z}/3\mathbb{Z}$, if we think of $[0]_3$ as 0. Indeed, we could list *all* products ab where $a \neq 0$ and $b \neq 0$:

$$[1]_3 \cdot [1]_3 = [1]_3, \quad [1]_3 \cdot [2]_3 = [2]_3, \quad [2]_3 \cdot [1]_3 = [2]_3, \quad [2]_3 \cdot [2]_3 = [1]_3$$

and it just does not happen that $[0]_3$ ever appears as the product of two nonzero elements. So from the point of view of this property, $\mathbb{Z}$ and $\mathbb{Z}/3\mathbb{Z}$ are similar. We say, to anticipate a definition we will officially introduce soon (Definition 3.23), that $\mathbb{Z}$ and $\mathbb{Z}/3\mathbb{Z}$ are both 'integral domains'. But this property is *not* true in all $\mathbb{Z}/n\mathbb{Z}$: for example,

$$[2]_4 \cdot [2]_4 = [4]_4 = [0]_4$$

---

[2] This is a rather common abuse of language.

even though $[2]_4 \neq [0]_4$. In terms of the tables drawn above, I am just observing that there is a 0 in the middle of the second table, away from the first row and the first column. Is this a rare phenomenon? Not really: for example, $[2]_6 \cdot [3]_6 = [6]_6 = [0]_6$, $[3]_{12} \cdot [4]_{12} = [0]_{12}$, etc. We can say that $\mathbb{Z}/4\mathbb{Z}, \mathbb{Z}/6\mathbb{Z}, \mathbb{Z}/12\mathbb{Z}$ are not 'integral domains'. It is very easy to find plenty more examples of this kind. By the end of the section, we will understand very thoroughly what is going on.

On the other hand, there are a bunch of properties of these operations which do hold for $\mathbb{Z}$ and for *all* $\mathbb{Z}/n\mathbb{Z}$. Take a minute to verify that the following properties hold in $\mathbb{Z}$. First of all, $a + b \in \mathbb{Z}$ and $a \cdot b \in \mathbb{Z}$ for all $a, b \in \mathbb{Z}$ (that is, $\mathbb{Z}$ is 'closed' with respect to these operations); then

$$
\begin{array}{ll}
\forall a, b, c & (a + b) + c = a + (b + c) \\
\forall a & a + 0 = 0 + a = a \\
\forall a \, \exists a' & a + a' = a' + a = 0 \\
\forall a, b & a + b = b + a \\
\forall a, b, c & (a \cdot b) \cdot c = a \cdot (b \cdot c) \\
\forall a & a \cdot 1 = 1 \cdot a = a \\
\forall a, b, c & a \cdot (b + c) = a \cdot b + a \cdot c \\
& (a + b) \cdot c = a \cdot c + b \cdot c
\end{array}
\tag{2.3}
$$

and further

$$
\forall a, b \qquad a \cdot b = b \cdot a.
$$

This looks like a long list, but it should not take you long to go through it. In the third property, the element a' is what we usually call $-a$.

Many of the properties listed in (2.3) have names: for example, the first one is called 'associativity' (of addition), the second is the 'existence of an additive identity', and so on. We will come back to these properties soon, in the appropriately general context (§3.1). For now the point I want to make is the following.

THEOREM 2.13   *For every positive integer n, all of these properties hold for $\mathbb{Z}/n\mathbb{Z}$ with $0 = [0]_n$ and $1 = [1]_n$.*

Remark 2.14   In view of this, I will often write 0 for $[0]_n$ and 1 for $[1]_n$. This is potentially confusing, because the two 0 in '$0 = [0]_n$' mean different things: the first is an element that we choose to call 0 in $\mathbb{Z}/n\mathbb{Z}$; the second is the integer 0. Spend one moment getting used to these conventions.   ⌐

With the terminology we will introduce in §3.1, Theorem 2.13 simply tells us that $\mathbb{Z}/n\mathbb{Z}$ is a 'ring', and in fact a 'commutative ring'. This theorem proves itself! Definition 2.10 prescribes that $[a]_n + [b]_n$ and $[a]_n \cdot [b]_n$ are elements of $\mathbb{Z}/n\mathbb{Z}$, so 'closure' is trivially satisfied. Every single other item in the list can be easily reduced to the corresponding property in $\mathbb{Z}$. For example, if we want to prove the 'distributivity' law in $\mathbb{Z}/n\mathbb{Z}$,

$$\forall a, b, c \in \mathbb{Z}/n\mathbb{Z} \qquad a \cdot (b + c) = a \cdot b + a \cdot c,$$

we can argue as follows. For all $a, b, c \in \mathbb{Z}/n\mathbb{Z}$, there are integers $k, \ell, m \in \mathbb{Z}$ such that

$$a = [k]_n, \quad b = [\ell]_n, \quad c = [m]_n.$$

Then using Definition 2.10 we have

$$
\begin{aligned}
a \cdot (b + c) &= [k]_n \cdot ([\ell]_n + [m]_n) \\
&= [k]_n \cdot [\ell + m]_n \\
&= [k \cdot (\ell + m)]_n \\
&\stackrel{!}{=} [k \cdot \ell + k \cdot m]_n \\
&= [k \cdot \ell]_n + [k \cdot m]_n \\
&= [k]_n \cdot [\ell]_n + [k]_n \cdot [m]_n \\
&= a \cdot b + a \cdot c,
\end{aligned}
$$

where the equality $\stackrel{!}{=}$ holds because the same 'distributivity' law holds in $\mathbb{Z}$. Every property in the list can be proved to hold in $\mathbb{Z}/n\mathbb{Z}$ by precisely the same method. This is rather tedious, and very easy—make sure you agree with me by trying a few other items in the list.

The long list given above (without the last, lone item) will be our starting point to define *rings*. In case you are wondering, I separated the last item from the rest because it is an additional property ('commutativity of multiplication') which is not part of the definition of *ring*. When it holds, we say that the ring is *commutative*. Therefore, $\mathbb{Z}$ and all $\mathbb{Z}/n\mathbb{Z}$ are 'commutative rings'. Incidentally, you know other examples: the sets $\mathbb{Q}$, $\mathbb{R}$, and $\mathbb{C}$ (denoting respectively the sets of *rational, real,* and *complex* numbers, along with the ordinary operations $+$ and $\cdot$) all satisfy the properties listed in (2.3), including the lone item following the longer list. They are all 'commutative rings'.

We will see that many other properties can be deduced from the ones listed in (2.3), and all such properties will hold for all rings. For the moment we will stay with $\mathbb{Z}/n\mathbb{Z}$, and examine how some features of the integer n get magically encoded in properties of the ring $\mathbb{Z}/n\mathbb{Z}$.

I have already pointed out that one property that holds in $\mathbb{Z}$, that is, the implication

$$ab = 0 \implies a = 0 \text{ or } b = 0,$$

holds in $\mathbb{Z}/n\mathbb{Z}$ for some, but not all, n. For example, $[2]_{10} \cdot [5]_{10} = [10]_{10} = [0]_{10}$, even though $[2]_{10} \neq [0]_{10}$ and $[5]_{10} \neq [0]_{10}$.

On the other hand, there are 'nice' properties that hold for some $\mathbb{Z}/n\mathbb{Z}$ and do not hold for $\mathbb{Z}$. You should have noticed that the properties listed in (2.3) include the existence of an 'additive inverse':

$$\forall a \, \exists a' \qquad a + a' = a' + a = 0.$$

For example, the additive inverse of 2 is -2, since $2 + (-2) = (-2) + 2 = 0$. You may also have noticed that I did *not* list a property prescribing the existence of 'multiplicative inverses', which would look like this:

$$\forall a \neq 0 \, \exists a' \qquad a \cdot a' = a' \cdot a = 1. \tag{2.4}$$

It is natural to require a to not equal 0, since '1/0' does not exist in any reasonable context. However, even with this natural requirement, (2.4) is simply not true in $\mathbb{Z}$: for example, $2 \neq 0$ and yet there is no integer n such that $2n = 1$. By contrast, (2.4) *is* verified if you allow 'fractions': if $\frac{p}{q}$ is a rational number (that is, p and q are integers, and $q \neq 0$), and $p \neq 0$, then the rational number $\frac{q}{p}$ is its multiplicative inverse: $\frac{p}{q} \cdot \frac{q}{p} = 1$. So 2 does *not* have a multiplicative inverse in $\mathbb{Z}$, but it does in $\mathbb{Q}$ (that is, $\frac{1}{2}$). More generally: (2.4) *is* satisfied for $\mathbb{Q}$, but it is not for $\mathbb{Z}$.

What about $\mathbb{Z}/n\mathbb{Z}$? It is easy to check whether (2.4) holds by examining a multiplication table: this property just says that every row of the table, with the exception of the row corresponding to 0, must contain a 1. (Why?) For example, (2.4) does not hold in $\mathbb{Z}/4\mathbb{Z}$: look back at the multiplication table, and note that there is no 1 in the row corresponding to 2. However, below are the tables for $\mathbb{Z}/5\mathbb{Z}$, as you may verify (and again neglecting the parentheses in the notation $[a]_5$ to avoid cluttering the picture).

+	0	1	2	3	4
0	0	1	2	3	4
1	1	2	3	4	0
2	2	3	4	0	1
3	3	4	0	1	2
4	4	0	1	2	3

·	0	1	2	3	4
0	0	0	0	0	0
1	0	1	2	3	4
2	0	2	4	1	3
3	0	3	1	4	2
4	0	4	3	2	1

It just so happens that every row in the multiplication table, other than the row headed by 0, does have a 1.

For conciseness, when a structure satisfying the properties listed in (2.3), including the commutativity of multiplication, also satisfies (2.4), we say that structure is a 'field'. (We will give a more official definition in §3.3.) Thus, $\mathbb{Z}$ and $\mathbb{Z}/4\mathbb{Z}$ are *not* fields, while $\mathbb{Z}/5\mathbb{Z}$ *is* a field; and the reader can verify that $\mathbb{Q}, \mathbb{R}, \mathbb{C}$ are also fields.

It would seem that $\mathbb{Z}/n\mathbb{Z}$ satisfies (2.4) for some values of n, and not for others. Is this just a random fact defying any sensible explanation, or is there a method to this madness? If you want to gain some personal intuition for what is going on, put away these notes and construct the multiplication tables for $\mathbb{Z}/n\mathbb{Z}$ for several positive integers n, say $n = 2, \ldots, 10$. List the numbers n for which (2.4) is satisfied, and see if you can detect a pattern. If you do, test it on some larger case, and then see if you can prove it in general. And then come back here and continue reading.

THEOREM 2.15 *Let $n > 1$. Then the following assertions are equivalent:*

(i) The integer n is prime.

> *(ii) In $\mathbb{Z}/n\mathbb{Z}$, property (2.2) is satisfied: $\forall a, b$, $ab = 0 \implies a = 0$ or $b = 0$.*
> *(iii) In $\mathbb{Z}/n\mathbb{Z}$, property (2.4) is satisfied: if $a \neq 0$, then a has a multiplicative inverse.*

Proof We are going to argue that (i) $\implies$ (iii) $\implies$ (ii) $\implies$ (i).

(i) $\implies$ (iii): Assume that $n = p > 1$ is a prime (hence irreducible) integer, and let $a \in \mathbb{Z}/p\mathbb{Z}$, $a \neq 0$. We have to prove that a has a multiplicative inverse in $\mathbb{Z}/p\mathbb{Z}$.

Let $\ell \in \mathbb{Z}$ be a representative for a: $a = [\ell]_p$. Since $a \neq 0$, we have $[\ell]_p \neq [0]_p$, that is, $p \nmid \ell$. By Lemma 1.18, this implies that $\gcd(p, \ell) = 1$; therefore, by Theorem 1.8, there exist integers q and k such that

$$1 = qp + k\ell.$$

Now 'read this identity modulo p': if $1 = qp + k\ell$, then $[1]_p = [qp + k\ell]_p$, and Definition 2.10 gives

$$[1]_q = [qp]_p + [k\ell]_p = [k]_p \cdot [\ell]_p,$$

since $[qp]_p = [0]_p$ as qp is a multiple of p. Letting $a' = [k]_p$, we have that $a'a = 1$ in $\mathbb{Z}/p\mathbb{Z}$ (and $aa' = 1$ follows by commutativity). Therefore a does have a multiplicative inverse, as we had to verify.

(iii) $\implies$ (ii): Assume that every nonzero $a \in \mathbb{Z}/n\mathbb{Z}$ has a multiplicative inverse. We have to verify that if $ab = 0$ in $\mathbb{Z}/n\mathbb{Z}$, then $a = 0$ or $b = 0$. If $a = 0$, then there is nothing to prove. If $a \neq 0$, then a has a multiplicative inverse a', by assumption. But then

$$b = 1 \cdot b = (a' \cdot a) \cdot b = a' \cdot (a \cdot b) = a' \cdot 0 \overset{!}{=} 0$$

as needed.

Take a moment to understand why each of these equalities is true: two of them will be a direct application of two of the properties listed in (2.3); the last equality, marked $\overset{!}{=}$, is left as an exercise (Exercise 2.11).

(ii) $\implies$ (i): Now we are assuming that $ab = 0 \implies a = 0$ or $b = 0$, for all $a, b \in \mathbb{Z}/n\mathbb{Z}$, and we have to prove that n is prime. We use the very definition of prime integer, Definition 1.19: we will assume that n divides a product km, and show that n must divide k or m. This is straightforward if we consider the elements $a = [k]_n$ and $b = [m]_n$ determined by k and m in $\mathbb{Z}/n\mathbb{Z}$. Indeed, then

$$ab = [k]_n[m]_n = [km]_n = [0]_n = 0,$$

since n divides km. By our assumption, we have that $a = 0$ or $b = 0$, that is,

$$[k]_n = 0 \quad \text{or} \quad [m]_n = 0,$$

and this says precisely that n divides k or n divides m. $\quad\square$

Using the fancy terminology I mentioned earlier, with which we will deal more extensively in Chapter 3, Theorem 2.15 says that $n > 1$ is prime if and only if $\mathbb{Z}/n\mathbb{Z}$ is an *integral domain,* if and only if $\mathbb{Z}/n\mathbb{Z}$ is a *field.*

The key to the argument proving (i) $\implies$ (iii) gives the following more general useful result.

PROPOSITION 2.16   *Let $n > 1$ be an integer. Then a class $[\ell]_n$ has a multiplicative inverse in $\mathbb{Z}/n\mathbb{Z}$ if and only if $\gcd(n, \ell) = 1$.*

Proof   If $\gcd(n, \ell) = 1$, then there exist integers q and k such that $qn + k\ell = 1$. It follows that

$$[k]_n[\ell]_n = [k\ell]_n = [1 - qn]_n = [1]_n .$$

Since multiplication is commutative in $\mathbb{Z}/n\mathbb{Z}$, this also implies $[\ell]_n[k]_n = [1]_n$, and the conclusion is that $[k]_n$ is a multiplicative inverse of $[\ell]_n$.

Conversely, if $[\ell]_n$ has a multiplicative inverse in $\mathbb{Z}/n\mathbb{Z}$, then $\gcd(n, \ell) = 1$. Indeed, let $[k]_n$ be an inverse to $[\ell]_n$, so that $[k]_n[\ell]_n = [1]_n$; then there is a q such that $1 - k\ell = qn$, so that $1 = qn + k\ell$. This shows that 1 is a linear combination of n and ℓ, and it follows that $\gcd(n, \ell) = 1$ by Corollary 1.9.   □

Remark 2.17   A convenient consequence of the fact that $\mathbb{Z}/p\mathbb{Z}$ is an integral domain when p is prime is that a 'multiplicative cancellation' law holds: if $a \neq 0$ in $\mathbb{Z}/p\mathbb{Z}$, then $ab = ac \implies b = c$. Indeed, if $ab = ac$, then $a(b - c) = 0$, and if p is prime it follows that $b - c = 0$ by part (ii) of the theorem.

This cancellation does *not* hold in general. For example, $[2]_6 \cdot [2]_6 = [4]_6 = [2]_6 \cdot [5]_6$, and yet $[2]_6 \neq [5]_6$ (you should have come up with something like this when you did Exercise 2.4).   ⌟

The proof of Theorem 2.15 is rather straightforward, but quite interesting nevertheless. You may have noticed that the proof of '(iii) $\implies$ (ii)' did not really use modular arithmetic (that is, considerations having specifically to do with $\mathbb{Z}/n\mathbb{Z}$): it looks like a formal manipulation that only relies on some of the basic properties listed at (2.3). And indeed, the same argument will work in a substantially more general setting.

There are two cases that are *not* contemplated in Theorem 2.15: $n = 0$ and $n = 1$. For $n = 0$, I pointed out in Remark 2.8 that $\mathbb{Z}/0\mathbb{Z}$ is 'a copy' of $\mathbb{Z}$; under this identification, the operations on $\mathbb{Z}/0\mathbb{Z}$ are clearly the same as those in $\mathbb{Z}$. We know that (2.2) holds in $\mathbb{Z}$, and (2.4) does not: in other words, $\mathbb{Z}$ is an integral domain but it is not a field. In this case, the equivalence of (ii) and (iii) in Theorem 2.15 simply does not hold. Looking carefully at the proof of the theorem, you will realize that this has to do with the fact that 0 *is* a prime number according to Definition 1.19, but it is *not* irreducible. As for $n = 1$: by Theorem 2.7, $\mathbb{Z}/1\mathbb{Z}$ consists of a single element, $[0]_1$. In particular, $[0]_1 = [1]_1$: in this strange place, $0 = 1$. This makes (ii) and (iii) vacuously true as stated. As it happens, this 'trivial' ring is not considered to be an integral domain or a field: the hypothesis that 0 and 1 are different will be part of the official definition of these types of structure, as we will see in §3.3.

2.5   Fermat's Little Theorem, and the RSA Encryption System

The lesson we learn from Theorem 2.15 is that rather sophisticated features of a number n are reflected in algebraic properties of the structure $(\mathbb{Z}/n\mathbb{Z}, +, \cdot)$ that we have been exploring. The following pretty statement should be seen as a manifestation of the same motive. We adopt the usual 'power' notation: so, $[a]_n^e$ stands for the product of $[a]_n$ by itself e times, which according to Definition 2.10 must equal $[a^e]_n$.

> THEOREM 2.18 (Fermat's little theorem)   *Let a, p be integers, with p positive and prime. Then $[a]_p^p = [a]_p$.*

In other words, if p is prime, then $a^p \equiv a \bmod p$. This is simply not (necessarily) the case if p is not prime: for example, $2^4 = 16$ is *not* congruent to 2 modulo 4. But it is true for primes, as Theorem 2.18 states and as we will prove below. The result looks quite impressive in any given example with largish numbers. For instance, take $a = 100$, $p = 13$: then (by the theorem) no computation is needed to conclude that

$$100000000000000000000000000 \equiv 100 \bmod 13 ;$$

and yet, if you feel like working this out, you do discover that indeed

$$100000000000000000000000000 - 100 = 7692307692307692307692300 \cdot 13 ,$$

confirming that the left-hand side is a multiple of 13.

Fermat's little theorem is often stated in the following equivalent formulation.

> THEOREM   *Let a, p be integers, with p positive and prime, and assume that $p \nmid a$. Then $a^{p-1} \equiv 1 \bmod p$.*

This is essentially the way Fermat himself stated it, in a letter to a friend in 1640. The equivalence of the two statements is a straightforward exercise using Theorem 2.15 (Exercise 2.18).

Fermat included the statement of the theorem in his letter, but did not include the proof; allegedly, it was too long.[3] By contrast, the argument we will see is sharp and quick; it probably was not what Fermat had in mind. In a rather distant future (§12.1) we will learn a bit of group theory that will make the statement essentially *evident;* but for now we do not have that luxury, so we will have to be a little clever.

Proof of Theorem 2.18   If $p \mid a$, then $[a]_p = [0]_p$, and the statement is that $[0]_p^p = [0]_p$. This is true. So we may assume that $p \nmid a$, and it is enough to prove that $[a]_p^{p-1} = [1]_p$ in this case, since the stated formula follows then by multiplying both sides by $[a]_p$.

Assume then that $p \nmid a$, that is, $[a]_p \neq [0]_p$. The clever step consists of considering the $p - 1$ classes

$$[a]_p, \quad [2a]_p, \quad \ldots, \quad [(p-2)a]_p, \quad [(p-1)a]_p . \tag{2.5}$$

These are all *different from one another* (we can say they are 'distinct'). Indeed, since $[a]_p \neq [0]_p$ and p is prime,

$$[\ell a]_p = [ka]_p \implies [\ell]_p = [k]_p$$

---

[3] Unlike in the case of the more famous 'Fermat's Last Theorem' (cf. §7.1), we can be quite confident that he *did* have a proof.

by cancellation (Remark 2.17). By the contrapositive of this statement, $[\ell]_p \neq [k]_p$ implies $[\ell a]_p \neq [ka]_p$. By the second part of Lemma 2.6 the classes $[1]_p, \ldots, [p-1]_p$ are all different, and then so are the classes $[a]_p, \ldots, [(p-1)a]_p$.

But then (this is the cleverest part) the classes listed in (2.5) must be *the same as the classes* $[1]_p, \ldots, [p-1]_p$, *just in a different order.* (Work out Exercise 2.19 to see this in action.) It follows that the products of the two sets of classes must be equal:

$$[a]_p \cdot [2a]_p \cdots [(p-1)a]_p = [1]_p \cdot [2]_p \cdots [p-1]_p .$$

Therefore $[1 \cdot 2 \cdots (p-1) \cdot a^{p-1}]_p = [1 \cdot 2 \cdots (p-1)]_p$, that is,

$$[(p-1)!]_p \cdot [a]_p^{p-1} = [(p-1)!]_p .$$

Finally, it suffices to observe that $[(p-1)!]_p \neq [0]_p$ (since p is prime and does not divide any of $1, \ldots, p-1$), so applying cancellation again gives

$$[a]_p^{p-1} = [1]_p$$

as needed.   □

I mentioned the statement of Fermat's little theorem at the end of §1.4, pointing out that it is actually useful in practice, for applications to encryption. I would like to take the time to explain how this works, even if it is not necessary for the development of the main material covered in this text.

If we are handed an integer n, and we find an integer a, which may be assumed to be in the range $1 < a < n$, such that $[a]_n^{n-1} \neq [1]_n$, then Theorem 2.18 guarantees that n is *not* prime, hence it is composite. Such an integer a is said to 'witness' the fact that n is composite. For example, 2 is a witness of the fact that 4 is composite, since $2^{4-1} = 8 \not\equiv 1 \bmod 4$. If n is large, this is remarkable: factoring large integers takes an unpractically long time, while the computation of $[a]_n^{n-1}$ is relatively fast. (You understand this particularly well if you worked out Exercise 2.8.) So we may be able to tell that a large integer n is *not* prime very quickly, without having to factorize it.

Further, finding a witness is relatively easy: it can be shown that most composite numbers n have at least (roughly) $n/2$ witnesses. (See Exercise 2.20 for an idea of how this may be shown.) Therefore, if n is one of these composite numbers, and we test (for example) 10 random integers between 1 and n, then we are going to find a witness to n with a probability of $1 - 1/2^{10} = 0.999\ldots$ Bottom line: it is not too computationally expensive to tell with a high degree of confidence whether a number is prime or composite.

These considerations lead to practical ways to produce large prime integers. We can generate a random integer with, say, 500 digits, and then use refinements of these methods to essentially determine whether this integer is prime; if it is not, we throw it away and generate another random integer, and keep going until we get one that is prime. As a consequence of the Prime Number Theorem (also briefly mentioned at the end of §1.4) it can be shown that the probability of finding a prime less than n is $1/\ln n$, so several hundred tries at most will be enough to find a 500-digit prime. With current technology, doing all this takes a fraction of a second.

Why do we want to construct large primes? This is a basic ingredient in the 'RSA encryption' system, which is common place as a way to encode information transmitted through the Internet. Your laptop is likely using it when you do your online banking, for example. RSA stands for 'Rivest, Shamir, and Adleman', who came up with it back in the 1970s. It should not strike you as a clever use of sophisticated and modern mathematical tools; it should strike you as a clever use of *extremely elementary, ancient* mathematics. Indeed, at this point we have already covered enough material in these notes to understand how it works, and we do not even really know yet what a ring is! (Of course, the fact that it only uses elementary mathematics makes RSA all the more impressive.)

RSA is a 'public key' encryption system, which means that everybody can encode a message: the 'key' to encode a message is published and available to whomever cares; so is the encoded message. Nevertheless, *decoding* the message can in practice only be done by someone who has some extra information. Thus, your laptop can send your bank a key that allows the bank to encode some sensitive information and send it back to you; someone can intercept the key and the encoded message, but will not be able (with current technology) to use this information to recover the message.

How does this work? Above I gave a rough explanation of how one can produce large prime integers. To show you that everything I say can be implemented, I will use this very laptop (on which I am writing these notes) to produce two largish primes, using the method I explained above. My laptop promptly complies:

$$p = 818428743210989683024895252563, \quad q = 644677944074338849892311039879$$

(I am not pushing all the way to 500 digits for typesetting reasons). I can also ask my laptop to multiply them:

$$n = pq = 527622959544605838546815712346082387453782597046540869959877.$$

This takes no time at all. And yet, no one knows how to go equally fast from n to the individual p and q: *this* is why this system works so well, as we will see. There is no harm in 'publishing' n; p and q will remain secret. Next, I note that the number $k = (p-1)(q-1)$ happens to equal

$$k = 527622959544605838546815712344619280766497268513623663667436$$

and by using the Euclidean algorithm it is easy to verify that $\gcd(e, k) = 1$ if

$$e = 23847938479384928749387.$$

I again chose this number randomly: I just typed very fast a bunch of random digits. This is the 'encryption exponent', and it will be used to encode messages.[4] We can publish

---

[4] In fact, in practice one usually chooses $e = 2^{16} + 1 = 65537$ (this is a prime number; if it happens to be a factor of k, one just starts from scratch and chooses different primes p and q). I have chosen a different random number to illustrate that this choice is irrelevant to the method.

the two integers

$$n = 5276229595446058385468157123460823874537825970465408699598772$$

$$e = 23847938479384928749387 .$$

Now your bank, or anyone else in fact, can send you information securely. It suffices to write this information as one or more integers a, $1 \le a < n$. Each a will be encoded by computing $[a]_n^e$, which is again something that can be done quickly. For example, if my Very Secret message to you is

$$a = 1010101010101010101010101010 ,$$

then I will send you

$$b = 1636053974647888640902635251477072421368139149061258364703122$$

since (my computer assures me) $[a]_n^e = [b]_n$. To this day, no one has found an efficient way to recover a from b, given only the knowledge of n and e. So b can be transmitted in the open, and it does not matter if someone intercepts it.

So you receive b. How do *you* recover my message? As we have seen in Proposition 2.16, since $\gcd(e, k) = 1$, then e has a multiplicative inverse modulo k. Chasing the Euclidean algorithm reveals that this inverse is

$$d = 7244491894555964353301587780975315291080320758295158853369522 .$$

(Go ahead and verify that $[d]_k \cdot [e]_k = [1]_k$ if you don't trust me.) Since you know d (the 'decryption' exponent), it is a very easy matter for *you* to recover the message: lo and behold,

$$[b]_n^d = [1010101010101010101010101010]_n ,$$

as my computer confirms in a fraction of a second.

This should look amazing, but of course it is no coincidence, and we have all the tools needed to understand how it works.

> THEOREM 2.19 (RSA algorithm)   *Let $p \ne q$ be positive prime integers, and let*
>
> - $n = pq$
> - $k = (p - 1)(q - 1)$
> - *e be an integer that is relatively prime to k*
> - *d be an integer such that $de \equiv 1 \bmod k$*
> - *a be a given integer*
> - *b be a representative of $[a]_n^e$.*
>
> *Then $a \equiv b^d \bmod n$.*

Proof   Note that $b^d \equiv a^{de} \bmod n$; so we have to prove that $a^{de} \equiv a \bmod n$. First, I note that it suffices to prove that $a^{de} \equiv a \bmod p$ *and* $a^{de} \equiv a \bmod q$. Indeed, if this is the case, then

$$p \mid (a^{de} - a) \quad \text{and} \quad q \mid (a^{de} - a) ,$$

and this implies that $n \mid (a^{de} - a)$ since p and q are relatively prime. (You worked this out in Exercise 1.10. It is particularly transparent if you think in terms of prime factorizations.)

Focus on the prime p. If $p \mid a$, then both a and a^{de} are 0 mod p, and in particular $a^{de} \equiv a$ mod p. Therefore we may assume that $p \nmid a$.

Since $de \equiv 1$ mod k, k divides $de - 1$. Therefore there is a positive integer ℓ such that $de = 1 + k\ell$. Then

$$a^{de} = a^{1+k\ell} = a \cdot a^{(p-1)(q-1)\ell} = a \cdot (a^{p-1})^{(q-1)\ell}.$$

By Fermat's little theorem, $a^{p-1} \equiv 1$ mod p, since p is prime and $p \nmid a$. Therefore

$$a^{de} = a \cdot (a^{p-1})^{(q-1)\ell} \equiv a \cdot 1^{(q-1)\ell} = a \text{ mod } p,$$

as needed. The proof for q is of course entirely similar, and we are done. □

You should work out a concrete example by hand, using small primes of course, to see the cogs turning as they should. The beauty of the mechanism is that an eavesdropper is welcome to find out what n, e, and the encoded message $[a]_n^e$ are; without knowing d, the eavesdropper cannot apply Theorem 2.19 and recover the message. But *if* the eavesdropper *can* factor n, discovering p and q, then all is lost: once p and q are known, so is k, and the decryption exponent d can be recovered quickly by finding the multiplicative inverse of e mod k.

No practical methods to factor a large integer (on a classical, as opposed to quantum, computer) are known, and this question has received a lot of attention. There is a possibility that some fiendishly clever method exists to factor an integer quickly. If such a method were discovered, the security of communications over the Internet would be compromised.

Some history: In the August 1977 issue of the *Scientific American,* a message was given in its encoded form using RSA, with a 129-digit n. The estimate then was that it would take 40 quadrillion years to decoded it on a fast supercomputer of the time. This particular message was cracked in 1994 by Arjen Lenstra, using new factoring algorithms and 600 computers linked over the Internet, taking 'only' 8 months. To read about this, look at the July 1994 issue of the *Scientific American,* page 17.

In Exercises 2.1–2.6, a, b, c, k, n are integers, and k and n are positive.

Exercises

2.1 Prove that if $a \equiv b$ mod n and $k \mid n$, then $a \equiv b$ mod k.

2.2 Prove that if $a \equiv b$ mod q for all positive irreducible integers q, then $a = b$.

2.3 Prove that if $q \geq 5$ and q is irreducible, then $[q]_6 = [1]_6$ or $[q]_6 = [5]_6$.

2.4 ▷ Find a proof or a counterexample of the following statement: If $[a]_n \neq [0]_n$, and $[a]_n[b]_n = [a]_n[c]_n$, then $[b]_n = [c]_n$.

2.5 Find a class $[a]_{99} \neq [0]_{99}$ in $\mathbb{Z}/99\mathbb{Z}$ such that there is *no* integer b for which $[a]_{99}[b]_{99} = [1]_{99}$.

2.6 Assume that $[a]_n = [1]_n$. Prove that $\gcd(a, n) = 1$. Find an example showing that the converse is not true.

2.7 Prove that every positive integer is congruent modulo 9 to the sum of its digits.

2.8 ▷ Find the last digit of $7^{1000000}$.

2.9 ▷ Find a congruence class $[a]_7$ in $\mathbb{Z}/7\mathbb{Z}$ such that every class $[b]_7$ except $[0]_7$ equals a power of $[a]_7$.

2.10 ▷ Let $n > 0$ be an odd integer that may be written as the sum of two perfect squares. Prove that $[n]_4 = [1]_4$.

2.11 ▷ Prove that for all $a \in \mathbb{Z}/n\mathbb{Z}$, $a \cdot 0 = 0$ (where 0 stands for $[0]_n$).

2.12 Let $p > 0$ be a prime integer, and let $a \in \mathbb{Z}/p\mathbb{Z}$, $a \neq 0$. Prove that the multiplicative inverse of a, whose existence is guaranteed by Theorem 2.15, is *unique*.

2.13 (i) Find all the solutions of the equation $[9]_{12} x = [6]_{12}$ in $\mathbb{Z}/12\mathbb{Z}$. That is, find all $[a]_{12}$ such that $[9]_{12} \cdot [a]_{12} = [6]_{12}$.

(ii) Does the equation $[9]_{12} x = [8]_{12}$ have solutions?

2.14 Let a, b, n be integers, with $n > 0$, and assume $\gcd(a, n) = 1$. Prove that the equation $[a]_n x = [b]_n$ has a unique solution in $\mathbb{Z}/n\mathbb{Z}$.

2.15 Let $p > 0$ be a prime integer. Prove that the equation $x^2 + x = 0$ has exactly two solutions in $\mathbb{Z}/p\mathbb{Z}$, and find those solutions.

2.16 Let $n > 0$ be an integer, and let $a, b \in \mathbb{Z}/n\mathbb{Z}$. Assume that both a and b have multiplicative inverses. Prove that ab has a multiplicative inverse.

2.17 Let $n > 0$ be an integer, and let $a \in \mathbb{Z}/n\mathbb{Z}$. Prove that either a has a multiplicative inverse or there exists a $b \in \mathbb{Z}/n\mathbb{Z}$, $b \neq 0$, such that $ab = 0$.

2.18 ▷ Prove that the two formulations of Fermat's little theorem are equivalent.

2.19 ▷ (i) Compute the classes $[1 \cdot 3]_{11}, [2 \cdot 3]_{11}, [3 \cdot 3]_{11}, \ldots, [10 \cdot 3]_{11}$ and confirm that these are the classes $[1]_{11}, \ldots, [10]_{11}$, in a different order. (A generalization of this fact is used in the proof of Theorem 2.18.)

(ii) Compute the classes $[1 \cdot 3]_{12}, [2 \cdot 3]_{12}, [3 \cdot 3]_{12}, \ldots, [11 \cdot 3]_{12}$ and observe that these do *not* equal the classes $[1]_{12}, \ldots, [11]_{12}$, in any order.

2.20 ▷ Let $n, a, b_1, \ldots, b_r$ be integers, with $n > 1$ and $1 < a < n$, $1 < b_i < n$. Assume that a is a witness to the fact that n is composite (that is, $a^{n-1} \not\equiv 1 \bmod n$) and that $\gcd(a, n) = 1$.

• Prove that if b_i is *not* a witness, then ab_i is a witness.
• Prove that if $[b_1]_n, \ldots, [b_r]_n$ are all different, then so are $[ab_1]_n, \ldots, [ab_r]_n$.
• Deduce that in this case n has fewer than $n/2$ non-witnesses.

2.21 A composite number n is a *Carmichael number* if $a^{n-1} \equiv 1$ modulo n for all a relatively prime to n.

• Prove that 1729 is a Carmichael number.
• Prove that if n is a square-free integer (that is, n is not a multiple of a perfect square) such that $p - 1$ divides $n - 1$ for all prime factors p of n, then n is a Carmichael number.

(*Hint:* Do the second part before doing the first.) It is known that there exist infinitely many Carmichael numbers, but this was only proved in 1994.

2.22 Construct a concrete example of RSA encryption, using 1-digit primes.

3 Rings

3.1 Definition and Examples

If you have made it this far, congratulations: this is where the text really begins. Up to this point, we have only given a good look at objects with which you were already likely rather familiar. This is useful in itself, and has the added advantage of showing you that you already know a large collection of examples of the notion that I am about to define.

A 'binary operation' on a set R is a function $R \times R \to R$; if the name of the operation is $*$, we usually write $a * b$ for the image of the pair (a, b). More generally, we may consider binary operations with values in another set S, maybe containing R; if $a * b \in R$ for all a, b in R, we would then say that R is 'closed' with respect to $*$. If no other set S is mentioned, the set R is assumed to be closed with respect to the operations. This is the case in the following definition.

DEFINITION 3.1 A 'ring' $(R, +, \cdot)$ consists of a set R along with two binary operations $+$ and $\cdot$ (called 'addition' and 'multiplication'), satisfying the following properties.

(i) Associativity for addition:

$$\forall a, b, c \in R \qquad (a + b) + c = a + (b + c)$$

(ii) Existence of additive identity:

$$\exists 0_R \in R \; \forall a \in R \qquad a + 0_R = 0_R + a = a$$

(iii) Existence of additive inverses:

$$\forall a \in R \; \exists z \in R \qquad a + z = z + a = 0_R$$

(iv) Commutativity for addition:

$$\forall a, b \in R \qquad a + b = b + a$$

(v) Associativity for multiplication:

$$\forall a, b, c \in R \qquad (a \cdot b) \cdot c = a \cdot (b \cdot c)$$

(vi) Existence of multiplicative identity:

$$\exists 1_R \in R \; \forall a \in R \qquad\qquad a \cdot 1_R = 1_R \cdot a = a$$

(vii) Distributivity:

$$\forall a, b, c \in R \qquad\qquad a \cdot (b + c) = a \cdot b + a \cdot c$$
$$(a + b) \cdot c = a \cdot c + b \cdot c$$

The operations are often understood in the notation, and we refer to a ring $(R, +, \cdot)$ simply by the name of the set R. The elements 0_R, 1_R are often simply denoted 0 and 1, if this does not lead to ambiguities. By the same token, the operations $+$, $\cdot$ may also be denoted $+_R$, $\cdot_R$.

Very soon we will show (Proposition 3.14) that the elements called 0 ($= 0_R$) and 1 ($= 1_R$) in Definition 3.1 are *unique:* in a ring, *only one* element can play the role of an 'additive identity', and we call that element 0; *only one* element can play the role of a 'multiplicative identity', and we call that element 1. This may seem natural, but it will require a proof: uniqueness is not explicitly listed among the axioms in Definition 3.1, so we will have to understand why it follows from those axioms. The same applies to the 'additive inverse' mentioned in property (iii).

Caveat. Not everybody agrees that the definition of 'ring' should include the existence of the multiplicative identity, axiom (vi) in the list in Definition 3.1. There are good reasons for both choices, and each is standard in a particular context; the reader should be aware that the choice taken in this text reflects my own personal bias. The distinction is relatively minor; what I call a *ring,* others may call a *ring with identity.* All the rings we will encounter will be 'rings with identity'. Below (Example 3.5) I will give an example of a structure satisfying all the axioms except (vi). From my point of view this is a 'ring without identity', and some have tried to adopt the term 'rng' (as in 'r*i*ng without *i*dentity') for this notion.

You should note that the list of properties given in Definition 3.1 is *exactly* the same as the list (2.3) presented in §2.4. These properties are called the 'ring axioms'. In §2.4 we observed that $\mathbb{Z}$ and $\mathbb{Z}/n\mathbb{Z}$ (with the operations $+$, $\cdot$ introduced in Definition 2.10) satisfy all the ring axioms; we can now simply say that $\mathbb{Z}$ and all $\mathbb{Z}/n\mathbb{Z}$ are *rings.*

Example 3.2 $\mathbb{Z}$ is a ring.

Example 3.3 The sets $\mathbb{Q}$, $\mathbb{R}$, and $\mathbb{C}$ of rational, real, and complex numbers form rings with the usual operations.

Example 3.4 For every nonnegative integer n, $\mathbb{Z}/n\mathbb{Z}$ is a ring.

Example 3.5 A 'non-example': The set $5\mathbb{Z}$ of multiples of 5: $\{\ldots, -5, 0, 5, 10, \ldots\}$. With the usual operations of addition and multiplication, this set satisfies all the axioms listed above with the exception of (vi): there is no multiplicative identity. Indeed, this would be the ordinary number 1, and $1 \notin 5\mathbb{Z}$. As mentioned above, some authors *would* consider this a ring (and may be appalled at the suggestion that it isn't!); from the perspective of this text, it is a 'rng'. Of course the same applies to every $n\mathbb{Z}$ for $n \neq \pm 1$. According to the convention adopted in this text, the set $2\mathbb{Z}$ of even integers is a rng, not a ring.

Example 3.6 Another non-example, for more serious reasons: Consider the set $\mathbb{N} = \mathbb{Z}^{\geq 0}$ of natural numbers. With the usual operations $+$, $\cdot$, $\mathbb{N}$ satisfies all the ring axioms *with the exception of* the existence of additive inverses, axiom (iii): for instance, there is no $a' \in \mathbb{N}$ such that $1 + a' = 0$. (Such a' would be -1, and -1 is not an element of $\mathbb{N}$.) Thus, $\mathbb{N}$ is *not* a ring. As it happens, axiom (iii) is so fundamental that nobody on this planet calls $\mathbb{N}$ a ring.

Example 3.7 Another non-example! The set $\mathbb{Z}^{\leq 0}$ of 'nonpositive numbers' is even worse off than $\mathbb{N}$, since $(-1) \cdot (-1) = 1 \notin \mathbb{Z}^{\leq 0}$: the set $\mathbb{Z}^{\leq 0}$ is not closed with respect to (ordinary) multiplication. Thus, we do not even know how to interpret the last three axioms in the list, for this set (at least if we stick to the ordinary operations).

Several of the examples or non-examples presented above have something to do with 'numbers', but you should not take this as a given. In fact, it is important that you dissociate symbols that you are used to, such as 0, 1, $+$, $\cdot$, from their usual meaning as 'numbers' or 'operations among numbers'. We have already studied the examples $\mathbb{Z}/n\mathbb{Z}$, where 0 really means $[0]_n$, etc.: elements of the rings $\mathbb{Z}/n\mathbb{Z}$ are not numbers. What $+$ and $\cdot$ mean in the context of $\mathbb{Z}/n\mathbb{Z}$ was explained in Definition 2.10, and required a careful discussion. In general, elements of a ring may have *nothing to do* with numbers, and the operations $+$, $\cdot$ will have to be defined anew in each instance. The reason why we recycle notation such as 0, 1, $+$, $\cdot$ is in order to avoid a proliferation of symbols, and because (as we will see) these symbols behave in many ways similarly to their 'ordinary' namesakes. But this is a good time to make a shift in perspective: notwithstanding uninformed platitudes such as 'mathematics is the science of numbers', it turns out that $\mathbb{Z}$ is just one object in the category of 'rings'; it is just one (important) example among a large infinity of structures all satisfying the axioms listed in Definition 3.1.

A very sensible question now would be: *why* do we study rings? The short answer is that they 'arise in nature'. There are many, many examples arising naturally in several branches of mathematics, which turn out to be rings. Studying rings as abstract entities will give us information on all these specific instances of this notion. If you find this a bit hard to appreciate at this stage, it is because you do not have a large catalogue of examples. Let's see a few more.

Example 3.8 A ring R cannot be *empty*: axioms (ii) and (vi) prescribe the existence of some elements in R, so $R = \emptyset$ is not an option. How small can R be? One simple example that defies our experience with numbers is the ring consisting of a *single* element, say r, with operations

$$r + r := r, \quad r \cdot r := r.$$

You can scan quickly the ring axioms and verify that they are all trivially satisfied; and indeed, the corresponding ring is called the 'trivial' ring. In the trivial ring 0_R and 1_R both equal r: there is no room for anything else. So $0 = 1$ in the trivial ring. We have run across a concrete realization of this ring: $\mathbb{Z}/1\mathbb{Z}$ consists of a single class $[0]_1$, and indeed $[0]_1 = [1]_1$. You should check that in fact *if* $0 = 1$ in a ring R, then necessarily R is trivial, in the sense that it consists of a single element as in this example (Exercise 3.5).

Example 3.9   You have likely taken a course in linear algebra, and in that case you are already familiar with *matrices*. Consider the set $M_{2,2}(\mathbb{R})$ of 2×2 matrices with entries in $\mathbb{R}$. In linear algebra you discuss operations among matrices, defined as follows:

$$\begin{pmatrix} a' & b' \\ c' & d' \end{pmatrix} + \begin{pmatrix} a'' & b'' \\ c'' & d'' \end{pmatrix} = \begin{pmatrix} a' + a'' & b' + b'' \\ c' + c'' & d' + d'' \end{pmatrix}$$

for addition, and

$$\begin{pmatrix} a' & b' \\ c' & d' \end{pmatrix} \cdot \begin{pmatrix} a'' & b'' \\ c'' & d'' \end{pmatrix} = \begin{pmatrix} a'a'' + b'c'' & a'b'' + b'd'' \\ c'a'' + d'c'' & c'b'' + d'd'' \end{pmatrix}$$

for multiplication. Then I claim that $(M_{2,2}(\mathbb{R}), +, \cdot)$ is a ring. You should check this yourself; you will let

$$0 = 0_{M_{2,2}(\mathbb{R})} = \begin{pmatrix} 0 & 0 \\ 0 & 0 \end{pmatrix}, \quad 1 = 1_{M_{2,2}(\mathbb{R})} = \begin{pmatrix} 1 & 0 \\ 0 & 1 \end{pmatrix},$$

figure out what the 'additive inverse' should be, and verify the other properties. The manipulations may turn out to be a little annoying; for example, to verify associativity of multiplication requires the following computations:

$$\left(\begin{pmatrix} a' & b' \\ c' & d' \end{pmatrix} \cdot \begin{pmatrix} a'' & b'' \\ c'' & d'' \end{pmatrix} \right) \cdot \begin{pmatrix} a''' & b''' \\ c''' & d''' \end{pmatrix} = \begin{pmatrix} a'a'' + b'c'' & a'b'' + b'd'' \\ c'a'' + d'c'' & c'b'' + d'd'' \end{pmatrix} \cdot \begin{pmatrix} a''' & b''' \\ c''' & d''' \end{pmatrix}$$

$$= \begin{pmatrix} (a'a'' + b'c'')a''' + (a'b'' + b'd'')c''' & (a'a'' + b'c'')b''' + (a'b'' + b'd'')d''' \\ (c'a'' + d'c'')a''' + (c'b'' + d'd'')c''' & (c'a'' + d'c'')b''' + (c'b'' + d'd'')d''' \end{pmatrix},$$

$$\begin{pmatrix} a' & b' \\ c' & d' \end{pmatrix} \cdot \left(\begin{pmatrix} a'' & b'' \\ c'' & d'' \end{pmatrix} \cdot \begin{pmatrix} a''' & b''' \\ c''' & d''' \end{pmatrix} \right) = \begin{pmatrix} a' & b' \\ c' & d' \end{pmatrix} \cdot \begin{pmatrix} a''a''' + b''c''' & a''b''' + b''d''' \\ c''a''' + d''c''' & c''b''' + d''d''' \end{pmatrix}$$

$$= \begin{pmatrix} a'(a''a''' + b''c''') + b'(c''a''' + d''c''') & a'(a''b''' + b''d''') + b'(c''b''' + d''d''') \\ c'(a''a''' + b''c''') + d'(c''a''' + d''c''') & c'(a''b''' + b''d''') + d'(c''b''' + d''d''') \end{pmatrix},$$

and then we need to work out the individual entries in the two results to verify that they agree. For example,

$$(a'a'' + b'c'')a''' + (a'b'' + b'd'')c''' = a'a''a''' + b'c''a''' + a'b''c''' + b'd''c''',$$

$$a'(a''a''' + b''c''') + b'(c''a''' + d''c''') = a'a''a''' + a'b''c''' + b'c''a''' + b'd''c'''$$

agree as promised, because addition is commutative in $\mathbb{R}$. Perform a similar comparison on the other three entries, and this will conclude the verification that whenever A', A'', A''' are in $M_{2 \times 2}(\mathbb{R})$, we have

$$(A' \cdot A'') \cdot A''' = A' \cdot (A'' \cdot A'''), \tag{3.1}$$

as needed in order to verify axiom (v).

Phew, this was not fun to typeset. Distributivity is similarly notation-intensive. But you may have already done all this work in Linear Algebra: surely (3.1) was discussed in such a course. It is in fact just as straightforward to verify that the set of $n \times n$ matrices with entries in $\mathbb{R}$, $M_{n \times n}(\mathbb{R})$, forms a ring with the usual matrix operations, for every $n \geq 1$.

So you do have a large collection of new examples of rings to contemplate, and again the elements of these rings are not 'numbers'. You can construct even more exotic examples, since you could consider a ring of square matrices $M_{n \times n}(R)$ with entries *in any ring,* and this would give you a new ring. The verification of associativity I worked out above boiled down to the fact that $\cdot$ is associative among the entries, and $+$ is commutative; both facts are true in every ring. If you are very patient, you could write out the multiplication table for the ring $M_{2 \times 2}(\mathbb{Z}/2\mathbb{Z})$, whose elements are things like this:

$$\begin{pmatrix} [0]_2 & [1]_2 \\ [1]_2 & [1]_2 \end{pmatrix},$$

or write out a formula for multiplication for the ring $M_{2 \times 2}(M_{2 \times 2}(\mathbb{Z}))$, whose elements are things like this:

$$\begin{pmatrix} \begin{pmatrix} 1 & 0 \\ -1 & 5 \end{pmatrix} & \begin{pmatrix} 3 & 7 \\ -11 & 4 \end{pmatrix} \\ \begin{pmatrix} 0 & -3 \\ 17 & 0 \end{pmatrix} & \begin{pmatrix} 21 & -3 \\ 57 & 8 \end{pmatrix} \end{pmatrix}.$$

These are all fine rings, and some of them are quite useful.

Example 3.10 Another class of examples with which you are already familiar is given by *polynomials*. A 'polynomial in one indeterminate x', with coefficients in (say) $\mathbb{R}$, is a gadget like this:

$$\alpha = \sum_{i \geq 0} a_i x^i = a_0 + a_1 x + \cdots + a_{n-1} x^{n-1} + a_n x^n,$$

where all the coefficients a_i are real numbers and *only finitely many of the coefficients are not equal to 0* ($a_i = 0$ for $i > n$ in this example). You have plenty of experience with polynomials, for example from calculus courses. You can add and multiply polynomials: if

$$\beta = \sum_{i \geq 0} b_i x^i = b_0 + b_1 x + \cdots + b_{m-1} x^{m-1} + b_m x^m,$$

then

$$\alpha + \beta := \sum_{i \geq 0} (a_i + b_i) x^i = (a_0 + b_0) + (a_1 + b_1) x + (a_2 + b_2) x^2 + \cdots$$

and

$$\alpha \cdot \beta := (a_0 + b_0) + (a_1 b_0 + a_0 b_1)x + (a_2 b_0 + a_1 b_1 + a_0 b_2)x^2 + \cdots .$$

Well, you probably expect the next punch line: if $\mathbb{R}[x]$ denotes the set of polynomials in an indeterminate x and with coefficients in $\mathbb{R}$, then $(\mathbb{R}[x], +, \cdot)$ satisfies all the axioms listed in Definition 3.1, so it is a ring. This requires a verification, which I am happily leaving to you.

You may also expect further punch lines. First, there is nothing special about $\mathbb{R}$, and we could replace it with *any* ring R, getting a new ring $R[x]$. Second, we could consider rings of polynomials in two indeterminates: $R[x, y]$, or however many you feel like: $R[x_1, \ldots, x_n]$. For example, an element of $R[x, y]$ will be a finite sum of the form $\sum_{i,j} a_{i,j} x^i y^j$, with $a_{i,j} \in R$.

Again, this gives you a huge collection of new examples of rings. We will explore polynomial rings in some detail later in these notes, particularly in Chapter 7.

Example 3.11   In calculus, polynomials are used as good examples of *functions*. In fact, polynomials detemine functions over *any* ring. If

$$f = a_0 + a_1 x + \cdots + a_n x^n$$

is a polynomial in $R[x]$, where R is any ring, then f defines a function $R \to R$, which I would also denote by f: this would be obtained by setting

$$\forall r \in R \qquad f(r) := a_0 + a_1 r + \cdots + a_n r^n ,$$

that is, the result of 'plugging in r for x'. We will come back to this below, in Example 4.22.

Remark 3.12   Note that different polynomials may correspond to the same function: for example, $x(x-1) \in (\mathbb{Z}/2\mathbb{Z})[x]$ determines the 0-function, but it is not the 0-polynomial. In the following, I may use the notation $f(x)$ to denote a polynomial in $R[x]$. This is just done to record the name of the indeterminate, and should not lead you into thinking that polynomials are some 'special kinds of function'. They are really different gadgets.   ⌟

Can we make rings by using more general functions? Of course we can. Let $\mathcal{F}$ denote the set of (for example) real-valued functions of one real variable x. If f, g are elements of $\mathcal{F}$, then we can define two new functions $f + g$, $f \cdot g$, by letting $\forall x$

$$(f + g)(x) := f(x) + g(x), \quad (f \cdot g)(x) = f(x) \cdot g(x). \tag{3.2}$$

The resulting structure $(\mathcal{F}, +, \cdot)$ is a ring, as you may verify. And again we may consider many, many variations on this theme. We could use only *continuous* functions, and we would get a new ring: in Calculus we prove—or at least mention—that the sum and product of two continuous functions is continuous, so this set is closed with respect to the operations; and the ring axioms are easy to verify. Or we could consider functions defined on a subset of $\mathbb{R}$, for example an interval; or $\mathbb{C}$-valued instead of $\mathbb{R}$-valued functions; and so on.

We could even just consider the set of functions $f: S \to R$, where S is *any* set and

R is *any* ring, with operations defined precisely as in (3.2). This structure is also a ring, as you may verify easily enough. The '0' in this ring is the function that sends every element of S to the 0 in R; the '1' is likewise the function that sends every element of S to the element 1 in R.

Example 3.13   For an example that may sound a little fancier, let V be a vector space and consider the set of *linear transformations* $V \to V$. (You have probably learned about all this in Linear Algebra, maybe with slightly different notation.) You can define addition and multiplication as follows. If α, β are linear transformations $V \to V$, define $\alpha + \beta$ by setting

$$\forall v \in V \quad (\alpha + \beta)(v) := \alpha(v) + \beta(v) \, ;$$

and $\alpha\beta$ by setting

$$\forall v \in V \quad (\alpha\beta)(v) := \alpha(\beta(v)) \, .$$

Once more, verifying the ring axioms for these operations should be your business, not mine; I promise that they all check just fine. So you have a 'ring of linear transformations of a vector space'. If you know your linear algebra, then you see that this is a natural generalization of the ring of matrices. We will not have any immediate use for this ring, so I won't discuss it further now. But it is in a sense more representative of what a 'general' ring may look like; examples such as $\mathbb{Z}$, $\mathbb{Z}/n\mathbb{Z}$ are much more special, and therefore not as representative.

Bottom line: the notion of 'ring' is ubiquitous. There is a technical reason why this is the case: in a distant future (Remark 10.7) we will see that every *abelian group* gives rise to a ring in a very natural way; but we have to cover quite a bit of material before I can even state this fact precisely, so you can safely ignore this comment for now. In the rest of this chapter we are going to explore consequences that can be drawn from the ring axioms, and further constructions we can perform whenever we have a ring. Such properties and constructions will apply to *every* example reviewed above, as well as to any other ring you may ever come across. We will also focus on a few key examples illustrating certain features of the theory; *polynomial rings* will play a prominent role. Such examples are special enough that we can analyze them in detail, and at the same time general enough that they are the basic objects of study in entire areas of mathematics, such as number theory or algebraic geometry.

3.2   Basic Properties

Now we know what a ring is, and we have looked at several examples. We will next establish some general facts about rings, deriving them as logical consequences of the ring axioms. Most of them will look very reasonable, if not outright 'obvious'; this is because the notation in ring theory has been chosen so as to evoke familiar facts, and many of them turn out to be true. But keep in mind that 'obvious' is an essentially forbidden word in mathematics: if something is obvious, then it should be very easy to explain. If you can't find a simple, logical explanation of something that you think is 'obvious', think again.

I have mentioned (and you can prove if you work out Exercise 3.5) that as soon as R contains two or more elements, then two of these must be 0_R and 1_R, respectively. Let's verify that these elements 0_R and 1_R are *unique*, that is, that you cannot satisfy (for example) ring axiom (ii) with two different elements playing the role of 0_R. In fact, this should have been checked even before giving these elements a name: if *two* elements in a ring satisfied ring axiom (ii), which one should be called 0_R? Giving the same name to more than one element in a set is not a viable option. But in this case the axioms do imply that 0_R and 1_R are unique, so this abuse of language is forgivable.

PROPOSITION 3.14   *Let R be a ring. Then*

- *If $0'_R$ and $0''_R$ both satisfy the requirement on 0_R in property (ii), then $0'_R = 0''_R$.*
- *If $1'_R$ and $1''_R$ both satisfy the requirement on 1_R in property (vi), then $1'_R = 1''_R$.*

Proof   More generally, we say that an element e is an 'identity' for an operation $*$ if

$$\forall a \qquad a * e = e * a = a. \qquad (3.3)$$

I claim that the identity of *any* operation is necessarily unique, that is: if e' and e'' both satisfy (the role of e in) (3.3), then $e' = e''$. To see this, simply note that since e'' satisfies (3.3), then

$$e' = e' * e'' ;$$

and since e' satisfies (3.3), then

$$e' * e'' = e''.$$

Therefore $e' = e' * e'' = e''$ as needed. (In fact, this proves that $e' = e''$ as soon as e' is an identity 'on the left' and e'' is an identity 'on the right'.) Applying this observation to the operations $+$ and $\cdot$ in a ring gives the statement.   □

A similar observation applies to the *additive inverse,* whose existence is required by ring axiom (iii). You might have expected me to call the additive inverse of a something like '$-a$', given that it should give[1] 0 when added to a. I refrained from doing so because calling it $-a$ implies that the inverse is *determined* by a, that is, that there is a unique element z satisfying requirement (iii) for a given element $a \in R$; we have not verified this yet. However, it *is* the case for us, because it can be proved from the ring axioms.

PROPOSITION 3.15   *Let R be a ring, and let a be an element of R. Then the additive inverse of a is unique, that is: if $z', z'' \in R$ are such that*

$$a + z' = z' + a = 0 \quad and \quad a + z'' = z'' + a = 0, \qquad (3.4)$$

---

[1]   I am now going to ignore the subscript R and call 0_R, 1_R simply 0 and 1. Usually this does not lead to confusion, and you get used to it very quickly.

then necessarily $z' = z''$.

Proof The key is to consider the sum of z', a, and z'', and use associativity. The whole argument can be compressed into one line:

$$z' \overset{(a)}{=} z' + 0 \overset{(b)}{=} z' + (a + z'') \overset{(c)}{=} (z' + a) + z'' \overset{(d)}{=} 0 + z'' \overset{(e)}{=} z'' :$$

identity (a) is true by ring axiom (ii); (b) is true by the first assumption in (3.4); (c) is true by the associativity of addition (ring axiom (i)); (d) is true by the second assumption in (3.4); and (e) is again true by ring axiom (ii). □

Now that we know that the additive inverse of a is indeed determined by a, I feel comfortable calling it $-a$, and axiom (iii) may be rewritten as follows:

$$a + (-a) = (-a) + a = 0. \tag{3.5}$$

This is surely as you expected. What is $-(-a)$? Call this element b for a moment, then b should satisfy

$$(-a) + b = b + (-a) = 0;$$

comparing with (3.5) we see that a *already* satisfies this requirement. And then $a = b = -(-a)$ because of the uniqueness of additive inverses. So $-(-a) = a$ as you might have expected. But even something like this is not 'obvious' in the general context of rings: it follows from the ring axioms by a simple argument, which you are well advised to understand and remember.

Notation: $b - a$ is short for $b + (-a)$. This operation behaves just like ordinary 'subtraction'.

A convenient shorthand is to avoid the use of parentheses where possible. The operation $+$ is a *binary* operation: it tells us how to add *two* elements of the ring; something like the addition of five elements a_1, a_2, a_3, a_4, a_5, which we would like to denote by $a_1 + a_2 + a_3 + a_4 + a_5$, is simply *not* defined *a priori*. If we want to define it, in principle I should indicate in which order I want to perform the operations, and make a distinction between (for example)

$$a_1 + ((a_2 + a_3) + (a_4 + a_5)) \quad \text{and} \quad (a_1 + a_2) + (a_3 + (a_4 + a_5)).$$

Again, $+$ is a *binary* operation, so in principle I only know what the addition of *two* elements should be. According to the first expression, I should first add a_2 and a_3; then a_4 and a_5; then add these two results; and finally add a_1 to the result of this last operation. The second expression specifies a different order in which these operations should be performed. Is it clear that I would get the same element as a result? Luckily, the associativity of addition ensures that these two elements *are* equal: indeed,

$$a_1 + \underline{((a_2 + a_3) + (a_4 + a_5))} = (a_1 + (a_2 + a_3)) + \underline{(a_4 + a_5)}$$
$$= ((a_1 + \underline{(a_2 + a_3)}) + a_4) + a_5$$
$$= (((a_1 + a_2) + a_3) + a_4) + a_5$$

and

$$(a_1 + a_2) + \underline{(a_3 + (a_4 + a_5))} = \underline{((a_1 + a_2) + a_3)} + (a_4 + a_5)$$
$$= (((a_1 + a_2) + a_3) + a_4) + a_5 \,,$$

with the same result. At each step I am *only* using one application of the associative law for addition, as stated in the ring axioms. (I am underlining the parentheses I am moving at each step by applying the associative law.) It can be shown (by induction) that this can always be done: you can always reduce any meaningful arrangement of parentheses in the addition or multiplication of any number of elements of a ring to a standard arrangement, as I did above. As a result, the specific order in which you perform the operations[2] in (for instance) the multiplication of finitely many elements is not important, and no parentheses need be included.

We will also adopt the usual shorthand of replacing $\cdot$ with simple juxtaposition when this does not lead to ambiguity or particularly ugly notation: for example, I could state the associativity of multiplication by simply writing $(ab)c = a(bc)$. We also use the usual power notation, in every ring: so a^n stands for a product $a \cdots a$ with n factors, whenever n is a positive integer; and $a^0 = 1$ by convention. Similarly, we adopt the 'multiple' notation: for example, $18a$ is shorthand for

$$a + a + a + a + a + a + a + a + a + a + a + a + a + a + a + a + a + a \,. \qquad (3.6)$$

By convention, $0a$ equals 0 and 'negative' multiples are defined through additive inverses: thus, $(-5)a$ means $-(a + a + a + a + a)$. Of course this equals $(-a) + (-a) + (-a) + (-a) + (-a)$, as you should verify (Exercise 3.7).

Remark 3.16   The 'multiple' notation may lead you astray, if you don't parse it carefully. In an arbitrary ring R, it makes sense to write '$2a$' for an element a. In general, this does *not* mean the multiplication of an element $2 \in R$ with the element $a \in R$; it may in fact make no sense whatsoever to think of '2' as an element of R. The notation $2a$ simply stands for $a + a$.   ⌐

One important consequence of the ring axioms is the following; it may be called 'additive cancellation', or simply 'cancellation'.

PROPOSITION 3.17   *Let R be a ring, and let $a, b, c \in R$. Assume that $a + b = a + c$. Then $b = c$.*

Proof   Here is a one-line proof:

$$b = 0 + b = ((-a) + a) + b = (-a) + (a + b) = (-a) + (a + c) = ((-a) + a) + c = 0 + c = c \,.$$

You should be able to justify each $=$ in this string of identities, for example by referring to a specific ring axiom. Please do so.   □

---

[2] *Note:* I am talking about the order *of the operations*, not *of the factors*. In a general ring, there is no reason why $a \cdot b$ and $b \cdot a$ should be the same element.

Notice that, in general, 'multiplicative cancellation' should *not* be expected to hold in a ring: we have already run across examples of rings where it does not hold (see Remark 2.17). So the statement 'if $ab = ac$, then $b = c$' is simply *not true* in general in a ring (even if we assume $a \neq 0$), no matter how 'obvious' it may look. No matter how clever you are, you will not be able to prove this statement by using the ring axioms. The statement *will* be true in special types of rings, as we will see in a moment (Proposition 3.26).

Additive cancellation is handy. For example, the following observation is an immediate consequence.

> COROLLARY 3.18 *Let R be a ring, and let $a \in R$. Then $0 \cdot a = a \cdot 0 = 0$.*

Proof To prove the identity $a \cdot 0 = 0$:

$$a \cdot 0 + a \cdot 0 = a \cdot (0 + 0) = a \cdot 0 = a \cdot 0 + 0$$

by distributivity and axiom (ii), and

$$a \cdot 0 = 0$$

follows then by cancellation (Proposition 3.17). The proof of the identity $0 \cdot a = 0$ is entirely analogous. □

There are other easy consequences of the ring axioms, which make the operations $+$ and $\cdot$ behave as you may expect, given your familiarity with examples such as $\mathbb{Z}$. It would be futile to try and compile a complete list; the list would likely be too long, and each individual statement is usually very easy to prove from the ring axiom, if it is true. Several examples are covered in the exercises.

3.3 Special Types of Rings

The notation $+, \cdot, 0, 1$ used in the ring axioms serves you well in the sense that these symbols guide your intuition in your expectations of what may be true in a ring, based on your experience with friendly rings such as $\mathbb{Z}$, $\mathbb{Q}$, and others. For example, although '$a \cdot 0 = 0$' is *not* one of the axioms listed in Definition 3.1, it turned out to be an easy consequence of these axioms, as we have seen in Corollary 3.18. Therefore, it must be true in *every* ring R, for *every* element $a \in R$. This is a pleasant state of affairs, but you should not completely trust your intuition, because examples such as $\mathbb{Z}$ are *very* special, so they do not really capture the essence of the notion of 'ring'. More exotic examples such as the rings we have encountered in Example 3.13 are more representative, but then, we are much less familiar with such examples. If you run into a promising-looking property, you should ask yourself whether it also is a consequence of the axioms. Unless you see how, you should remain skeptical.

For example, you may have noticed that the ring axioms include the requirement that the addition be *commutative:* ring axiom (iv) states

$$\forall a, b \in R \qquad a + b = b + a.$$

On the other hand, for $R = \mathbb{Z}$ it so happens that

$$\forall a, b \in R \qquad ab = ba.$$

That is, multiplication is *also* commutative in $\mathbb{Z}$.

Might there be some very clever way to combine the ring axioms and deduce this property in every ring? To convince the most skeptical of readers that this is not the case, I only need to show an example of one ring R and of two elements $a, b \in R$ for which $ab \neq ba$.

Example 3.19   Consider the ring of matrices $M_{2 \times 2}(\mathbb{R})$ (cf. Example 3.9). Since

$$\begin{pmatrix} 1 & 1 \\ 0 & 1 \end{pmatrix} \cdot \begin{pmatrix} 1 & 0 \\ 1 & 1 \end{pmatrix} = \begin{pmatrix} 2 & 1 \\ 1 & 1 \end{pmatrix}$$

while

$$\begin{pmatrix} 1 & 0 \\ 1 & 1 \end{pmatrix} \cdot \begin{pmatrix} 1 & 1 \\ 0 & 1 \end{pmatrix} = \begin{pmatrix} 1 & 1 \\ 1 & 2 \end{pmatrix},$$

we see that the multiplication in this ring of matrices does *not* satisfy the commutativity law.

DEFINITION 3.20   A 'commutative ring' is a ring R in which multiplication is commutative, that is, in which the property

$$\forall a, b \in R \qquad ab = ba$$

holds.

As we have just seen (Example 3.19), *not every* ring is commutative. You may go through the lists of examples given in §3.1 and see which of the rings shown there are commutative. Many will be. Commutative rings abound in nature, even though in principle a 'random' ring should not be expected to be commutative. *Commutative algebra* is a branch of mathematics of its own, and it is devoted to the study of commutative rings.[3] Results in commutative algebra are crucial to the fields of *number theory*[4] and *algebraic geometry.*[5] Most of the examples we will use in this text will be commutative, mostly because I earn a living by (teaching and) doing research in algebraic geometry. However, noncommutative rings are extremely important. A basic example of a noncommutative ring is the ring of 'linear transformations of a vector space' V, discussed

---

[3] In the useful 'Mathematics Subject Classification' (MSC) compiled by the American Mathematical Society, *commutative algebra* is item 13; 'algebra' is too big to have its own item.

[4] Item 11.

[5] Item 14.

in Example 3.13: it is noncommutative as soon as $\dim V > 1$. Rings of matrices are ubiquitous in physics, particularly quantum mechanics. Methods of *noncommutative geometry*[6] derive the Lagrangian of the standard model of particle physics from a space constructed starting from rings of 1×1, 2×2, and 3×3 complex matrices.

For the rest of this section, let's assume that our rings are commutative: you know examples such as $\mathbb{Z}$, $\mathbb{Z}/n\mathbb{Z}$, polynomial rings such as $\mathbb{Q}[x]$, and so on. Here is another property we are used to, based on our familiarity with $\mathbb{Z}$:

$$\forall a, b \qquad ab = 0 \implies a = 0 \text{ or } b = 0. \tag{3.7}$$

If you have paid attention, you already know that this property does *not* hold in every ring, even if commutative.

Example 3.21 In the ring $\mathbb{Z}/10\mathbb{Z}$, $[2]_{10} \cdot [5]_{10} = [0]_{10}$ even though $[2]_{10} \neq [0]_{10}$ and $[5]_{10} \neq [0]_{10}$.

Elements such as $[2]_{10}$ and $[5]_{10}$ in this example are called *zero-divisors*, somewhat unfortunately since after all *every* element divides 0.

DEFINITION 3.22 Let R be a ring. We say that $a \in R$ is a 'zero-divisor' if there exists a *nonzero* $b \in R$ such that $ab = 0$ or $ba = 0$.

Therefore 0 is always a zero-divisor in a nontrivial ring (since $0 \cdot 1 = 0$ and $1 \neq 0$), while there are rings such as $\mathbb{Z}/10\mathbb{Z}$ where there are *nonzero* zero-divisors: $[2]_{10}$ is a zero-divisor. If an element is not a zero-divisor, we say that it is a '*non-zero-divisor*', often abbreviated *nzd*.

Thus, a is a non-zero-divisor precisely when $ab = 0$ implies $b = 0$ and $ba = 0$ implies $b = 0$. There is a name for the nontrivial commutative rings in which *every nonzero element is a non-zero-divisor*, that is, nontrivial rings satisfying (3.7); the following terminology was already mentioned in §2.4.

DEFINITION 3.23 Let R be a commutative ring. Then R is an 'integral domain' if

$$\forall a, b \in R \qquad ab = 0 \implies a = 0 \text{ or } b = 0$$

and $0 \neq 1$ in R.

There are good reasons for requiring $0 \neq 1$, i.e., for requiring that R not be trivial, essentially boiling down to the fact that we don't want to view 1 as irreducible (which was briefly discussed after Remark 1.23). This will become clearer when we discuss the general relation between integral domains and 'prime ideals' in §6.1.

Example 3.24 In fact, we have seen an inkling of this relation already. We have proved that (3.7) holds for the ring $\mathbb{Z}/n\mathbb{Z}$ if and only if n is a *prime* integer: see Theorem 2.15 for $n > 1$; and $\mathbb{Z}/0\mathbb{Z}$ is an integral domain because $\mathbb{Z}$ is.

---

[6] Items 58B34, 81R60 in the MSC.

Thus, $\mathbb{Z}/p\mathbb{Z}$ in an integral domain precisely when p is a prime integer. (Keep in mind that according to my somewhat nonstandard convention, 0 is a prime integer!)

Example 3.25 You should also go through the examples in §3.1 and decide whether each given example is or is not an integral domain. Most will turn out to be integral domains, even if a 'random' commutative ring should not be expected to be an integral domain. For example, the polynomial rings $\mathbb{Z}[x]$, $\mathbb{Q}[x]$, $\mathbb{R}[x]$, $\mathbb{C}[x]$ all are (cf. Exercise 3.14).

Integral domains are a more comfortable environment than arbitrary commutative rings. Here is one reason why.

PROPOSITION 3.26 *Let R be a ring, and let $a \in R$. If a is a non-zero-divisor, then multiplicative cancellation holds for a, in the sense that*

$$\forall b, c \in R \qquad ab = ac \implies b = c \quad and \quad ba = ca \implies b = c.$$

In particular, if R is an integral domain, then multiplicative cancellation holds for every nonzero *element of R.*

(Cf. Remark 2.17, where this property was already discussed in the context of the rings $\mathbb{Z}/p\mathbb{Z}$.)

Proof Let a be a non-zero-divisor, and assume $ab = ac$. Then $a(b - c) = 0$; since a is a non-zero-divisor, this implies $b - c = 0$, and hence $b = c$. The argument for 'cancellation on the right' is analogous. □

So commutative rings are a special kind of rings, and integral domains are a special kind of commutative rings. There are sophisticated ways in which an integral domain may be 'even more special', and we will encounter a few in due time, particularly in Chapter 6. One property which produces a very interesting class of special integral domains is the existence of multiplicative inverses.

DEFINITION 3.27 An element b of a ring R is a 'multiplicative inverse' of an element $a \in R$ if $ab = ba = 1_R$. If $a \in R$ has a multiplicative inverse (in R), then it is said to be 'invertible', or a 'unit'. The set of units in a ring R is denoted R^*.

If the multiplicative inverse of a exists, then it is unique (argue as in Proposition 3.15), and it is usually denoted a^{-1}. Of course in this case a^{-n} will stand for $(a^{-1})^n$, and then the usual formula $a^{m+n} = a^m a^n$ holds for all integers m and n.

A random element of a random ring should not be expected to be invertible: $2 \in \mathbb{Z}$ is not invertible in $\mathbb{Z}$, while it *is* invertible in $\mathbb{Q}$; $[2]_n$ is invertible in $\mathbb{Z}/n\mathbb{Z}$ precisely if $\gcd(2, n) = 1$, i.e., if n is odd. Indeed, if $n = 2k + 1$, then

$$[2]_n[-k]_n = [-2k]_n = [1 - n]_n = [1]_n.$$

In fact we already know much more:

PROPOSITION 3.28   *Let n > 0 be an integer. Then*

$$(\mathbb{Z}/n\mathbb{Z})^* = \{[a]_n \in \mathbb{Z}/n\mathbb{Z} \mid \gcd(a, n) = 1\}.$$

We do not need to prove this statement, since it is simply Proposition 2.16 rephrased using the notation introduced in Definition 3.27.

The element 0 is not invertible in any nontrivial ring (Exercise 3.17); in particular, 0 is not invertible in integral domains. There are wonderful rings in which *every* nonzero element is invertible.

DEFINITION 3.29   A 'field' is a nontrivial commutative ring R in which every nonzero element of R is invertible, that is:

$$\forall a \in R \qquad a \neq 0 \implies (\exists b \in R \quad ab = ba = 1).$$

Thus, a field is a nontrivial commutative ring R such that $R^* = R \smallsetminus \{0\}$.

This terminology was also already introduced in §2.4.

Example 3.30   The reader knows many examples already: $\mathbb{Q}, \mathbb{R}, \mathbb{C}$ are all fields, and we have proved (Theorem 2.15) that $\mathbb{Z}/p\mathbb{Z}$ is a field whenever p is a positive prime integer, that is, a positive irreducible number.

Thus $\mathbb{Z}/2\mathbb{Z}, \mathbb{Z}/3\mathbb{Z}, \mathbb{Z}/5\mathbb{Z}, \mathbb{Z}/7\mathbb{Z}, \mathbb{Z}/11\mathbb{Z}$, etc. are all fields.

In fact, let me reiterate what we know about $\mathbb{Z}/n\mathbb{Z}$ at this point, in terms of the language we have developed. For *all* integers n:

- $\mathbb{Z}/n\mathbb{Z}$ is an integral domain if and only if n is prime;
- $\mathbb{Z}/n\mathbb{Z}$ is a field if and only if n is irreducible.

This is just a sharp restatement of Theorem 2.15, including the case $n = 0$ for completeness (cf. Example 3.24).

As promised, fields are special cases of integral domains.

PROPOSITION 3.31   *Fields are integral domains.*

Proof   Let R be a field. Then $0 \neq 1$ in R since R is nontrivial. To prove that R is an integral domain, assume that $a, b \in R$ and $ab = 0$; we have to show that $a = 0$ or $b = 0$. If $a = 0$, there is nothing to prove. If $a \neq 0$, then a has a multiplicative inverse a^{-1} in R, since R is a field. But then

$$b = 1 \cdot b = (a^{-1}a)b = a^{-1}(ab) = a^{-1} \cdot 0 = 0$$

(by Corollary 3.18) as needed.   □

Remark 3.32   The careful reader will probably notice that what the argument really shows is that invertible elements are necessarily non-zero-divisors. So in a field every nonzero element is a non-zero-divisor, and this makes it an integral domain.   ⌐

By Proposition 3.31, *field* $\implies$ *integral domain*. The converse does *not* hold: always carry a copy of $\mathbb{Z}$ in your pocket to remind you of this fact, since $\mathbb{Z}$ is an integral domain that is not a field. Another example is the polynomial ring $\mathbb{Q}[x]$: even though $\mathbb{Q}$ is a field, the ring $\mathbb{Q}[x]$ is *not* a field, because x does not have an inverse. (This would be a polynomial $\alpha(x) = a_0 + q_1 x + \cdots + q_n x^n$ such that $x\alpha(x) = 1$, and there is no such polynomial.)

From this perspective, it is almost amusing that the concepts of field and of integral domain *do* coincide for the rings $\mathbb{Z}/n\mathbb{Z}$ when $n > 1$: we have proved in Theorem 2.15 that n is prime if and only if $\mathbb{Z}/n\mathbb{Z}$ is an integral domain, if and only if $\mathbb{Z}/n\mathbb{Z}$ is a field. Here is a different explanation for the same observation.

PROPOSITION 3.33   *Let R be a* finite *integral domain. Then R is a field.*

Proof   Let R be a finite integral domain, and let $a \in R$ be a nonzero element. We have to prove that a has a multiplicative inverse. Let

$$r_1, \quad r_2, \quad \ldots, \quad r_m$$

be all the elements of R, without repetitions. (So one of these elements is a, and one of these elements is 1.) Consider the products

$$ar_1, \quad ar_2, \quad \ldots, \quad ar_m. \tag{3.8}$$

I claim that these are *also* all the elements of R, without repetitions. Indeed, since $a \neq 0$ and R is an integral domain, then multiplicative cancellation holds for R (Proposition 3.26). If $ar_i = ar_j$, then $r_i = r_j$ by cancellation, hence $i = j$. Therefore there are no repetitions in the list at (3.8). Since the list contains m elements of R with no repetitions, and R itself has m elements, all elements of R must appear in this list. But then $ar_i = 1$ for one of the elements r_i, and $r_i a = 1$ follows since R is commutative. This shows that a has a multiplicative inverse (that is, r_i).   □

Thus, 'field' and 'integral domain' happen to be the same notion for the rings $\mathbb{Z}/n\mathbb{Z}$ when $n \neq 0$ simply because these are *finite*. As soon as a ring R is infinite, this argument no longer applies and R may well be an integral domain and not a field. Once more, $\mathbb{Z}$ is an example of this phenomenon. Incidentally, can you dream of a finite field which is *not* of the form $\mathbb{Z}/p\mathbb{Z}$ for p prime? There are such things, and in due time we will learn a lot about them. Believe it or not, by the time we reach §14.5 we will be able to construct explicitly *all* finite fields. For now, work out Exercise 3.22.

Later on in this text we will focus on the study of fields: the whole of Part III is devoted to this topic. it may not seem so right now, since you do not have a large catalogue of examples, but this is an exceptionally rich environment and extremely beautiful and powerful mathematics has been developed to explore it.

Exercises

3.1 Consider the set $5\mathbb{Z}$ of integer multiples of 5. We have seen that this is not a ring with the usual operations (cf. Example 3.5; it is at best a *rng*). However, consider the usual addition and the following different multiplication:

$$a \odot b := \frac{ab}{5}.$$

Verify that $5\mathbb{Z}$ is closed with respect to this operation, and that $(5\mathbb{Z}, +, \odot)$ is a ring according to Definition 3.1.

3.2 Define a new 'multiplication' operation on $\mathbb{Z}$, by setting $a \odot b = -ab$. Is $(\mathbb{Z}, +, \odot)$ a ring according to Definition 3.1? (Prove it is, or explain why it is not.)

3.3 Define new operations $\oplus, \odot$ on $\mathbb{Z}$, by setting

$$a \oplus b = a + b - 1, \qquad a \odot b = a + b - ab.$$

Prove that $(\mathbb{Z}, \oplus, \odot)$ is a ring. What are 0 and 1 in this new ring?

3.4 Let $R = \mathcal{P}(S)$ denote the *set of subsets* of a set S. Define two operations on R: for all $A, B \subseteq S$, set

$$A + B := (A \setminus B) \cup (B \setminus A),$$
$$A \cdot B := A \cap B.$$

Prove that $(R, +, \cdot)$ is a ring.

3.5 ▷ Let R be a ring, and assume that $0 = 1$ in R. Prove that R consists of a single element.

3.6 ▷ Let R be a ring. Prove that $\forall a \in R, (-1) \cdot a = -a$.

3.7 ▷ Let R be a ring, and $a, b \in R$. Prove that $-(a + b) = (-a) + (-b)$. Deduce that $(-n)a = -(na)$ equals $n(-a)$ for every nonnegative integer n.

3.8 ▷ Let R be a ring. Prove that $(2(1_R)) \cdot (3(1_R)) = 6(1_R)$. More generally, give an argument showing that $(m1_R) \cdot (n1_R) = (mn)1_R$. (Show that for all $a \in R, (m1_R) \cdot a = ma$. You may assume that m is nonnegative. Why?)

3.9 Prove that $(-a) \cdot b = a \cdot (-b) = -a \cdot b$.

3.10 What is $(a + b)(a - b)$ in a ring? What is $(ab)^3$?

3.11 A 'Boolean ring' is a ring R such that $\forall a \in R, a^2 = a$.
- Prove that for all a in a Boolean ring, $2a = 0$.
- Prove that Boolean rings are *commutative,* that is, $ab = ba$ for all a, b.
(*Hint:* Consider $(a + a)^2, (a + b)^2$.)

3.12 ▷ The 'characteristic' of a ring R is the smallest positive integer n such that $n(1_R) = 0_R$, or 0 if there is no such positive integer.
- Show that the characteristic of $\mathbb{Z}$ is 0, and the characteristic of $\mathbb{Z}/m\mathbb{Z}$ is m.
- Show that if R has characteristic n, then $na = 0_R$ for all elements a of R.

3.13 ▷ Prove that the characteristic of an integral domain is a prime integer. (The characteristic of a ring is defined in Exercise 3.12.)

3.14 ▷ Prove that if R is an integral domain, then $R[x]$ is also an integral domain.

3.15   Consider the ring $C(\mathbb{R})$ of continuous real-valued functions of a real variable (cf. Example 3.11). Is this ring an integral domain?

3.16   If you have had complex analysis: consider the ring $\mathcal{H}(\mathbb{C})$ of analytic complex-valued functions of a complex variable. (Realize that this is a ring!) Is $\mathcal{H}(\mathbb{C})$ an integral domain?

3.17   ▷ Let R be a ring, and assume that 0_R is invertible in R. Prove that R is necessarily a trivial ring.

3.18   Consider the set $R = \mathbb{R}^+$ of positive real numbers. Define new operations $\oplus$, $\odot$ on this set, by setting $x \oplus y := x \cdot y$, $x \odot y = x^{\log y}$ where log denotes logarithm in base e.

- Prove that $(R, \oplus, \odot)$ is a ring, with additive identity 0_R equal to 1 (!) and multiplicative identity 1_R equal to e.
- Is the multiplication commutative in this ring?
- Assume that $x \odot y = 0_R$. Does it follow that $x = 0_R$ or $y = 0_R$?
- Does every $r \in R$, $r \neq 0_R$ have a multiplicative inverse in R? That is, is there an $s \in R$ such that $r \odot s = 1_R$?

3.19   Consider the ring $R = M_{2\times2}(\mathbb{Z}/2\mathbb{Z})$ (cf. Example 3.9).

- Let A, B be elements of R, and assume that $A \cdot B = 0$. Does it necessarily follow that $A = 0$ or $B = 0$?
- Does every $A \in R$, $A \neq 0_R$ have a multiplicative inverse in R? That is, is there a matrix $B \in R$ such that $A \cdot B = 1_R$?

3.20   ▷ An element a in a ring R is *nilpotent* if $a^k = 0$ for some $k \geq 1$.

- Prove that if a is nilpotent, then it is a zero-divisor.
- Find an example of a nonzero nilpotent element, in one of the rings $\mathbb{Z}/n\mathbb{Z}$.
- Prove that if a is nilpotent, then $1 + a$ is invertible.

(*Hint:* For the third part, think about geometric series.)

3.21   Let $n > 0$ be an integer and $R = \mathbb{Z}/n\mathbb{Z}$. Prove that every element of R is either invertible or a zero-divisor. Find a ring R for which this is not true.

3.22   ▷ Suppose the set R consists of four elements: $R = \{0, 1, x, y\}$, and two operations $+$ and $\cdot$ are defined on R, with the following multiplication tables:

+	0	1	x	y
0	0	1	x	y
1	1	0	y	x
x	x	y	0	1
y	y	x	1	0

+	0	1	x	y
0	0	0	0	0
1	0	1	x	y
x	0	x	y	1
y	0	y	1	x

Prove that R is a field. (It would take too long to verify completely the associativity of $+$ and $\cdot$ and the distributivity axiom. They happen to work in this case; please illustrate this fact by choosing an example for each of these properties. On the other hand, make sure you give a complete verification for the other axioms needed to prove that R is a field.)

4 The Category of Rings

4.1 Cartesian Products

You are probably wondering what's going to happen next; if you aren't, you should be. Think back to your first encounter with *set theory:* after you learned the definition of 'set', you likely introduced notions such as 'subset', and studied various operations such as union, intersection, Cartesian product, quotient by an equivalence relation, and possibly others. These are ways to construct new sets from old ones. Very importantly, you introduced and studied the notion of *functions* between sets, which allow you to transfer information from a set to another set. Functions are the underlying texture of the collection of all sets, placing each set in relation with all other sets. I mentioned briefly at the very beginning of these notes that studying 'sets, and functions between sets' means studying the *category* of sets. You may not be fully aware of this, but your encounter with sets has amounted to defining[1] the category of sets, and studying some interesting features of this category.

Now that we know what *rings* are, we want to essentially carry out the same program: we would like to find ways to construct new rings from the ones we know already, and we should understand what is the sensible notion of 'function' from a ring to another ring. Thus, we would like to understand what the category of rings should be, and some natural operations among rings. In fact, we will first come up with some natural operations, and then use them to get hints on how the most natural notion of 'function' should be defined.

DEFINITION 4.1   Let R, S be rings. The *Cartesian product* of R and S is the ring $(R \times S, +, \cdot)$, where

$$R \times S = \{(r, s) \mid r \in R, s \in S\}$$

is the ordinary Cartesian product of sets, and the operations $+, \cdot$ are defined *componentwise:*

$$(r_1, s_1) + (r_2, s_2) := (r_1 + r_2, s_1 + s_2),$$
$$(r_1, s_1) \cdot (r_2, s_2) := (r_1 \cdot r_2, s_1 \cdot s_2).$$

As usual, pay attention to what the symbols mean: for example, $+$ has three different

---

[1] Maybe not too rigorously. Actual 'set theory' is a more serious business.

meanings in this definition—as the addition operation in R, the addition operation in S, and the addition operation being defined in $R \times S$.

It is an uninspiring exercise to verify that Definition 4.1 indeed defines a ring. The additive and multiplicative identities are

$$0_{R \times S} = (0_R, 0_S), \quad 1_{R \times S} = (1_R, 1_S),$$

and are of course denoted $(0,0)$, $(1,1)$ for short. You should take the few minutes necessary to convince yourself that the ring axioms in Definition 3.1 are satisfied for $R \times S$.

Example 4.2   You know of one ring with four elements: $\mathbb{Z}/4\mathbb{Z}$. (Or maybe two, if you have worked out Exercise 3.22.) Well, now you know another one: $(\mathbb{Z}/2\mathbb{Z}) \times (\mathbb{Z}/2\mathbb{Z})$. These two rings look rather different. In $(\mathbb{Z}/2\mathbb{Z}) \times (\mathbb{Z}/2\mathbb{Z})$, adding any element to itself gives the additive identity: for example,

$$([1]_2, [1]_2) + ([1]_2, [1]_2) = ([1]_2 + [1]_2, [1]_2 + [1]_2) = ([0]_2, [0]_2).$$

This is not the case in $\mathbb{Z}/4\mathbb{Z}$: adding $[1]_4$ to itself does *not* give the additive identity:

$$[1]_4 + [1]_4 = [2]_4 \neq [0]_4.$$

Using a piece of terminology we will introduce soon, we would say that these two rings are not *isomorphic*.

Example 4.3   Similarly, you can contemplate two rings with six elements: $(\mathbb{Z}/2\mathbb{Z}) \times (\mathbb{Z}/3\mathbb{Z})$ and $\mathbb{Z}/6\mathbb{Z}$. Can you find some distinguishing feature that is different in these two rings? (*Hint:* No. This will become crystal clear when we get to Example 4.28.)

Example 4.4   Every complex number $a + bi$ is determined by a pair of real numbers a, b; and indeed, $\mathbb{C}$ is usually represented as a (real) plane $\mathbb{R}^2$. Now you have *another* way to think of $\mathbb{R}^2$ as a ring: the Cartesian product $\mathbb{R} \times \mathbb{R}$. You could spend a moment looking for similarities and differences between $\mathbb{C}$ and $\mathbb{R} \times \mathbb{R}$; here is one difference. In $\mathbb{R} \times \mathbb{R}$ there are nonzero zero-divisors: $(1,0)$ and $(0,1)$ are not zero in $\mathbb{R} \times \mathbb{R}$, and yet $(1,0) \cdot (0,1) = (0,0)$ *is* 0. On the other hand, $\mathbb{C}$ is a *field*, and in particular it is an integral domain (Proposition 3.31). So there are *no* nonzero zero-divisors in $\mathbb{C}$. The rings $\mathbb{C}$ and $\mathbb{R} \times \mathbb{R}$ will not be 'isomorphic'.

What can Cartesian products teach us about functions? Among all functions between a set like $R \times S$ and one of its 'factors', say R, there is a particularly compelling one, the 'projection' $\pi \colon R \times S \to R$. This is defined by

$$\forall (r, s) \in R \times S \qquad \pi(r, s) := r. \tag{4.1}$$

There of course are many others: if R and S have two elements each (as in Example 4.2), there are $2^4 = 16$ different functions from the set $R \times S$ to the set R, and π is just one of them. This function plays a special role in set theory. If we do not really know anything about R and S, it is the 'only' function that we can write out as explicitly as we do in (4.1).

Well, the next observation is that this function π is quite special from the point of view of ring theory. Compare the result of performing an operation (for example, $+$) *and then* projecting with π, vs. projecting first *and then* performing the operation. Starting from two elements $(r_1, s_1), (r_2, s_2)$ of $R \times S$, the first way to do this gives

$$(r_1, s_1) + (r_2, s_2) = (r_1 + r_2, s_1 + s_2) \mapsto r_1 + r_2 \,;$$

the second way first extracts the components r_1, r_2 and then adds them:

$$(r_1, s_1), (r_2, s_2) \mapsto r_1, r_2 \mapsto r_1 + r_2 \,.$$

Remarkably, we get the same result. We can write this as follows:

$$\pi((r_1, s_1) + (r_2, s_2)) = r_1 + r_2 = \pi((r_1, s_1)) + \pi((r_2, s_2)) \,.$$

Similarly,

$$\pi((r_1, s_1) \cdot (r_2, s_2)) = r_1 \cdot r_2 = \pi((r_1, s_1)) \cdot \pi((r_2, s_2)) \,.$$

Further, note that $\pi((1, 1)) = 1$: π sends the multiplicative identity of $R \times S$ to the multiplicative identity of R.

This projection is one prototype of the type of function that we want to adopt as giving us the 'texture' underlying the collection of rings.

4.2 Subrings

Summarizing, 'Cartesian products', with which you were likely familiar in set theory, have a direct analogue for rings, and the natural projections onto the factors turn out to have a special property from the point of view of rings. What about 'subsets'? Do they also have an analogue in ring theory?

DEFINITION 4.5   Let R be a ring, and let S be a subset of R. Assume $0_R \in S$, $1_R \in S$, and that S is closed with respect to both operations $+, \cdot$ in R. We say that S is a 'subring' of R if all axioms in Definition 3.1 are satisfied, with $0_S = 0_R$ and $1_S = 1_R$.

In other words, a subset S of R is a subring if it is made into a ring of its own by the operations defined in R, with the same 0 and 1 as in R.

Example 4.6   The ring $\mathbb{Z}$ is a subring of $\mathbb{Q}$, which is a subring of $\mathbb{R}$, which is a subring of $\mathbb{C}$.

Example 4.7   The ring $\mathbb{Z}/2\mathbb{Z}$ is NOT a subring of $\mathbb{Z}$. If you are tempted to interpret $[0]_2$ as $0 \in \mathbb{Z}$ and $[1]_2$ as $1 \in \mathbb{Z}$, please realize that this will not satisfy the requirements in Definition 4.5. Indeed, applying the operations *in* $\mathbb{Z}$ as required will give $1 + 1 = 2$, while $[1]_2 + [1]_2$ should be $[0]_2$.

By the same token, $\mathbb{Z}/n\mathbb{Z}$ is not a subring of $\mathbb{Z}$ for any $n > 0$. I have already made this point fairly strongly in §2.2: you should not think of 'congruence classes of numbers' as 'numbers'.

Example 4.8 Look back at Example 4.2, and consider the ring $R = \mathbb{Z}/2\mathbb{Z} \times \mathbb{Z}/2\mathbb{Z}$. Let Δ be the 'diagonal' subset,[2]

$$\Delta = \{(a, a) \mid a \in \mathbb{Z}/2\mathbb{Z}\}.$$

This set consists of exactly two elements: $([0]_2, [0]_2)$ and $([1]_2, [1]_2)$. It is closed under the operations, as is easy to verify; for example,

$$([1]_2, [1]_2) + ([1]_2, [1]_2) = ([0]_2, [0]_2)$$

is an element of Δ. Spend the couple of minutes necessary to verify that Δ is indeed a subring of R.

If you feel that this new ring Δ 'looks a lot like' $\mathbb{Z}/2\mathbb{Z}$ itself, you are right. These rings will turn out to be 'isomorphic'.

Example 4.9 In fact, the same example can be constructed from every ring R: the 'diagonal' subset in $R \times R$, $\Delta = \{(a, a) \mid a \in R\}$, is a subring of $R \times R$, and it will be a doppelgänger of R itself.

Example 4.10 By contrast, the subset $S = \{(a, 0) \mid a \in R\}$ may also look a lot like R, but it is *not* a subring according to our definition: indeed, it does not contain the multiplicative identity $(1, 1)$ (if R is not trivial).

Example 4.11 Let R be a ring, and consider the ring of polynomials $R[x]$ over R (cf. Example 3.10). We may view R itself as a subring of $R[x]$, by associating with each $r \in R$ the 'constant' polynomial $a + 0 \cdot x + 0 \cdot x^2 + \cdots$.

Example 4.12 For a fancier example of a subset that turns out to be a subring, consider the subset of $\mathbb{C}$ defined as follows:

$$\mathbb{Z}[i] := \{a + bi \in \mathbb{C} \mid a \in \mathbb{Z}, b \in \mathbb{Z}\}.$$

Once more, spend a moment to realize that $\mathbb{Z}[i]$ is a subring of $\mathbb{C}$. The main point is that if $a_1 + b_1 i$ and $a_2 + b_2 i$ are elements of $\mathbb{Z}[i]$, so that a_1, b_1, a_2, b_2 are integers, then

$$(a_1 + b_1 i) \cdot (a_2 + b_2 i) = (a_1 a_2 - b_1 b_2) + (a_1 b_2 + a_2 b_1)i$$

also has integer real and complex parts; therefore $\mathbb{Z}[i]$ is closed with respect to $\cdot$. (It is even easier to verify that it is closed with respect to $+$.) This new ring $\mathbb{Z}[i]$ is called the ring of 'Gaussian integers' and it plays an important role in number theory. We will get a taste of one application of this ring in §10.4.

---

[2] Why 'diagonal'? Well, for example, the 45° diagonal in the ordinary plane consists of points of the form (x, x). We are doing the same here.

If you want to explore this example a little more, compare $\mathbb{Z}[i]$ with $\mathbb{Z} \times \mathbb{Z}$: in what ways are these two rings similar, and in what ways are they different? Do you think they will turn out to be 'isomorphic'? (Exercise 4.4.)

Example 4.13   One more fancy-looking example: let R be the subset of $M_{2\times2}(\mathbb{R})$ consisting of matrices of the form

$$\begin{pmatrix} a & -b \\ b & a \end{pmatrix}.$$

Go ahead and verify that this is a *subring* of $M_{2\times2}(\mathbb{R})$. The main point now is that the product of two matrices of this form is again a matrix of the same form:

$$\begin{pmatrix} a_1 & -b_1 \\ b_1 & a_1 \end{pmatrix} \cdot \begin{pmatrix} a_2 & -b_2 \\ b_2 & a_2 \end{pmatrix} = \begin{pmatrix} a_1 a_2 - b_1 b_2 & -(a_1 b_2 + b_1 a_2) \\ a_1 b_2 + b_1 a_2 & a_1 a_2 - b_1 b_2 \end{pmatrix}.$$

Peek ahead at Proposition 4.14 if you want to streamline the verification that this subset of $M_{2\times2}(\mathbb{R})$ indeed determines a ring.

Verifying that a subset S of a ring R determines a subring according to Definition 4.5 may seem an annoying deal, since the definition instructs us to verify that all the ring axioms are verified for S. In practice, most of this work is unnecessary, due to the following observation.

> PROPOSITION 4.14   *Let R be a ring, and let $S \subseteq R$ be a subset containing 0_R and 1_R. Then S is a subring of R if and only if (i) it is closed under $+$ and $\cdot$, and (ii) it contains the additive inverse of every element of S.*

Proof If S is a subring, then the stated conditions hold by Definition 4.5 and since the existence of the additive inverse is one of the ring axioms.

Conversely, I claim that if these conditions are satisfied, then S satisfies all of the ring axioms. The point here is that all ring axioms other than the existence of 0, 1, and additive inverses only depend on the 'universal quantifier' $\forall$. (Look back at the list of axioms!) In other words, these axioms are verified for *all* choices of elements in R; and then they must in particular be satisfied for choices of elements in S. That takes care of axioms (i), (iv), (v), and (vii). Axioms (ii) and (iii) are verified since $0 \in S$ and we are specifically requiring S to contain additive inverses. Axiom (vi) is verified since $1 \in S$.

Therefore all of the ring axioms are satisfied for S, and S is a subring as prescribed in Definition 4.5. □

In fact, we could be more efficient: (i) and the requirement that $-1 \in S$ would be enough (use Exercise 3.6). Alternative: in order to prove that S is a subring, it suffices to show that it contains 1 and is closed under 'subtraction' and multiplication. (Exercise 4.5.)

The notion of 'subring' also comes with a distinguished function associated with it. If S is a subset of R, there will in general be many functions from S to R: for example, if

S has two elements and R has four elements (as in Example 4.8), then there are $4^2 = 16$ different functions from S to R. Among these, there is a 'special' one: the function which simply sends an element $s \in S$ to *the same element s,* in R. I call this the 'inclusion' function, $i \colon S \to R$, defined by

$$\forall s \in S \qquad i(s) = s.$$

For example, if $S = \mathbb{Z}$ is viewed as a subring of $R = \mathbb{Q}$, then $i(2) = \frac{2}{1}$ would be the number 2, viewed as a *rational* number rather than as an integer.

The next observation is that this inclusion function *also* satisfies the properties we detected in the case of projections from a product. That is, $\forall s_1, s_2 \in S$:

$$i(s_1 + s_2) = s_1 + s_2 = i(s_1) + i(s_2),$$
$$i(s_1 \cdot s_2) = s_1 \cdot s_2 = i(s_1) \cdot i(s_2).$$

Further, $i(1_S) = 1_R$, since we have required the identity of S to be the same as the identity of R.

All of this may seem almost too obvious to deserve being pointed out, but in mathematics that is often how important notions manifest themselves. The examples we have run across will prompt us to make the important definition of 'ring homomorphism'.

4.3   Ring Homomorphisms

The examples we have reviewed in the previous sections lead us to the following definition.

DEFINITION 4.15   Let R and S be rings. A function $f \colon R \to S$ is a 'homomorphism of rings' (or 'ring homomorphism') if $\forall a, b \in R$,

$$f(a + b) = f(a) + f(b),$$
$$f(a \cdot b) = f(a) \cdot f(b),$$

and further $f(1_R) = 1_S$.

We say that a ring homomorphism 'preserves' addition and multiplication: this is what the two formulas in Definition 4.15 tell us. We will get a notion of 'homomorphism' for every structure we will encounter in this text: so at some point we will deal with 'homomorphisms of modules', 'group homomorphisms', etc. Again, these notions will be functions which preserve appropriate operations.

If the context is clear, one can just write 'homomorphism'. For example, since in these chapters we are just dealing with rings, we could drop the qualifier 'ring'.

There are easy properties of (ring) homomorphisms, which you should check for yourself (the exercises give a small sample). For example, Definition 4.15 requires that homomorphisms send 1 to 1. What about 0?

PROPOSITION 4.16 *Let $f: R \to S$ be a ring homomorphism. Then $f(0_R) = 0_S$.*

Proof Indeed,

$$f(0_R) = f(0_R + 0_R) = f(0_R) + f(0_R),$$

from which $f(0_R) = 0_S$ by cancellation. □

Let's look at a few examples of homomorphisms.

Example 4.17 Let R be a ring, and let T be a trivial ring—so T consists of a single element t. Then the 'constant' function $R \to T$, sending every $r \in R$ to t, is a homomorphism. We could call this the 'trivial homomorphism'.

Note that if S is *not* a trivial ring, then the function $R \to S$ sending every $r \to R$ to 0_S is *not* a ring homomorphism, since according to Definition 4.15 a ring homomorphism must send 1_R to 1_S, and $1_S \neq 0_S$ in nontrivial rings.

Example 4.18 As we have seen, the projection $\pi: R \times S \to R$, $(r, s) \mapsto r$ is a ring homomorphism. We have also seen that the inclusion $i: S \to R$ of a subring into a ring is a ring homomorphism.

Example 4.19 Surprise! You know many more examples. Let $n \geq 0$ be an integer, and define a function $\varphi: \mathbb{Z} \to \mathbb{Z}/n\mathbb{Z}$ by setting $\forall a \in \mathbb{Z}$,

$$\varphi(a) = [a]_n .$$

Then $\varphi(1) = [1]_n$, and

$$\varphi(a + b) = [a + b]_n = [a]_n + [b]_n = \varphi(a) + \varphi(b),$$
$$\varphi(a \cdot b) = [a \cdot b]_n = [a]_n \cdot [b]_n = \varphi(a) \cdot \varphi(b).$$

Indeed, these identities are true by the very definition of $+$ and $\cdot$ in $\mathbb{Z}/n\mathbb{Z}$; look back at Definition 2.10.

Incidentally, now I can clarify a comment I made back in §2.2, after defining $\mathbb{Z}/n\mathbb{Z}$ (Definition 2.5). When I warned you *not* to think of $\mathbb{Z}/n\mathbb{Z}$ as a 'set of numbers', I was really alerting you to the fact that you cannot realize $\mathbb{Z}/n\mathbb{Z}$ as a subring of $\mathbb{Z}$ if $n > 0$ (cf. Example 4.7). This really means that there is no *injective* ring homomorphism $\mathbb{Z}/n\mathbb{Z} \to \mathbb{Z}$ if $n > 0$; even more is true, as we will see below in Example 4.25. On the other hand, I promised that $\mathbb{Z}$ and $\mathbb{Z}/n\mathbb{Z}$ would have a different kind of close kinship, and now we know what that is: there is a *surjective* ring homomorphism the other way around, from $\mathbb{Z}$ to $\mathbb{Z}/n\mathbb{Z}$.

Example 4.20 We can generalize Example 4.19: let R be any ring, and define a function $\sigma: \mathbb{Z} \to R$ by setting $\forall n \in \mathbb{Z}$,

$$\sigma(n) = n \, 1_R .$$

Then this is a ring homomorphism. Indeed, $\sigma(1) = 1_R$, and $\forall m, n \in \mathbb{Z}$,

$$\sigma(m + n) = (m + n)1_R = (m1_R) + (n1_R), \quad \sigma(m \cdot n) = (m \cdot n)1_R = (m1_R) \cdot (n1_R).$$

These formulas are particularly transparent if m and n are positive; it is not hard to verify them for arbitrary m and n. (You already did most of this when you did Exercise 3.8.)

Example 4.21   In fact, you know even more examples. Define a function $\psi \colon \mathbb{Z}/12\mathbb{Z} \to \mathbb{Z}/4\mathbb{Z}$ by setting $\forall [a]_{12} \in \mathbb{Z}/12\mathbb{Z}$,

$$\psi([a]_{12}) = [a]_4 \,.$$

Wait… this seems to depend on the choice of the representative a for the class $[a]_{12}$. Don't we have to check that it is well-defined, that is, independent of this representative? You bet we do. Check it now.

OK, OK: Let's check it together. If a and b represent the same class mod 12, i.e., $[a]_{12} = [b]_{12}$, then $a \equiv b$ mod 12, that is, 12 divides $b - a$. Now, if 12 divides $b - a$, then 4 divides $b - a$. It follows that $a \equiv b$ mod 4, and that says precisely that $[a]_4 = [b]_4$, as we need.

Now please verify that ψ is a ring homomorphism.

Example 4.22   And more! You know about the polynomial ring $\mathbb{Z}[x]$ (look at Example 3.10). Define a function $e_2 \colon \mathbb{Z}[x] \to \mathbb{Z}$ by setting $\forall \alpha(x) \in \mathbb{Z}[x]$,

$$e_2(\alpha(x)) = \alpha(2) \,.$$

(Or replace 2 by your favorite number.) Then e_2 is a ring homomorphism. Verify this! More generally, if $f \in R[x]$ is a polynomial with coefficients in a ring R, and r is any element of R, then we can define a function e_r 'evaluating' the polynomial at r:

$$e_r(f(x)) = f(r) \,.$$

We had already considered this back in Example 3.11. All of these functions $R[x] \to R$ are ring homomorphisms.

Example 4.23   For an example with a different flavor, consider *complex conjugation:* the function $\mathbb{C} \to \mathbb{C}$ sending $z = a + bi$ to $\bar{z} = a - bi$, where $a, b \in \mathbb{R}$. Then $\bar{1} = 1$, and more generally $\bar{a} = a$ if $a \in \mathbb{R}$; and for $z_1 = a_1 + b_1 i$, $z_2 = a_2 + b_2 i$,

$$\overline{z_1} + \overline{z_2} = \overline{a_1 + b_1 i} + \overline{a_2 + b_2 i} = a_1 - b_1 i + a_2 - b_2 i = (a_1 + a_2) - (b_1 + b_2)i = \overline{z_1 + z_2}$$

and

$$\overline{z_1} \cdot \overline{z_2} = (a_1 - b_1 i) \cdot (a_2 - b_2 i) = (a_1 a_2 - b_1 b_2) - (a_1 b_2 + a_2 b_1)i = \overline{z_1 \cdot z_2} \,.$$

Complex conjugation preserves both addition and multiplication.

Thus, it is a ring homomorphism.

Example 4.24 One non-example: Consider the function $f \colon \mathbb{Z} \to \mathbb{Z}$ defined by $f(a) = 2a$. There are a couple of reasons why this is not a *ring* homomorphism. (When the time comes, we will see it is a fine *module* homomorphism, but that is a different story: see Example 8.16.) For example, $f(1) = 2$, while Definition 4.15 requires the image of 1 to be 1. In any case, this function does not preserve multiplication:

$$f(a) \cdot f(b) = 2a \cdot 2b = 4ab,$$

while

$$f(a \cdot b) = 2(a \cdot b) = 2ab,$$

and $4ab \neq 2ab$ if $ab \neq 0$.

Example 4.25 Here is another non-example. We have seen that there is a homomorphism $\mathbb{Z} \to \mathbb{Z}/2\mathbb{Z}$. On the other hand, I claim that there are *no* homomorphisms from $\mathbb{Z}/2\mathbb{Z}$ to $\mathbb{Z}$. Indeed, let f be a function $\mathbb{Z}/2\mathbb{Z} \to \mathbb{Z}$ and assume that f is a homomorphism. By definition of homomorphism, $f([1]_2) = 1$; and by Proposition 4.16, $f([0]_2) = 0$. But then we would have

$$2 = 1 + 1 = f([1]_2) + f([1]_2) = f([1 + 1]_2) = f([0]_2) = 0,$$

which is nonsense.

By the same token, there are *no* ring homomorphisms from $\mathbb{Z}/n\mathbb{Z}$ to $\mathbb{Z}$ for any $n > 0$.

Example 4.26 Yet another non-example: Consider the *determinant*, defined (for example) on the ring $M_{2\times 2}(\mathbb{R})$ of 2×2 matrices with real entries:

$$M = \begin{pmatrix} a & b \\ c & d \end{pmatrix} \mapsto \det(M) = ad - bc.$$

The multiplicative identity in $M_{2\times 2}(\mathbb{R})$ is the identity matrix I, and $\det(I) = 1$, so this condition for homomorphisms is verified. The determinant also preserves multiplication: $\det(AB) = \det(A)\det(B)$. This is true for square matrices of any size, as you certainly have verified in Linear Algebra. For 2×2 matrices, it is no big deal:

$$\begin{pmatrix} a_1 & b_1 \\ c_1 & d_1 \end{pmatrix} \cdot \begin{pmatrix} a_2 & b_2 \\ c_2 & d_2 \end{pmatrix} = \begin{pmatrix} a_1 a_2 + b_1 c_2 & a_1 b_2 + b_1 d_2 \\ c_1 a_2 + d_1 c_2 & c_1 b_2 + d_1 d_2 \end{pmatrix};$$

the determinant of this matrix is

$$(a_1 a_2 + b_1 c_2)(c_1 b_2 + d_1 d_2) - (a_1 b_2 + b_1 d_2)(c_1 a_2 + d_1 c_2)$$
$$= a_1 a_2 d_1 d_2 - a_1 b_2 c_2 d_1 - a_2 b_1 c_1 d_2 + b_1 b_2 c_1 c_2,$$

and it magically agrees with

$$(a_1 d_1 - b_1 c_1)(a_2 d_2 - b_2 c_2) = a_1 a_2 d_1 d_2 - a_1 b_2 c_2 d_1 - a_2 b_1 c_1 d_2 + b_1 b_2 c_1 c_2.$$

So this condition is also satisfied. However,

$$\begin{pmatrix} 1 & 0 \\ 0 & 0 \end{pmatrix} + \begin{pmatrix} 0 & 0 \\ 0 & 1 \end{pmatrix} = \begin{pmatrix} 1 & 0 \\ 0 & 1 \end{pmatrix},$$

and the determinants of both matrices on the left equal 0. Since $0 + 0 \neq 1$, the function 'determinant' does *not* preserve addition. It is not a ring homomorphism. (It will be a 'group' homomorphism, cf. Example 11.16.)

The collection of all rings, taken along with the notion of ring homomorphism we just introduced, forms the 'category' of rings. In categorical parlance, we would say that rings are the 'objects' of the category of rings, and ring homomorphisms are the 'morphisms' in the category of rings. (Similarly, sets are the objects of the category of sets, and the morphisms in the category of sets are ordinary functions.)

Checking that some given notions of 'objects' and 'morphisms' do form a category amounts to verifying a few reasonable-looking axioms, which you are welcome to look up if you care. For rings, the content of these axioms is summarized in the following proposition.

PROPOSITION 4.27   *Let R, S, T, U be rings.*

- *The identity function* $\mathrm{id}_R \colon R \to R$ *is a ring homomorphism.*
- *Let* $f \colon R \to S$, $g \colon S \to T$ *be ring homomorphisms. Then the composition* $g \circ f \colon R \to T$ *is a ring homomorphism.*

Composition is associative: for all ring homomorphisms $f \colon R \to S$, $g \colon S \to T$, *and* $h \colon T \to U$, *we have*

$$h \circ (g \circ f) = (h \circ g) \circ f.$$

Proof   All statements prove themselves. The first is in fact a particular case of the observation that inclusions of subrings are homomorphisms (view R as a subring of itself). For the second statement, let a, b be elements of R. Then

$$(g \circ f)(a + b) = g(f(a + b)) = g(f(a) + f(b)) = g(f(a)) + g(f(b))$$
$$= (g \circ f)(a) + (g \circ f)(b)$$

and

$$(g \circ f)(a \cdot b) = g(f(a \cdot b)) = g(f(a) \cdot f(b)) = g(f(a)) \cdot g(f(b))$$
$$= (g \circ f)(a) \cdot (g \circ f)(b),$$

where I have repeatedly used the fact that f and g preserve the operations. Further,

$$(g \circ f)(1_R) = g(f(1_R)) = g(1_S) = 1_T,$$

and this concludes the proof that $g \circ f$ is a ring homomorphism.

Composition is associative because it already is for set-functions. Explicitly, for all $r \in R$ we have

$$((h \circ g) \circ f)(r) = (h \circ g)(f(r)) = h(g(f(r))) = h((g \circ f)(r)) = (h \circ (g \circ f))(r).$$

Since this holds $\forall r \in R$, it follows that $(h \circ g) \circ f = h \circ (g \circ f)$ as stated.   □

4.4 Isomorphisms of Rings

Whenever we have a category, we become interested in establishing when different 'objects' (rings, in this case) may be 'identified', that is, have the same properties. For example, we have called *trivial* any ring with only one element (this was Example 3.8). Surely you will agree that two trivial rings constructed from *different* sets with one element are actually just different manifestations of the 'same' concept. This already happens in set theory: the sets {○} and {●} are different, but any 'set-theoretic property' one of the these two sets may have, the other one will also have. We just have a set with one element, and we are choosing to label that one element by the symbol ○ in one case, and by the symbol ● in the other.

Similarly, while the sets {a, b, c} and {0, 1, 2} are different sets, again you can see the difference just as a different choice of labels for the elements of a set with three elements. The technical term for 'labeling' is 'bijection' (aka 'one-to-one correspondence'). We can define a function from the set {○, ●, ∗} to the set {0, 1, 2} by specifying

$$\circ \mapsto 0, \quad \bullet \mapsto 1, \quad \ast \mapsto 2,$$

and this function is rather special: it is both *injective* (i.e., 'one-to-one') and *surjective* (i.e., 'onto'). It gives an exact match between the elements of one set with the elements of the other one. We say it is *bijective,* or a *bijection* (or a 'one-to-one correspondence'). This is not the only function that has this property: the function

$$\circ \mapsto 1, \quad \bullet \mapsto 2, \quad \ast \mapsto 0$$

is also a bijection. In fact, you have likely learned in your previous exposure to this material (and/or you should be able to see by just thinking about it for a moment) that if two sets have n elements, then there are $n!$ bijections from one set to the other.

One (easy) fact to keep in mind is that bijections are precisely those functions that have an *inverse:* a function $f : A \to B$ between two sets is a bijection if and only if there is a function $g : B \to A$ such that $g \circ f = \mathrm{id}_A$ and $f \circ g = \mathrm{id}_B$. If it exists, this function g is usually denoted f^{-1}.

But you know all this. Coming back to rings, we ask the same question: when are two rings R, S different just because of a 'relabeling' of the elements?

Example 4.28 Take the two rings $\mathbb{Z}/6\mathbb{Z}$ and $(\mathbb{Z}/2\mathbb{Z}) \times (\mathbb{Z}/3\mathbb{Z})$ we encountered in Example 4.3. They have the following tables for addition and multiplication:

+	0	1	2	3	4	5
0	0	1	2	3	4	5
1	1	2	3	4	5	0
2	2	3	4	5	0	1
3	3	4	5	0	1	2
4	4	5	0	1	2	3
5	5	0	1	2	3	4

·	0	1	2	3	4	5
0	0	0	0	0	0	0
1	0	1	2	3	4	5
2	0	2	4	0	2	4
3	0	3	0	3	0	3
4	0	4	2	0	4	2
5	0	5	4	3	2	1

for $\mathbb{Z}/6\mathbb{Z}$ and

+	(0,0)	(1,1)	(0,2)	(1,0)	(0,1)	(1,2)
(0,0)	(0,0)	(1,1)	(0,2)	(1,0)	(0,1)	(1,2)
(1,1)	(1,1)	(0,2)	(1,0)	(0,1)	(1,2)	(0,0)
(0,2)	(0,2)	(1,0)	(0,1)	(1,2)	(0,0)	(1,1)
(1,0)	(1,0)	(0,1)	(1,2)	(0,0)	(1,1)	(0,2)
(0,1)	(0,1)	(1,2)	(0,0)	(1,1)	(0,2)	(1,0)
(1,2)	(1,2)	(0,0)	(1,1)	(0,2)	(1,0)	(0,1)

·	(0,0)	(1,1)	(0,2)	(1,0)	(0,1)	(1,2)
(0,0)	(0,0)	(0,0)	(0,0)	(0,0)	(0,0)	(0,0)
(1,1)	(0,0)	(1,1)	(0,2)	(1,0)	(0,1)	(1,2)
(0,2)	(0,0)	(0,2)	(0,1)	(0,0)	(0,2)	(0,1)
(1,0)	(0,0)	(1,0)	(0,0)	(1,0)	(0,0)	(1,0)
(0,1)	(0,0)	(0,1)	(0,2)	(0,0)	(0,1)	(0,2)
(1,2)	(0,0)	(1,2)	(0,1)	(1,0)	(0,2)	(1,1)

for $(\mathbb{Z}/2\mathbb{Z}) \times (\mathbb{Z}/3\mathbb{Z})$. (Of course I am not writing all the parentheses that I should: 4 in the first two tables stands for $[4]_6$, and $(0,1)$ in the last two stands for $([0]_2, [1]_3)$.) If you are careful and very patient, you can check that these tables match on the nose if you apply the following bijection:

$$0 \mapsto (0,0), \quad 1 \mapsto (1,1), \quad 2 \mapsto (0,2), \quad 3 \mapsto (1,0), \quad 4 \mapsto (0,1), \quad 5 \mapsto (1,2).$$

This is why I chose to list the elements in the last two tables in the funny order you see: it makes it easier to see that the tables match. For example, the entry in row 5 and column 3 in the second table for $(\mathbb{Z}/2\mathbb{Z}) \times (\mathbb{Z}/3\mathbb{Z})$ is 3; correspondingly, the entry in the row marked $(1,2)$ and the column marked $(1,0)$ in the second table for $(\mathbb{Z}/2\mathbb{Z}) \times (\mathbb{Z}/3\mathbb{Z})$ is $(1,0)$.

This is a very special requirement! Most of the $720 = 6!$ bijections between $\mathbb{Z}/6\mathbb{Z}$ and $(\mathbb{Z}/2\mathbb{Z}) \times (\mathbb{Z}/3\mathbb{Z})$ would *not* satisfy this property.

Of course, the fact that the tables match in this example simply means that the bijection listed above is a homomorphism.

DEFINITION 4.29   Let $f: R \to S$ be a ring homomorphism. We say that f is an 'isomorphism' if it is a bijection.

Thus, an isomorphism is a 'match of labels' between elements of a ring and elements of another, performed in a way which preserves the operations. There is an important observation to be made here. If $f: R \to S$ is an isomorphism, then in particular it is a bijection, so that (as recalled above) f has an inverse: there is a function $f^{-1}: S \to R$ such that $f^{-1} \circ f = \mathrm{id}_R$, $f \circ f^{-1} = \mathrm{id}_S$. The observation is that this inverse is then *also* a homomorphism!

PROPOSITION 4.30   *Let $f: R \to S$ be an isomorphism of rings. Then the inverse $f^{-1}: S \to R$ is also an isomorphism.*

Proof   The inverse of a bijection is a bijection, so we have to prove that f^{-1} is a homomorphism. Since f is a homomorphism, $f(1_R) = 1_S$. This implies that $f^{-1}(1_S) = 1_R$, which is one of the requirements for a ring homomorphism. Next, let b_1, b_2 be elements of S, and let $a_1 = f^{-1}(b_1)$, $a_2 = f^{-1}(b_2)$. Equivalently, $b_1 = f(a_1)$ and $b_2 = f(a_2)$. Then

$$f^{-1}(b_1 + b_2) = f^{-1}(f(a_1) + f(a_2)) = f^{-1}(f(a_1 + a_2))$$

since f is a homomorphism

$$= (f^{-1} \circ f)(a_1 + a_2) = a_1 + a_2$$
$$= f^{-1}(b_1) + f^{-1}(b_2).$$

Thus, f^{-1} preserves addition. The same type of argument shows that f^{-1} also preserves multiplication, and it follows that f^{-1} is a homomorphism, as needed.   □

The following corollary is an easy consequence of Proposition 4.30 (make sure you could write a proof if you had to!), and it is conceptually important.

COROLLARY 4.31   *Let $f: R \to S$ be a homomorphism of rings. Then f is an isomorphism if and only if there exists a homomorphism $g: S \to R$ such that $g \circ f = \mathrm{id}_R$ and $f \circ g = \mathrm{id}_S$.*

Thus, ring isomorphisms really play a role for rings analogous to bijections for sets: they are precisely the function-like gadgets that have an inverse.

Isomorphisms can be used to define a nice relation in the collection of rings.

DEFINITION 4.32   We say that two rings R, S are 'isomorphic', and write $R \cong S$, if there exists a ring isomorphism $f: R \to S$.

This relation is 'nice' in the sense that it determines an equivalence relation in any family of rings; this is a direct consequence of Propositions 4.27 and 4.30.

COROLLARY 4.33   *The isomorphism relation is reflexive, symmetric, and transitive.*

Proof   Let R, S, T be rings.
• The identity $\mathrm{id}_R: R \to R$ is an isomorphism, so $R \cong R$.
• Assume $R \cong S$. Then there exists an isomorphism $f: R \to S$. By Proposition 4.30, $f^{-1}: S \to R$ is an isomorphism; therefore $S \cong R$.
• Assume $R \cong S$ and $S \cong T$. Then there are isomorphisms $f: R \to S$ and $g: S \to T$. By Proposition 4.27, $g \circ f: R \to T$ is a homomorphism; and it is a bijection since compositions of bijections are bijections. Therefore $g \circ f$ is an isomorphism. It follows that $R \cong T$, as needed.   □

Remark 4.34 By Corollary 4.31, I could have *defined* ring isomorphisms as those ring homomorphisms f for which there is a ring homomorphism g such that both $f \circ g$ and $g \circ f$ are the identity. This is the general notion of isomorphism in a category: a 'morphism' f is an isomorphism if and only if there is a 'morphism' g such that both $f \circ g$ and $g \circ f$ are the identity. Analogously to Definition 4.32, we say that two objects A, B of a category are 'isomorphic' if there exists an isomorphism $A \to B$. For example, in the category of sets, 'isomorphisms' are bijections, so two objects (that is, two sets) are isomorphic in the category of sets precisely when they have the same cardinality. Two rings may well be isomorphic *as sets* but not isomorphic *as rings:* see Example 4.38 below.   ⌟

Ring isomorphisms preserve all 'ring-theoretic' properties, that is, all the properties which we can define by only referring to the ring operations.[3]

Example 4.35 Assume that $R \cong S$, and let $f : R \to S$, $g : S \to R$ be ring isomorphisms. Then a and b commute in R, i.e., $ab = ba$, if and only if $f(ab) = f(ba)$ in S, if and only if $f(a)f(b) = f(b)f(a)$ in S, if and only if $f(a)$ and $f(b)$ commute in S. By the same token, c and d commute in S if and only if $g(c)$ and $g(d)$ commute in R.

In particular, R is a *commutative* ring if and only if S is a commutative ring. 'Commutativity' is a ring-theoretic property, so it is preserved by ring isomorphisms.

The same applies to the other properties and definitions we have encountered. Here is another example.

PROPOSITION 4.36 *Let $f : R \to S$ be a ring isomorphism, and let $a \in R$. Then a is a zero-divisor if and only if $f(a)$ is a zero-divisor, and a is a unit if and only if $f(a)$ is a unit.*

Proof Let $a \in R$ be a zero-divisor. Then there exists a nonzero $b \in R$ such that $ab = 0$ or $ba = 0$. Dealing with the first possibility (the second is analogous) we have $f(a)f(b) = f(ab) = f(0) = 0$ (by Proposition 4.16), and $f(b) \neq 0$ since f is an isomorphism (and in particular it must be injective). Therefore $f(a)$ is a zero divisor. Applying the same argument to f^{-1} gives the converse implication.

Let $a \in R$ be a unit. Then there exists an element $b \in R$ such that $ab = ba = 1$. We have $f(a)f(b) = f(ab) = f(1) = 1$ (by definition of homomorphism), and similarly $f(b)f(a) = 1$. Therefore $f(a)$ has an inverse in S, so it is a unit. Again, applying the same argument to f^{-1} gives the converse implication.   □

COROLLARY 4.37 *Let R, S be isomorphic rings. Then R is an integral domain if and only if S is an integral domain, and R is a field if and only if S is a field.*

---

[3] As a non-example, the fact that elements of $\mathbb{Z}$ are 'numbers' is *not* a ring-theoretic property.

This follows from Proposition 4.36 (and Example 4.35, and the observation that if a ring R is nontrivial, then all rings isomorphic to R are also nontrivial).

The foregoing considerations should not be surprising—two isomorphic rings only differ in the way we 'label' their elements, therefore it should be clear that they will share properties which do not depend on such labeling. The properties of 'being commutative' or 'being a field' have nothing to do with the names we assign to the elements, so of course such properties are preserved by isomorphisms. This general principle will be applied without detailed verifications in the rest of this text.

Example 4.38 (Cf. Example 4.2.) There are $4! = 24$ bijections $f\colon \mathbb{Z}/4\mathbb{Z} \to (\mathbb{Z}/2\mathbb{Z}) \times (\mathbb{Z}/2\mathbb{Z})$. We do not need any big machinery to see that *none* of these bijections is an isomorphism. Indeed, $x + x = 0$ for every x in $(\mathbb{Z}/2\mathbb{Z}) \times (\mathbb{Z}/2\mathbb{Z})$, while, e.g., $[1]_4 + [1]_4 = [2]_4 \neq 0$ in $\mathbb{Z}/4\mathbb{Z}$. An isomorphism would preserve such a property.

Example 4.39 On the other hand, if you have worked out the details of Example 4.28, then you have proved 'by hand' that the rings $\mathbb{Z}/6\mathbb{Z}$ and $\mathbb{Z}/2\mathbb{Z} \times \mathbb{Z}/3\mathbb{Z}$ *are* isomorphic.

Example 4.40 Complex conjugation (Example 4.23) is an *isomorphism* of the field $\mathbb{C}$ to itself. Indeed, we have verified that it is a ring homomorphism, and it has an inverse, namely, itself: if $z = a + bi$ with $a, b \in \mathbb{R}$, then

$$\overline{(\overline{z})} = \overline{(\overline{a + bi})} = \overline{a - bi} = a + bi = z\,.$$

'Galois theory' deals with *automorphisms* of fields, that is, isomorphisms of a field to itself. It turns out that every polynomial determines a whole group of automorphisms, which encode sophisticated information about the polynomial. For example, complex conjugation has to do with the polynomial $x^2 + 1$. We will scratch the surface of this deep subject in Chapter 15.

Example 4.41 Consider the subring R of $M_{2\times2}(\mathbb{R})$ given in Example 4.13, and define a function $f\colon \mathbb{C} \to R$ as follows:

$$f(a + bi) = \begin{pmatrix} a & -b \\ b & a \end{pmatrix}.$$

Then f is a ring homomorphism: $f(1)$ is the identity matrix, f clearly preserves addition, and (maybe a little more surprisingly) it preserves multiplication. Indeed,

$$f((a_1 + b_1 i) \cdot (a_2 + b_2 i)) = f((a_1 a_2 - b_1 b_2) + (a_1 b_2 + b_1 a_2)i)$$
$$= \begin{pmatrix} a_1 a_2 - b_1 b_2 & -(a_1 b_2 + b_1 a_2) \\ a_1 b_2 + b_1 a_2 & a_1 a_2 - b_1 b_2 \end{pmatrix},$$

while

$$f(a_1 + b_1 i) \cdot f(a_2 + b_2 i) = \begin{pmatrix} a_1 & -b_1 \\ b_1 & a_1 \end{pmatrix} \cdot \begin{pmatrix} a_2 & -b_2 \\ b_2 & a_2 \end{pmatrix} = \begin{pmatrix} a_1 a_2 - b_1 b_2 & -b_1 a_2 - a_1 b_2 \\ a_1 b_2 + b_1 a_2 & -b_1 b_2 + a_1 a_2 \end{pmatrix}$$

with the same result. So f is a ring homomorphism. It is clearly bijective, so it is an isomorphism.

Thus, we see that $M_{2 \times 2}(\mathbb{R})$ contains an 'isomorphic copy' of $\mathbb{C}$ (that is, the subring R of $M_{2 \times 2}(\mathbb{R})$).

In retrospect, this is clear from a linear algebra point of view. The matrix $f(z)$ corresponding to a complex number z is nothing but the matrix expressing the linear map $\mathbb{R}^2 \to \mathbb{R}^2$ determined by 'multiplication by z' after we identify $\mathbb{C}$ with $\mathbb{R}^2$. For example, multiplication by i is expressed by a counterclockwise rotation by $90°$, which has matrix

$$\begin{pmatrix} 0 & -1 \\ 1 & 0 \end{pmatrix}$$

and this matrix is nothing but $f(i)$.

Example 4.42   In Example 4.9 we also observed that if R is a ring, then the 'diagonal' $\Delta = \{(a, a) \mid a \in R\}$ is a subring of $R \times R$. I claimed in passing that this will turn out to be isomorphic to R. Now we can verify this claim. Define a function $\delta \colon R \to \Delta$ by setting $\delta(a) = (a, a)$ for all $a \in R$. It takes a moment to verify that δ is a ring homomorphism: $\delta(1) = (1, 1)$ is the identity in Δ, and

$$\delta(a + b) = (a + b, a + b) = (a, a) + (b, b) = \delta(a) + \delta(b) ,$$

and similarly for multiplication. Further, δ has an inverse, defined by sending (a, a) back to a. So δ is an isomorphism, and this verifies that $R \cong \Delta$.

Example 4.43   In Example 3.10 I have briefly mentioned the ring $R[x, y]$ of polynomials in two indeterminates x, y with coefficients in a ring R. This ring is isomorphic to the ring $R[x][y]$ of polynomials in *one* indeterminate with coefficients in $R[x]$.

Indeed, we obtain a function $\varphi R[x, y] \to R[x][y]$ by simply collecting the coefficients of the powers of y: in proper notation, I could write

$$\varphi \left(\sum_{i,j} a_{ij} x^i y^j \right) = \sum_j \left(\sum_i a_{ij} x^i \right) y^j \in R[x][y] .$$

It should be immediate that this is a bijection (what is its inverse?), and it clearly preserves addition and multiplication. So φ is a ring isomorphism. Of course the role of x and y can be reversed: $R[x, y]$ is also isomorphic to $R[y][x]$.

By the same token and iterating, we obtain the following useful interpretation of a polynomial ring in any (finite) number of indeterminates:

$$R[x_1, \ldots, x_n] \cong R[x_1, \ldots, x_{n-1}][x_n] \cong R[x_1][x_2] \cdots [x_n] .$$

Exercises

4.1 Let R and S be commutative rings. Prove that $R \times S$ is a commutative ring.

4.2 ▷ Let R, S be nontrivial rings. Prove that $R \times S$ has nonzero zero-divisors.

4.3 ▷ Let R, S be rings. Prove that $(R \times S)^* = R^* \times S^*$.

4.4 ▷ Compare the rings $\mathbb{Z} \times \mathbb{Z}$ and $\mathbb{Z}[i]$. Elements of both rings are determined by the choice of two integers (a, b), so these rings 'look the same' *as sets*. Find differences between them *as rings*.

4.5 ▷ Let R be a ring, and let S be a subset of R containing 0_R and 1_R. Assume further that S satisfies the following requirements:

- $\forall a, b \in S \qquad a \cdot b \in S$,
- $\forall a, b \in S \qquad a - b \in S$.

That is, assume that S is closed under multiplication and under 'subtraction'. Prove that S is a subring of R.

4.6 Consider the subset $S = \{[0]_6, [3]_6\}$ of $\mathbb{Z}/6\mathbb{Z}$. Note that S is closed with respect to the operations $+$ and $\cdot$ in $\mathbb{Z}/6\mathbb{Z}$. Does $(S, +, \cdot)$ satisfy the ring axioms? Is it a subring of $\mathbb{Z}/6\mathbb{Z}$?

4.7 Note that every rational number can be written with an *even* denominator: for example, 1 may be written as $\frac{2}{2}$. Give an example of a rational number which cannot be written with an *odd* denominator. Prove that the set of rational numbers $\frac{a}{b}$ which *can* be written with b odd forms a subring of $\mathbb{Q}$.

4.8 Let S be a subring of an integral domain. Prove that S is an integral domain.

4.9 A *subfield* of a field is a subring which is a field. Prove that the subset of $\mathbb{R}$ consisting of numbers $a + b\sqrt{2}$, with $a, b \in \mathbb{Q}$, is a subfield of $\mathbb{R}$.

4.10 Let R, S be rings, and let $f : R \to S$ be a *surjective* function such that $\forall a, b \in R$: $f(a + b) = f(a) + f(b)$ and $f(a \cdot b) = f(a) \cdot f(b)$. Prove that f is a ring homomorphism. That is, prove that in this case $f(1_R)$ necessarily equals 1_S.

4.11 ▷ Let $f : R \to S$ be a ring homomorphism. Prove that for all $a, b \in R$ we have $f(b - a) = f(b) - f(a)$. (In particular, $f(-a) = -f(a)$.)

4.12 Let $f : R \to S$ be a ring homomorphism. Prove that for all $a \in R$ and all $n \in \mathbb{Z}$ we have $f(na) = nf(a)$.

4.13 ▷ Let R be a ring. Prove that the homomorphism $\sigma : \mathbb{Z} \to R$ defined in Example 4.20 is the *unique* ring homomorphism $\mathbb{Z} \to R$.

4.14 ▷ By Exercise 4.13, the object $\mathbb{Z}$ is very special in the category of rings: for all objects R there is *exactly one* morphism $\mathbb{Z} \to R$. We say that $\mathbb{Z}$ is 'initial' in the category of rings.

Figure out what it should mean for an object to be 'final' in a category, and find out whether the category of rings has a final object.

(Also: Does the category of sets have an initial object? A final object?)

4.15 Let T be a trivial ring, and let $f : T \to R$ be a ring homomorphism. Prove that R is also a trivial ring.

4.16   Let $f: R \to S$ be a ring homomorphism. If $a \in R$ is a zero-divisor, does it follow that $f(a)$ is a zero-divisor? If $a \in R$ is a unit, does it follow that $f(a)$ is a unit?

4.17   Let $f: R \to S$ be an *onto* ring homomorphism. Prove that if R is commutative, then S is commutative.

4.18   Prove that there is a ring homomorphism from $\mathbb{Z}/m\mathbb{Z}$ to $\mathbb{Z}/n\mathbb{Z}$ if and only if m is a multiple of n.

4.19   ▷ Prove that if m and n are positive integers and are not relatively prime, then $\mathbb{Z}/mn\mathbb{Z}$ is *not* isomorphic to $\mathbb{Z}/m\mathbb{Z} \times \mathbb{Z}/n\mathbb{Z}$.

5 Canonical Decomposition, Quotients, and Isomorphism Theorems

5.1 Rings: Canonical Decomposition, I

Let's set the stage for the next development. There is an interesting observation concerning ordinary functions between sets, which I will recall below. If it sounds unfamiliar, I recommend that you review the material in §A.9, where this is discussed in some detail.

Let A and B be sets, and let $\varphi\colon A \to B$ be a function. Then I claim that φ may be written as a composition of an injective function, a bijective function, and a surjective function. More precisely, there are sets A' and B', and functions $\iota\colon B' \to B$, $\tilde{\varphi}\colon A' \to B'$, and $\pi\colon A \to A'$ such that ι is injective, $\tilde{\varphi}$ is bijective, π is surjective, and

$$\varphi = \iota \circ \tilde{\varphi} \circ \pi. \tag{5.1}$$

That sounds like a mouthful, and I prefer to think in terms of the following diagram:

$$A \xrightarrow{\quad\pi\quad} A' \xrightarrow[\tilde{\varphi}]{\;\sim\;} B' \xrightarrow[\iota]{} B.$$

with φ the curved arrow from A to B.

We can travel from the leftmost point of this diagram to the rightmost point by following the arrows in two different ways: along the arrows marked by π, $\tilde{\varphi}$, ι on the bottom side, or along the single curved arrow marked by φ. The bottom side corresponds to the composition $\iota \circ \tilde{\varphi} \circ \pi$, while the curved side represents the function φ. We say that a diagram of functions is 'commutative' if the result of traveling through it from any point to any other point along the arrows does not depend on the way you choose to travel. Thus, saying that the above diagram 'commutes' or 'is commutative' is equivalent to stating the equality in formula (5.1), $\varphi = \iota \circ \tilde{\varphi} \circ \pi$.

The funny decorations on the bottom arrows tell us that π is supposed to be surjective, i.e., onto (that's what the double tip of the arrow signifies), ι is injective, i.e., one-to-one (that's what the hook stands for), and $\tilde{\varphi}$ is a bijection (the $\sim$ over the arrow is reminding us of the notation $\cong$).

This may be called the 'canonical decomposition' of φ. It may sound fancy, but it is actually a very simple business: B' is simply the image $\mathrm{im}(\varphi) = \varphi(A)$ of φ, and A' is obtained as the quotient $A/\sim$ with respect to the equivalence relation $\sim$ defined on A by setting $a_1 \sim a_2 \iff \varphi(a_1) = \varphi(a_2)$. The bijection $\tilde{\varphi}$ is defined by

$$\tilde{\varphi}([a]_\sim) := \varphi(a).$$

I had already recalled the gist of these observations in the last bullet point in §2.1. Again, a more extensive discussion may be found in §A.9. Familiarity with the way this decomposition works is advisable, because it will give you a solid context for what we are going to do. In any case, full details will be provided.

Essentially, we are going to verify that the situation summarized above also occurs in the world of *rings* and *ring homomorphisms*: if $f : R \to S$ is a ring homomorphism, we are going to find rings R', S', and a surjective ring homomorphism $\pi : R \to R'$, an injective ring homomorphism $\iota : S' \to S$, and a ring isomorphism $\tilde{f} : R' \to S'$, such that $f = \iota \circ \tilde{f} \circ \pi$. That is, we will determine a commutative diagram

$$R \xrightarrow[\pi]{} R' \xrightarrow[\tilde{f}]{\sim} S' \overset{f}{\underset{\iota}{\hookrightarrow}} S \,, \tag{5.2}$$

where now R, R', S, S' are rings and all the arrows represent ring homomorphisms. This is a diagram 'in the category of rings'.

This may all sound very abstract, but I promise that we will extract out of it some very concrete tools to deal with rings and ring homomorphisms.

First, let's deal with $\iota : S' \to S$. This hinges on the following simple observation.

PROPOSITION 5.1　*Let $f : R \to S$ be a ring homomorphism. Then the image $S' = \operatorname{im} f := f(R)$ of f is a subring of S.*

Proof Since homomorphisms send 1 to 1 (by definition) and 0 to 0 (by Proposition 4.16), $f(R)$ contains both 0 and 1. To verify that $f(R)$ is closed under the operations, let $s_1, s_2 \in f(R)$; then $\exists r_1, r_2 \in R$ such that $s_1 = f(r_1)$ and $s_2 = f(r_2)$, by definition of $f(R)$. But then

$$s_1 + s_2 = f(r_1) + f(r_2) = f(r_1 + r_2) \quad \text{and} \quad s_1 \cdot s_2 = f(r_1) \cdot f(r_2) = f(r_1 \cdot r_2) \,,$$

since f preserves the operations, and this shows that both $s_1 + s_2$ and $s_1 \cdot s_2$ belong to $f(R)$. Finally, to show that $f(R)$ contains the additive inverses of elements of $f(R)$, let $s \in f(R)$; then $\exists r \in R$ such that $s = f(r)$, and

$$-s = -f(r) = f(-r)$$

(as you hopefully verified in Exercise 4.11), proving that $-s \in f(R)$.

By Proposition 4.14, $f(R)$ is a subring of S. □

Remark 5.2　Of course the same argument shows that if R' is a subring of R, then $f(R')$ is a subring of S. ⌐

Thus, we are going to take $S' = f(R)$, and we will let $\iota : S' \to S$ be the inclusion function. We already know this is a homomorphism (cf. §4.2) and ι is injective. So we already have a piece of our diagram:

$$R \dashrightarrow \overset{??}{\dashrightarrow} S' \overset{f}{\underset{\iota}{\hookrightarrow}} S \,, \tag{5.3}$$

and we have to figure out what to do with the ?? part.

This is a slightly more involved business. Again, remember from the analogous situation for sets (recalled above) that we want to mod out by the equivalence relation $\sim$ defined on R by setting $a_1 \sim a_2 \iff f(a_1) = f(a_2)$. We are going to take a longish detour motivated by looking at this relation; eventually we will come back to the task of figuring out how to fill the rest of diagram (5.3), but for now this goal will have to move to the back burner.

5.2   Kernels and Ideals

One motivating observation for what comes next is the fact that if $f : R \to S$ is a ring homomorphism, then the relation introduced above may be expressed as follows:

$$a_1 \sim a_2 \iff f(a_1) = f(a_2) \iff f(a_2 - a_1) = 0.$$

Thus, the relation $\sim$ is closely related to the requirement '$f(r) = 0$'. We need a name for the set of elements satisfying this condition.

DEFINITION 5.3   Let $f : R \to S$ be a ring homomorphism. The 'kernel' of f, denoted ker f, is the subset of R

$$\ker f = \{r \in R \mid f(r) = 0\}.$$

Thus, $r \in \ker f$ if and only if $f(r) = 0$.

Example 5.4   Let $n \geq 0$ be an integer, and let $\varphi : \mathbb{Z} \to \mathbb{Z}/n\mathbb{Z}$ be the homomorphism defined by $\varphi(a) = [a]_n$ (cf. Example 4.19). Since

$$\varphi(n) = 0 \iff [a]_n = 0 \iff a \in n\mathbb{Z},$$

we have

$$\ker \varphi = \{a \in \mathbb{Z} \mid f(a) = 0\} = n\mathbb{Z}.$$

Example 5.5   Let $e_0 : R[x] \to R$ be the homomorphism defined by $e_0(f(x)) = f(0)$ (cf. Example 4.22). If $f(x) = a_0 + a_1 x + \cdots + a_n x^n$, then

$$e_0(f(x)) = f(0) = a_0.$$

Therefore ker e_0 consists of all polynomials $a_0 + a_1 x + \cdots + a_n x^n$ with $a_0 = 0$, i.e., the polynomials with 'no constant term'.

If we adopt Definition 5.3, then we can reinterpret the equivalence relation $\sim$ in the following way:

$$a_1 \sim a_2 \iff a_2 - a_1 \in \ker f.$$

Indeed, $a_2 - a_1 \in \ker f$ says precisely that $f(a_2 - a_1) = 0$.

What kind of object is this new set ker f? It is not a subring if f is not a trivial homomorphism, since then $1 \notin \ker f$ (because $f(1) = 1 \neq 0$ if f is not trivial). However, it is rather close to being a ring—here is an example of a 'rng'! For instance, it is easy to check that ker f satisfies the ring axioms having to do with *addition*.

> PROPOSITION 5.6　*Let $f: R \to S$ be a ring homomorphism. Then* (ker f, +) *satisfies axioms (i)–(iv) from Definition 3.1.*

Proof　First of all, ker f is closed with respect to +: indeed, $\forall a, b \in \ker r$,

$$f(a + b) = f(a) + f(b) = 0 + 0 = 0.$$

Further, $0 \in \ker f$ since $f(0) = 0$ (Proposition 4.16), and if $a \in \ker f$, then $-a \in \ker f$ since $f(-a) = -f(a) = -0 = 0$.

This is all we have to show in order to prove the proposition, since associativity and commutativity of + are universal requirements (so they hold for all choices of elements in ker f since they hold for all choices of elements in R). □

This is already interesting. When the time comes we will learn a name for structures satisfying these axioms: they are called 'abelian groups'. Chapter 10 is entirely devoted to abelian groups. But the set ker f is more special than this: it satisfies the following somewhat bizarre property.

> PROPOSITION 5.7　*Let $f: R \to S$ be a ring homomorphism. Then for all $a \in \ker f$ and all $r \in R$, we have*
>
> $$ra \in \ker f, \quad ar \in \ker f.$$

Proof　Indeed,

$$f(ar) = f(a)f(r) = 0 \cdot f(r) = 0$$

by Corollary 3.18, and $f(ra) = 0$ by the same token. □

I will refer to this property as the 'absorption property': it tells us that ker f 'absorbs' multiplications by all elements of R, both on the left and on the right. The absorption property and the property spelled out in Proposition 5.6 are so important that we give a name to all subsets of R satisfying both of them.

DEFINITION 5.8　A subset I of a ring R is an 'ideal' if it is closed under + and satisfies axioms (i)–(iv) of Definition 3.1 and the absorption property:

$$(\forall a \in I)(\forall r \in R) \quad ar \in I \text{ and } ra \in I.$$

These should actually properly be named *two-sided* ideals. Subsets that satisfy (i)–(iv) and absorption of multiplication on the *left* by elements of R are called *left-ideals;* and, similarly, *right-ideals* satisfy absorption on the right. We will only use ideals which satisfy both absorption properties. (For commutative rings, the distinction is immaterial.)

Just as the verification that a given subset of a ring is a subring may be streamlined (Proposition 4.14, Exercise 4.5), so can the verification that a given subset is an ideal.

PROPOSITION 5.9 *A nonempty subset I of a ring R is an ideal if and only if it is closed with respect to addition and satisfies the absorption property.*

Proof If I is an ideal, then it is in particular closed with respect to $+$ and satisfies absorption by definition. Conversely, assume I is closed with respect to $+$ and satisfies the absorption property; in order to show that it is an ideal, we just have to verify axioms (i)–(iv). Associativity and commutativity of $+$ (i.e., axioms (i) and (iv)) are satisfied because these properties are universal requirements and they hold in R, and I is closed with respect to addition. Since I is nonempty, it contains at least one element a, and then it contains $0 \cdot a = 0$ by absorption; so axiom (ii) holds. For $a \in I$, $-a = (-1) \cdot a$ is in I by absorption, and this implies axiom (iii). □

Here are a few examples of ideals.

Example 5.10 Propositions 5.6 and 5.7 say that ker f is an ideal of R, for every ring homomorphism $f \colon R \to S$. *Preview of coming attractions:* We will soon discover that *all* ideals are of this form (Remark 5.36).

Example 5.11 The ring R itself is an ideal in R: this should be clear from the definition, and also follows from the fact that R is the kernel of the homomorphism from R to a trivial ring.

Example 5.12 For every ring R, the subset $\{0_R\}$ is an ideal of R. Again, this should be clear from the definition, and also follows from the fact that $\{0_R\}$ is the kernel of the identity homomorphism $\mathrm{id}_R \colon R \to R$.

Example 5.13 A non-example: $\mathbb{Z}$ is *not* an ideal in $\mathbb{Q}$. Indeed, it is not true that if a is an integer and q is a rational number, then qa is an integer. For instance, take $a = 3$, $q = \frac{1}{2}$: the number $\frac{1}{2} \cdot 3 = \frac{3}{2}$ is not an integer.

Example 5.14 The set $n\mathbb{Z}$ of multiples of an integer n is an ideal of $\mathbb{Z}$. Indeed, we have seen in Example 5.4 that $n\mathbb{Z}$ is a kernel. (Note that $(-n)\mathbb{Z} = n\mathbb{Z}$, so it is not restrictive to assume $n \geq 0$.) It is also straightforward to check this independently by using Proposition 5.9: $0 \in n\mathbb{Z}$, so $n\mathbb{Z}$ is not empty; the sum of two multiples of n is a multiple of n, so $n\mathbb{Z}$ is closed under addition; and multiplying a multiple of n by *any* integer gives a multiple of n, so $n\mathbb{Z}$ satisfies absorption.

Example 5.15 The set of polynomials $f(x, y) \in \mathbb{C}[x, y]$ that have no constant term is an ideal of $\mathbb{C}[x, y]$. (Recall that $\mathbb{C}[x, y]$ is the ring of polynomials in two indeterminates

x and y with coefficients in $\mathbb{C}$; cf. Examples 3.10 and 4.43.) Check this as an exercise.

There are useful and straightforward generalizations of these examples, encapsulated below in Definition 5.17. They rely on the following observation.

> PROPOSITION 5.16   Let R be a commutative *ring, and let $a \in R$. Then the subset*
>
> $$(a) = \{ra \mid r \in R\},$$
>
> *i.e., the set of 'multiples' of a in R, is an ideal of R.*

Proof   We use Proposition 5.9 again. Since $a \in (a)$, the set (a) is nonempty. Next, let $r_1 a, r_2 a$ be multiples of a; then

$$r_1 a + r_2 a = (r_1 + r_2)a$$

is also a multiple of a. Therefore, (a) is closed with respect to $+$. Further, if $b = ca$ is a multiple of a, and $r \in R$, then

$$rb = r(ca) = (rc)a$$

by associativity of multiplication, and this implies that $rb \in (a)$. Since R is commutative, it also follows that $br \in (a)$. Therefore (a) satisfies the absorption property. By Proposition 5.9, (a) is an ideal.   □

The commutativity of R is necessary for this statement. (See Exercise 5.4.)

More generally, if R is commutative, then the set

$$(a_1, \ldots, a_n) = \{r_1 a_1 + \cdots + r_n a_n \mid r_1, \ldots, r_n \in R\}$$

of 'linear combinations' of $a_1, \ldots, a_n$ is an ideal of R. This is another straightforward verification, and good practice for you (Exercise 5.5).

The notions introduced in Proposition 5.16 and in the previous paragraph are useful and commonplace, so they deserve their own terminology.

> DEFINITION 5.17   Let R be a commutative ring, and $a \in R$. Then (a) (defined above) is the 'principal ideal' generated by a. More generally, for $a_1, \ldots, a_n \in R$, the ideal $(a_1, \ldots, a_n)$ is called the 'ideal generated by $a_1, \ldots, a_n$'.

Most of the examples reviewed above are principal (for commutative rings). Indeed, $R = (1)$; $\{0\} = (0)$; and $n\mathbb{Z}$ is the principal ideal (n) generated by n in $\mathbb{Z}$. Example 5.15 is not principal; it is the ideal (x, y) with two generators x, y in the ring $\mathbb{C}[x, y]$.

Remark 5.18   If $a_1, \ldots, a_n$ belong to an ideal I, then every linear combination of $a_1, \ldots, a_n$ belongs to I: this follows from absorption and closure with respect to addition. (Right?) Therefore, $(a_1, \ldots, a_n)$ is in fact the *smallest* ideal containing $a_1, \ldots, a_n$.

More generally, if A is any subset of a ring R, then the 'ideal generated by A' is by definition the smallest ideal of R containing A. Such an ideal always exists, as you can verify (Exercise 5.10).

5.3 Quotient Rings

Bottom line: examples of ideals abound. What do we do with them?

Rewind a bit and look where we started. We were given a ring homomorphism $f: R \to S$, and for some reasons we were looking at the relation $\sim$ defined on R by the prescription $a_1 \sim a_2 \iff f(a_1) = f(a_2)$. We had realized that this relation may be interpreted in terms of the kernel of f:

$$a_1 \sim a_2 \iff a_2 - a_1 \in \ker f. \tag{5.4}$$

Then we realized that $\ker f$ is an 'ideal' of R. If you stare at this long enough, you may come up with the idea of defining a relation in the style of (5.4) for *every* ideal I of R.

DEFINITION 5.19 ' Let R be a ring, and let I be an ideal of R. Define a relation $\sim_I$ on R by setting $\forall a, b \in R$

$$a \sim_I b \iff b - a \in I.$$

We say that 'a is congruent to b modulo the ideal I'.

Thus, if $f: R \to S$ is a ring homomorphism and $I = \ker f$, then $\sim_I$ is precisely the same relation $\sim$ we were considering earlier in connection with the puzzling diagram (5.2) we are trying to construct.

A more popular notation for this relation is '$a \equiv b \bmod I$'; I am going to use $\sim_I$ because it makes it a little easier to state the basic facts that will follow. In a short while I'll dispose of this notation altogether.

Whenever you run into a new relation, you should ask yourself *what kind* of relation it is, and you should be happy if it turns out to be an equivalence relation.

PROPOSITION 5.20 *Let R be a ring, let I be an ideal of R. Then the relation $\sim_I$ is an equivalence relation.*

Proof • $\sim_I$ *is reflexive:* Indeed, $a \sim_I a$ since $a - a = 0$ and $0 \in I$ for every ideal I.

• $\sim_I$ *is symmetric:* Indeed, assume $a \sim_I b$; then $b - a \in I$, and since $a - b = -(b - a)$ and I is an ideal, it follows that $a - b \in I$. Therefore $b \sim_I a$.

• $\sim_I$ *is transitive:* Indeed, let $a, b, c \in R$, and assume $a \sim_I b$ and $b \sim_I c$. Then $b - a \in I$ and $c - b \in I$, and since

$$c - a = (c - b) + (b - a)$$

and I is an ideal, it follows that $c - a \in I$. This shows that $a \sim_I c$, as needed. □

The careful reader will note that the proof of this proposition only used axioms (i)–(iv) in the definition of an ideal; the absorption property is not needed here.

You should now have a feeling of *déjà-vu:* where have we seen something like this already? Flip back to the discussion leading to the definition of $\mathbb{Z}/n\mathbb{Z}$ in Chapter 2, and you will see a clear parallel. You already have a key example in hand of what we are in the process of developing.

Example 5.21   Let $n \geq 0$ be an integer, and let $I = n\mathbb{Z}$ be the ideal generated by n. Then $\forall a, b \in \mathbb{Z}$,

$$a \sim_I b \iff a \equiv b \bmod n.$$

That is, in this case the relation $\sim_I$ is nothing but congruence modulo n.

As you may have surmised at this point, we are going to carry out for any ideal I in a ring R the same program we carried out for the special case of the ideals $n\mathbb{Z}$ in $\mathbb{Z}$. First, we need a good name for the set of equivalence classes modulo $\sim_I$, and for the equivalence classes themselves.

DEFINITION 5.22   Let R be a ring, I an ideal of R, and $\sim_I$ the relation introduced above. We will denote by $a + I$ the equivalence class of $a \in R$ with respect to $\sim_I$; $a + I$ is called the 'coset of a (modulo I)'. We will denote by R/I the set of cosets; R/I is called the 'quotient of R modulo I'.

Remark 5.23   The notation $a + I$ is sensible: this set consists of all $b \in R$ such that $a \sim_I b$, that is, for which $b - a \in I$. Writing $b \in a + I$ for this condition is a good mnemonic device. We can in fact interpret the notation $a + I$ rather literally, i.e.,

$$a + I = \{a + r \mid r \in I\} \subseteq R.$$

This allows us to view $a + I$ as a *subset* of R, which it is: it is an equivalence class. (Note that I itself is a coset: $I = 0 + I$ as subsets of R. So we may view I as an element of R/I.) However, we also have to get used to viewing $a + I$ as an *element* of R/I: it is an equivalence class, therefore an element of the corresponding partition.

This issue is precisely the same one I raised concerning the notation $[a]_n$: a congruence class can be viewed both as a subset of $\mathbb{Z}$ and as an element of $\mathbb{Z}/n\mathbb{Z}$. Perhaps unfortunately, we do not use a different notation for these different points of view. With the notation introduced in Definition 5.22, we would now write $a + n\mathbb{Z}$ for the class $[a]_n$. (In fact, we occasionally did do this back in §2.2.)

The notation R/I is consistent with this motivating example: if $R = \mathbb{Z}$ and $I = n\mathbb{Z}$, we indeed wrote $\mathbb{Z}/n\mathbb{Z}$ for R/I, as early as Chapter 2 (Definition 2.5).

Next, we are going to obtain an analogue of Definition 2.10: that is, we will introduce two operations $+, \cdot$ on the set R/I.

DEFINITION 5.24   Let I be an ideal of a ring R, and let $a + I$, $b + I$ be elements of R/I (i.e., cosets). We define

$$(a + I) + (b + I) := (a + b) + I,$$
$$(a + I) \cdot (b + I) := (a \cdot b) + I.$$

Of course, you should immediately react to this prescription by asking yourself whether these operations are well-defined. This is the case, and the proof will follow closely the pattern of the proof given in Chapter 2 for the analogous fact concerning $\mathbb{Z}/n\mathbb{Z}$. Here is the analogue of Lemma 2.9. Its proof *will* use the absorption property.

> LEMMA 5.25   *Let I be an ideal of R, and let a, a', b, b' be elements of R.*
> *If $a' - a \in I$ and $b' - b \in I$, then*
>
> *(i) $(a' + b') - (a + b) \in I$,*
> *(ii) $a' \cdot b' - a \cdot b \in I$.*

Proof (i) follows immediately from the fact that ideals satisfy ring axioms (i)–(iv). (Work this out!)

(ii) Since $a' - a \in I$ and $b' - b \in I$, there are elements $i \in I$, $j \in I$ such that $a' - a = i$, $b' - b = j$. Then

$$a'b' - ab = (a + i)(b + j) - ab = (ab + aj + ib + ij) - ab = aj + ib + ij.$$

Now the three elements aj, ib, ij are all in I since I is an ideal, by the absorption property. Since ideals are closed with respect to addition, we can conclude that $aj + ib + ij \in I$, and this shows that $a'b' - ab \in I$ as needed.   □

The well-definedness of the operations introduced in Definition 5.24 is a straightforward consequence of Lemma 5.25. Indeed, if $a + I = a' + I$ and $b + I = b' + I$, then we have $a' - a \in I$ and $b' - b \in I$, and Lemma 5.25 guarantees then that the definitions for $(a + I) + (b + I)$ and $(a + I) \cdot (b + I)$ proposed in Definition 5.24 agree with the definitions for $(a' + I) + (b' + I)$ and $(a' + I) \cdot (b' + I)$, respectively.[1]

Good! Then our set R/I has two nice operations $+$, $\cdot$. The reader should now fully expect the next result.

> THEOREM 5.26   *Let I be an ideal of a ring R. Then the structure $(R/I, +, \cdot)$ is a ring, with $0_{R/I} = 0 + I = I$ and $1_{R/I} = 1 + I$.*

Proof   This is straightforward! All the hard work went into the definition. Proving that the ring axioms (from Definition 3.1) are satisfied is routine work, which I gladly leave to the reader. For example, to verify axiom (iii), let $a + I$ be any coset. Then we can consider the coset $(-a) + I$, and note that

$$(a + I) + ((-a) + I) = (a - a) + I = 0 + I, \quad ((-a) + I) + (a + I) = (-a + a) + I = 0 + I$$

as needed.   □

Example 5.27   The ring $\mathbb{Z}/n\mathbb{Z}$, with which we are already familiar, is a brilliant example of this construction: we can now see it as the ring R/I in the particular case $R = \mathbb{Z}$,

---

[1] If this sentence looks confusing, please work through it as long as needed for it to make sense. This is a super-important point.

$I = n\mathbb{Z}$. We do not need to place any restriction on n, since $(-n)\mathbb{Z} = n\mathbb{Z}$, so there is no need to assume that n is nonnegative (as we did in Chapter 2).

If you work out Exercise 5.14, then you will also discover that the rings $\mathbb{Z}/n\mathbb{Z}$ are *all* the quotient rings of $\mathbb{Z}$, since there are no other ways to construct an ideal of $\mathbb{Z}$.

Example 5.28   Let R be any ring, and let $I = R$; as we have seen, R is an ideal of R. Then R/R is a trivial ring. Indeed, it consists of a single coset, that is, R itself.

Example 5.29   At the other extreme, consider $R/(0)$, where (0) is the ideal $\{0_R\}$. We can define a function $R \to R/(0)$ by sending a to $a + (0)$. A moment's thought should reveal that this is a bijective homomorphism. For example: if $a + (0) = b + (0)$, then $b - a$ is a multiple of 0, so $b - a = 0$, i.e., $a = b$. This shows that the function is injective, and it is just as easy to verify that it is surjective and that it is a homomorphism. That is, $R/(0) \cong R$ is just an isomorphic copy of R.

Example 5.30   Let R be a *commutative* ring, and let I be an ideal of R. Then R/I is a commutative ring. Indeed, for every $a, b \in R$,

$$(a + I)(b + I) = ab + I = ba + I = (b + I)(a + I).$$

Example 5.31   On the other hand, we already know that if R is an integral domain, then R/I is not necessarily an integral domain. Indeed, let $R = \mathbb{Z}, I = 4\mathbb{Z}$; then $R = \mathbb{Z}$ is an integral domain and $R/I = \mathbb{Z}/4\mathbb{Z}$ is not an integral domain.

Example 5.32   Let $R = \mathbb{Z}[x]$, the polynomial ring with coefficients in $\mathbb{Z}$. As we have seen, the set (x) of multiples of x is an ideal of R. What does the quotient $\mathbb{Z}[x]/(x)$ look like?

Every polynomial in $\mathbb{Z}[x]$ may be written as $f(x) = a_0 + a_1 x + \cdots + a_n x^n$ for some integers $a_0, \ldots, a_n$. Note that $a_0 = f(0)$, and

$$(a_0 + a_1 x + \cdots + a_n x^n) - a_0 = (a_1 + a_2 x + \cdots + a_n x^{n-1}) \cdot x \in (x).$$

That is, $f(x) - f(0) \in (x)$, giving the equality of cosets

$$f(x) + (x) = f(0) + (x). \tag{5.5}$$

That is, *every* element of the quotient $\mathbb{Z}[x]/(x)$ is of the form $a + (x)$, where a is an integer. We can view this fact a little differently: define a function

$$\varphi: \mathbb{Z} \to \mathbb{Z}[x]/(x)$$

by setting $\varphi(a) = a + (x)$ for all $a \in \mathbb{Z}$. Then the fact we just verified tells us that this function φ is *surjective*. On the other hand, assume that $\varphi(a) = \varphi(b)$, for $a, b \in \mathbb{Z}$. Then

$$a + (x) = b + (x),$$

that is, $b - a$ is a multiple of x. But $b - a$ is an integer, and the only integer that is a multiple of x is 0. Therefore $a = b$. This shows that φ is also *injective*.

Further, φ is a homomorphism! This should be clear: $\varphi(1) = 1 + (x)$ is the identity in $\mathbb{Z}[x]/(x)$, and

$$\varphi(a + b) = (a + b) + (x) = (a + (x)) + (b + (x)) = \varphi(a) + \varphi(b).$$

This shows that φ preserves addition, and it is just as straightforward to verify that it preserves multiplication.

Conclusion: this new ring $\mathbb{Z}[x]/(x)$ turns out to be *isomorphic* to $\mathbb{Z}$.

There is nothing special about $\mathbb{Z}$ here: the same argument shows that

$$R[x]/(x) \cong R$$

for every[2] ring R.

Example 5.33 In fact, let R be a ring and let $r \in R$. Assume R is commutative and consider the ideal $(x - r)$ of $R[x]$. What can we say about $R[x]/(x - r)$?

Perform a 'change of variables', letting $x = t + r$. Then $g(t) := f(t + r)$ is a polynomial in t, and arguing as we just did a moment ago we see that $g(t) - g(0)$ is a multiple of t. That is, $f(x) - f(r)$ is a multiple of $x - r$:

$$f(x) - f(r) \in (x - r),$$

giving the equality of cosets

$$f(x) + (x - r) = f(r) + (x - r).$$

This shows that *every* coset in $R/(x - r)$ is of the form $a + (x - r)$ for some $a \in R$. We can then define a function $\varphi : R \to R[x]/(x - r)$ by setting $\varphi(a) = a + (x - r)$; we have just checked that φ is *surjective,* and you can verify as before that φ is also injective and is a homomorphism. (Please do this.)

Conclusion: $R[x]/(x - r) \cong R$ for every commutative ring R and every $r \in R$. This fact is actually used quite frequently in 'concrete' examples, so you are well advised to remember it.

Example 5.34 That was fun. Let's try something even more fun. Can we 'identify' the quotient ring $\mathbb{R}[x]/(x^2 + 1)$, i.e., find a famous ring isomorphic to it? (Recall that $\mathbb{R}$ stands for the field of *real numbers;* it is not just a random ring R.)

First, essentially the same technique we employed in the previous examples shows that every coset in $\mathbb{R}[x]/(x^2 + 1)$ may be written uniquely as

$$a + bx + (x^2 + 1); \tag{5.6}$$

we will come back to this point later on (in §7.2 to be precise), so let's not dwell on it now. Thus, every coset determines and is determined by a pair (a, b) of real numbers. (This says that $\mathbb{R}[x]/(x^2 + 1) \cong \mathbb{R} \times \mathbb{R}$ *as sets.*) How do the operations work, in terms of

---

[2] Fine print: since x commutes with every element in $R[x]$, we do not even need to assume that R is commutative for (x) to be an ideal of $R[x]$.

these pairs? If we want to add (a_1, b_1) and (a_2, b_2), we can add the corresponding cosets, and then translate the result back in terms of a pair. This is actually not too exciting:

$$(a_1, b_1) + (a_2, b_2) \rightsquigarrow (a_1 + b_1 x + (x^2 + 1)) + (a_2 + b_2 x + (x^2 + 1))$$
$$= (a_1 + a_2) + (b_1 + b_2)x + (x^2 + 1) \rightsquigarrow (a_1 + a_2, b_1 + b_2).$$

We see that the addition is the usual componentwise addition:

$$(a_1, b_1) + (a_2, b_2) = (a_1 + a_2, b_1 + b_2).$$

Multiplication is more interesting. Working just as above, we get

$$(a_1, b_1) \cdot (a_2, b_2) \rightsquigarrow (a_1 + b_1 x + (x^2 + 1)) \cdot (a_2 + b_2 x + (x^2 + 1))$$
$$= (a_1 a_2) + (a_1 b_2 + a_2 b_1)x + b_1 b_2 x^2 + (x^2 + 1) \rightsquigarrow \; ??$$

The coset we obtain is not of the form given in (5.6). But note that

$$b_1 b_2 x^2 + (x^2 + 1) = -b_1 b_2 + (x^2 + 1):$$

indeed,

$$b_1 b_2 x^2 - (-b_1 b_2) = b_1 b_2 \cdot (x^2 + 1)$$

is a multiple of $x^2 + 1$. Therefore,

$$(a_1 a_2) + (a_1 b_2 + a_2 b_1)x + b_1 b_2 x^2 + (x^2 + 1) = (a_1 a_2 - b_1 b_2) + (a_1 b_2 + a_2 b_1)x + (x^2 + 1)$$

and we conclude that the multiplication, in terms of pairs, works like this:

$$(a_1, b_1) \cdot (a_2, b_2) = (a_1 a_2 - b_1 b_2, a_1 b_2 + a_2 b_1).$$

If you are still with me, you will recognize that this is the same prescription defining the multiplication of *complex numbers:*

$$(a_1 + b_1 i) \cdot (a_2 + b_2 i) = (a_1 a_2 - b_1 b_2) + (a_1 b_2 + a_2 b_1) i.$$

Addition also works as in $\mathbb{C}$.

We can summarize this whole discussion by saying that the function

$$\mathbb{C} \to \mathbb{R}[x]/(x^2 + 1)$$

defined by $a + bi \mapsto a + bx + (x^2 + 1)$ is an isomorphism. Therefore, we successfully identified our quotient: we have discovered that

$$\mathbb{R}[x]/(x^2 + 1) \cong \mathbb{C}.$$

The verifications in the last few items were straightforward, but lengthy and a little cumbersome. We are on our way to developing a tool that will make such verifications much easier to carry out.

5.4   Rings: Canonical Decomposition, II

The quotient ring comes endowed with a natural 'projection' function:

$$\pi: R \to R/I, \quad r \mapsto r + I.$$

This is just the map that sends an element $r \in R$ to the corresponding equivalence class $[r]_{\sim_I}$; this equivalence class is called $r + I$ (Definition 5.22; and this is where we abandon the notation $\sim_I$, since we no longer need it).

PROPOSITION 5.35   *Let R be a ring, and let I be an ideal of R. Then the natural projection $\pi: R \to R/I$ is a surjective ring homomorphism, and its kernel is I.*

Proof   The first claim follows immediately from the definition of $+$ and $\cdot$ in the quotient:

$$\pi(a) + \pi(b) = (a + I) + (b + I) = (a + b) + I = \pi(a + b),$$

and similarly for multiplication; and of course $\pi(1) = 1 + I$ is the identity in R/I. Thus π is a ring homomorphism. It is clearly surjective, since every coset is the coset of *some* element of R: $a + I = \pi(a)$.

For the second claim, observe that

$$\pi(a) = 0 + I \iff a + I = 0 + I \iff a = a - 0 \in I.$$

This says precisely that $\ker \pi = I$, as stated.   □

Remark 5.36   Thus, we have now verified a claim I made in passing in Example 5.10: *every* ideal of a ring R is the kernel of some homomorphism.

I hope you find this somewhat striking. Look at the definitions of 'kernel' (Definition 5.3) and 'ideal' (Definition 5.8). Do they even look alike? To me, they do not. And yet, we have been able to establish that *they express precisely the same concept!*

We have now developed enough 'language' to get back to the issue that prompted this discussion. Recall that we were in the process of trying to understand a 'canonical decomposition' for a given ring homomorphism $f: R \to S$. We were looking for a commutative diagram

$$R \xrightarrow[\pi]{} R' \xrightarrow[\tilde{f}]{\sim} S' \xhookrightarrow{\iota} S ,$$

where R', S' are also rings, and the arrows denote ring homomorphisms. We had guessed right away that we could take $f(R)$ for S', and let $\iota: f(R) \to S$ be the inclusion. The quotient $R/\ker f$ is the missing piece in the puzzle: it will play the role of R' in our diagram. To define the function

$$\tilde{f}: R/\ker f = R' \to S' = f(R),$$

let $a + \ker f$ be an arbitrary coset in $R/\ker f$, and set

$$\tilde{f}(a + \ker f) := f(a). \qquad (5.7)$$

To verify that the result of this prescription is well-defined, suppose $a' + \ker f = a + \ker f$, i.e., $a' - a \in \ker f$. Since f is a homomorphism, this implies that

$$f(a') - f(a) = f(a' - a) = 0$$

by definition of kernel. This says that $f(a') = f(a)$, so indeed the proposed definition of $\tilde{f}(a + \ker f)$ does not depend on the representative chosen for the coset.

We say that $\tilde{f}$ is 'induced' by f. It should be clear that $\tilde{f}$ is a ring homomorphism:

$$\tilde{f}(a + \ker f) + \tilde{f}(b + \ker f) = f(a) + f(b) = f(a + b) = \tilde{f}((a + b) + \ker f)$$
$$= \tilde{f}((a + \ker f) + (b + \ker f));$$

the same token gives you $\tilde{f}(a + \ker f) \cdot \tilde{f}(b + \ker f) = \tilde{f}((a + \ker f) \cdot (b + \ker f))$, and $\tilde{f}(1 + \ker f) = f(1) = 1$ as it should.

We are finally ready to wrap up this lengthy discussion into one neat statement.

> THEOREM 5.37 (Canonical decomposition) *Let $f: R \to S$ be a ring homomorphism. Then we have a commutative diagram*
>
> $$R \xrightarrow[\pi]{} (R/\ker f) \xrightarrow[\tilde{f}]{} f(R) \xrightarrow{\iota} S \,, \quad \overset{f}{\frown}$$
>
> *where π is the natural projection, ι is the inclusion homomorphism, and $\tilde{f}$ is the induced homomorphism (defined above). Further: π is surjective, ι is injective, and $\tilde{f}$ is an* isomorphism *of rings.*

Proof We already know that π is a surjective homomorphism and ι is an injective homomorphism. To prove that the diagram is commutative, let a be an arbitrary element of R; then

$$(\iota \circ \tilde{f} \circ \pi)(a) = \iota(\tilde{f}(\pi(a))) = \iota(\tilde{f}(a + I)) = \iota(f(a)) = f(a).$$

This shows that $\iota \circ \tilde{f} \circ \pi = f$, as needed.

Thus, all we have left to do is prove that the homomorphism $\tilde{f}$ is an isomorphism. To see that $\tilde{f}$ is surjective, let s be an element of $f(R)$; then by definition of $f(R)$ there exists an element $a \in R$ such that $f(a) = s$, and we have

$$\tilde{f}(a + \ker f) = f(a) = s.$$

This shows that s has a preimage in $R/\ker f$, proving that $\tilde{f}$ is surjective. To verify that $\tilde{f}$ is injective, suppose $\tilde{f}(a + \ker f) = \tilde{f}(b + \ker f)$. Then $f(a) = f(b)$, i.e., $f(b - a) = 0$. This says that $b - a \in \ker f$, and shows that $a + \ker f = b + \ker f$, as needed.

We conclude that $\tilde{f}$ is an isomorphism, and we are done. □

5.5   The First Isomorphism Theorem

The canonical decomposition of a homomorphism is a very useful tool. The following consequence is often highlighted, and even gets a name.

THEOREM 5.38 (First Isomorphism Theorem)   *Let $f: R \to S$ be a surjective ring homomorphism. Then $S \cong R/\ker f$.*
   In fact, f induces an isomorphism $\tilde{f}: R/\ker f \xrightarrow{\sim} S$, defined by $\tilde{f}(a + \ker f) = f(a)$.

Proof By the canonical decomposition, $(R/\ker f) \cong f(R)$, and the isomorphism $\tilde{f}$ is defined by $a + \ker f \mapsto f(a)$. Since f is surjective, $f(R) = S$. The statement follows immediately.   □

Beyond technical details, the main item that makes Theorem 5.37 (and hence the First Isomorphism Theorem) work is the definition of the 'induced' homomorphism, (5.7). The same definition can be given in a slightly more general case, and this mild generalization is quite useful.

PROPOSITION 5.39   *Let $f: R \to S$ be a ring homomorphism, and assume that $I \subseteq \ker f$. Then f induces a unique ring homomorphism $\tilde{f}: R/I \to S$ making the diagram*

commute.

In case you are curious, Proposition 5.39 spells out the 'universal property of quotients', another item in the categorical language that underlies everything we are covering in these notes.

The commutativity of the diagram actually determines the definition of $\tilde{f}$. Indeed, since $a + I \in R/I$ equals $\pi(a)$, the desired property that $f = \tilde{f} \circ \pi$ forces us to define

$$\tilde{f}(a + I) = \tilde{f}(\pi(a)) = \tilde{f} \circ \pi(a) = f(a) :$$

there is no choice here. In this sense $\tilde{f}$ is 'unique': only one function $\tilde{f}$ can make the diagram commute. The content of the proposition is that this definition works, i.e., this prescription for $\tilde{f}$ is well-defined, and $\tilde{f}$ is then a ring homomorphism. The proof of these facts is essentially identical to the case worked out in detail in the case leading to the proof of Theorem 5.37, so it does not seem necessary to go over it again. You should now grab a piece of paper and just do it yourself—it should not give you any trouble. (If it does, that means you have not absorbed the proof of Theorem 5.37 as well as you may want to.)

Of course, in general the induced homomorphism will neither be injective nor surjective. It will be injective precisely when $I = \ker \varphi$. It will be surjective precisely when f is surjective to begin with.

What are all these considerations good for? The First Isomorphism Theorem has two main types of applications. First, it often allows us to quickly 'recognize' a quotient, that is, determine that it is isomorphic to a known ring. Second, it also allows us to determine the image of a homomorphism: if $f : R \to S$ is a homomorphism, then f determines a *surjective* homomorphism $f : R \to f(R)$, and applying the First Isomorphism Theorem gives some kind of realization of the image $f(R)$. In both cases, all that is required is an explicit 'computation' of $\ker f$.

To see the first type of application in action, we will revisit below the 'harder' items we worked out in Examples 5.27–5.34. In §5.7 we will see an important application of the same strategy, and in §5.6 we will look at an application of the second type.

Example 5.40 Suppose we have to identify $\mathbb{Z}[x]/(x)$ (cf. Example 5.32). Consider the function $f : \mathbb{Z}[x] \to \mathbb{Z}$ defined by sending $\alpha(x)$ to $\alpha(0)$, for all $\alpha(x) \in \mathbb{Z}[x]$. This is a surjective homomorphism. (Make sure you see why!) What is $\ker f$?

$$\ker f = \{\alpha(x) \in \mathbb{Z}[x] \mid \alpha(0) = 0\} = \{\alpha(x) \in \mathbb{Z}[x] \mid \text{the constant term of } \alpha(x) \text{ is } 0\}.$$

This is the set of multiples of x, and therefore $\ker f = (x)$. By the first isomorphism theorem, we can conclude

$$\mathbb{Z}[x]/(x) \cong \mathbb{Z}$$

without any further computation.

Example 5.41 More generally, let R be any commutative ring and consider the surjective ring homomorphism

$$e_r : R[x] \to R$$

obtained by 'plugging-in $x = r$': that is, sending $f(x) \in R[x]$ to $f(r)$. Then

$$\ker(e_r) = \{f(x) \in R[x] \mid f(r) = 0\}.$$

In Example 5.33 we have verified that for all polynomials $f(x)$, $f(x) - f(r)$ is a multiple of $(x - r)$; that is, there exists a polynomial $q(x)$ such that

$$f(x) = (x - r)q(x) + f(r).$$

This is true for all $f(x)$. But then $f(r) = 0$ if and only if $f(x) = (x - r)q(x)$ for some polynomial $q(x)$, i.e., if and only if $f(x) \in (x - r)$. This shows that

$$\ker(e_r) = (x - r),$$

and by the First Isomorphism Theorem we recover that

$$R[x]/(x - r) \cong R. \tag{5.8}$$

This is a surprisingly useful observation: it will occasionally allow you to realize seemingly complicated quotients without having to perform any work at all. For instance,

$$\mathbb{C}[x, y]/(x - y^2 - y^{17}) \cong \mathbb{C}[y] :$$

for this, just apply (5.8) with $R = \mathbb{C}[y]$, $r = y^2 + y^{17}$. (Recall that the polynomial ring $\mathbb{C}[x, y]$ is isomorphic to $\mathbb{C}[y][x]$, cf. Example 4.43.)

Example 5.42 What about $\mathbb{R}[x]/(x^2+1)$? This time, consider the function $f: \mathbb{R}[x] \to \mathbb{C}$ obtained by sending $\alpha(x)$ to $\alpha(i)$. This is a surjective ring homomorphism, and

$$\ker f = \{\alpha(x) \in \mathbb{R}[x] \mid \alpha(i) = 0\} = (x^2 + 1).$$

This is straightforward if you think in terms of 'long division of polynomials'. By long division, you can find unique $a, b \in \mathbb{R}$ such that

$$\alpha(x) = \beta(x)(x^2 + 1) + a + bx$$

(this is the same observation leading to (5.6); we will talk about it in more detail later on, see Example 7.4). Therefore

$$\alpha(i) = \beta(i)(i^2 + 1) + a + bi = a + bi,$$

and this shows that $\alpha(i) = 0$ if and only if $a = b = 0$; that is, if and only if $\alpha(x)$ is a multiple of $x^2 + 1$; that is, if and only if $\alpha(x) \in (x^2 + 1)$.

This may look like a mouthful, but realize that *no more work is needed* in order to conclude that

$$\mathbb{R}[x]/(x^2 + 1) \cong \mathbb{C} :$$

this now follows immediately by applying the First Isomorphism Theorem. In passing, precisely the same argument shows that

$$\mathbb{Z}[x]/(x^2 + 1) \cong \mathbb{Z}[i],$$

the ring of 'Gaussian integers' we run across in Example 4.12.

We will use the First Isomorphism Theorem quite frequently. In fact, one soon gets used to thinking of *every* 'homomorphic image' of a ring R, i.e., any ring S such that there is a surjective ring homomorphism $R \to S$, as a quotient of R by an ideal, up to a natural identification. One may do so because of the First Isomorphism Theorem.

5.6 The Chinese Remainder Theorem

Here is a good extended example of the second type of application of the First Isomorphism Theorem, in which it is used to 'identify the image' of a homomorphism. The main result here goes under the name of 'Chinese Remainder Theorem'. One example of application of this theorem (in the form of a problem) may be found in the *Sūnzǐ*

Suànjīng, a Chinese mathematical text dating back to the fourth century AD. A proof of the theorem (over[3] $\mathbb{Z}$) was given by Qín Jiǔsháo, around 1250.

Let R be a ring, and let I and J be ideals of R. There is a natural function

$$\rho: R \to R/I \times R/J$$

sending $r \in R$ to the pair $(r + I, r + J)$, and it will take you just a moment to check that this function is a ring homomorphism.

Notice that ρ need not be surjective. For example, let $R = \mathbb{Z}$, $I = J = \mathbb{Z}/2\mathbb{Z}$; then $\rho(0) = (0,0), \rho(1) = (1,1), \rho(2) = (0,0)$, etc. (dropping a bit of extra notation as usual). It should be clear that $(1,0)$ (for example) is not in the image of ρ. In fact, it should be clear that the image of ρ is the 'diagonal' subring of Example 4.9 in this case.

On the other hand, sometimes ρ *is* surjective: you can quickly verify for fun that the analogous function $\mathbb{Z} \to \mathbb{Z}/2\mathbb{Z} \times \mathbb{Z}/3\mathbb{Z}$ is surjective.

Regardless of whether ρ is surjective, we can ask what the ring $\rho(R)$ looks like. This is a subring of $R/I \times R/J$. Can we 'understand' what it is?

The First Isomorphism Theorem directly answers that question, by telling us that $\rho(R) \cong R/\ker\rho$. (Indeed, ρ restricts to a surjective function $R \to \rho(R)$.) So all we need to do is 'compute' the kernel $\ker(\rho)$ of ρ. This is easily done:

$$\ker\rho = \{r \in R \,|\, \rho(r) = 0\} = \{r \in R \,|\, (r + I, r + J) = (0,0)\}$$
$$= \{r \in R \,|\, r + I = 0 + I \text{ and } r + J = 0 + J\}$$
$$= \{r \in R \,|\, r \in I \text{ and } r \in J\}$$
$$= I \cap J.$$

(Incidentally, since kernels are ideals, this shows that $I \cap J$ is an ideal if I and J are ideals; you can also check this 'by hand', and I encourage you to do so in Exercise 5.9.) The conclusion is that the First Isomorphism Theorem tells us that $\rho(R) \cong R/(I \cap J)$. In this case, the 'canonical decomposition' looks like this:

$$R \xrightarrow[\pi]{\rho} R/(I \cap J) \xrightarrow[\tilde{\rho}]{\sim} \rho(R) \xhookrightarrow{\iota} R/I \times R/J.$$

For example, if $R = \mathbb{Z}$ and $I = J = 2\mathbb{Z}$, then $I \cap J = 2\mathbb{Z}$, so that $\rho(\mathbb{Z}) \cong \mathbb{Z}/2\mathbb{Z}$ has two elements. This matches what we verified by hand a moment ago. Work out what happens for $R = \mathbb{Z}$, $I = 2\mathbb{Z}$, $J = 3\mathbb{Z}$, for fun.

This is a good occasion to mention a few operations we can perform on ideals. (You have already seen these if you already worked out Exercise 5.9.) Given ideals I and J of a ring R, we can create new ideals in several ways.

DEFINITION 5.43   The 'sum' of I and J is

$$I + J = \{a + b \,|\, a \in I, b \in J\}.$$

---

[3] People had not come up with rings and ideals yet back then.

It only takes a moment to verify that this is an ideal of R. It may be described as the *smallest* ideal of R containing both I and J. The *largest* ideal contained in both I and J is clearly the intersection $I \cap J$: since $I \cap J$ is the largest *subset* contained in both I and J, and since it is an ideal, it is indeed the largest ideal with this property. There is another naturally defined ideal contained in both I and J, which in general is *not* the largest ideal with this property.

DEFINITION 5.44 The 'product' of I and J is

$$IJ = \left\{ \sum_i a_i b_i \mid a_i \in I, b_i \in J \right\}.$$

Watch out: IJ does *not* consist of the products of elements of I by elements of J; it is the set of *sums* of such products. Again, it is easy to verify that this is an ideal of R. To see that it is contained in I, note that each product $a_i b_i$ is contained in I by absorption, since $a_i \in I$; and then every element of IJ must be contained in I, since it is a sum of elements of I. Essentially the same argument shows that $IJ \subseteq J$. It follows that $IJ \subseteq I \cap J$; this is true for all ideals I and J.

Nothing prevents us from taking $I = J$. For example, let $R = \mathbb{Z}[x]$, and $I = (x)$. Then $I \cdot I = I^2$ is the ideal (x^2), $I^3 = (x^3)$, etc. This case gives a good example showing that $IJ \subsetneq I \cap J$ in general: if $I = J$, then $IJ = I^2$, while $I \cap J = I$ and, e.g., $(x^2) \subsetneq (x)$.

It is easy to see what these definitions amount to for $R = \mathbb{Z}$ and $I = (a) = a\mathbb{Z}$ and $J = (b) = b\mathbb{Z}$.

Example 5.45 I claim that $(a) + (b) = (\gcd(a,b))$. Indeed, let $d = \gcd(a,b)$. Since $d \mid a$, we see that $a \in (d)$, that is, $(a) \subseteq (d)$. By the same token, $(b) \subseteq (d)$. Since $(a) + (b)$ is the smallest ideal containing both (a) and (b), we see that $(a) + (b) \subseteq (d)$. On the other hand, we know that d is a linear combination of a and b: this is our old friend Theorem 1.8. This means precisely that $d \in (a) + (b)$, and therefore $(d) \subseteq (a) + (b)$. Since $(a) + (b) \subseteq (d)$ and $(d) \subseteq (a) + (b)$, we can conclude that $(d) = (a) + (b)$, which was my claim.

Example 5.46 Next, I claim that $(a) \cap (b) = (\text{lcm}(a,b))$ (i.e, it equals the principal ideal generated by the *least common multiple* of a and b). Indeed, $n \in (a) \cap (b)$ if and only if n is a multiple of both a and b; since every multiple of both a and b is a multiple of $\text{lcm}(a,b)$, this shows that $(a) \cap (b) \subseteq (\text{lcm}(a,b))$; on the other hand, $(\text{lcm}(a,b)) \subseteq (a) \cap (b)$ since the lcm of a and b is itself a multiple of a and b.

Example 5.47 It should be clear that $(a) \cdot (b) = (ab)$.

Thus, in this setting the difference between IJ and $I \cap J$ is transparent: IJ corresponds to the product, and $I \cap J$ corresponds to the least common multiple. These ideals coincide in $\mathbb{Z}$ if and only if the least common multiple of a and b equals ab, that is (Exercise 1.23) if and only if a and b are relatively prime. In Lemma 5.51 below we will prove a substantial generalization of this fact.

Now that we have gotten familiar with this language, let's go back to the situation at the beginning of the section. We were given *two* ideals I and J of a ring R, and we looked at the homomorphism

$$\rho: R \to (R/I) \times (R/J)$$

defined in the natural way—by sending $r \in R$ to $\rho(r) = (r + I, r + J)$. As an application of the First Isomorphism Theorem, we discovered that

$$\rho(R) \cong R/(I \cap J).$$

We also observed that ρ may or may not be surjective, that is, $\rho(R)$ may or may not be the whole of $(R/I) \times (R/J)$. It is useful to identify a condition which guarantees that ρ *is* surjective. (The reader will prove that this condition is equivalent to the surjectivity of ρ.)

PROPOSITION 5.48   *Let I, J be ideals in a ring R, and assume that $I + J = (1)$. Then the homomorphism ρ is surjective.*

The condition that $I + J = (1)$ means that the *smallest* ideal containing both I and J is the ring itself. In the example $R = \mathbb{Z}$, with $I = (a)$ and $J = (b)$, as we have seen this would mean that $\gcd(a, b) = 1$, that is, the integers a and b are relatively prime. We may take the condition $I + J = (1)$ as a natural generalization of the 'relative prime' condition.

Proof   Assume that $I + J = (1)$ and let $(a + I, b + J) \in R/I \times R/J$. We have to find $r \in R$ such that $(a + I, b + J) = \rho(r)$, that is, such that

$$(a + I, b + J) = (r + I, r + J).$$

That is, such that $r - a \in I$, $r - b \in J$. By hypothesis, there exist $i \in I$, $j \in J$ such that $1 = i + j$. Let then

$$r = bi + aj.$$

With this definition,

$$r - a = r - a \cdot 1 = (bi + aj) - (ai + aj) = (b - a)i \in I,$$
$$r - b = r - b \cdot 1 = (bi + aj) - (bi + bj) = (a - b)j \in J,$$

by the absorption properties. So this r works, and this concludes the proof.   □

Remark 5.49   The key to the proof is the idea of letting $r = bi + aj$, where $i \in I$ and $j \in J$ are such that $i + j = 1$. This may seem a bit out of the blue. It isn't really, if you think along the following lines. Given that $i + j = 1$, we have $i = 1 - j$; therefore

$$\rho(i) = (i + I, i + J) = (i + I, (1 - j) + J) = (0 + I, 1 + J).$$

By the same token, $\rho(j) = (1 + I, 0 + J)$. If we want to cook up the pair $(a + I, b + J)$, then it makes sense to view it as

$$(a + I, b + J) = (0 + I, b \cdot 1 + J) + (a \cdot 1 + I, 0 + J) = \rho(bi) + \rho(aj) = \rho(bi + aj).$$

This is where the chosen r comes from. ⌐

COROLLARY 5.50   *Let R be a ring, and let I, J be ideals of R such that $I + J = (1)$.
Then*

$$R/(I \cap J) \cong (R/I) \times (R/J).$$

Proof   We used the First Isomorphism Theorem earlier to show that $R/(I \cap J) \cong \rho(R)$; and $\rho(R) = (R/I) \times (R/J)$ by the proposition, since $I + J = (1)$.   □

Corollary 5.50 is a version of the so-called 'Chinese Remainder Theorem'. I am making the arbitrary choice to give this name to a slightly different version of the result, which assumes that the ring is *commutative*. Many variations on the same theme may be called 'Chinese Remainder Theorem' in other sources. If R is commutative, we have the following further handy observation.

LEMMA 5.51   *Let R be a commutative ring, and let I, J be ideals of R such that $I + J = (1)$. Then $I \cap J = IJ$.*

Again, in the case $R = \mathbb{Z}$ we know this already (cf. Example 5.47): it says that the lcm of two integers a and b equals ab if a and b are relatively prime.

Proof   The inclusion $IJ \subseteq I \cap J$ is always true, as pointed out earlier. The interesting inclusion is $I \cap J \subseteq IJ$, which is *not* necessarily true for arbitrary I and J.
Assume that $I + J = (1)$. Then there exist $i \in I$, $j \in J$ such that $i + j = 1$. For $r \in I \cap J$,

$$r = r \cdot 1 = r \cdot (i + j) = ri + rj = ir + rj.$$

Note that I have used the commutativity of R in the last equality, to perform the switch $ri = ir$. This proves the statement, since $ir \in IJ$ as $i \in I$, $r \in J$, and $rj \in IJ$ as $r \in I$, $j \in J$.   □

THEOREM 5.52 (Chinese Remainder Theorem)   *Let R be a commutative ring, and let I, J be ideals of R such that $I + J = (1)$. Then*

$$R/IJ \cong (R/I) \times (R/J).$$

Proof   This is just Corollary 5.50 again, with Lemma 5.51 taken into consideration.   □

The statement of the Chinese Remainder Theorem should strike you as pleasantly general. In the (very!) particular case in which $R = \mathbb{Z}$, $I = (m) = m\mathbb{Z}$, $J = (n) = n\mathbb{Z}$, it says that

$$\mathbb{Z}/mn\mathbb{Z} \cong \mathbb{Z}/m\mathbb{Z} \times \mathbb{Z}/n\mathbb{Z}$$

if m and n are relatively prime. The case $m = 2$, $n = 3$ has been with us since our first encounter with isomorphisms back in §4.4. (And if you have worked out Exercise 4.19, then you know that the condition that a and b are relatively prime is not only sufficient, but also necessary.)

The following seemingly more impressive statement is an easy consequence. Prove it yourself (by induction).

> COROLLARY 5.53 *Let $n_1, \ldots, n_r$ be pairwise relatively prime positive integers, and let $N = n_1 \cdots n_r$. Then*
>
> $$\mathbb{Z}/N\mathbb{Z} \cong (\mathbb{Z}/n_1\mathbb{Z}) \times \cdots \times (\mathbb{Z}/n_r\mathbb{Z}).$$

In particular,

$$\mathbb{Z}/N\mathbb{Z} \cong (\mathbb{Z}/p_1^{a_1}\mathbb{Z}) \times \cdots \times (\mathbb{Z}/p_r^{a_r}\mathbb{Z})$$

if $N = p_1^{a_1} \cdots p_r^{a_r}$ is the prime factorization of N.

Corollary 5.53 tells us that we can solve 'congruence equations'. If $n_1, \ldots, n_r$ are pairwise relatively prime integers, then given arbitrary integers $m_1, \ldots, m_r$, there must exist some integer M such that

$$\begin{cases} M \equiv m_1 \bmod n_1 \\ M \equiv m_2 \bmod n_2 \\ \quad \cdots \\ M \equiv m_r \bmod n_r \end{cases} \tag{5.9}$$

and further M is uniquely determined modulo the product $N := n_1 \cdots n_r$. This is now clear, since it is a restatement of Corollary 5.53. Was it clear to you before? It was not clear to me! Chasing the proof of the Chinese Remainder Theorem (particularly Proposition 5.48, which we proved along the way) would even show how to come up with M if the numbers m_i and n_i are given.

The fact that we can solve arbitrary congruence equations of the type (5.9), if the numbers n_i are pairwise relatively prime, is the form of the Chinese Remainder Theorem that appears in the early literature on the subject. The related problem in the *Sūnzǐ Suànjīng* mentioned at the beginning of the section is a problem of this type.

Example 5.54 Suppose that you have to count a largish number of items—for example, the number of people in a group, and you know it is in the order of a couple of hundred, but you need the exact number, and need it fast. Tell your friends to assemble in rows of 17 people each; you find that the last row only has 8 people in it. Then tell them to regroup in rows of 19 people each; and the last row only has 1 person. You think about it a second, and conclude that the group consists of exactly 229 people.

How did you do it? Well, you happen to know that

$$1 = -10 \cdot 17 + 9 \cdot 19 :$$

that is, you happen to know two numbers $i = -170 = -10 \cdot 17 \in (17)$ and $j = 171 =$

$9 \cdot 19 \in (19)$ such that $1 = i + j$. Look back at the proof of Proposition 5.48: you will see that if you need a number which is $\equiv 8 \bmod 17$ and $\equiv 1 \bmod 19$, you can then take

$$1 \cdot (-170) + 8 \cdot 171 = 1198.$$

Next you note that the unique number between 0 and $17 \cdot 19 = 323$ which is congruent to 1198 modulo 323 is $1198 - 3 \cdot 17 \cdot 19 = 229$, and you are done.

OK, so this is not the greatest example: with such small numbers we could just as well count the rows and get the exact number quickly enough. But this method becomes practical if the numbers are larger; it was allegedly used to count, e.g., soldiers in cohorts. Spend a moment appreciating that it works precisely because of the Chinese Remainder Theorem.

The key point of Corollary 5.53 is that it allows us to do arithmetic modulo large numbers N by reducing it to the more manageable case of, for example, the prime power factors of N. This is useful in practice, in more serious applications than what may transpire from Example 5.54.

Example 5.55 For an example with a different flavor, look at Exercise 5.20. The question concerns the ring $\mathbb{R}[x]/(x^2 - 1)$, and the problem asks you to construct by hand an explicit isomorphism from this ring to $\mathbb{R} \times \mathbb{R}$. You are now able to verify that the two rings are isomorphic without performing any messy computation. Since

$$(x^2 - 1) = (x + 1)(x - 1)$$

as ideals, and $(x + 1)$, $(x - 1)$ are relatively prime (indeed, $1 = \frac{x+1}{2} - \frac{x-1}{2}$), Theorem 5.52 tells us that

$$\mathbb{R}[x]/(x^2 - 1) \cong \mathbb{R}[x]/(x + 1) \times \mathbb{R}[x]/(x - 1)$$

for free; and $\mathbb{R}[x]/(x - r) \cong \mathbb{R}$ for any real number r by the First Isomorphism Theorem (as we have seen more generally in Example 5.41).

Here is one more application of the Chinese Remainder Theorem. This is rather pleasant, since it can be formulated without any reference to the more abstract language we have developed, and yet the *proof* uses this language very naturally.

Leonhard Euler is credited with introducing the 'totient function': for a positive integer n, the 'totient' $\phi(n)$ is the number of positive integers k, $1 \le k \le n$, that are relatively prime to n. The first few values of this function are

$$\phi(1) = 1, \quad \phi(2) = 1, \quad \phi(3) = 2, \quad \phi(4) = 2, \quad \phi(5) = 4, \quad \phi(6) = 2, \quad \phi(7) = 6, \ldots.$$

This function has important applications, and in fact we will encounter it again later on in this text (§10.2, §15.6). It seems to jump around a lot, and it may seem difficult to detect any pattern. Well, clearly $\phi(p) = p - 1$ if p is irreducible, since in that case *every* positive integer less than p is relatively prime to p. But what is, e.g., $\phi(105)$?

You can patiently work this out by hand if you want, but by now you have figured out that mathematicians are *not* patient people. What we have learned makes it very easy to handle this function.

PROPOSITION 5.56   *If a and b are relatively prime, then $\phi(ab) = \phi(a)\phi(b)$.*

Proof The key observation is that $\phi(n)$ equals the number of *units* in $\mathbb{Z}/n\mathbb{Z}$: indeed, $[\ell]_n$ is a unit in $\mathbb{Z}/n\mathbb{Z}$ precisely when $\gcd(n, \ell) = 1$. This was observed a while back, in Proposition 2.16. A common notation for the set of units in a ring R is R^*; so the number $\phi(n)$ equals the cardinality of $(\mathbb{Z}/n\mathbb{Z})^*$. Now assume that a and b are relatively prime. By the Chinese Remainder Theorem,

$$(\mathbb{Z}/ab\mathbb{Z}) \cong (\mathbb{Z}/a\mathbb{Z}) \times (\mathbb{Z}/b\mathbb{Z}).$$

It follows that

$$(\mathbb{Z}/ab\mathbb{Z})^* \cong (\mathbb{Z}/a\mathbb{Z})^* \times (\mathbb{Z}/b\mathbb{Z})^* : \tag{5.10}$$

indeed, (r, s) is a unit in $R \times S$ if and only if r is a unit in R and s is a unit in S. (Exercise 4.3!) Since the number of elements in a Cartesian product of sets is the product of the numbers of elements of the factors, (5.10) implies $\phi(ab) = \phi(a)\phi(b)$ as stated.   □

For instance, $\phi(105) = \phi(3 \cdot 5 \cdot 7) = \phi(3)\phi(5)\phi(7) = 2 \cdot 4 \cdot 6 = 48$.

5.7   The Third Isomorphism Theorem

Where there is a First Isomorphism Theorem, you may suspect a Second Isomorphism Theorem is also lurking, and maybe a third one as well. I'll leave the second one alone, at least for now: the context of rings is not really appropriate for this result. (We will come back to it in later chapters, when we deal with modules and with groups.) For now, we can take a look at the Third Isomorphism Theorem.

Recall that the quotient R/I of a ring by an ideal I is the set of all cosets $r + I$ with $r \in R$ (Definition 5.22). I will more generally use the notation A/I for the set of all cosets $r + I$ with $r \in A$, where A is any subset of R containing I. That is, $A/I = \pi(A)$, where $\pi: R \to R/I$ is the natural projection. We have seen that R/I (that is, the case $A = R$) is itself a ring, and we have already learned a few things about it. The next question is: what are the ideals of the ring R/I? They should be sets of cosets, that is, sets of the form A/I for some $A \subseteq R$.

THEOREM 5.57 (Third Isomorphism Theorem)   *Let R be a ring, and let I be an ideal of R. Let J be an ideal of R containing I. Then J/I is an ideal of R/I, and all ideals of R/I may be realized in this way. Further,*

$$(R/I)/(J/I) \cong R/J.$$

Proof Let $\pi\colon R \to R/I$ be the natural projection. If J is an ideal of R containing I, then $\pi(J) = J/I$ is an ideal of R/I since π is surjective[4] (Exercise 5.7).

If $\bar{J}$ is an ideal of R/I, let $J = \pi^{-1}(\bar{J})$. Then J is an ideal of R (Exercise 5.23), and J contains $I = \pi^{-1}(0)$ since $\bar{J}$ contains 0. It is clear that $J/I = \pi(J) = \bar{J}$, and this verifies that all ideals of R/I are of the form J/I as stated. (Also note that $J = \pi^{-1}(J/I)$; we'll use this fact in a moment, at the very end of the proof.)

The last part is a good application of the First Isomorphism Theorem, as you know already if you have worked out Exercise 5.23. We have the composition ρ of the natural projections, which I will call π and π':

$$R \xrightarrow{\ \pi\ } R/I \xrightarrow{\ \pi'\ } (R/I)/(J/I)\,.$$

with ρ the composite arrow $R \to (R/I)/(J/I)$.

Both natural projections are surjective, therefore ρ is surjective. By the First Isomorphism Theorem,

$$(R/I)/(J/I) \cong R/\ker\rho\,.$$

What is $\ker\rho$?

$$\ker\rho = \{r \in R \mid \rho(r) = 0\} = \rho^{-1}(0)$$

by definition. Since $\rho = \pi' \circ \pi$,

$$\rho^{-1}(0) = \pi^{-1}(\pi'^{-1}(0)) = \pi^{-1}(J/I) = J$$

and this concludes the proof. $\qquad\qquad\qquad\qquad\qquad\qquad\qquad\Box$

In order to handle J/I in applications of this theorem, it may be useful to observe that if R is commutative, and $J = (a)+I$, then J/I will simply be the ideal $(a+I)$ generated by $a + I$. Prove this! And in fact, since you are at it, prove a version with more generators. (Exercise 5.34.)

Example 5.58 Recall that $(2, x) \subseteq \mathbb{Z}[x]$ stands for the ideal consisting of linear combinations of 2 and x (Definition 5.17). Concretely, it is the set of all polynomials with *even* constant term. In order to identify the quotient $\mathbb{Z}[x]/(2, x)$, we can argue as follows. By the Third Isomorphism Theorem, $\mathbb{Z}[x]/(2, x) \cong (\mathbb{Z}[x]/(x))/((2, x)/(x))$. We already know that $\mathbb{Z}[x]/(x) \cong \mathbb{Z}$ (Example 5.40). Since $(2, x) = (2) + (x)$, $(2, x)/(x)$ is the ideal of $\mathbb{Z}[x]/(x)$ generated by the coset $2 + (x)$. This coset corresponds to the integer 2 via the isomorphism $\mathbb{Z}[x]/(x) \cong \mathbb{Z}$. The conclusion is that

$$\mathbb{Z}[x]/(2, x) \cong (\mathbb{Z}[x]/(x))/((2, x)/(x)) \cong \mathbb{Z}/2\mathbb{Z}\,.$$

---

[4] In case you have not worked out Exercise 5.7, just note that $\pi(I)$ is clearly closed under addition, and it satisfies the absorption property: if $r + I$ is any coset, and $a + I \in J/I$, then

$$(r + I)(a + I) = ra + I \in (J/I)$$

since $ra \in J$ as J is an ideal; and similarly for $ar + I$. Therefore J/I is indeed an ideal of R/I.

Statements like Theorem 5.57 become really useful when one's catalogue of rings is more extensive than what you may have at this point. But we already know enough to draw the general conclusion that there is a dictionary between ring-theoretic features of the quotient R/I and specific features of the ideal I. For instance, Theorem 5.57 tells us that the set of ideals of the quotient R/I is in one-to-one correspondence with the set of ideals of R containing I. Mod-ing out by I is a way to throw away all the ideals of R which do *not* contain I.

Example 5.59   The diligent reader knows that all ideals of $\mathbb{Z}$ are of the form $n\mathbb{Z}$ for some n (Exercise 5.14), and we will discuss this fact soon enough anyway (Example 6.27). Given this, what are the ideals of $\mathbb{Z}/12\mathbb{Z}$? It would not take too long to work this out by hand, but the Third Isomorphism Theorem dispenses us from having to do any work. By Theorem 5.57, the ideals of $\mathbb{Z}/12\mathbb{Z}$ correspond to the ideals of $\mathbb{Z}$ containing $12\mathbb{Z}$, that is, the ideals $n\mathbb{Z}$ containing $12\mathbb{Z}$. An ideal $n\mathbb{Z}$ contains $12\mathbb{Z}$ if and only if $12 \in n\mathbb{Z}$, that is, if and only if 12 is a multiple of n. That is, if and only if

$$n = \pm 1, \pm 2, \pm 3, \pm 4, \pm 6, \pm 12 \,.$$

We can conclude that $\mathbb{Z}/12\mathbb{Z}$ contains exactly six ideals: denoting the class $[a]_{12}$ by $\bar{a}$ for short, these six ideals are

$$(\bar{1}), \quad (\bar{2}), \quad (\bar{3}), \quad (\bar{4}), \quad (\bar{6}), \quad \text{and} \quad (\overline{12}) = (\bar{0}) \,.$$

Inclusions between these ideals also reflect divisibility: for example, $3 \mid 6$, and correspondingly

$$(\bar{6}) = \{\bar{0}, \bar{6}\} \quad \subseteq \quad (\bar{3}) = \{\bar{0}, \bar{3}, \bar{6}, \bar{9}\} \,.$$

We will run into this general picture in other contexts, e.g., cf. Example 10.11.

From this point of view, when is $\mathbb{Z}/p\mathbb{Z}$ a field? We know that a commutative ring is a field if and only if its only ideals are (1) and (0). (Well, we know this if we have worked out Exercise 5.16.) Arguing as we just did in Example 5.59, that is, using the Third Isomorphism Theorem, we see that this is the case if and only if the only divisors of p are ± 1 and $\pm p$. Aha, so this is the case if and only if p is irreducible. This is something we proved back in §2.4 (Theorem 2.15), at the price of what seemed somewhat hard work back then.

The reason we don't need to do any work now to recover the same fancy statement is that the steps in that proof have been absorbed by elements of the language we have learned, so that we don't really need to think about the proof anymore—the language itself does the legwork for us. This is how things should be. Rather than devising clever proofs, one tries to devise a clever language, and then coming up with proofs is more or less the same as coming up with (grammatically correct) sentences in that language—something our brains are evolutionarily adept at doing. In a nutshell, this is in my opinion the main worth of developing 'abstract' mathematics.

Exercises

5.1 Verify that ker f is a rng for every homomorphism $f \colon R \to S$. What can you say about S if ker f is actually a *ring?*

5.2 Prove that a ring homomorphism $f \colon R \to S$ is *injective* if and only if ker $f = (0)$.

5.3 Prove that if I is an ideal in a ring R, and I contains a unit u, then $I = R$. In particular, the only subring of R that is an ideal is R itself.

5.4 ▷ Consider the set of 2×2 real matrices of the form

$$\begin{pmatrix} a & b \\ c & d \end{pmatrix} \cdot \begin{pmatrix} 1 & 0 \\ 0 & 0 \end{pmatrix},$$

that is, the set of 'multiples' of the matrix $\begin{pmatrix} 1 & 0 \\ 0 & 0 \end{pmatrix}$. Prove that this is *not* an ideal of $M_{2\times2}(\mathbb{R})$.

5.5 ▷ Let R be any *commutative* ring, and let $a_1, \dots, a_n \in R$. Prove that the subset

$$(a_1, \dots, a_n) = \{r_1 a_1 + \dots + r_n a_n \mid r_1, \dots, r_n \in R\}$$

of 'linear combinations' of $a_1, \dots, a_n$ is an ideal of R.

5.6 Let R, S be rings, and let I be an ideal of R and J be an ideal of S. Prove that $I \times J$ is an ideal of $R \times S$.

5.7 ▷ Let $f \colon R \to S$ be a surjective homomorphism of rings, and let I be an ideal of R. Prove that $f(I)$ is an ideal of S. Find an example showing that this is not necessarily the case if f is not surjective.

5.8 Let $f \colon R \to S$ be a ring homomorphism, and let J be an ideal of S. Prove that $f^{-1}(J)$ is an ideal of R. (Recall that $f^{-1}(J)$ stands for the set $\{r \in R \mid f(r) \in J\}$. This is defined even if f is not a bijection.)

5.9 ▷ Let I, J be ideals of a ring R.
- Prove that $I \cap J$ is an ideal.
- Prove that the subset

$$IJ = \left\{ \sum_i a_i b_i \mid a_i \in I, b_i \in J \right\}$$

is an ideal. Prove that $IJ \subseteq I \cap J$.
- Prove that the subset

$$I + J = \{a + b \mid a \in I \text{ and } b \in J\}$$

is an ideal of R. Prove that $I + J$ is the smallest ideal of R containing both I and J.

5.10 ▷ Generalizing part of Exercise 5.9, let $\{I_\alpha\}_\alpha$ be any (possibly infinite) set of ideals of a ring R. Prove that the intersection $\cap_{\alpha \in A} I_\alpha$ is an ideal of R.

As a consequence, prove that for every subset $A \subseteq R$ of R there is a smallest ideal I such that $A \subseteq I$ (cf. Remark 5.18).

5.11 Find an example showing that if I and J are ideals of a ring R, $I \cup J$ satisfies the absorption property, but is not necessarily an ideal of R.

5.12 ▷ Let

$$I_1 \subseteq I_2 \subseteq I_3 \subseteq I_4 \subseteq \cdots$$

be a chain of ideals in a ring R, and let

$$I := \bigcup_{j \geq 1} I_j = \{r \in R \mid \exists j, \, r \in I_j\}.$$

Prove that I is an ideal of R.

5.13 ▷ Let R be a commutative ring, and let $a, b, q, r \in R$. Assume $a = bq + r$. Prove that $(a, b) = (r, b)$.

5.14 ▷ Let I be an ideal of $\mathbb{Z}$. Prove that $I = n\mathbb{Z}$ for some n. (*Hint:* If $I \neq (0)$, argue that it must contain some positive number, and let n be the smallest such number. Prove that $I = n\mathbb{Z}$.)

5.15 ▷ Let I be an ideal of a field F. Prove that $I = (0)$ or $I = (1)$.

5.16 ▷ Let R be a nontrivial commutative ring, and assume that the only ideals of R are (0) and (1). Prove that R is a field.

5.17 Let R be a ring, and let I be an ideal of R. Prove that R/I is commutative if and only if $\forall a, b \in R$, $ab - ba \in I$.

5.18 ▷ Let F be a field, and let $\varphi \colon F \to R$ be a ring homomorphism, where $0 \neq 1$ in R. Prove that φ is necessarily injective.

5.19 Every coset in $\mathbb{R}[x]/(x^2 - 1)$ can be written uniquely as $a + bx + (x^2 - 1)$ (you don't need to verify this). Therefore we can identify cosets in $\mathbb{R}[x]/(x^2 - 1)$ with pairs (a, b) of real numbers. Determine the result of adding and multiplying pairs via this identification (cf. Example 5.34, where this is worked out for $\mathbb{R}[x]/(x^2 + 1)$).

5.20 ▷ Prove that $\mathbb{R}[x]/(x^2 - 1) \cong \mathbb{R} \times \mathbb{R}$, as follows. Every coset in $\mathbb{R}[x]/(x^2 - 1)$ can be written uniquely as $a + bx + (x^2 - 1)$ (you don't need to verify this). Define a function $\varphi \colon \mathbb{R}[x]/(x^2 - 1) \to \mathbb{R} \times \mathbb{R}$ by $\varphi(a + bx + (x^2 - 1)) = (a + b, a - b)$. Prove that φ is an isomorphism. (Working out Exercise 5.19 first may be helpful.)

5.21 What does the canonical decomposition theorem say, when $f \colon R \to S$ is an *injective* ring homomorphism?

5.22 Let $f \colon R \to S$ be a ring homomorphism. Let I be an ideal of R, and J an ideal of S, such that $f(I) \subseteq J$. Prove that f induces a ring homomorphism $\tilde{f} \colon R/I \to S/J$ making the diagram

$$
\begin{array}{ccc}
R & \xrightarrow{\ f\ } & S \\
\downarrow & & \downarrow \\
R/I & \xrightarrow{\ \tilde{f}\ } & S/J
\end{array}
$$

commute, where the vertical functions are the natural projections.

5.23 ▷ Let $f \colon R \to S$ be a ring homomorphism, and let J be an ideal of S. Therefore $I = f^{-1}(J)$ is an ideal of R, by Exercise 5.8. Prove that if f is surjective, then $R/I \cong S/J$.

5.24 Let R, S be rings, and let I, J be ideals of R, S, respectively. Prove that

$$(R \times S)/(I \times J) \cong (R/I) \times (S/J).$$

5.25 ▷ Let R be a ring, I an ideal of R, and consider the subset $I[x]$ of $R[x]$ consisting of those polynomials whose coefficients belong to I. Prove that $I[x]$ is an ideal of $R[x]$, and that

$$R[x]/I[x] \cong (R/I)[x].$$

5.26 ▷ Show that the definition of gcd, Definition 1.5, may be paraphrased as follows: A nonnegative integer d is the gcd of two integers a and b if the ideal (d) is the smallest principal ideal containing both a and b.

5.27 Let R be a commutative ring, and let $I = (a)$, $J = (b)$ be two principal ideals. Prove that $I + J = (a, b)$ (see Definition 5.17 for the notation). Prove that $IJ = (ab)$.

5.28 Prove the converse to Proposition 5.48: if I and J are ideals of R, and the function $\rho : R \to R/I \times R/J$ defined by $r \mapsto (r + I, r + J)$ is surjective, then $I + J = (1)$.

5.29 Prove Corollary 5.53 as a consequence of the Chinese Remainder Theorem. (If you want to give a formal argument, you will likely need an induction.)

5.30 (i) Find a positive integer $M < 100$ such that

$$\begin{cases} M \equiv 2 \bmod 3 \\ M \equiv 3 \bmod 5 \\ M \equiv 2 \bmod 7. \end{cases}$$

(ii) Prove that there is no integer M such that

$$\begin{cases} M \equiv 2 \bmod 3 \\ M \equiv 3 \bmod 6 \\ M \equiv 2 \bmod 7. \end{cases}$$

(i) is the problem appearing in the *Sūnzĭ Suànjīng*. Of course your task is to find out how it can be solved by using (the proof of) the Chinese Remainder Theorem, *not* by trial and error.

5.31 (a) Prove that $\mathbb{Z}/15\mathbb{Z} \times \mathbb{Z}/7\mathbb{Z} \cong \mathbb{Z}/3\mathbb{Z} \times \mathbb{Z}/35\mathbb{Z}$.

(b) Prove that $\mathbb{Z}/4\mathbb{Z} \times \mathbb{Z}/5\mathbb{Z} \not\cong \mathbb{Z}/2\mathbb{Z} \times \mathbb{Z}/10\mathbb{Z}$.

5.32 Prove that $\mathbb{R}[x]/(x^2 - 3x + 2) \cong \mathbb{R}[x]/(x^2 + x)$, doing as little work as possible.

5.33 ▷ Find an explicit formula for the totient function of n, given the prime factorization $n = p_1^{a_1} \cdots p_r^{a_r}$ of n.

5.34 ▷ Let R be a commutative ring, $a \in R$, and let I, J be ideals, with $J = (a) + I$. Prove that $J/I = (a + I)$. More generally, prove that if $J = (a_1, \ldots, a_n) + I$, then $J/I = (a_1 + I, \ldots, a_n + I)$. (Look at Exercise 5.5 if you need a reminder on the notation.)

5.35 Let R be a commutative ring, and let $r_1, \ldots, r_n$ be elements of R. Prove that $R[x_1, \ldots, x_n]/(x_1 - r_1, \ldots, x_n - r_n) \cong R$. (*Hint:* Example 5.41.)

5.36 ▷ Prove that $\mathbb{C}[x, y]/(y - x^2) \cong \mathbb{C}[x]$, by finding explicitly an isomorphism from one ring to the other. (*Hint:* Example 5.41.)

6 Integral Domains

6.1 Prime and Maximal Ideals

One common underlying theme in several examples we have analyzed so far is that properties of the ring R/I reflect properties of the ideal I. For instance, we know (Theorem 2.15, and see the comments following Example 5.59) that $\mathbb{Z}/n\mathbb{Z}$ is a field precisely when n is *irreducible*. If you read Example 5.55 with the right mindset, you will see that the fact that the quotient $\mathbb{R}[x]/(x^2 - 1) \cong \mathbb{R} \times \mathbb{R}$ has nonzero zero-divisors has to do with the fact that $(x^2 - 1) = (x - 1)(x + 1)$ can be factored—this is the reason why we can apply the Chinese Remainder Theorem in this example.

These observations guide us in the following two definitions.

DEFINITION 6.1 Let R be a commutative ring, and let I be an ideal of R. Then I is a 'prime' ideal if the quotient R/I is an integral domain.

DEFINITION 6.2 Let R be a commutative ring, and let I be an ideal of R. Then I is a 'maximal' ideal if the quotient R/I is a field.

These are extremely important definitions in commutative algebra, number theory, and algebraic geometry. (The rings mainly used in these fields are commutative, and that is why I am only giving these definitions for commutative rings.) In algebraic geometry, certain loci defined by polynomial equations correspond to prime ideals in a ring, and their geometric points correspond to maximal ideals. To really understand this, you will have to turn to a different book; but you could try Exercise 6.8 if you are curious.

Here is an immediate but useful consequence of the definitions.

PROPOSITION 6.3 *Maximal ideals are prime.*

Proof Fields are integral domains (Proposition 3.31). □

Both notions of 'prime' and 'maximal' ideals admit alternative equivalent formulations. In fact, in other texts you are just as likely to see these alternative formulations given as definitions. Let's first deal with prime ideals.

THEOREM 6.4   *Let R be a commutative ring, and let I be an ideal of R. Then I is a prime ideal if and only if $I \neq (1)$ and the following condition holds:*

$$\forall a, b \in R \quad ab \in I \iff (a \in I) \vee (b \in I). \tag{6.1}$$

Proof   Assume that I is prime in R. According to Definition 6.1, R/I is an integral domain, and in particular $0 \neq 1$ in R/I. Then R/I is not a trivial ring, and in particular $I \neq (1)$. Next, let a, b be such that $ab \in I$. Then $ab + I = 0$ in R/I. Since $ab + I = (a + I)(b + I)$, this says that $(a + I)(b + I) = 0$ in R/I. By definition of integral domain, this implies that $a + I = 0$ or $b + I = 0$; and this means that $a \in I$ or $b \in I$. Therefore (6.1) holds for I.

The converse implication is obtained by reversing this argument. Explicitly, assume that the stated condition holds for I; we have to show that R/I is an integral domain. Since R is commutative, so is R/I. Since $I \neq (1)$, R/I is not trivial, and hence $0 \neq 1$ in R/I. Next, let $(a + I)$, $(b + I)$ be elements of R/I and assume $(a + I)(b + I) = 0$ in R/I. This says that $ab + I = 0$ in R/I, that is, $ab \in I$. Condition (6.1) implies that $a \in I$ or $b \in I$, and this in turn shows that $a + I = 0$ or $b + I = 0$ in R/I, verifying the defining condition for integral domains.   □

Example 6.5   As a direct application of Theorem 6.4, a commutative ring R is an integral domain if and only if the ideal (0) is prime in R. Indeed, $R/(0) \cong R$ (Example 5.29).

Example 6.6   Let's consider the case $R = \mathbb{Z}$ of integers. If p is an integer, then by Theorem 6.4, the ideal $(p) = p\mathbb{Z}$ is prime if and only if $p \neq \pm 1$ and

$$\forall b, c \in \mathbb{Z} \quad p \mid bc \iff p \mid b \vee p \mid c.$$

Look back at Definition 1.19! This is *precisely* the definition of prime *integer*. Therefore, the notion of prime *ideal* is a direct generalization of the notion of prime *integer*. In the case of integers, the fact that p is prime if and only if $\mathbb{Z}/p\mathbb{Z}$ is an integral domain was verified 'by hand' in Theorem 2.15, (i) $\iff$ (ii) for $p > 0$, and observed in Example 3.24 for $p = 0$. This fact may now be seen as a very, very particular case of Theorem 6.4, which works for *all* prime ideals in *all* commutative rings. So we have made some progress.

Incidentally, this is the reason why I choose to view 0 as a prime number: after all, everybody agrees that (0) is a prime *ideal* in $\mathbb{Z}$; and then demoting 0 to the status of a non-prime *integer* would create a clash in terminology. But beware that many (probably most) authors simply disagree with me, and do not count 0 among the prime integers.

Example 6.7   The ideal (x) in $\mathbb{Z}[x]$ is prime, since $\mathbb{Z}[x]/(x) \cong \mathbb{Z}$ (Example 5.40) and $\mathbb{Z}$ is an integral domain. The ideals (y) and (x, y) are prime in $\mathbb{Z}[x, y]$, since $\mathbb{Z}[x, y]/(y) \cong \mathbb{Z}[x]$ and $\mathbb{Z}[x, y]/(x, y) \cong \mathbb{Z}$ are both integral domains. (Use the First Isomorphism Theorem to verify these isomorphisms.)

Example 6.8 In algebraic geometry, one would associate the parabola $y = x^2$ with the ideal $(y - x^2)$ in $\mathbb{C}[x, y]$. (For some reasons, having to do with the 'Fundamental Theorem of Algebra', Theorem 7.17, algebraic geometers like me prefer to work over the complex numbers.) The ideal $(y - x^2)$ is prime in $\mathbb{C}[x, y]$: indeed, it so happens that $\mathbb{C}[x, y]/(y - x^2) \cong \mathbb{C}[x]$ (Exercise 5.36) and $\mathbb{C}[x]$ is an integral domain. This makes the parabola 'irreducible', in the language of algebraic geometry.

Moving on to maximal ideals, we have the following alternative characterization.

THEOREM 6.9   *Let R be a commutative ring, and let I be an ideal of R. Then I is a maximal ideal if and only if $I \neq (1)$ and there is no ideal J such that $I \subsetneq J \subsetneq R$.*

The term 'maximal' is a good description of the condition appearing in Theorem 6.9: the condition says that I is a maximal ideal if it is indeed 'maximal', i.e., it cannot be made bigger, among proper ideals (i.e., ideals $\neq R$). The ideal I is maximal if and only if whenever J is an ideal such that $I \subseteq J \subseteq R$, then necessarily $I = J$ or $J = R$. There is no room 'in between'.

Proof If I is maximal according to Definition 6.2, then R/I is a field, and in particular R/I is not a trivial ring. This shows that $I \neq (1)$. To show that the stated condition holds, we can work contrapositively: assume that there exists an ideal J such that $I \subsetneq J \subsetneq R$, and prove that R/I is not a field. This is a nice consequence of the Third Isomorphism Theorem. Indeed, if $I \subsetneq J \subsetneq R$, then (by Theorem 5.57) J/I is an ideal in R/I and $(0) \subsetneq J/I \subsetneq (1)$ in R/I. This implies that R/I is not a field, since the only ideals in a field are (0) and (1) (Exercise 5.15).

Conversely, we have to show that if the stated conditions hold, then R/I is a field. First of all, R/I is not a trivial ring since $I \neq (1)$, and hence $0 \neq 1$ in R/I. To show that R/I is a field, again argue contrapositively. If R/I is *not* a field, then R/I has some ideal $\bar{J}$ different from (0) and (1) (Exercise 5.16): $(0) \subsetneq \bar{J} \subsetneq (1)$. By Theorem 5.57, $\bar{J}$ corresponds to some ideal J containing I, and necessarily $I \subsetneq J \subsetneq R$. This shows that the stated condition does not hold, concluding the proof.   □

Example 6.10 By Theorem 6.9, a commutative ring R is a field if and only if (0) is maximal in R, if and only if R contains exactly two ideals: (0) and (1). This is a restatement of Exercises 5.15 and 5.16 (which, however, were used in the proof).

Example 6.11 An ideal is maximal in $\mathbb{Z}$ if and only if it is a nonzero prime ideal.

Indeed, let I be a maximal ideal in $\mathbb{Z}$. Then $I \neq (0)$ since $\mathbb{Z}/(0) \cong \mathbb{Z}$ is not a field; and I is prime since $\mathbb{Z}/I$ is a field, hence an integral domain.

Conversely, let I be a nonzero prime ideal. Then $I = n\mathbb{Z}$ for some $n \neq 0$ (Exercise 5.14), so $\mathbb{Z}/I = \mathbb{Z}/n\mathbb{Z}$ is finite, and it is an integral domain since I is prime. Finite integral domains are fields (Proposition 3.33), so this shows that I is maximal.

Combining with Example 6.6, we see that $\mathbb{Z}/p\mathbb{Z}$ is a field if and only if p is irreducible (that is, a nonzero prime integer, cf. Theorem 1.21). Of course this recovers (i) $\Longleftrightarrow$ (iii) from Theorem 2.15, cf. Example 3.30.

Example 6.12 The ideal $(2, x)$ is maximal in $\mathbb{Z}[x]$. Indeed, $\mathbb{Z}[x]/(2, x) \cong \mathbb{Z}/2\mathbb{Z}$ (Example 5.58), and $\mathbb{Z}/2\mathbb{Z}$ is a field. The ideal (x) is prime, but not maximal: indeed $\mathbb{Z}[x]/(x) \cong \mathbb{Z}$ is an integral domain, but not a field. And indeed we have $(x) \subsetneq (2, x) \subsetneq \mathbb{Z}[x]$.

Example 6.13 If k is a field, the ideal (x) *is* maximal in $k[x]$. In $k[x, y]$, the ideal (x) is not maximal; the ideal (x, y) is maximal. We have $(x) \subsetneq (x, y) \subsetneq k[x, y]$.

Example 6.14 As we saw in Example 6.7, the ideal $(y - x^2)$ of $\mathbb{C}[x, y]$ is prime, and this ideal is associated with the parabola $y = x^2$ (in $\mathbb{C}^2$). The ideal (x, y) is maximal, and we have $(y - x^2) \subseteq (x, y)$. In algebraic geometry, this inclusion of ideals corresponds to the fact that the origin $(0, 0)$ (which is associated with the ideal (x, y) in the sense that it is the solution of the equations $x = y = 0$) is a point of the parabola $y = x^2$: 'points' correspond to 'maximal ideals' in the dictionary of algebraic geometry. Points of $y = x^2$ correspond to maximal ideals of $\mathbb{C}[x, y]$ containing the ideal $(y - x^2)$, that is (by the Third Isomorphism Theorem) to maximal ideals of the ring $\mathbb{C}[x, y]/(y - x^2)$.

The dictionary of algebraic geometry mentioned in Example 6.14 actually goes much further. If R is a commutative ring, the 'spectrum' Spec(R) denotes the set of prime ideals in R. This set may be viewed as a space determined by R, whose 'usual' points correspond to the *maximal* ideals.

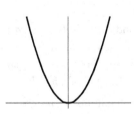

For example, the parabola $y = x^2$ (again viewed in $\mathbb{C}^2$) corresponds to Spec(R) for $R = \mathbb{C}[x, y]/(y - x^2)$. The ordinary points (a, b) of the parabola determine maximal ideals of R, in a way that you will understand rather well if you work out Exercise 6.8. Unbeknownst to you, when you drew a parabola in Calculus you were actually drawing certain maximal ideals of a certain ring.

In fact, *all* elements of a spectrum Spec(R) may be viewed as 'points' in a suitable sense. For every ideal I of R, denote by $V(I)$ the subset of Spec(R) consisting of the prime ideals of R containing I. Then you can verify that the family of subsets $V(I)$ satisfies the axioms for the *closed* sets of a topology on Spec(R). That is: the spectrum of a ring is in a natural way a topological space, whose points are the prime ideals of R. (This topology is called the *Zariski topology.*) The 'usual' points are those points which happen to be *closed* in this topological space.

If this seems obscure, don't worry—we will not use these facts in what follows. If it seems interesting, spend some quality time with a text in algebraic geometry.

6.2   Primes and Irreducibles

It is likely dawning on you that $\mathbb{Z}$ is a rather privileged ring. Beside being an integral domain, which is nice already, $\mathbb{Z}$ has properties that are not shared by more general rings. We have run across one such instance a moment ago: in Example 6.11 we have seen that an ideal is maximal in $\mathbb{Z}$ if and only if it is a nonzero prime ideal. This is not true in other perfectly respectable integral domains: for example, (x) is nonzero and prime in $\mathbb{Z}[x]$, and yet (x) is not maximal (Example 6.12).

Our general aim in this chapter is to investigate to what extent the idyllic situation we have in $\mathbb{Z}$ can be generalized to more general integral domains. Many nice properties of $\mathbb{Z}$ follow from the fact that we have decompositions into irreducibles: we proved the Fundamental Theorem of Arithmetic, Theorem 1.22, very early in the game. Does a 'fundamental theorem' hold in *every* integral domain?

To explore this question, we have to settle some terminology. First of all, we adopt for all commutative rings the same terminology introduced (for $\mathbb{Z}$) in the very first definition we gave in this text.

DEFINITION 6.15   Let R be a commutative ring, and let $a, b \in R$. We say that 'b divides a', and write $b \mid a$, if $a \in (b)$, i.e., $\exists c \in R$ such that $a = bc$.

We could also mimic precisely the definition of 'greatest common divisor' introduced for integers in §1.3: for a, b in an integral domain R, we could say that $d \in R$ is a gcd of a, b provided that (i) d is a common divisor of a and b (i.e., $d \mid a$ and $d \mid b$) and (ii) every common divisor c of a and b divides d. Look at Definition 1.5. The following is a sleek reformulation of this definition.

DEFINITION 6.16   Let R be an integral domain, and let $a, b \in R$. We say that $d \in R$ is a 'greatest common divisor' (gcd) of a and b if (d) is the smallest principal ideal of R containing both a and b.

Indeed (cf. Exercise 5.26), (d) contains both a and b precisely when d is a common divisor of a and b; if (d) is the smallest such ideal, then d must divide every c that is also a common divisor of a and b. Note that if d is a gcd of a and b, then so is du for every unit u, so it is not very appropriate to talk of 'the' gcd of a and b.

Remark 6.17   While this definition is natural, note that it raises a number of interesting questions. For example, does a gcd necessarily *exist*? No. Even if it exists, is it clear that a gcd of a and b is necessarily a linear combination of a and b? For example, if a and b are 'relatively prime' in the sense that 1 is a gcd of a and b (directly extending Definition 1.10), does it follow that 1 is a linear combination of a and b? No again. This happens in $\mathbb{Z}$ (Theorem 1.8, Corollary 1.9), but it does not hold in general: for example, 1 is a gcd of 2 and x in $\mathbb{Z}[x]$ according to the definition proposed above, but 1 is *not* a linear combination of 2 and x. Again, $\mathbb{Z}$ is a special place.   ⌐

Moving on to the notion of 'prime' element, the following definition is natural in view of Theorem 6.4.

DEFINITION 6.18   Let R be an integral domain, and let $p \in R$. We say that p is a 'prime' element of R if the principal ideal (p) is prime.

Remark 6.19   According to this definition, 0 is a prime element in an integral domain. As I pointed out in §6.1, some prefer to make 0 an exception and count it out. (I do not like making exceptions.)   ⌐

We could of course consider more general rings for such a definition; but we will only use this terminology in the case of integral domains.

By Theorem 6.4, and using the terminology introduced in Definition 6.15, this notion may be translated into a condition which generalizes word-for-word the condition defining prime *integers* (Definition 1.19): p is prime if and only if p is not a unit and whenever $p \mid ab$ we have that $p \mid a$ or $p \mid b$.

We can also generalize the definition of 'irreducible', as follows.

DEFINITION 6.20   Let R be an integral domain, and let $q \in R$. We say that q is 'irreducible' if q is not a unit and

$$\forall a, b \in R \quad q = ab \implies a \text{ is a unit or } b \text{ is a unit.} \tag{6.2}$$

(The definition easily implies that $q \neq 0$.) Note that the condition could be rephrased as saying that $q = ab$ implies $(q) = (a)$ or $(q) = (b)$ (Exercise 6.15). You should also go ahead and verify that an integer is irreducible if and only if it satisfies this condition in $\mathbb{Z}$ (Exercise 6.16). So Definition 6.20 is a direct generalization of Definition 1.17.

Next, recall that we have verified that a nonzero *integer* is prime if and only if it is irreducible (Theorem 1.21). Does this fact generalize to arbitrary integral domains? No!

Example 6.21   Consider the ring $R = \mathbb{C}[x, y]/(y^2 - x^3)$. One can check (Exercise 6.12) that R is an integral domain, that $\underline{x}$ is irreducible, and that $\underline{y} \notin (\underline{x})$, where I am underlining elements of $\mathbb{C}[x, y]$ to denote the corresponding cosets in R. In R, $\underline{y}^2 = \underline{x}^3$: indeed, $\underline{y}^2 - \underline{x}^3$ is the coset of $y^2 - x^3$, and this is 0 in the quotient. This shows that $(\underline{x})$ is not prime: indeed, $\underline{y} \cdot \underline{y} = \underline{y}^2 \in (\underline{x})$, and yet $\underline{y} \notin (\underline{x})$. So $\underline{x}$ is an irreducible element which is not prime.

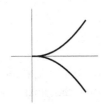

The 'algebro-geometric' reason for the phenomenon observed in Example 6.21 has to do with the fact that $y^2 = x^3$ defines a *singular* curve. If you draw its real graph, you will see that it looks 'cuspy' at the origin. By contrast, the parabola of Example 6.8 is *not* singular, and in its ring all irreducibles are prime. Indeed, we have seen that its ring is isomorphic to $\mathbb{C}[t]$, and we will be able to verify that all irreducible elements in $\mathbb{C}[t]$ are prime (because $\mathbb{C}[t]$ is a 'PID', cf. Proposition 6.33).

So once again $\mathbb{Z}$ turns out to be 'special'. We will see that nonzero prime elements are irreducible in every integral domain, but irreducible elements are necessarily prime only in a certain type of ring.

The following characterization of irreducible elements streamlines several arguments.

> LEMMA 6.22   *Let R be an integral domain, and let $q \in R$, $q \neq 0$, q not a unit. Then q is irreducible if and only if the ideal (q) is maximal among proper principal ideals of R: if $(q) \subseteq (a)$, then $(a) = (1)$ or $(a) = (q)$.*

Proof   First, let us assume that q is irreducible, and let $a \in R$, $(q) \subseteq (a)$. Then $q = ab$ for some $b \in R$, and since q is irreducible, then a is a unit or b is a unit. If a is a unit, then $(a) = (1)$. If b is a unit, then $(a) = (ab) = (q)$. This verifies that (q) is indeed maximal among proper principal ideals.

Conversely, let us assume that $q \neq 0$, q is not a unit, and that the stated maximality condition holds. Assume $q = ab$. Then $(q) \subseteq (a)$, so either $(a) = (1)$ or $(a) = (q)$ by maximality. In the first case, a is a unit. In the second, we have $a = qu$ for a unit u (Exercise 6.11); it follows that $q = qub$, therefore $1 = ub$ since R is an integral domain and $q \neq 0$, therefore b is a unit.   □

Armed with Lemma 6.22, we can prove that the implication 'nonzero prime $\implies$ irreducible' holds in every integral domain.

> THEOREM 6.23   *Let R be an integral domain, and let $p \in R$ be a nonzero prime element. Then p is irreducible.*

Proof   Let $p \in R$ be a nonzero prime element, and assume $(p) \subseteq (a)$. We have to prove that $(a) = (p)$ or $(a) = (1)$ (this implies that p is irreducible by Lemma 6.22).

Since $(p) \subseteq (a)$, there exists an element $b \in R$ such that $p = ab$. Since p is prime, $a \in (p)$ or $b \in (p)$. In the first case we have $(a) \subseteq (p)$; since we are assuming $(p) \subseteq (a)$, we can conclude that $(a) = (p)$. In the second case, there exists a $c \in R$ such that $b = cp$. Then $p = ab = a(cp) = (ac)p$. Since R is an integral domain and $p \neq 0$, we see that $1 = ac$, and this shows that $(a) = (1)$, concluding the proof.   □

The other implication, 'irreducible $\implies$ prime', may fail in an arbitrary ring: this is what Example 6.21 taught us. On the other hand, this implication does hold in $\mathbb{Z}$. Again, what makes $\mathbb{Z}$ so special?

6.3   Euclidean Domains and PIDs

Remember that in a distant past we derived the main properties of $\mathbb{Z}$ from the fact that we have a 'division with remainder' in $\mathbb{Z}$. We can axiomatize the situation as follows.

DEFINITION 6.24   Let R be an integral domain. We say that R is a 'Euclidean domain' if there is a function $v \colon R \setminus \{0\} \to \mathbb{Z}^{\geq 0}$ with the following property: for all $a \in R$ and all nonzero $b \in R$ there exist $q, r \in R$ such that

$$a = bq + r$$

with either $r = 0$ or $v(r) < v(b)$. The function v is called a 'Euclidean valuation'.

It is common to require that valuations satisfy more stringent requirements, but these will not be needed in what follows.

Definition 6.24 certainly is a mouthful, but it gets better if you realize that you already know a very good example of 'Euclidean domain': $\mathbb{Z}$. In $\mathbb{Z}$ we can define $v(n) = |n|$, and then it is indeed the case that for all a and for all $b \neq 0$, we can find q and r with $a = bq + r$ and $r = 0$ or $|r| < b$. This is essentially a restatement of the 'division with remainder' we discussed in Theorem 1.3; we are now less interested in the issue of uniqueness, but the gist of that statement is captured by Definition 6.24. Euclidean domains are integral domains endowed with a good notion of 'division with remainder'.

Are there *other* examples of Euclidean domains? Yes—in §7.2 we will study polynomial rings $k[x]$ with coefficients in a field k, and we will see that these are also Euclidean domains. We will also look at the ring $\mathbb{Z}[i]$ of 'Gaussian integers' (we had run across this in Example 4.12), which also turns out to be a Euclidean domain, cf. Proposition 10.30.

Euclidean domains are exceedingly convenient rings. Let us check that *every ideal in a Euclidean domain is principal*. In fact, this property is important enough that it has its own name.

DEFINITION 6.25   Let R be an integral domain. We say that R is a 'principal ideal domain', or 'PID', if every ideal in R is principal: for every ideal I of R, there exists an element $a \in R$ such that $I = (a)$.

You already know many rings that are *not* PIDs: for example, we have run across the ideal $(2, x)$ in $\mathbb{Z}[x]$ (Example 5.58), and this ideal is not principal (Exercise 6.10). The vast majority of integral domains are *not* PIDs. On the other hand, fields are PIDs. Indeed, the only ideals in a field are (0) and (1), and these are principal. This is not so exciting. The following theorem is much more interesting.

THEOREM 6.26   *Euclidean domains are PIDs.*

Proof   Let R be a Euclidean domain, and let I be an ideal of R. If $I = (0)$, then I is principal. So we may assume that $I \neq (0)$. Then consider the set $S \subseteq \mathbb{Z}^{\geq 0}$ consisting of values $v(s)$ for $s \in I$, $s \neq 0$. This is a nonempty set of nonnegative integers, therefore it contains a smallest element by the well-ordering principle. Let $b \in I$, $b \neq 0$, be an element of I such that $v(b)$ is this smallest element of S.

Then I claim that $I = (b)$. Indeed, it is clear that $(b) \subseteq I$, since $b \in I$; we have to verify the converse inclusion: for $a \in I$, we have to prove that a is a multiple of b. Since R is a Euclidean domain and $b \neq 0$, we know that there exist $q, r \in R$ such that $a = bq + r$, with $r = 0$ or $v(r) < v(b)$. Now we have

$$r = a - bq \in I.$$

Indeed, both a and bq are elements of I. If $r \neq 0$, then $v(r) < v(b)$: but this is forbidden, since $v(b)$ was selected as the *smallest* valuation of a nonzero element of I. Therefore, necessarily, $r = 0$, i.e., $a = bq$. But then a is a multiple of b, and this concludes the proof.   □

Example 6.27   By Theorem 6.26, $\mathbb{Z}$ is a PID: that is, all ideals of $\mathbb{Z}$ are of the form $n\mathbb{Z}$ for some $n \in \mathbb{Z}$. If you worked out Exercise 5.14, you knew this already; and we used this fact in Example 5.59. Theorem 6.26 is a substantial generalization of this observation.

Example 6.28   As I mentioned, we will prove (Theorem 7.2) that if k is a field, then $k[x]$ is a Euclidean domain. By Theorem 6.26, $k[x]$ is a PID. For instance, $\mathbb{Q}[x]$, $\mathbb{R}[x]$, $\mathbb{C}[x]$, and $(\mathbb{Z}/p\mathbb{Z})[x]$, with p a positive prime, are all PIDs. But keep in mind that $\mathbb{Z}[x]$ is *not* a PID: the ideal $(2, x)$ is not principal in $\mathbb{Z}[x]$.

'Greatest common divisors' are very easy to deal with in PIDs. If R is a PID and $a, b \in R$, then the ideal (a, b) is itself principal, because *every* ideal is principal in a PID. Therefore $(a, b) = (d)$ for some $d \in R$, and (d) is tautologically the 'smallest principal ideal containing a and b'. Therefore d is a greatest common divisor of a and b according to Definition 6.16. The issues raised in Remark 6.17 are transparent for PIDs: greatest common divisors exist, and further, in a PID a greatest common divisor of a and b is automatically a linear combination of a and b. Indeed, the ideal (a, b) consists of the set of linear combinations of a and b, so if $(d) = (a, b)$, then d is a linear combination of a and b. For example, a and b are relatively prime in a PID (i.e., 1 is a gcd of a and b or, equivalently, a and b have no common irreducible factor) if and only if $(a, b) = (1)$. This is not necessarily the case in more general rings, cf. Remark 6.17.

Remark 6.29   Since the situation is so neat in PIDs, it is just as neat in Euclidean domains, by Theorem 6.26. In fact, in a Euclidean domain, greatest common divisors are even easier to deal with: you should have no difficulty verifying that the Euclidean algorithm, described in §1.3 (see Theorem 1.14), generalizes to arbitrary Euclidean domains. (The role of Lemma 1.15 will be played by Exercise 5.13.) Therefore, computing greatest common divisors in a Euclidean domain is straightforward, if you can effectively perform 'division with remainder'.   ⌐

We are ready to provide a new explanation for the fact that nonzero prime ideals in $\mathbb{Z}$ are maximal. The 'real' reason this happens is that $\mathbb{Z}$ is a PID!

PROPOSITION 6.30   *Let R be a PID, and let I be a nonzero prime ideal. Then I is maximal.*

Proof Let I be a nonzero prime ideal of R. Since R is a PID, $I = (p)$ for some $p \in R$, $p \neq 0$; and p is prime according to Definition 6.18. By Theorem 6.23, p is irreducible. By Lemma 6.22, $I = (p)$ is maximal *among proper principal ideals in R*. But in R *every* ideal is principal, so this means that I is maximal among proper ideals of R. By Theorem 6.9, I is a maximal ideal of R.   □

Once again: since $\mathbb{Z}$ is a PID, as we now know, then the maximal ideals are precisely the nonzero prime ideals. We have been aware of this fact for a while, but the explanation we have now discovered is 'better' than the reason given earlier, in Example 6.11 (which depended on the fact that the quotients $\mathbb{Z}/n\mathbb{Z}$ are finite sets for $n \neq 0$), and more streamlined than the reasoning you could extract from Theorem 2.15. More importantly, we have proved Proposition 6.30 for *all* PIDs, and there are many interesting PIDs out there. Our friend $\mathbb{Z}$ is just one example.

Remark 6.31   Here is another message from our sponsor (that is, algebraic geometry). We can attempt to define a number for a given integral domain R, in the following way. We consider all *finite chains of prime ideals*

$$(0) \subsetneq I_1 \subsetneq I_2 \subsetneq \cdots \subsetneq I_n , \tag{6.3}$$

where each I_j is a prime ideal. Note that (0) is prime, since we have been assuming that R is an integral domain. The 'length' of a chain is the number of proper inclusions; so (6.3) has length n. If R has arbitrarily long such chains, we will leave it alone. If, on the contrary, there is a *longest* finite chain of prime ideals in R, then we say that R has 'Krull dimension' equal to n if the longest chain of prime ideals has length n.

Believe it or not, this notion of dimension matches precisely a 'geometric' notion of dimension. For example, $\mathbb{C}^2$ has (complex) dimension 2, and the ring algebraic geometry associates with it is $\mathbb{C}[x, y]$; it can be shown that this ring has Krull dimension 2. In fact, the chain

$$(0) \subsetneq (x) \subsetneq (x, y)$$

is as long as possible in $\mathbb{C}[x, y]$, There is no longer chain of prime ideals.[1]

Fields have Krull dimension 0, since the only ideals of a field are (0) and (1) (Exercise 5.15), so the only chain of prime ideals starts and ends with (0). (Remember that (1) is not prime!) And indeed, in algebraic geometry, fields correspond to isolated *points,* that is, dimension 0 geometric objects.

What Proposition 6.30 shows is that PIDs that are not fields have Krull dimension equal to 1 (Exercise 6.19). Thus, from the point of view of algebraic geometry these rings correspond to 'curves'. Our main example will be $k[x]$, where k is a field: we will soon see that this is a PID, so it has Krull dimension 1; and it is the ring that algebraic geometers associate with a 'line'. The 'parabola' showing up in Example 6.8 should also be a curve, and indeed (as we have seen) its ring turned out to be isomorphic to $\mathbb{C}[t]$, which has Krull dimension 1.

Maybe a little more surprisingly, this tells us that $\mathbb{Z}$ may be thought of as a 'curve', in a suitably general setting.   ⌙

---

[1] It is not too easy to prove this, however.

Several other properties of $\mathbb{Z}$ upgrade to all PIDs. Here is one also having to do with chains.

> PROPOSITION 6.32   *Let R be a PID. Then in R every chain of ideals stabilizes.*
> *That is, if*
>
> $$I_1 \subseteq I_2 \subseteq I_3 \subseteq I_4 \subseteq \cdots$$
>
> *is a chain of ideals in R, then there is an index m such that $I_m = I_{m+1} = I_{m+2} = \cdots$.*

The condition expressed in Proposition 6.32 is known by the catchy name of 'ascending chain condition', and it defines 'Noetherian' rings. (Named after Emmy Noether, one of the most powerful mathematicians of the twentieth century.) This is a very important notion, which we are only going to touch upon now; we will come back to it in §9.3. Proposition 6.32 states that 'PIDs are Noetherian'. As it happens, commutative rings in which every ideal admits *finitely many* generators are Noetherian; you will check this yourself when the time comes (Exercise 9.10). Since in a PID every ideal admits a single generator, PIDs are Noetherian in a very strong sense.

To get a feel of why this hypothesis is true in $\mathbb{Z}$, try to construct an 'infinitely ascending chain' of ideals: it will start with $(0) \subseteq (n)$ for some n; choose your favorite n, and you will see that your chain will have to stop, no matter how you try to construct it. For example,

$$(0) \subsetneq (80) \subsetneq (40) \subsetneq (20) \subsetneq (10) \subsetneq (2) \subsetneq (1)$$

can't go any further. Choosing a different n may give you a longer chain, but that also will reach the ceiling and stop there. In an arbitrary ring, this may not necessarily happen (Exercise 6.20). According to Proposition 6.32, it necessarily does if the ring is a PID.

Proof of Proposition 6.32   Let

$$I_1 \subseteq I_2 \subseteq I_3 \subseteq I_4 \subseteq \cdots$$

be a chain of ideals in a PID R. Consider the subset of R defined by the *union* of all the ideals in the chain:

$$I := \bigcup_{j \geq 1} I_j = \{r \in R \mid \exists j, r \in I_j\}.$$

Then I is an ideal of R (Exercise 5.12). Since I is an ideal and R is a PID, $I = (r)$ for some $r \in R$. By definition of I, there exists some index m such that $r \in I_m$. But then we have

$$I = (r) \subseteq I_m \subseteq I_{m+1} \subseteq \cdots \subseteq I$$

and it follows that $I_m = I_{m+1} = \cdots$. So the sequence stabilizes, as stated.   □

What is Proposition 6.32 good for? It will allow us to generalize to all PIDs nothing less than the Fundamental Theorem of Arithmetic! We will see how in the next section.

6.4 PIDs and UFDs

We are ready to clarify the relation between prime and irreducible elements. We have shown that nonzero primes are automatically irreducible in any integral domain (Theorem 6.23), and we have noticed that the converse does not necessarily hold, even if it does hold in $\mathbb{Z}$ (Theorem 1.21). Well, this fact holds in *every* PID.

> PROPOSITION 6.33   *Let R be a PID, and let $q \in R$ be an irreducible element. Then (q) is a maximal ideal; in particular, q is prime.*

Proof   By Lemma 6.22, (q) is maximal among proper principal ideals. All ideals are principal in a PID, therefore (q) is a maximal ideal (Theorem 6.9). It follows that (q) is a prime ideal (Proposition 6.3), so q is prime, as stated. □

Remark 6.34   Conversely, if R is a PID *but not a field,* then all maximal ideals of R are of the form (q), for q irreducible. This is simply because maximal ideals are prime, and nonzero primes are irreducible (Proposition 6.3 and Theorem 6.23). If R is a field, then (0) is maximal, so this case has to be excluded. ⌐

Theorem 6.23 and Proposition 6.33 upgrade Theorem 1.21, that is, the case of $\mathbb{Z}$, to all PIDs. And now we can formulate our generalization of the Fundamental Theorem of Arithmetic to all PIDs.

> THEOREM 6.35   *Let R be a PID, and let $a \in R$; assume that $a \neq 0$ and a is not a unit. Then there exist finitely many irreducible elements $q_1, \ldots, q_r$ such that*
>
> $$a = q_1 \cdots q_r.$$
>
> *Further, this factorization is unique in the sense that if*
>
> $$a = q_1 \cdots q_r = p_1 \cdots p_s, \tag{6.4}$$
>
> *with all q_i, p_j irreducible, then $s = r$ and after reordering the factors we have $(p_1) = (q_1)$, $(p_2) = (q_2)$, ..., $(p_r) = (q_r)$.*

If you look back at Theorem 1.22, you will see that I copied it here almost word-for-word. In the particular case $R = \mathbb{Z}$, this *is* the Fundamental Theorem of Arithmetic.

Also note that with due care we could also extend the statement to units, by allowing the case $r = 0$ of no factors.

Proof   First, let's deal with the existence of a factorization. We will work by contradiction, assuming that there exists an element $a \in R$, $a \neq 0$, a not a unit, which does *not* admit a factorization into irreducible as in (6.4). In particular, a is itself not irreducible, so we can write $a = a_1 b_1$, with neither a_1 nor b_1 a unit; note that then $(a) \subsetneq (a_1)$ and $(a) \subsetneq (b_1)$. If both a_1 and b_1 admitted factorizations as in (6.4), then so would a. Therefore, we may assume that a_1 or b_1 does *not* have such a factorization; without loss of generality we may assume that a_1 does not have a factorization. Therefore, we have found $a_1 \in R$ (with $a_1 \neq 0$, a_1 not a unit) such that

$$(a) \subsetneq (a_1)$$

and such that a_1 does not admit a factorization. By precisely the same argument, we can find $a_2 \in R$ (and $a_2 \neq 0$, a_2 not a unit) such that

$$(a) \subsetneq (a_1) \subsetneq (a_2),$$

and such that a_2 does not admit a factorization. The process continues, generating an infinite chain[2]

$$(a) \subsetneq (a_1) \subsetneq (a_2) \subsetneq \cdots \subsetneq (a_m) \subsetneq \cdots .$$

This is a contradiction: we have verified in Proposition 6.32 that in a PID there are *no* infinitely ascending chains of ideals. This contradiction proves the existence of factorizations into irreducibles for every nonzero, nonunit $a \in R$.

To prove uniqueness, we will again argue by contradiction: we will assume that the factorization is not necessarily unique, and show that this leads to a contradiction.

If the factorization is not unique, then (by the well-ordering principle) there must be a smallest positive r such that there is an element a with at least two *different* factorizations:

$$a = q_1 \cdots q_r = p_1 \cdots p_s, \qquad (6.5)$$

with all q_i, p_j irreducible. We must have $r > 1$: if $q_1 = p_1 \cdots p_s$, then $s = 1$ since q_1 is irreducible, and $q_1 = p_1$; this would contradict the assumption that the factorizations are different. So $r \geq 2$. We have

$$q_1 \cdots q_r \in (p_1). \qquad (6.6)$$

Since p_1 is irreducible and R is a PID, p_1 is prime by Proposition 6.33. Therefore, (6.6) implies that one of the elements q_i belongs to (p_1). After rearranging the factors q_i, we may assume that $q_1 \in (p_1)$.

Next, q_1 is irreducible; by Lemma 6.22, it is maximal among proper principal ideals. Since $(q_1) \subseteq (p_1)$, and $(p_1) \neq R$, necessarily $(q_1) = (p_1)$. Therefore we may assume $q_1 = p_1$ (by multiplying p_2 by a unit if necessary), and we discover

$$a = p_1 q_2 \cdots q_r = p_1 p_2 \cdots p_s.$$

Since R is a domain, we may cancel p_1, and we get

$$q_2 \cdots q_r = p_2 \cdots p_s. \qquad (6.7)$$

Note that the factorization on the left-hand side has $r - 1 < r$ factors: since r was the *smallest* number of possible factors of an element with different factorizations, the two factorizations (6.7) must coincide. That is, $r-1 = s-1$, and the factors can be reordered so that $(q_i) = (p_i)$. But then $r = s$, and $(q_i) = (p_i)$ for all i after a reordering (we had

---

[2] A more rigorous version of this argument would use induction.

already arranged for (q_1) to agree with (p_1)). This contradicts the assumption that the factorizations (6.5) were different, and this contradiction completes the proof. □

Example 6.36 Once again, the case $R = \mathbb{Z}$ is simply the Fundamental Theorem of Arithmetic, Theorem 1.22. As I mentioned in Example 6.28, you will soon add several other rings to your palette of PIDs: for instance, all the rings $k[x]$, where k is your favorite field, are PIDs. By Theorem 6.35, each of these rings will have its own 'Fundamental Theorem': every nonzero, nonunit element in each of these rings will have a unique factorization into irreducibles.

To summarize the content of Theorem 6.35, we can say that 'unique factorization' holds for PIDs, or that 'PIDs are UFDs'.

DEFINITION 6.37 An integral domain R is a 'unique factorization domain' (UFD) if the conclusion of Theorem 6.35 holds for R. That is, every nonzero, nonunit element of R admits a factorization into irreducibles, and this factorization is unique in the sense explained in Theorem 6.35.

Remark 6.38 We have just proved that PIDs are UFDs, but the class of UFDs is *much* larger than the class of PIDs: there are many, many UFDs that are not PIDs. For example, $\mathbb{Z}[x]$ and $\mathbb{C}[x, y]$ are such rings. We will understand this a little better in a short while (cf. Remark 7.28). ⌐

Remark 6.39 I wrote out the statement of Theorem 6.35 in terms of *elements* in order to stress the parallel with the Fundamental Theorem of Arithmetic. A perhaps even more elegant formulation would be in terms of *ideals:* if $I = (a) \neq (0)$ is an ideal of a PID R, then there exist prime ideals (q_i), $i = 1, \ldots, r$, such that

$$I = (q_1) \cdots (q_r),$$

and this collection of prime ideals is unique (up to reordering).

Make sure you understand that this is just a reformulation of Theorem 6.35. It includes the case in which a is a unit: then $r = 0$ and the right-hand side should be interpreted as (1). It follows (see Exercise 6.27) that every nonzero ideal I in a PID may be 'decomposed' as an intersection of rather special ideals:

$$I = (q_1^{m_1}) \cap \cdots \cap (q_r^{m_r}),$$

where the elements q_i are irreducible and the exponents m_i are positive integers. The ideals (q^m) appearing in this statement are 'primary' ideals. This notion is a generalization of the notion of *prime* ideal; in a PID, a primary ideal is an ideal of the form (q^m) for some irreducible element q. The definition is more involved for more general rings; the above observation generalizes to the statement that in a *Noetherian* ring every ideal admits a 'primary decomposition'. You will have to delve into commutative algebra, algebraic geometry, or number theory to see how this works. ⌐

I will give one example of a consequence of 'unique factorization', for use in a remarkable application in the next chapter (§7.1).

Example 6.40   Say that an element is an 'nth power up to a unit' in a ring R if it can be written as ua^n for some $a \in R$ and a unit $u \in R$.

> LEMMA 6.41   *Let R be a UFD. Let $b, c, d \in R$, n a positive integer, and assume that*
> (1) $d = bc$,
> (2)   *b and c have no common irreducible factors, and*
> (3)   *d is an nth power in R.*
> *Then b and c are nth powers in R up to units.*

Proof   By (3), $d = a^n$ for some $a \in R$. Let $a = q_1 \cdots q_r$ be an irreducible factorization (by hypothesis, R is a UFD, so factorizations exist in R). Then

$$d = q_1^n \cdots q_r^n$$

is an irreducible factorization of d. On the other hand, b and c have irreducible factorizations, and combining them gives *another* irreducible factorization of $d = bc$. By uniqueness of factorizations in R, these factorizations must coincide up to multiplication by units and reordering. By (2), b and c have no common irreducible factors. Therefore, up to reordering there must be an i such that

$$b = uq_1^n \cdots q_i^n, \quad c = vq_{i+1}^n \cdots q_r^n,$$

where u and v are units. It follows that $b = u(q_1 \cdots q_i)^n$ and $c = v(q_{i+1} \cdots q_r)^n$ are nth powers up to units, as claimed.   □

The result clearly extends to arbitrarily many factors: if $d = b_1 \cdots b_s$ is an nth power and the elements b_i pairwise have no common factors, then each b_i must be an nth power up to a unit.

The conclusion of Lemma 6.41 may well fail over rings that are *not* UFDs. For example, in the ring $\mathbb{C}[x, y, z]/(y^2 - xz)$ we have $\underline{y}^2 = \underline{xz}$ (where I am underlining an element to denote the corresponding coset); and yet $\underline{x}$ and $\underline{z}$ are not squares. This ring is not a UFD. (As it happens, irreducible factorizations do exist in this ring. What is missing is the U in UFD.) Geometrically speaking, it is the ring of a 'cone' in space, and the fact that the ring is not a UFD has to do with the fact that cones have a 'singular' point, the origin in this case.

6.5   The Field of Fractions of an Integral Domain

There is a tight kinship between $\mathbb{Z}$ and $\mathbb{Q}$, which has transpired occasionally and will come to the fore later on (for example in §7.4). It turns out that this is again a situation in which $\mathbb{Z}$ may be replaced by more general integral domains: in fact, *every* integral domain comes equipped with a field in which it sits just as tightly as $\mathbb{Z}$ sits in $\mathbb{Q}$.

We can loosely say that $\mathbb{Q}$ is the 'smallest' field containing $\mathbb{Z}$. What this means is that if k is a field, and $\mathbb{Z} \subseteq k$, then there is an isomorphic copy of $\mathbb{Q}$ within k, containing $\mathbb{Z}$. Here is a diagram to visualize this situation:

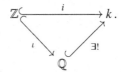

We have the inclusion homomorphism $\iota : \mathbb{Z} \hookrightarrow \mathbb{Q}$ (just sending a number a to the fraction $\frac{a}{1}$). Given another inclusion $i : \mathbb{Z} \to k$ of $\mathbb{Z}$ into a field k, I claim that there is a unique inclusion of $\mathbb{Q}$ into k which makes the diagram commute: that is, we have $\mathbb{Z} \subseteq \mathbb{Q} \subseteq k$. You cannot fit a 'smaller' field between $\mathbb{Z}$ and $\mathbb{Q}$.

Why is this true? Just send a fraction $\frac{p}{q}$ to $i(p)i(q)^{-1} \in k$: this can be done since $q \neq 0$, so $i(q) \neq 0$ (we are assuming that i is injective), so $i(q)$ has an inverse in k (because k is a field), and this allows us to define the element $i(p)i(q)^{-1}$ in k. This defines an injective homomorphism $\mathbb{Q} \hookrightarrow k$: in a moment we will see why, in a much more general setting.

What I have just described is another example of a *universal property,* something that is best understood in the context of categories. Universal properties are efficient ways to describe requirements that a construction must satisfy. For *every* integral domain R, the *field of fractions F* of R will be a field satisfying the 'same' universal property summarized above for $\mathbb{Q}$ w.r.t. $\mathbb{Z}$. In other words, we want F to be 'the smallest field containing R', and what that means is that every field k containing R must contain an isomorphic copy of F. The diagram visualizing this requirement is the same diagram shown above, *mutatis mutandis:*

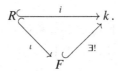

Part of the 'universal property' package guarantees that *if such a field F exists, then it is unique up to isomorphism.* I encourage you to prove this fact on your own (Exercise 6.28), and you will see that you do not need to know anything about a specific construction of F in order to do it.

Still, does a field F as above exist? How do we construct this field?

Let $\hat{F}$ be the set defined as follows:

$$\hat{F} = \{(r, s) \in R \times R \mid s \neq 0\} .$$

This is just the subset of $R \times R$ consisting of pairs whose second element is not zero. I will define a relation on $\hat{F}$:

$$(r_1, s_1) \sim (r_2, s_2) \iff r_1 s_2 = r_2 s_1 .$$

LEMMA 6.42   *The relation $\sim$ on $\hat{F}$ is an* equivalence *relation.*

Proof   Reflexivity: $(r, s) \sim (r, s)$ since $rs = rs$.

Symmetry: If $(r_1, s_1) \sim (r_2, s_2)$, then $r_1 s_2 = r_2 s_1$; therefore $r_2 s_1 = r_1 s_2$, and this shows that $(r_2, s_2) \sim (r_1, s_1)$.

Transitivity: Assume that $(r_1, s_1) \sim (r_2, s_2)$ and $(r_2, s_2) \sim (r_3, s_3)$. Then

$$r_1 s_2 = r_2 s_1 \quad \text{and} \quad r_2 s_3 = r_3 s_2 .$$

It follows that

$$(r_1 s_3)s_2 = (r_1 s_2)s_3 = (r_2 s_1)s_3 = (r_2 s_3)s_1 = (r_3 s_2)s_1 = (r_3 s_1)s_2 .$$

Since we are assuming that $s_2 \neq 0$, and R is an integral domain, we can deduce that $r_1 s_3 = r_3 s_1$. Therefore $(r_1, s_1) \sim (r_3, s_3)$, concluding the proof of transitivity and of the lemma.   □

Note that the proof of transitivity used two important hypotheses: for the proof to work, we need R to be an integral domain, and we need that the second element of each pair is not zero. The process we are going through *can* be generalized substantially, but that requires some care, and is beyond the scope of this text. (See Remark 6.46.)

Now that we have an equivalence relation $\sim$ on the set $\hat{F}$, we can consider the quotient $\hat{F}/\sim$, that is, the set of equivalence classes.

DEFINITION 6.43   Let R be an integral domain. The *field of fractions* of R is the quotient $F = \hat{F}/\sim$ endowed with the two operations $+$ and $\cdot$ described below.

Here are the operations. Denote by $[(r, s)] \in F$ the equivalence class of (r, s). (We will soon come up with a much friendlier notation.) The operations $+$ and $\cdot$ on F are defined as follows.

$$[(r_1, s_1)] + [(r_2, s_2)] := [(r_1 s_2 + r_2 s_1, s_1 s_2)] ,$$
$$[(r_1, s_1)] \cdot [(r_2, s_2)] := [(r_1 r_2, s_1 s_2)] .$$

Alas, there is work to do. Since these definitions appear to depend on the choices of representatives, we have to prove that these choices are immaterial; only then will we have well-defined operations.[3]

CLAIM 6.44   *The proposed operations $+$, $\cdot$ on F are indeed well-defined.*

Proof   This is routine work, but it needs to be done. Suppose we replace (r_1, s_1) by an equivalent pair (r_1', s_1'): that is, assume that $r_1 s_1' = r_1' s_1$. We have to prove that the result of the operations does not change if we replace r_1 by r_1' and s_1 by s_1'. (The same reasoning will show that we can also replace (r_2, s_2) by an equivalent pair.) That is, we have to show that if $r_1 s_1' = r_1' s_1$, then

---

[3] We have run into this problem several times before; look at the discussion preceding Claim 2.12 if this seems at all foggy.

(i) $(r_1 s_2 + r_2 s_1, s_1 s_2) \sim (r_1' s_2 + r_2 s_1', s_1' s_2)$ and

(ii) $(r_1 r_2, s_1 s_2) \sim (r_1' r_2, s_1' s_2)$.

To verify (i), we need to prove that

$$(r_1 s_2 + r_2 s_1)(s_1' s_2) = (r_1' s_2 + r_2 s_1')(s_1 s_2),$$

and this just amounts to expanding the two sides: the left-hand side gives

$$r_1 s_2 s_1' s_2 + r_2 s_1 s_1' s_2$$

and the right-hand side gives

$$r_1' s_2 s_1 s_2 + r_2 s_1' s_1 s_2.$$

These two elements agree, because we are assuming that $r_1 s_1' = r_1' s_1$ (and since R is commutative).

The proof of (ii) is even more straightforward, so it is left to the reader. □

Now we have a set F and two operations $+$ and $\cdot$. If I were (even) more committed to giving a complete account of the story, I would now proceed to verify that $(F, +, \cdot)$ satisfies the ring axioms from Definition 3.1. This is essentially straightforward, so it is left to you (Exercise 6.30). Just for fun, let's verify the associativity of addition. Let $[(r_1, s_1)], [(r_2, s_2)], [(r_3, s_3)]$ be three elements of F. We have

$$([(r_1, s_1)] + [(r_2, s_2)]) + [(r_3, s_3)] = [(r_1 s_2 + r_2 s_1, s_1 s_2)] + [(r_3, s_3)]$$
$$= [((r_1 s_2 + r_2 s_1)s_3 + r_3(s_1 s_2), (s_1 s_2)s_3)]$$
$$= [(r_1 s_2 s_3 + r_2 s_1 s_3 + r_3 s_1 s_2, s_1 s_2 s_3)],$$
$$[(r_1, s_1)] + ([(r_2, s_2)] + [(r_3, s_3)]) = [(r_1, s_1)] + [(r_2 s_3 + r_3 s_2, s_2 s_3)]$$
$$= [(r_1(s_2 s_3) + (r_2 s_3 + r_3 s_2)s_1, s_1(s_2 s_3))]$$
$$= [(r_1 s_2 s_3 + r_2 s_3 s_1 + r_3 s_2 s_1, s_1 s_2 s_3)].$$

The results of these two operations agree, again because R is commutative. The verification of the other axioms is just as (un)inspiring. In the process you will find that $[(0, 1)]$ is the 0 element in this new ring, and $[(1, 1)]$ is the element 1. Indeed, for example,

$$[(0, 1)] + [(r, s)] = [0 \cdot s + r \cdot 1, 1 \cdot s] = [(r, s)].$$

Now we have a new ring F. It is immediately seen to be commutative, as a consequence of the fact that R is commutative. Since R is an integral domain, we have $0 \neq 1$ *in R*. Then $0 \cdot 1 \neq 1 \cdot 1$ in R, which implies $(0, 1) \not\sim (1, 1)$, therefore $[(0, 1)] \neq [(1, 1)]$ in F. That is, $0 \neq 1$ *in F*.

We are almost done verifying that F is a field: we now know that it is a commutative ring in which $0 \neq 1$, so we just need to verify that every nonzero element has an inverse. For this, let $[(r, s)] \neq [(0, 1)]$. Then $(r, s) \not\sim (0, 1)$, therefore $r \cdot 1 \neq 0 \cdot s$, that is $r \neq 0$. But then $(s, r) \in \hat{F}$, so we can consider $[(s, r)] \in F$; and

$$[(r, s)] \cdot [(s, r)] = [(rs, sr)].$$

Since $rs \cdot 1 = 1 \cdot sr$ (once again we use the fact that R is commutative), we see that

$(rs, sr) \sim (1, 1)$, and we conclude that $[(r, s)] \cdot [(s, r)] = 1_F$: the multiplicative inverse of $[(r, s)]$ is $[(s, r)]$.

Hurray! This concludes the definition of the field F. In the process you have probably gotten annoyed at the notation $[(r, s)]$, and indeed there is a much better notation for the corresponding element of F: we denote $[(r, s)]$ by the 'fraction' $\frac{r}{s}$. (This is why F is called the 'field of fractions' of R!) Thus, we have

$$\frac{r_1}{s_1} = \frac{r_2}{s_2} \iff r_1 s_2 = r_2 s_1 , \tag{6.8}$$

which is probably not unexpected.[4] In terms of fractions, the operations given above work like this:

$$\frac{r_1}{s_1} + \frac{r_2}{s_2} := \frac{r_1 s_2 + r_2 s_1}{s_1 s_2} \quad \text{and} \quad \frac{r_1}{s_1} \cdot \frac{r_2}{s_2} := \frac{r_1 r_2}{s_1 s_2},$$

surprise surprise.

We are not quite done. I had promised that this field would satisfy the *universal property* discussed at the beginning of the section. First of all, there is an injective homomorphism $\iota : R \to F$, defined by

$$\forall r \in R : \quad \iota(r) = \frac{r}{1} .$$

This is indeed a ring homomorphism:

$$\iota(r_1) + \iota(r_2) = \frac{r_1}{1} + \frac{r_2}{1} = \frac{r_1 + r_2}{1} = \iota(r_1 + r_2),$$

and similarly for multiplication; and $\iota(1_R) = \frac{1}{1} = 1_F$. This homomorphism allows us to view R as a subring of F. (Just as we are used to viewing $\mathbb{Z}$ as a subring of $\mathbb{Q}$.) Finally, the universal property can be spelled out as in the following statement.

THEOREM 6.45   *Let R be an integral domain, and let F and $\iota : R \hookrightarrow F$ be the field of fractions of R and the injective ring homomorphism defined above. Let k be a field, and let $i : R \hookrightarrow k$ be an injective ring homomorphism. Then there exists a unique ring injective homomorphism $j : F \hookrightarrow k$ making the diagram*

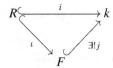

commute, that is, such that $i = j \circ \iota$.

Proof   How do we go about proving such a statement? Maybe a little surprisingly, the statement itself provides us with a crucial hint. (This is typical when you encounter a universal property.) Suppose that we can find a function $j : F \to k$ satisfying all the

---

[4] It is not unexpected because we are used to handling fractions of integers; but now we are starting from an arbitrary integral domain, and identity (6.8) amounts to an explicit expression of the equivalence relation *defining* fractions in this more general context.

requirements in this statement. Then since j has to be a ring homomorphism, necessarily we must have

$$j\left(\frac{r}{s}\right) = j\left(\frac{r}{1} \cdot \frac{1}{s}\right) = j\left(\frac{r}{1}\right) \cdot j\left(\frac{1}{s}\right) = j\left(\frac{r}{1}\right) \cdot j\left(\left(\frac{s}{1}\right)^{-1}\right) = j\left(\frac{r}{1}\right) \cdot j\left(\frac{s}{1}\right)^{-1}$$

$$= j(\iota(r)) \cdot j(\iota(s))^{-1}.$$

On the other hand, j must satisfy the requirement $i = j \circ \iota$. Therefore

$$j(\iota(r)) \cdot j(\iota(s))^{-1} = (j \circ \iota)(r) \cdot (j \circ \iota)(s)^{-1} = i(r)i(s)^{-1}.$$

This shows that *if a ring homomorphism j as required exists, then it must necessarily be given by*

$$j\left(\frac{r}{s}\right) := i(r)i(s)^{-1}. \tag{6.9}$$

The fact that this formula is forced upon us says that j, if it exists, is *unique*. This already proves part of the statement of Theorem 6.45. It also tells us what we have to do in order to complete the proof: we have to show that the prescription (6.9) *does* define a ring homomorphism $j\colon F \to k$ satisfying the requirement that $i = j \circ \iota$. If we do this, we will have successfully proved the theorem.

First of all, note that by definition the 'denominator' of a fraction is nonzero; since i is injective, $i(s) \neq 0$ in (6.9), and then $i(s)^{-1}$ exists since k is a field. Therefore, at least we *can* write the product $i(r)i(s)^{-1}$ appearing in (6.9).

Is j well-defined? Suppose $\frac{r_1}{s_1} = \frac{r_2}{s_2}$. Then $r_1 s_2 = r_2 s_1$, therefore

$$i(r_1)i(s_2) = i(r_1 s_2) = i(r_2 s_1) = i(r_2)i(s_1),$$

since i is a ring homomorphism. It then follows that

$$i(r_1)i(s_1)^{-1} = i(r_2)i(s_2)^{-1},$$

and this indeed shows that j is well-defined.

Is j a ring homomorphism? Yes:

$$j\left(\frac{r_1}{s_1}\right) + j\left(\frac{r_2}{s_2}\right) = i(r_1)i(s_1)^{-1} + i(r_2)i(s_2)^{-1} = i(r_1 s_2)i(s_1 s_2)^{-1} + i(r_2 s_1)i(s_1 s_2)^{-1}$$

$$= (i(r_1 s_2) + i(r_2 s_1))i(s_1 s_2)^{-1} = i(r_1 s_2 + r_2 s_1)i(s_1 s_2)^{-1}$$

$$= j\left(\frac{r_1 s_2 + r_2 s_1}{s_1 s_2}\right)$$

$$= j\left(\frac{r_1}{s_1} + \frac{r_2}{s_2}\right),$$

$$j\left(\frac{r_1}{s_1}\right) \cdot j\left(\frac{r_2}{s_2}\right) = i(r_1)i(s_1)^{-1} \cdot i(r_2)i(s_2)^{-1} = i(r_1)i(r_2)i(s_1)^{-1}i(s_2)^{-1}$$

$$= i(r_1 r_2)i(s_1 s_2)^{-1} = j\left(\frac{r_1 r_2}{s_1 s_2}\right)$$

$$= j\left(\frac{r_1}{s_1} \cdot \frac{r_2}{s_2}\right),$$

and of course $j(\frac{1}{1}) = i(1)i(1)^{-1} = 1$ as needed.

Is j injective? Well, *every* homomorphism from a field to a ring is injective (this is old fare: Exercise 5.18), so this is automatic.

Does j make the diagram commute? Yes: for all $r \in R$,

$$j \circ \iota(r) = j\left(\frac{r}{1}\right) = i(r)i(1)^{-1} = i(r),$$

and this verifies that $j \circ \iota = i$ as homomorphisms. This concludes the proof of the theorem. $\square$

Somehow the proof of Theorem 6.45 looks imposing. However, as I tried to empha-size, the statement itself tells you how you *must* define the required homomorphism; and then proving that the definition works is completely routine work, even if it looks complicated once you dot all the i's and cross all the t's as I have done above. In the scheme of things, the proof of Theorem 6.45 is so straightforward that in more advanced texts you should expect it to be omitted entirely, just left to the reader.

Remark 6.46 Summarizing, the field of fractions F of an integral domain R is a field obtained by 'making all nonzero elements of R invertible': we have constructed F by including by hand the inverse $\frac{1}{s}$ of every nonzero element $s \in R$.

This construction is a particular case of a more general process, called 'localization'. Using localization, one can start from an arbitrary commutative ring R and a 'multi-plicative subset' S of R (a subset satisfying certain properties: $1 \in S$, $0 \notin S$, and $s, t \in S \implies st \in S$) and construct a new ring, denoted $S^{-1}R$ (the 'localization of R at S'), in which every element of S becomes a unit. Further, this ring $S^{-1}R$ satisfies a universal property making it the 'smallest' ring in which elements of S are units. The field of fractions of an integral domain R is the localization of R at the set $S = R \setminus \{0\}$.

Localization is an important tool in number theory and algebraic geometry. In al-gebraic geometry, it is a way to extract 'local' information, that is, information in a neighborhood of a given point, concerning objects such as the parabola discussed at the end of §6.1 or the cuspidal curve appearing in Example 6.21.

If you are curious about this, take a course in algebraic geometry!   ⌟

Exercises

6.1   Let $f : R \to S$ be a homomorphism of rings, and let J be a prime ideal of S. Prove that $I = f^{-1}(J)$ is a prime ideal of R.

6.2   Consider the set I of polynomials in $\mathbb{Z}[x]$ whose constant term is a multiple of p, where $p \neq 0$ is a prime integer. Prove that I is a maximal ideal of $\mathbb{Z}[x]$.

6.3   Let I be a prime ideal of a commutative ring R. Is $I \times I$ a prime ideal of $R \times R$? (*Hint:* Exercises 5.24 and 4.2.)

6.4   Let $f : R \to S$ be a homomorphism of rings, and let J be a maximal ideal of S. Find an example showing that $I = f^{-1}(J)$ is not necessarily a maximal ideal of R. Under what additional condition on f is I necessarily a maximal ideal?

6.5 Let R be a finite commutative ring, and let I be a prime ideal of R. Prove that I is maximal.

6.6 Prove that if R is an integral domain, then (x) is a prime ideal in $R[x]$.

6.7 Prove that $\mathbb{Z}/1024\mathbb{Z}$ has exactly one maximal ideal.

6.8 Let (a, b) be a point in the plane $\mathbb{C}^2$. Prove that the ideal $(x - a, y - b)$ is maximal in $\mathbb{C}[x, y]$, and that (a, b) is a point of the parabola $y = x^2$ if and only if the ideal $(y - x^2)$ is contained in the ideal $(x - a, y - b)$. Conclude that every point (a, b) of the parabola determines a maximal ideal in the ring $\mathbb{C}[x, y]/(y - x^2)$. (As it happens, *every* maximal ideal in $\mathbb{C}[x, y]/(y - x^2)$ arises in this way, but it is not easy to show that this is the case.)

6.9 Prove that the ideal $(y + 1, x^2 + 1)$ is maximal in $\mathbb{R}[x, y]$ and contains the ideal $(y - x^2)$. This ideal does *not* correspond to a point with real coordinates on the parabola $y - x^2$. (This phenomenon is one reason why algebraic geometry is 'easier' over the complex numbers.)

6.10 ▷ Prove that the ideal $(2, x)$ of $\mathbb{Z}[x]$ is not principal. (Therefore $\mathbb{Z}[x]$ is not a PID.)

6.11 ▷ Let R be an integral domain, and let $a, b \in R$. Prove that $(a) = (b)$ if and only if $a = bu$, with u a unit in R.

6.12 ▷ Consider the ring homomorphism $\varphi \colon \mathbb{C}[x, y] \to \mathbb{C}[t]$ defined by $\varphi(f(x, y)) = f(t^2, t^3)$. (It should be clear that this is a homomorphism.)

- Prove that $\ker(\varphi) = (y^2 - x^3)$. (*Hint:* You may assume that for every polynomial $f(x, y)$ there exist polynomials $g(x, y)$ and $h(x), k(x)$ such that $f(x, y) = g(x, y)(y^2 - x^3) + h(x)y + k(x)$.)
- Prove that $\mathbb{C}[x, y]/(y^2 - x^3)$ is an integral domain.
- Prove that the coset $\underline{x}$ of x in $\mathbb{C}[x, y]/(y^2 - x^3)$ is irreducible.

6.13 Let $R = \mathbb{C}[x, y]/(y^2 - x^3)$, as in Example 6.21, and denote by $\underline{x}$ the coset of x in R. Determine the ring $R/(\underline{x})$ up to isomorphism, and show that it is not an integral domain. (This will just be another way to see that $\underline{x}$ is not prime in R. It may be helpful to note that $(x, y^2 - x^3) = (x, y^2)$, by Exercise 5.13.)

6.14 Let R be a commutative ring, and let $p \in R$. Prove that p is prime if and only if $(ab) \subseteq (p)$ implies $(a) \subseteq (p)$ or $(b) \subseteq (p)$.

6.15 ▷ Let R be an integral domain, and let $q \in R$. Prove that q is irreducible if and only if $(ab) = (q)$ implies $(a) = (q)$ or $(b) = (q)$. Equivalently: q is *not* irreducible precisely when there exist a, b in R such that $q = ab$ and $(q) \subsetneq (a)$, $(q) \subsetneq (b)$.

6.16 ▷ Prove that an integer is irreducible in the sense of Definition 6.20 if and only if it is irreducible in the sense of Definition 1.17.

6.17 Prove that fields are Euclidean domains. (That is: if k is a field, find a function $v \colon k^{\neq 0} \to \mathbb{Z}^{\geq 0}$ satisfying the property given in Definition 6.24.)

6.18 ▷ Prove that every ring contains exactly one subring isomorphic to $\mathbb{Z}/n\mathbb{Z}$, for a well-defined nonnegative integer n. (This integer is the *characteristic* of the ring, defined in Exercise 3.12. *Hint:* Use the First Isomorphism Theorem and the fact that $\mathbb{Z}$ is a PID.)

6.19 ▷ Prove that fields have Krull dimension 0 and PIDs that are not fields have Krull dimension 1.

6.20 ▷ You can consider a ring $\mathbb{Z}[x_1, x_2, x_3, \dots]$ of polynomials in 'infinitely many indeterminates'. An element of this ring is just an ordinary polynomial in any (finite) number of indeterminates; addition and multiplication are defined as usual. Prove that in this ring there *does* exist an infinitely ascending chain of ideals. (Therefore, this ring is not 'Noetherian'.)

6.21 Let R be a PID. Prove that every proper ideal of R is contained in a maximal ideal. (*Note:* This fact is true in every ring, but the proof in the general case is quite a bit subtler.)

6.22 Let R be a UFD, and let a and b be nonzero, nonunit elements of R. Prove that there are pairwise distinct irreducible elements $q_1, \ldots, q_r$, nonnegative integers $m_1, \ldots, m_r$, and units u, v such that

$$a = uq_1^{m_1} \cdots q_r^{m_r}, \quad b = vq_1^{n_1} \cdots q_r^{n_r}.$$

6.23 With notation as in Exercise 6.22, prove that the ideal (a, b) equals the principal ideal $(q_1^{\min(m_1,n_1)} \cdots q_r^{\min(m_r,n_r)})$.

Thus, the generator of this ideal is a gcd of a and b. In particular, greatest common divisors exist in UFDs.

6.24 Let R be a UFD, and let p, q be irreducible elements such that $(p) \neq (q)$. Prove that for all positive integers m, n the gcd of p^m and q^n is 1.

6.25 Let R be a PID, and let a and b be elements of R with no common irreducible factor. Prove that $(a) + (b) = (1)$.

6.26 ▷ Let R and S be PIDs, and assume that R is a subring of S. Let $a, b \in R$. Prove that if a and b are *not* relatively prime in S, then they are not relatively prime in R.

6.27 ▷ Let R be a PID, and let I be a nonzero ideal. Prove that there exist irreducible elements $q_1, \ldots, q_r$ and positive integers $m_1, \ldots, m_r$ such that

$$I = (q_1^{m_1}) \cap \cdots \cap (q_r^{m_r}).$$

(*Hint:* Lemma 5.51 will likely be helpful.)

6.28 ▷ Prove that a field satisfying the requirement explained at the beginning of §6.5 is unique up to isomorphism. That is, prove that if two fields F_1, F_2 satisfy this requirement for a given integral domain R, then necessarily $F_1 \cong F_2$. (Therefore, if you find a different way to construct a field for which you can prove the analogue of Theorem 6.45, then your field will be isomorphic to mine.)

6.29 Let k be a field. What is the field of fractions of k?

6.30 ▷ Verify (at least some of) the ring axioms for $(F, +, \cdot)$, with F as in Definition 6.43 and $+, \cdot$ defined in the discussion following that definition.

6.31 Recall from Remark 6.46 that a *multiplicative subset* S of a commutative ring is a set such that $0 \notin S$, $1 \in S$, and $s, t \in S \implies st \in S$.

- Prove that if I is a prime ideal of a ring R, then $R \setminus I$ is a multiplicative subset of R.
- Prove that if $f \in R$ is not nilpotent, then the set of powers f^i, $i \geq 0$, is a multiplicative subset of R.

Both of these cases lead to localizations that are of fundamental importance in (e.g.) algebraic geometry.

6.32 If you are a little ambitious, see if you can construct the ring $S^{-1}R$ mentioned in Remark 6.46, at least if R is an integral domain. What is the universal property satisfied by this 'localization'?

7 Polynomial Rings and Factorization

7.1 Fermat's Last Theorem for Polynomials

You may well wonder whether knowing about PIDs and all that is in any sense 'useful'. Are we able to do now *anything* that we weren't able to do before learning this material? In §7.2 we will prove that *if k is a field, then $k[t]$ is a Euclidean domain;* and then we will know it must be a PID, hence a UFD, by what we learned in Chapter 6. In particular, $\mathbb{C}[t]$ is a UFD. Even before we prove this, I would like to illustrate why knowing something of this sort is helpful, by using it to solve an 'easy' version of a very famous problem. It is fun to see these abstract ideas put to good use.

The famous *Fermat's Last Theorem* states that there are *no* solutions to the equation $a^n + b^n = c^n$ for $n \geq 3$ if we require a, b, c to be positive integers. Pierre de Fermat (a lawyer!) wrote this statement in the margin of a math book he was reading, in 1637. It took a mere 358 years to prove it—Andrew Wiles, a Princeton mathematician, proved in 1995 a substantial part of the 'Taniyama–Shimura–Weil' conjecture, which was known to imply Fermat's Last Theorem. (Richard Taylor collaborated with Wiles in an essential step in the proof.)

Are we going to prove Fermat's Last Theorem here? Fat chance. Wiles' proof uses extremely sophisticated mathematics, that goes light years beyond what we have learned so far. *However,* we can consider 'the same question' over a different ring. Let's ask whether there can exist *polynomials $f(t)$, $g(t)$, $h(t)$* in $\mathbb{C}[t]$, not all constant, such that

$$f(t)^n + g(t)^n = h(t)^n, \tag{7.1}$$

for $n \geq 3$. By clearing common factors we may assume that $g(t)$ and $h(t)$ are relatively prime. We may also assume that the sum of the degrees of $f(t)$, $g(t)$, $h(t)$, a positive integer, is as small as possible subject to the condition that (7.1) holds.[1] (I am counting on the fact that you remember what the 'degree' of a polynomial is; if not, glance now at Definition 7.1.)

Our tools to explore this question are the facts that $\mathbb{C}[t]$ is a UFD (because it is a Euclidean domain, hence a PID, as we will prove in §7.2), that we can take nth roots of every complex number for every n, and in particular that the polynomial $1 - t^n$ factorizes completely in $\mathbb{C}[t]$:

$$1 - t^n = (1 - t)(1 - \zeta t)(1 - \zeta^2 t) \cdots (1 - \zeta^{n-1} t), \tag{7.2}$$

---

[1] Hopefully you realize that here I am once again using the well-ordering principle.

where $\zeta = e^{2\pi i/n}$. Even if you are not very familiar with complex numbers, this fact should not surprise you too much: for example, for $n = 4$ we have $e^{2\pi i/n} = e^{\pi i/2} = i$, giving $\zeta^2 = -1$, $\zeta^3 = -i$, so the statement is that

$$1 - t^4 = (1 - t)(1 - it)(1 + t)(1 + it),$$

something you can check in no time at all. In general, the powers ζ^i are the vertices of a regular n-gon in the complex plane. For $n = 7$, they look like this:

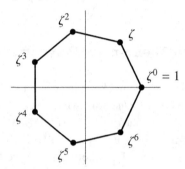

Now replace t by $g(t)/h(t)$ in (7.2):

$$1 - \frac{g(t)^n}{h(t)^n} = \left(1 - \frac{g(t)}{h(t)}\right)\left(1 - \zeta\frac{g(t)}{h(t)}\right)\left(1 - \zeta^2\frac{g(t)}{h(t)}\right)\cdots\left(1 - \zeta^{n-1}\frac{g(t)}{h(t)}\right)$$

and clear denominators:

$$h(t)^n - g(t)^n = (h(t) - g(t))(h(t) - \zeta g(t))\left(h(t) - \zeta^2 g(t)\right)\cdots\left(h(t) - \zeta^{n-1}g(t)\right). \quad (7.3)$$

We are assuming that (7.1) holds, so the left-hand side equals $f(t)^n$. What can we say about the factors on the right-hand side? Consider two distinct factors $\alpha(t) = h(t) - \zeta^i g(t)$, $\beta(t) = h(t) - \zeta^j g(t)$. Note that

$$\frac{\alpha(t) - \beta(t)}{\zeta^j - \zeta^i} = g(t) \quad \text{and} \quad \frac{\zeta^j\alpha(t) - \zeta^i\beta(t)}{\zeta^j - \zeta^i} = h(t) \quad\quad (7.4)$$

(check my algebra!). It follows that if $\alpha(t)$ and $\beta(t)$ had a common irreducible factor, then $g(t)$ and $h(t)$ would have a common irreducible factor. We are assuming that this is not the case, so we see that *the factors on the right-hand side of (7.3) are pairwise relatively prime.*

Now use the fact that $\mathbb{C}[t]$ is a PID (as we will prove soon), and therefore unique factorization holds in $\mathbb{C}[t]$ (Theorem 6.35). Since

$$f(t)^n = (h(t) - g(t))(h(t) - \zeta g(t))\left(h(t) - \zeta^2 g(t)\right)\cdots\left(h(t) - \zeta^{n-1}g(t)\right)$$

is an nth power and the factors on the right are relatively prime, by Lemma 6.41 we can deduce that *each factor is an nth power* up to a unit. In particular, there exist polynomials $a(t)$, $b(t)$, $c(t)$, such that

$$h(t) - g(t) = a(t)^n, \quad h(t) - \zeta g(t) = b(t)^n, \quad h(t) - \zeta^2 g(t) = c(t)^n.$$

(This is where we use the fact that $n \geq 3$: this implies that we have at least three factors.

Also note that I have absorbed units in $a(t)$, $b(t)$, $c(t)$, since units are nonzero complex numbers and are therefore themselves nth powers: unlike for real numbers, one can extract *every* root of a complex number. The field $\mathbb{C}$ is *algebraically closed*, as we will discuss in §7.3.) Another bit of mindless algebra gives:

$$(-\zeta)a(t)^n + (1 + \zeta)b(t)^n = (-\zeta)(h(t) - g(t)) + (1 + \zeta)(h(t) - \zeta g(t))$$
$$= h(t) - \zeta^2 g(t)$$
$$= c(t)^n.$$

Letting λ and μ be complex numbers such that $\lambda^n = -\zeta, \mu^n = 1 + \zeta$, we get

$$(\lambda a(t))^n + (\mu b(t))^n = c(t)^n.$$

But this is an identity of the same type as (7.1), and the sum of the degrees of the three polynomials $a(t)$, $b(t)$, $c(t)$ is less than the sum for (7.1) (because $f(t)$, $g(t)$, $h(t)$ were assumed to not all be constant). This yields a contradiction: we had assumed that this sum was the *least* possible for which there could be such an identity.

Thus, a solution to identity (7.1) by nonconstant polynomials would lead to a contradiction. Therefore there is no such solution.

The moral of the story is that Fermat's Last Theorem for polynomials is a relatively simple business. (When I teach this material to graduate students, I assign this result as homework, with a short hint.) You may wonder whether this argument may be cleverly adapted to work over $\mathbb{Z}$: after all, $\mathbb{Z}$ is also a PID. The problem is that the ring where the argument would run is not $\mathbb{Z}$, rather a ring which contains $\mathbb{Z}$ *and* ζ, and possibly nth roots of other elements as needed. Why should *this* ring be a PID, or even a UFD? The ring $\mathbb{Z}[\zeta]$ (the smallest subring of $\mathbb{C}$ containing $\mathbb{Z}$ and $\zeta = e^{2\pi i/n}$) is *not* a UFD for most values of n.

7.2   The Polynomial Ring with Coefficients in a Field

Recall (from Example 3.10) that if R is a ring, then we can construct a new ring $R[x]$, whose elements are polynomials in the indeterminate x with coefficients in R. The operations $+$, $\cdot$ are defined in the usual way.

DEFINITION 7.1   Let $f(x) \in R[x]$ be a nonzero polynomial: $f(x) = a_0 + a_1 x + \cdots + a_d x^d$, with $a_d \neq 0$. The integer d is called the 'degree' of f, denoted $\deg f$; a_d is the 'leading coefficient' of f; and a_0 is the 'constant term' of f. A polynomial is 'constant' if it equals its constant term. A polynomial is 'monic' if its leading coefficient is 1.

(The polynomial 0 does not have a degree or a leading coefficient; it is taken to be 'constant'.)

The constant polynomials form a subring of $R[x]$ isomorphic to R (Example 4.11).

I will try to state a few of the results that follow for polynomial rings over arbitrary rings R; but notions such as 'degree' are better behaved if we assume R to be an integral domain (Exercise 7.1). This assumption is also convenient in that it leads to a reasonable description of the invertible elements of a polynomial ring: if R is an integral domain, then the units of $R[x]$ are simply the units of R, viewed as constant polynomials (Exercise 7.2). In particular, if k is a field, then the units of $k[x]$ are all the nonzero constant polynomials.

Coming back to more general rings, we have observed in Example 3.11 that polynomials in $R[x]$ determine *functions* $R \to R$. (However, in general the function corresponding to a polynomial does not determine it, as you will verify in the exercises.) Thus, it makes sense to talk about the element $f(a)$ of R, for any polynomial $f(x) \in R[x]$ and any $a \in R$. If $f(x)$ is a *constant* polynomial in the sense of Definition 7.1, then this function is a constant function; this justifies the terminology for polynomials.

Here is a rather open-ended, but interesting question: Which properties of R are 'inherited' by $R[x]$? For example, if R is commutative, then $R[x]$ is also commutative. You have even proved already (Exercise 3.14) that if R is an integral domain, then $R[x]$ is an integral domain. An interesting property that is inherited by R has to do with the 'ascending chain condition', which we have verified for PIDs in Proposition 6.32. There I mentioned that rings satisfying this condition are called 'Noetherian'. It can be shown that if R is Noetherian, then $R[x]$ is Noetherian; this is the so-called *Hilbert's basis theorem*. This fact is very important in, for example, algebraic geometry, and its proof is actually not difficult, but I will show some restraint and not discuss it here.

What about the PID property? If R is a PID, is $R[x]$ a PID? Not necessarily: $\mathbb{Z}$ is a PID, but $\mathbb{Z}[x]$ is *not* a PID. The ideal $(2, x)$ is not principal (Exercise 6.10).

What about fields? A polynomial ring $R[x]$ is never a field: x is not 0, and yet it has no multiplicative inverse in $R[x]$. Even if k is a field, $k[x]$ is not a field.

This seems a bit discouraging, but something very interesting happens: if k is a field, then $k[x]$ is a PID. In fact, something stronger is true: we will show that if k is a field, then $k[x]$ is a *Euclidean domain*. Once we prove this, the whole package of properties of PIDs will be available to us when we deal with the polynomial ring $k[x]$. In §7.1 we have seen how handy this observation can be.

THEOREM 7.2 *Let k be a field. Then $k[x]$ is a Euclidean domain.*

To prove this theorem, we have to find a function $v \colon (k[x] \setminus 0) \to \mathbb{Z}^{\geq 0}$ satisfying the requirement specified in Definition 6.24, that is, endowing $k[x]$ with a suitable 'division with remainder'. This function will be the *degree;* the process of 'long division' will show that the property in Definition 6.24 is satisfied by $k[x]$ if k is a field. In fact, long division can be performed over any ring, so long as we are dividing by polynomials with *invertible* leading coefficient. More precisely, the following holds.

PROPOSITION 7.3 ('Long division') *Let R be a ring, and let $f(x), g(x) \in R[x]$ be polynomials, with $f(x) \neq 0$. Assume that the leading coefficient of $f(x)$ is a unit. Then there exist polynomials $q(x), r(x) \in R[x]$ such that*

$$g(x) = q(x)f(x) + r(x),$$

with $r(x) = 0$ or $\deg r(x) < \deg f(x)$.

Proof   If $g(x) = 0$ or $\deg g(x) < \deg f(x)$, then choosing $q(x) = 0$, $r(x) = g(x)$ gives

$$g(x) = 0 \cdot f(x) + g(x),$$

satisfying the requirements in the statement. We have to prove that the result holds if $g(x) \neq 0$ and $n = \deg g(x) \geq m = \deg f(x)$. Now suppose that there is a polynomial $g(x)$ of degree $n \geq m$ for which the statement does *not* hold. We will be done if we show that this leads to a contradiction. By the well-ordering principle, we may choose a polynomial $g(x)$ with this property and for which $n = \deg g(x)$ is as small as possible; so the statement is satisfied for all polynomials of degree $< n$. Write

$$f(x) = a_0 + a_1 x + \cdots + a_m x^m, \quad g(x) = b_0 + b_1 x + \cdots + b_n x^n,$$

and note that a_m is a unit by hypothesis, so there exists an element $u = a_m^{-1} \in R$ such that $u a_m = 1$. The coefficient of x^n in

$$g_1(x) := g(x) - b_n u f(x) x^{n-m} = (b_0 + \cdots + b_n x^n) - (b_n u a_0 x^{n-m} + \cdots + b_n u a_m x^n)$$

is $b_n - b_n u a_m = b_n - b_n = 0$; it follows that $\deg g_1(x) < n$ or $g_1(x) = 0$. By our hypothesis on $g(x)$, the statement holds for $g_1(x)$: there exist polynomials $q_1(x)$, $r_1(x)$, with $r_1(x) = 0$ or $\deg r_1(x) < \deg f(x)$, such that

$$g(x) - b_n u f(x) x^{n-m} = g_1(x) = q_1(x) f(x) + r_1(x).$$

It follows that

$$g(x) = (q_1(x) + b_n u x^{n-m}) f(x) + r_1(x),$$

and this shows that the statement *does hold* for $g(x)$, with $q(x) = q_1(x) + b_n u x^{n-m}$ and $r(x) = r_1(x)$. This is a contradiction: $g(x)$ had been chosen so that the statement *would not hold* for it. It follows that the statement is true for all $g(x) \in R[x]$.   □

Example 7.4   Let R be a ring, and let $\alpha(x) \in R[x]$. Then there exists $\beta(x) \in R[x]$ and $a, b \in R$ such that

$$\alpha(x) = \beta(x)(x^2 + 1) + a + bx.$$

Indeed, the leading coefficient of $x^2 + 1$ is 1, and every polynomial of degree < 2 may be written as $a + bx$. This fact was used (for $R = \mathbb{R}$) in Example 5.42.

Example 7.5   If $a \in R$, then we can always divide a polynomial $g(x)$ by $f(x) = x - a$: again, the leading coefficient of $x - a$ is 1. The remainder $r(x)$ is then a constant $r \in R$, possibly equal to 0. It is useful to remember that this constant is $g(a)$: indeed, write

$$g(x) = q(x) f(x) + r(x) = q(x)(x - a) + r,$$

then

$$g(a) = q(a)(a - a) + r = r.$$

In particular, $x - a$ is a factor of $g(x)$ if and only if $g(a) = 0$.

The observation in Example 7.5 is streamlined by the following definition.

DEFINITION 7.6   Let R be a ring, and let $g(x) \in R[x]$ be a polynomial. An element $a \in R$ is a 'root' of $g(x)$ if $g(a) = 0$.

With this terminology, we have that $x - a$ is a factor of $g(x)$ if and only if a is a root of $g(x)$. This fact will be used below.

Theorem 7.2 is a consequence of Proposition 7.3.

Proof of Theorem 7.2   Since k is a field, it is an integral domain, and it follows that $k[x]$ is an integral domain. We define a valuation $v : (k[x] \smallsetminus 0) \to \mathbb{Z}^{\geq 0}$ by letting $v(f(x)) =$ deg $f(x)$. Since every nonzero element of a field is a unit, the leading coefficient of *every* nonzero polynomial $f(x)$ is a unit, thus the hypothesis of Proposition 7.3 is always satisfied. Therefore, for *all* polynomials $f(x), g(x) \in k[x]$ with $f(x) \neq 0$ there exist $q(x)$ and $r(x)$ such that

$$g(x) = q(x)f(x) + r(x)$$

with $r(x) = 0$ or $v(r(x)) < v(f(x))$. This fact verifies the requirement of Definition 6.24, proving that $k[x]$ is a Euclidean domain.   □

COROLLARY 7.7   *Let k be a field. Then $k[x]$ is a PID.*

Proof   Every Euclidean domain is a PID, by Theorem 6.26.   □

COROLLARY 7.8   *Let k be a field. Then unique factorization holds in $k[x]$.*

Proof   Every PID is a UFD, by Theorem 6.35.   □

The conclusion is that in many respects—for example, for what concerns gcds, or general remarks about irreducibility—polynomial rings over a field behave just like $\mathbb{Z}$. It is 'as if' $\mathbb{Z}$ itself were a ring of 'polynomials' over something like a field. This is not literally the case, but there is some active current research aimed at identifying a type of structure, which people have been calling[2] '$\mathbb{F}_1$' (the 'field with one element'), and which may play the role of a ring of coefficients that would allow us to view $\mathbb{Z}$ as a polynomial ring.

Before we delve further into polynomial rings over a field, it is useful to make a few more general observations. Proposition 7.3 works for every ring; if the ring is more special, then one can go a little further.

PROPOSITION 7.9   *Let R be an integral domain, and let $f(x) \neq 0$, $g(x)$ be polynomials in $R[x]$. Assume that the leading coefficient of $f(x)$ is a unit. Then the quotient $q(x)$ and remainder $r(x)$ obtained in Proposition 7.3 are uniquely determined by $f(x)$ and $g(x)$.*

---

[2] This terminology is essentially a joke. As we know, $0 \neq 1$ in a field, so every field has *at least two* elements.

Proof Assume that

$$g(x) = q_1(x)f(x) + r_1(x) = q_2(x)f(x) + r_2(x),$$

with both $r_1(x)$, $r_2(x)$ either 0 or of degree less than deg $f(x)$. Then

$$(q_1(x) - q_2(x))f(x) + (r_1(x) - r_2(x)) = g(x) - g(x) = 0,$$

and therefore $(q_1(x) - q_2(x))f(x) = r_2(x) - r_1(x)$. If $r_2(x) - r_1(x) \neq 0$, we have a contradiction: indeed, since R is an integral domain, the degree of the left-hand side would be at least as large as the degree of $f(x)$ (Exercise 7.1), while the degree of the right-hand side is necessarily smaller than the degree of $f(x)$. Therefore

$$(q_1(x) - q_2(x))f(x) = r_2(x) - r_1(x) = 0,$$

and it follows that $r_1(x) = r_2(x)$ and $q_1(x) = q_2(x)$ (since R is an integral domain). $\qquad\square$

> COROLLARY 7.10 *Let R be an integral domain, and let $f(x) \neq 0$. Assume that the leading coefficient of $f(x)$ is a unit. Then every coset in $R[x]/(f(x))$ has a unique representative of the form $r(x) + (f(x))$, with either $r(x) = 0$ or deg $r(x) <$ deg $f(x)$.*

Proof Let $g(x) + (f(x))$ be an element of $R[x]/(f(x))$. By Propositions 7.3 and 7.9, there exist unique $q(x)$ and $r(x)$ such that

$$g(x) = q(x)f(x) + r(x)$$

with $r(x) = 0$ or deg $r(x) <$ deg $f(x)$. But then

$$g(x) + (f(x)) = r(x) + (f(x)),$$

with $r(x)$ as in the statement and uniquely determined by $g(x)$. $\qquad\square$

Corollary 7.10 applies in particular if $R = k$ is a field, and for all nonzero polynomials $f(x) \in k[x]$: fields are integral domains, and nonzero elements in a field are units. Therefore, we obtain a very efficient description of the quotient $k[x]/(f(x))$ of $k[x]$ by the ideal generated by a nonzero element $f(x)$. In a sense, this is the analogue of the description of the quotient $\mathbb{Z}/n\mathbb{Z}$ obtained in Theorem 2.7. Using the language of linear algebra, I can state this result as follows.

> THEOREM 7.11 *Let k be a field, and let $f(x) \in k[x]$ be a nonzero polynomial of degree d. Then $k[x]/(f(x))$ is a d-dimensional k-vector space.*

All this means is that we can identify the quotient $k[x]/(f(x))$ with a set of d-tuples $(a_0, \ldots, a_{d-1})$ with entries in k, compatibly with addition and multiplication by scalars.

Proof This is a consequence of what we have established in the remarks following Corollary 7.10. Since k is an integral domain, and the leading coefficient of a nonzero polynomial is automatically a unit (as k is a field), *every* element of $k[x]/(f(x))$ may be written uniquely as

$$r(x) + (f(x)),$$

where $r(x) = a_0 + a_1x + \cdots + a_{d-1}x^{d-1}$ is 0 or a polynomial of degree $< \deg f(x)$. We can then associate the d-tuple of coefficients $(a_0, \ldots, a_{d-1})$ with the coset $g(x) + (f(x))$, and this identification preserves addition and multiplication by elements of k.   □

We have actually encountered a few instances of Theorem 7.11 along the way. In Example 5.34 we studied $\mathbb{R}[x]/(x^2+1)$, and we agreed that cosets can be represented by 'pairs of real numbers (a, b)'. That is, we can view $\mathbb{R}[x]/(x^2+1)$ (and hence $\mathbb{C}$) as a two-dimensional real vector space. This is nothing but a particular case of Theorem 7.11.

Example 7.12   The ring $(\mathbb{Z}/5\mathbb{Z})[x]/(x^3+1)$ has exactly 125 elements. Indeed, by Theorem 7.11 its elements may be described by triples (a_0, a_1, a_2) with entries in $\mathbb{Z}/5\mathbb{Z}$. Each of a_0, a_1, a_2 may take one of five values, so there are $5^3 = 125$ such triples. Concretely, we could list all the cosets: they are just all cosets $r(x) + (x^3+1)$ with $r(x)$ a polynomial with coefficients in $\mathbb{Z}/5\mathbb{Z}$ and of degree 0, 1, or 2.

7.3   Irreducibility in Polynomial Rings

Corollary 7.10 and Theorem 7.11 give us a good handle on quotients $k[x]/(f(x))$ when k is a field, and we can ask whether these quotients satisfy nice properties. Since $k[x]$ is a PID, we know (Proposition 6.33, Remark 6.34) that the ideal $(f(x))$ is maximal if and only if $f(x)$ is irreducible. Thus, $k[x]/(f(x))$ is a field precisely when $f(x)$ is irreducible in $k[x]$. The next natural question is: can we tell whether a polynomial $f(x)$ with coefficients in a field k is irreducible?

This question depends on the field k, and little can be said in general; I will mention a few simple but useful general observations, and then we will spend some time analyzing irreducibility over some familiar fields.

There is no big mystery concerning degree-1 polynomials.

Example 7.13   Degree-1 polynomials with coefficients in a field are *irreducible;* verify this now (Exercise 7.13).

Over more general rings, even integral domains, this is not necessarily the case. For example, the polynomial $2x \in \mathbb{Z}[x]$ has degree 1 and yet it factors as the product of two non-units: both 2 and x are non-units in $\mathbb{Z}[x]$.

In higher degree, the following result takes care of many cases.

PROPOSITION 7.14   *Let k be a field, and let $f(x) \in k[x]$ be a polynomial of degree at least 2. Assume $f(x)$ has a root. Then $f(x)$ is not irreducible.*

Proof   As observed in Example 7.5, if a is a root of $f(x)$, then $x - a$ is a factor of $f(x)$. If $\deg f(x) \geq 2$, this gives $f(x) = (x-a)\,g(x)$ with $\deg g(x) \geq 1$, and in particular $g(x)$ is then not a unit. This shows that $f(x)$ is not irreducible in this case.   □

One can be more precise if the degree of $f(x)$ is 2 *or* 3: then the irreducibility of $f(x)$ is in fact equivalent to the absence of roots.

> PROPOSITION 7.15 *Let k be a field, and let $f(x) \in k[x]$ be a nonzero polynomial of degree 2 or 3. Then $f(x)$ is irreducible if and only if $f(x)$ has no roots in k.*

Proof ($\Rightarrow$) If $f(x)$ has a root and degree ≥ 2, then $f(x)$ is reducible by Proposition 7.14. Therefore, if $f(x)$ is irreducible, then it has no roots.

($\Leftarrow$) Argue contrapositively again. If $f(x)$ is not irreducible, then $f(x) = q_1(x)q_2(x)$ for polynomials $q_1(x)$, $q_2(x)$, neither of which is a unit. Since k is a field, this implies that $q_1(x)$, $q_2(x)$ are not constant polynomials; since $\deg f(x) = 2$ or 3, one of $q_1(x)$, $q_2(x)$ has degree 1, so it equals $ax + b$ with $a, b \in k$, $a \neq 0$. But then $-ba^{-1}$ is a root of $f(x)$, and this concludes the proof. □

Remark 7.16 Note that we took the multiplicative inverse of a (nonzero) element of k in the second part of the proof. And indeed, the 'if' statement in the proposition is not true in general over rings that are not fields. For example, the polynomial $(2x - 1)^2$ in $\mathbb{Z}[x]$ is *reducible*, but it has no roots in $\mathbb{Z}$. (It has one root, $x = 1/2$, in $\mathbb{Q}$.)

Also note that the hypothesis that the degree be 3 or less is necessary. The polynomial $x^4 - 4 \in \mathbb{Q}[x]$ is *reducible,* and has no roots in $\mathbb{Q}$. Indeed, $x^4 - 4 = (x^2 - 2)(x^2 + 2)$; its roots are $\pm\sqrt{2}$, $\pm i\sqrt{2}$, and none of these is in $\mathbb{Q}$. ⌐

Now let's look at irreducibility over a few fields with which we are familiar. It turns out that irreducibility of *complex* polynomials is a very simple matter, due to the following sophisticated result.

> THEOREM 7.17 (Fundamental Theorem of Algebra) *Every nonconstant polynomial $f(x) \in \mathbb{C}[x]$ has a root in $\mathbb{C}$.*

What this statement says is that $\mathbb{C}$ is an 'algebraically closed' field. This is a hugely important fact in, e.g., algebraic geometry.

An efficient proof of Theorem 7.17 is usually presented in courses in complex analysis. After all, defining complex (or even real) numbers requires a limiting process, so it may not be surprising that analysis should turn out to be an essential ingredient of the proof of the Fundamental Theorem of Algebra. The gist of the standard proof is that if $f(x)$ did not have roots, then the function $1/f(x)$ would be analytic over the whole complex plane $\mathbb{C}$, and it is not hard to squeeze a contradiction from this by using (for example) the *maximum modulus principle*. There are more 'algebraic' arguments, and we will see one in the very last section of this text, §15.6; but even that argument will use a little analysis ('just' real analysis, though).

For now, I will simply record the following immediate consequences.

> COROLLARY 7.18 *A polynomial $f(x) \in \mathbb{C}[x]$ is irreducible if and only if it has degree 1.*

Proof Degree-1 polynomials are irreducible over any field, as you should have verified (Exercise 7.13). For the converse: if $f(x) \in \mathbb{C}[x]$ has degree 0, then it is a unit (Exercise 7.2), hence it is not irreducible. If it has degree ≥ 2, then it has a root a by Theorem 7.17. By Proposition 7.14, $f(x)$ is not irreducible: $x - a$ is an irreducible factor. □

> COROLLARY 7.19 *Every polynomial $f(x) \in \mathbb{C}[x]$ of degree d factors as a product of d factors of degree 1, determined up to units.*

Proof Every polynomial must factor as a product of irreducibles since unique factorization holds in $\mathbb{C}[x]$ (Corollary 7.8). The statement follows, since the irreducible polynomials are the polynomials of degree 1 by Corollary 7.18. The factors are determined up to units by the uniqueness of factorizations. □

Therefore, every degree-d polynomial $f(x) \in \mathbb{C}[x]$ may be written as

$$f(x) = u\,(x - a_1) \cdots (x - a_d),$$

where $u \neq 0$ and $a_1, \ldots, a_d$ are complex numbers uniquely determined by $f(x)$. In fact, the numbers $a_1, \ldots, a_d$ are the roots of $f(x)$.

The situation is a little more complicated over $\mathbb{R}$, but just a little. It is clear from calculus (!) considerations that every polynomial $f(x)$ of *odd* degree ≥ 3 in $\mathbb{R}[x]$ is *reducible*. Indeed, its limits as $x \to \pm\infty$ are $\pm\infty$, and both signs are obtained if the degree is odd; by the *Intermediate Value Theorem*,[3] the graph of $f(x)$ must cross the x-axis (since polynomials are continuous). This means that every such polynomial has a root, and hence it is not irreducible if its degree is at least 3, by Proposition 7.14.

This is a nice observation, but something much stronger is true.

> PROPOSITION 7.20 *Let $f(x) \in \mathbb{R}[x]$. Then $f(x)$ is irreducible if and only if*
>
> *(i)* $\deg f(x) = 1$, *or*
> *(ii)* $\deg f(x) = 2$ *and* $f(x) = ax^2 + bx + c$ *with* $b^2 - 4ac < 0$.

Proof ($\Leftarrow$) We have seen already that if $f(x)$ has degree 1, then it is irreducible (Example 7.3). If $f(x) = ax^2 + bx + c$ and $b^2 - 4ac < 0$, then $f(x)$ has *no* real roots (by the quadratic formula), and it is irreducible by Proposition 7.15.

($\Rightarrow$) Conversely, if $f(x) = ax^2 + bx + c$ and $b^2 - 4ac \geq 0$, then $f(x)$ has real roots, hence it is reducible (Proposition 7.14). If $\deg f(x) = 0$, then it is a unit (hence not irreducible). If $\deg f(x) \geq 3$, we can show that it must have a real root or a degree-2 real factor; in either case this will show that $f(x)$ is reducible, completing the proof.

To see that $f(x)$ must have real roots or degree-2 real factors, view $f(x)$ as a *complex*

---

[3] If you have taken calculus, you *have* run across this theorem, even though it is more of a topological statement than a calculus statement. Go back and look it up in your calculus book if you don't quite remember.

polynomial. By Theorem 7.17, $f(x)$ has a root $\alpha \in \mathbb{C}$. If α is real, we are done. In any case, note that if α is a root of $f(x)$, then so is the complex conjugate $\overline{\alpha}$: indeed,

$$f(\overline{\alpha}) \stackrel{!}{=} \overline{f(\alpha)} = \overline{0} = 0\,,$$

where $\stackrel{!}{=}$ holds because $f(x)$ has real coefficients (Exercise 7.18). If α is *not* real, then $\overline{\alpha} \neq \alpha$ is another root of $f(x)$, and it follows that both $x - \alpha$ and $x - \overline{\alpha}$ must appear in an irreducible factorization of $f(x)$. (Keep in mind that $\mathbb{C}[x]$ is a UFD!) Therefore

$$(x - \alpha)(x - \overline{\alpha}) \mid f(x)\,.$$

Now

$$(x - \alpha)(x - \overline{\alpha}) = x^2 - (\alpha + \overline{\alpha})x + \alpha\overline{\alpha} :$$

this is a polynomial of degree 2 with real coefficients, and we are done.   □

Example 7.21   The polynomial $x^4 + 4$ must be *reducible* over $\mathbb{R}$ (even if it does not have real roots!). We do not need any computation to know this, since *every* polynomial in $\mathbb{R}[x]$ of degree > 2 must be reducible, by Proposition 7.20. However, it is instructive to run through the argument proving Proposition 7.20 in this specific example, to see the gears turn.

Over $\mathbb{C}$, $x^4 + 4$ has four roots: $x^4 = -4$ implies that $x = \sqrt{2}\,\zeta$, where ζ is any of the four fourth roots of -1. These are $\zeta = e^{\pi i/4}$ for $i = 1, 3, 5, 7$; working out what these are shows that the four roots of $x^4 + 4$ in $\mathbb{C}$ are $\pm 1 \pm i$. As promised, the roots come in 'conjugate pairs': $1 \pm i$ and $-1 \pm i$. As in the proof, the first pair gives rise to the product

$$(x - (1 + i))(x - \overline{(1 + i)}) = ((x - 1) - i)((x - 1) + i) = (x - 1)^2 - (i^2) = x^2 - 2x + 2\,.$$

So this must be a factor of $x^4 + 4$, confirming that $x^4 + 4$ is reducible.

(Similarly, the other pair gives the factor $(x^2 + 2x + 2)$; and then indeed you may verify that $x^4 + 4 = (x^2 - 2x + 2)(x^2 + 2x + 2)$.)

Summary: Thanks to the Fundamental Theorem of Algebra, irreducibility in $\mathbb{C}[x]$ and $\mathbb{R}[x]$ is child's play. The situation is *very* different for $\mathbb{Q}[x]$: irreducibility in $\mathbb{Q}[x]$ is a deep and profound business. The surprise here will be that irreducibility in $\mathbb{Q}[x]$ is essentially the same as irreducibility in $\mathbb{Z}[x]$; I will proceed to make this statement more precise, and proving it will require a certain amount of cleverness.

7.4   Irreducibility in $\mathbb{Q}[x]$ and $\mathbb{Z}[x]$

There is one simple distinction between irreducibility over a field (for example, in $\mathbb{Q}[x]$) and irreducibility over a more general integral domain (for example, in $\mathbb{Z}[x]$). A polynomial $f(x) \in k[x]$, where k is a field, is reducible if and only if it *can* be written as a product of polynomials of lower degree: $f(x) = g(x)h(x)$ with $0 < \deg g(x) < \deg f(x)$,

$0 < \deg h(x) < \deg f(x)$. This is not quite so in a ring such as $\mathbb{Z}[x]$: for example, $2x + 2$ is *reducible* in $\mathbb{Z}[x]$ since it can be written as $2 \cdot (x + 1)$ and neither factor is a unit. The same polynomial $2x + 2$ is *irreducible* in $\mathbb{Q}[x]$ (the factor 2 is a unit in $\mathbb{Q}[x]$).

Maybe surprisingly, this turns out to be the *only* source of distinctions between irreducibility in $\mathbb{Z}[x]$ and in $\mathbb{Q}[x]$. This is not at all 'obvious'!

Let's get into the nuts and bolts of this particular question. First, note that if $f(x) \in \mathbb{Q}[x]$, then $af(x) \in \mathbb{Z}[x]$ for some $a \neq 0$ in $\mathbb{Q}$: this just amounts to clearing denominators. Since a is a unit in $\mathbb{Q}$, factoring $f(x)$ in $\mathbb{Q}[x]$ is the same as factoring $af(x)$; therefore it is not restrictive to assume that the polynomial we want to factor has *integer* coefficients.

> THEOREM 7.22 (Gauss's Lemma)   *Let $f(x) \in \mathbb{Z}[x]$, and let $g(x), h(x) \in \mathbb{Q}[x]$ be such that $f(x) = g(x)h(x)$. Then there exist rational numbers $a, b \in \mathbb{Q}$ such that $ag(x) \in \mathbb{Z}[x]$, $bh(x) \in \mathbb{Z}[x]$, and $f(x) = (ag(x))(bh(x))$.*

What this statement says is that every factorization of an integer polynomial $f(x)$ in $\mathbb{Q}[x]$ can be turned into a factorization of the same polynomial *in $\mathbb{Z}[x]$*. Conversely, of course, any factorization in $\mathbb{Z}[x]$ as a product of lower-degree factors is a factorization in $\mathbb{Q}[x]$. In this sense, factoring a polynomial in $\mathbb{Q}[x]$ is equivalent to factoring it in $\mathbb{Z}[x]$.

The key to proving Gauss's Lemma is the following observation.

> LEMMA 7.23   *Let $\alpha(x), \beta(x) \in \mathbb{Z}[x]$, and assume that an irreducible integer q divides all the coefficients of the product $\alpha(x)\beta(x)$. Then q divides all the coefficients of $\alpha(x)$ or all the coefficients of $\beta(x)$.*

Proof   Consider the ideal $I = q\mathbb{Z}$ of $\mathbb{Z}$ as well as the ideal $I[x]$ generated by q in $\mathbb{Z}[x]$. (Usually we would denote both ideals by (q), but this would be confusing here since we are dealing with two different rings: $\mathbb{Z}$ and $\mathbb{Z}[x]$.) Recall from Exercise 5.25 that we have an isomorphism

$$(\mathbb{Z}/I)[x] \cong \mathbb{Z}[x]/I[x] : \tag{7.5}$$

this is a standard application of the First Isomorphism Theorem. It implies that $I[x]$ is prime. Indeed, $I = (q)$ is prime in $\mathbb{Z}$ since q is irreducible, so $\mathbb{Z}/I$ is an integral domain; so $(\mathbb{Z}/I)[x]$ is an integral domain; so $\mathbb{Z}[x]/I[x]$ is an integral domain by (7.5); so $I[x]$ is a prime ideal.

Now, $I[x]$ consists of the polynomial multiples of q. These are precisely the polynomials whose coefficients are divisible by q. Since $I[x]$ is prime,

$$\alpha(x)\beta(x) \in I[x] \implies \alpha(x) \in I[x] \text{ or } \beta(x) \in I[x].$$

We are done, since this is precisely the statement of the lemma.   □

Proof of Gauss's Lemma   Let $f(x) \in \mathbb{Z}[x]$, and assume $f(x) = g(x)h(x)$ with $g(x), h(x)$ in $\mathbb{Q}[x]$ as in the statement. Choose any rational numbers a, b such that $ag(x)$ and $bh(x)$ are both in $\mathbb{Z}[x]$. Then we have

$$(ag(x))(bh(x)) = ab\, f(x),$$

and we have to show that we may choose a and b so that $ab = 1$. First, by multiplying a or b by large enough integers, and changing the sign of one of them if necessary, we see that there are positive integers of the form ab. We will then consider *all* of them: consider the set S of positive integers defined by

$$S = \{c = ab \mid a \in \mathbb{Q}, b \in \mathbb{Q}, ag(x) \in \mathbb{Z}[x], bh(x) \in \mathbb{Z}[x], ab \in \mathbb{Z}, ab > 0\}.$$

This set is nonempty, as we just observed. We will be done if we can prove that $1 \in S$.

By the well-ordering principle, S contains a smallest element c, corresponding to a specific choice of a and b. Now I will assume that $c \neq 1$, and derive a contradiction. If $c \neq 1$, then (by the Fundamental Theorem of Arithmetic!) c has some irreducible factor q. Then q divides all coefficients of $abf(x) = (ag(x))(bh(x))$, and by Lemma 7.23 we can conclude that q divides all the coefficients of $ag(x)$ or all the coefficients of $bh(x)$. Without loss of generality, assume that q divides all the coefficients of $ag(x)$. This implies that

$$q^{-1}ag(x)$$

has integer coefficients, and it follows that $\frac{c}{q} = (q^{-1}a)b$ is an element of S. But $\frac{c}{q} < c$, so this contradicts the choice of c as the smallest element of S. This contradiction shows that necessarily the smallest element of S is $c = 1$, and we are done.   □

Remark 7.24   The key contradiction step in this argument hinges on the 'size' of the smallest element of a set S. For the purpose of a generalization that I will mention soon (Remark 7.28), it is useful to recast the argument in terms of a different measure of 'how big' ab can be: the number of its positive irreducible factors, counting repetitions, works well. (For example, 20 has three irreducible factors: 2, counted twice, and 5.) You should try to construct a proof using this alternative notion. (Exercise 7.23.)   ⌐

As a consequence of Gauss's Lemma, we obtain a description of the irreducible elements in $\mathbb{Z}[x]$. It is clear that irreducible *integers* remain irreducible in $\mathbb{Z}[x]$ (Exercise 7.21). It is also clear that given a polynomial $f(x) \in \mathbb{Z}[x]$, we can factor out of $f(x)$ the gcd of its coefficients.

DEFINITION 7.25   The 'content' of a nonzero polynomial $f(x) \in \mathbb{Z}[x]$ is the gcd of its coefficients. A polynomial in $\mathbb{Z}[x]$ is 'primitive' if its content is 1.

Adopting this terminology, we can write

$$f(x) = d\, f_1(x),$$

where d is the content of $f(x)$ and $f_1(x) \in \mathbb{Z}[x]$ is primitive. The interesting question is: when is $f_1(x)$ irreducible?

COROLLARY 7.26   *Let $f(x) \in \mathbb{Z}[x]$, and assume that $f(x)$ is primitive. Then $f(x)$ is irreducible in $\mathbb{Z}[x]$ if and only if $f(x)$ is irreducible in $\mathbb{Q}[x]$.*

Proof   Assume that $f(x)$ is *not* irreducible in $\mathbb{Z}[x]$: so we can write $f(x) = g(x)h(x)$ where neither $g(x)$ nor $h(x)$ is a unit. Since we are assuming that the gcd of the coefficients of $f(x)$ is 1, neither $g(x)$ nor $h(x)$ can be a constant factor. It follows that $g(x)$ and $h(x)$ both have positive degree, and in particular they are not units in $\mathbb{Q}[x]$. Since $f(x)$ can be written as the product of two non-units in $\mathbb{Q}[x]$, it is not irreducible in $\mathbb{Q}[x]$.

The converse implication follows from Gauss's Lemma. Indeed, if $f(x)$ is *not* irreducible in $\mathbb{Q}[x]$, then we can write it as a product of two positive degree factors $g(x)h(x)$ with $g(x), h(x) \in \mathbb{Q}[x]$. By Gauss's Lemma, we can do the same in $\mathbb{Z}[x]$, so $f(x)$ is not irreducible in $\mathbb{Z}[x]$.   □

Summarizing, the irreducible elements of $\mathbb{Z}[x]$ are precisely the irreducible integers and the primitive polynomials that are irreducible in $\mathbb{Q}[x]$. The following important consequence is now essentially immediate.

COROLLARY 7.27   *Unique factorization holds in $\mathbb{Z}[x]$: the ring $\mathbb{Z}[x]$ is a UFD.*

Proof   Given $f(x) \in \mathbb{Z}[x]$, $f(x) \neq 0$, write $f(x) = d f_1(x)$, where d is the content of $f(x)$ and $f_1(x)$ is primitive. The content d has a unique factorization in $\mathbb{Z}$, hence in $\mathbb{Z}[x]$, by the Fundamental Theorem of Arithmetic. Viewing $f_1(x)$ as a polynomial in $\mathbb{Q}[x]$, it has a unique factorization as a product of irreducible polynomials in $\mathbb{Q}[x]$. By Gauss's Lemma, the factors may be taken to be with integer coefficients, and they will have to be primitive: a common factor of the coefficients of one of them would give a common factor of the coefficients of their product $f_1(x)$, which is primitive. These factors are then irreducible in $\mathbb{Z}[x]$ by Corollary 7.26.

Thus $f(x)$ can be written as a product of irreducible factors in $\mathbb{Z}[x]$, and the factors are uniquely determined (as the argument shows).   □

Remark 7.28   The argument we just went through can be upgraded from $\mathbb{Z}$ to any UFD, and it shows that if R is a UFD, then $R[x]$ is a UFD. Some texts would call *this* result 'Gauss's Lemma'.

To understand this generalization, it is useful to briefly summarize again the argument for $\mathbb{Z}$ and $\mathbb{Q}$. We have seen that finding a factorization into irreducibles of a polynomial $f(x) \in \mathbb{Z}[x]$ amounts to finding factorizations of its content and of its primitive part. The first can be done, and done uniquely, because $\mathbb{Z}$ is a UFD; the second can be done, and done uniquely, because $\mathbb{Q}[x]$ is a UFD. Well, the same strategy can be carried out with every UFD R, using the *field of fractions* F of R in place of $\mathbb{Q}$ (§6.5). The key result would be an analogue of Gauss's Lemma (Theorem 7.22); to upgrade its proof to this more general setting, you can use the number of irreducible factors to control how 'big' the key multiple ab may be (cf. Remark 7.24). This will lead to the conclusion that irreducibility in $R[x]$ is essentially equivalent to irreducibility in $F[x]$, by the same reasoning we used above for $\mathbb{Z}[x]$ and $\mathbb{Q}[x]$. The needed work does not involve any new tool and would solidify your understanding of these matters, so you are strongly encouraged to try your hand at it.

If you carry this out, you will have shown that if R is a UFD, then $R[x]$ is a UFD. For example, it follows that the polynomial ring $k[x_1, \ldots, x_n]$ in any number of indeterminates is a UFD if k is a field (Exercise 7.24). By the same token, $\mathbb{Z}[x_1, \ldots, x_n]$ is a UFD. These facts are very important in, e.g., algebraic geometry and number theory. ⌟

7.5   Irreducibility Tests in $\mathbb{Z}[x]$

Corollary 7.26 is useful as an ingredient in the proof that $\mathbb{Z}[x]$ is a UFD. It is also useful in that it gives us tools to test whether a polynomial $f(x) \in \mathbb{Q}[x]$ is irreducible *in* $\mathbb{Q}[x]$: after multiplying the polynomial by a rational number if necessary, we may assume that $f(x)$ has integer coefficients and content 1, and then its irreducibility in $\mathbb{Q}[x]$ is equivalent to irreducibility in $\mathbb{Z}[x]$ by Corollary 7.26. This allows us to use some tools that would not be available in $\mathbb{Q}[x]$. We will close this circle of ideas by looking at some of these tools. They all admit straightforward generalizations to UFDs, or even more general integral domains.

First, if a polynomial of degree ≥ 2 has a root, then it is not irreducible. Let $f(x) \in \mathbb{Z}[x]$. How do we tell whether $f(x)$ has a root *in* $\mathbb{Q}$?

> **PROPOSITION 7.29**   *Let $f(x) = a_0 + a_1 x + \cdots + a_d x^d \in \mathbb{Z}[x]$ be a polynomial of degree d, and let $c = \frac{p}{q} \in \mathbb{Q}$ be a root of $f(x)$, with p, q relatively prime integers. Then $p \mid a_0$ and $q \mid a_d$.*

Proof   Since $c = \frac{p}{q}$ is a root of $f(x)$,

$$a_0 + a_1 \frac{p}{q} + a_2 \frac{p^2}{q^2} + \cdots + a_d \frac{p^d}{q^d} = 0.$$

Clearing denominators, we see that

$$a_0 q^d + a_1 p q^{d-1} + a_2 p^2 q^{d-2} + \cdots + a_{d-1} p^{d-1} q + a_d p^d = 0,$$

and therefore

$$a_0 q^d = -(a_1 q^{d-1} + \cdots + a_d p^{d-1}) p,$$
$$a_d p^d = -(a_0 q^{d-1} + \cdots + a_{d-1} p^{d-1}) q.$$

This shows that

$$p \mid a_0 q^d, \quad q \mid a_d p^d;$$

since p and q are relatively prime, it follows that $p \mid a_0$ and $q \mid a_d$ as stated, by Corollary 1.11. □

Proposition 7.29 implies that finding the rational roots of an integer polynomial is a finite process: we can find all divisors of a_0 and a_d, and just test all fractions $\frac{p}{q}$ with $p \mid a_0$ and $q \mid a_d$. If a root is found and $\deg f(x) \geq 2$, we will know that $f(x)$ is not irreducible in $\mathbb{Q}[x]$. If a root is *not* found and $\deg f(x) \leq 3$, we will know that $f(x)$ *is* irreducible in $\mathbb{Q}[x]$. This is called the *rational root test*.

Example 7.30   The polynomial $f(x) = 4x^3 + 2x^2 - 5x + 3$ is irreducible in $\mathbb{Q}[x]$ (and hence in $\mathbb{Z}[x]$, since it is primitive, by Corollary 7.26). Indeed, $\deg f(x) = 3$, so to prove that $f(x)$ is irreducible it suffices to verify that it has no rational roots (Proposition 7.15). By Proposition 7.29, a root $c = \frac{p}{q}$ of $f(x)$ would necessarily have $p \mid 3$, $q \mid 4$. Therefore, c would have to be one of

$$\pm\frac{1}{1}, \quad \pm\frac{3}{1}, \quad \pm\frac{1}{2}, \quad \pm\frac{3}{2}, \quad \pm\frac{1}{4}, \quad \pm\frac{3}{4}.$$

It is tedious but completely straightforward to plug each of these 12 possibilities into $f(x)$; none of them turns out to be a root of $f(x)$, so we can conclude that $f(x)$ is irreducible.

Example 7.31   The polynomial $g(x) = 4x^3 + 2x^2 - 4x + 3$ is *reducible* in $\mathbb{Q}[x]$. Since the constant term and the leading coefficients of $g(x)$ are also 3 and 4, the list of candidates for a root are the same as those we found for $f(x)$ in the previous example. Plugging them in shows that $g(-\frac{3}{2}) = 0$, therefore $g(x)$ does have a root. Note that this implies (by Corollary 7.26) that $g(x)$ must be reducible *in* $\mathbb{Z}[x]$ as well (even if it does *not* have roots in $\mathbb{Z}$): in this example, $2x - 3$ is a factor of $g(x)$.

Example 7.32   The polynomial $h(x) = x^4 - 3x^2 + 1$ has no rational roots: according to Proposition 7.29, the only candidates are ±1, and these are not roots. It does *not* follow that $h(x)$ is irreducible: Proposition 7.15 only applies to polynomials of degree 2 and 3. As it happens,

$$x^4 - 3x^2 + 1 = (x^2 + x - 1)(x^2 - x - 1),$$

so $h(x)$ is *reducible* in $\mathbb{Z}[x]$ and $\mathbb{Q}[x]$.

Another general remark is perhaps more sophisticated. Let R be an integral domain, suppose $f(x) \in R[x]$, and let I be a prime ideal[4] of R. Via the natural homomorphism $R[x] \to (R/I)[x]$ (that is, taking the coefficients of $f(x)$ modulo I), $f(x)$ determines a polynomial $\underline{f}(x)$ in $(R/I)[x]$. Roughly speaking, if $f(x)$ can be factored in $R[x]$, then $\underline{f}(x)$ can also be factored in $(R/I)[x]$. As a consequence, if $\underline{f}(x)$ is irreducible, then we expect $f(x)$ to be irreducible. This is not quite right (do you see why?), but suitable hypotheses will make it work just fine. Since we are talking about $\mathbb{Z}$, let me take $R = \mathbb{Z}$ and $I = (p)$, where p is an irreducible integer; so I is a nonzero prime ideal. (Try your hand at a generalization: Exercise 7.26.)

> PROPOSITION 7.33   *Let $f(x) \in \mathbb{Z}[x]$ and let p be an irreducible integer. With notation as above, assume that the leading coefficient of $f(x)$ is not a multiple of p and that $\underline{f}(x)$ is irreducible in $(\mathbb{Z}/p\mathbb{Z})[x]$. Then $f(x)$ cannot be written as a product of polynomials of positive degree; hence it is irreducible in $\mathbb{Q}[x]$.*

---

[4] I am insisting on these hypotheses since I have only dealt with irreducible elements for rings that are integral domains.

> *In particular, if $f(x)$ is primitive, then under the stated hypotheses it must be irreducible in $\mathbb{Z}[x]$.*

Proof   Arguing contrapositively, assume that $f(x) = g(x) \cdot h(x)$ with $\deg g(x) = m$, $\deg h(x) = n$, both m and n positive; we will show that $\underline{f(x)}$ is then necessarily reducible in $(\mathbb{Z}/p\mathbb{Z})[x]$.

Since the leading coefficient of $f(x)$ is the product of the leading coefficients of $g(x)$ and $h(x)$, and it is not a multiple of p, it follows that the leading coefficients of $g(x)$ and $h(x)$ also are not multiples of p (as p is prime).

Therefore $\deg \underline{g(x)} = \deg g(x)$ and $\deg \underline{h(x)} = \deg h(x)$ are both positive, and we have $\underline{f(x)} = \underline{g(x)} \cdot \underline{h(x)}$. This shows that $\underline{f(x)}$ is reducible in $(\mathbb{Z}/p\mathbb{Z})[x]$ as needed.

Since $f(x) \in \mathbb{Z}[x] \subseteq \mathbb{Q}[x]$ cannot be written as a product of two positive-degree polynomials, it is irreducible in $\mathbb{Q}[x]$. Finally, irreducibility in $\mathbb{Q}[x]$ is equivalent to irreducibility in $\mathbb{Z}[x]$ for primitive polynomials by Corollary 7.26.   □

Example 7.34   Let $f(x) = 675x^3 - 23129x + 1573$. The leading coefficient and constant term have lots of divisors, so applying the rational root test would be impractical. However, the leading coefficient is odd, and reducing the polynomial mod $p = 2$ gives $x^3 + x + 1 \in (\mathbb{Z}/2\mathbb{Z})[x]$. This is irreducible, because it has no roots in $\mathbb{Z}/2\mathbb{Z}$ (it suffices to note that neither $x = 0$ nor $x = 1$ is a root). By Proposition 7.33, $f(x)$ is irreducible.

One consequence of Proposition 7.33 is that *there are irreducible polynomials in $\mathbb{Q}[x]$ of arbitrarily high degree:* indeed, for every irreducible p, there are irreducible polynomials in $(\mathbb{Z}/p\mathbb{Z})[x]$ of arbitrarily high degree (Exercise 7.11). We will see a possibly more direct proof of this fact below in Corollary 7.38.

Proposition 7.33 is pleasant, but it has limited applicability, for example because there exist polynomials that are reducible mod p for *all* positive primes p, and yet are *irreducible* in $\mathbb{Z}[x]$: an example is $x^4 + 1$. This polynomial is easily seen to be irreducible in $\mathbb{Z}[x]$; verifying that it is reducible modulo every prime $p > 0$ requires more sophisticated considerations.

These *caveats* should reinforce the impression that irreducibility in $\mathbb{Z}[x]$ is a deep issue. I am not aware of any truly efficient 'criterion' that would allow you to decide quickly whether a given polynomial in $\mathbb{Z}[x]$ is or is not irreducible. The last tool we will review is called *Eisenstein's criterion,* but this is a bit of a misnomer, since it also is *not* an 'if-and-only-if' statement. Again, I will give the version for $R = \mathbb{Z}$ and an irreducible number p, and leave you the pleasure of considering more general integral domains (Exercise 7.32).

> THEOREM 7.35 (Eisenstein's criterion)   *Let $f(x) \in \mathbb{Z}[x]$ be a polynomial of degree d:*
>
> $$f(x) = a_0 + a_1 x + \cdots + a_{d-1} x^{d-1} + a_d x^d .$$
>
> *Let p be an irreducible integer, and assume that*

- *p does not divide the leading coefficient a_d;*
- *p divides all the other coefficients a_i, $i = 0, \ldots, d-1$;*
- *p^2 does not divide the constant term a_0.*

Then $f(x)$ cannot be written as a product of two polynomials of positive degree, and hence it is irreducible in $\mathbb{Q}[x]$.

The following lemma will streamline the proof. Say that a polynomial is a 'monomial' if all its coefficients but one are 0, i.e., if it is of the form ax^r with $a \neq 0$.

LEMMA 7.36 *Let R be an integral domain, and let $g(x), h(x) \in R[x]$. If $g(x)h(x)$ is a monomial, then both $g(x)$ and $h(x)$ are monomials.*

Proof Let $g(x) = b_0 + b_1 x + \cdots + b_m x^m$, $h(x) = c_0 + c_1 x + \cdots + c_n x^n$, with $b_m \neq 0$, $c_n \neq 0$. Let $r \leq m$, resp. $s \leq n$, be the smallest exponents in $g(x)$, resp. $h(x)$, with nonzero coefficients. Therefore

$$g(x) = b_r x^r + \cdots + b_m x^m, \quad h(x) = c_s x^s + \cdots + c_n x^n,$$

and hence

$$g(x)h(x) = b_r c_s x^{r+s} + \cdots + b_m c_n x^{m+n}.$$

Working contrapositively, suppose $g(x)$ is *not* a monomial; then $r < m$, and therefore $r + s < m + n$. Since R is an integral domain, both $b_r c_s$ and $b_m c_n$ are nonzero; it follows that $g(x)h(x)$ has at least two nonzero coefficients, therefore it is not a monomial. The same conclusion is reached if $h(x)$ is not a monomial. $\square$

Proof of Eisenstein's criterion We have to prove that if the stated hypotheses are satisfied, then $f(x)$ cannot be written as a product $g(x)h(x)$ with $g(x), h(x) \in \mathbb{Z}[x]$ polynomials with positive degree. Argue by contradiction, and assume that such polynomials exist. Write

$$g(x) = b_0 + b_1 x + \cdots + b_m x^m, \quad h(x) = c_0 + c_1 x + \cdots + c_n x^n,$$

with $m > 0, n > 0, m + n = d$. Note that

$$\underline{f(x)} = \underline{g(x)} \cdot \underline{h(x)},$$

where I am underlining a term to mean its image modulo p; so this equality holds in $(\mathbb{Z}/p\mathbb{Z})[x]$. By the first hypothesis, $\underline{a_d} \neq 0$; by the second,

$$\underline{f(x)} = \underline{a_d} x^d$$

is a monomial: all the other coefficients of $f(x)$ are multiples of p, so their class modulo p is 0. The ring $\mathbb{Z}/p\mathbb{Z}$ is an integral domain, so $\underline{g(x)}$ and $\underline{h(x)}$ are both monomials, by Lemma 7.36: necessarily $\underline{g(x)} = \underline{b_m} x^m$, $\underline{h(x)} = \underline{c_n} x^n$. As $m > 0$ and $n > 0$, it follows that $\underline{b_0} = \underline{c_0} = 0$, that is, p divides both b_0 and c_0. But then p^2 divides $b_0 c_0 = a_0$, contradicting the third hypothesis. This contradiction proves the statement. $\square$

While not a characterization, Eisenstein's criterion is occasionally quite useful.

Example 7.37 Let p be an irreducible integer. By Eisenstein's criterion, the polynomial $x^2 - p$ is irreducible in $\mathbb{Q}[x]$. This gives a different proof of the result of Exercise 1.25: the real number $\sqrt{p}$ is *not* rational. Indeed, if it were, then $x^2 - p$ would have a rational root, so it would not be irreducible.

The same token gives quite a bit more: for any n, the polynomial $x^n - p$ satisfies the hypotheses of Eisenstein's criterion, hence it is irreducible in $\mathbb{Q}[x]$. In particular, $\sqrt[n]{p}$ is not rational. (And in fact the result is much stronger, since for $n > 3$ a polynomial may be reducible even if it does not have roots.)

COROLLARY 7.38 *In $\mathbb{Q}[x]$ there are irreducible polynomials of arbitrarily high degree.*

Proof We have just seen that $x^n - p$ is irreducible in $\mathbb{Q}[x]$ if p is irreducible. Here n can be any positive integer whatsoever. $\qquad\qquad\square$

Compare with the situation over $\mathbb{C}$, where irreducible polynomials have degree 1 (Corollary 7.18), or $\mathbb{R}$, where irreducible polynomials have degree ≤ 2 (Proposition 7.20). No such bound holds for irreducible polynomials in $\mathbb{Q}[x]$.

Sometimes applying Eisenstein's criterion successfully may require a shift in the variable. The following is a classical example, and it will be useful in the distant future, when we will study field theory more in depth. We will apply this result in §14.1.

Example 7.39 Let $p \in \mathbb{Z}$ be irreducible. Then the 'cyclotomic' polynomial $\gamma(x) = 1 + \cdots + x^{p-2} + x^{p-1}$ is irreducible in $\mathbb{Q}[x]$.

It wouldn't seem that Eisenstein's criterion can help here, since all the coefficients of this polynomial equal 1. However, $\gamma(x)$ is irreducible if and only if $\gamma(1+x)$ is irreducible. What is $\gamma(1 + x)$? Note that

$$\gamma(x) = 1 + \cdots + x^{p-2} + x^{p-1} = \frac{x^p - 1}{x - 1};$$

therefore

$$\gamma(1 + x) = \frac{(1 + x)^p - 1}{(1 + x) - 1} = \frac{(1 + x)^p - 1}{x} = p + \binom{p}{2}x + \cdots + \binom{p}{p-1}x^{p-2} + x^{p-1}$$

$$= \sum_{i=1}^{p} \binom{p}{i}x^{i-1},$$

where I used the binomial theorem. Now I claim that $p \mid \binom{p}{i}$ for $i = 1, \ldots, p-1$. Indeed, recall that

$$\binom{p}{i} = \frac{p(p - 1)\cdots(p - i + 1)}{i(i - 1)\cdots 1} = \left[\frac{(p - 1)\cdots(p - i + 1)}{i(i - 1)\cdots 1}\right]p.$$

This is an integer, by definition; and since p is irreducible, no non-unit term in the denominator (that is, none of $2, \ldots, i$) is a divisor of p for $i \le p - 1$. It follows that p divides this integer, as claimed.

By Eisenstein's criterion, $\gamma(1 + x)$ is irreducible, and then so must be $\gamma(x)$.

For instance,

$$1 + x + x^2 + x^3 + x^4 + x^5 + x^6 + x^7 + x^8 + x^9 + x^{10} + x^{11} + x^{12} + x^{13} + x^{14} + x^{15} + x^{16}$$

is irreducible in $\mathbb{Q}[x]$.

Believe it or not, this has to do with the fact that one can construct a regular polygon with 17 sides by just using straightedge and compass. If this sounds a little baffling to you, welcome to the club! We will understand this quite completely by the time we reach §15.6.

Exercises

7.1 ▷ Let R be an integral domain, and let $f(x), g(x) \in R[x]$. Prove that $\deg(fg) = \deg f + \deg g$. Find an example showing that this is not necessarily the case if R is not an integral domain.

7.2 ▷ Let R be an integral domain. Prove that a polynomial $f(x) \in R[x]$ is a unit in $R[x]$ if and only if $f(x) = a$ is constant, with a a unit in R. In particular, if k is a field, then the units in $k[x]$ are the nonzero constant polynomials.

7.3 Find a unit in $(\mathbb{Z}/4\mathbb{Z})[x]$ which is *not* a constant polynomial. Why does this not contradict the result of Exercise 7.2? Also, what does this have to do with Exercise 3.20?

7.4 We have observed in §7.2 that constant polynomials in $R[x]$ determine constant functions $R \to R$. Show that the converse does *not* hold, that is, there may be nonconstant polynomials which are constant if viewed as functions. (*Hint:* Try $R = \mathbb{Z}/2\mathbb{Z}$.)

7.5 Let k be a field, and let I be a nonzero ideal of $k[t]$. Prove that I is generated by a uniquely determined monic polynomial. (Use Exercise 6.11.)

7.6 Let k be a field. Prove that an ideal I of $k[x]$ is maximal if and only if I is a nonzero prime ideal.

7.7 Let k be a field, and let $f(x), g(x) \in k[x]$. Define a notion of 'greatest common divisor' of $f(x)$ and $g(x)$, and prove that it is a linear combination of $f(x)$ and $g(x)$ with coefficients in $k[x]$.

7.8 List all elements of $(\mathbb{Z}/2\mathbb{Z})[x]/(x^3 + x + 1)$. Can you tell whether this ring is a field?

7.9 Prove a converse to Corollary 7.7: If $R[x]$ is a PID, then R is a field.

7.10 Let k be a field. Prove that there are infinitely many irreducible polynomials in $k[x]$. (*Hint:* Adapt the argument used in Exercise 1.26.)

7.11 ▷ Prove that for every irreducible integer p and every integer $d > 0$ there exist irreducible polynomials of degree $> d$ in $(\mathbb{Z}/p\mathbb{Z})[x]$. (*Hint:* Use Exercise 7.10.)

7.12   Prove that algebraically closed fields are necessarily infinite.
(*Hint:* Use Exercise 7.10.)

7.13   ▷ Let k be a field, and let $f(x) \in k[x]$ be a polynomial of degree 1. Prove that $f(x)$ is irreducible. Is this fact true over arbitrary integral domains?

7.14   ▷ (i) Let $f(x)$ be a polynomial of degree d in $k[x]$, where k is a field. Prove that $f(x)$ has at most d roots.

(ii) Consider the polynomial $x^2 + x \in (\mathbb{Z}/6\mathbb{Z})[x]$. How many roots does it have in $\mathbb{Z}/6\mathbb{Z}$? Why doesn't this contradict part (i)?

7.15   Recall that polynomials determine *functions* (see Example 3.11): for a ring R, every polynomial $f(x) = a_0 + a_1 x + \cdots + a_d x^d \in R[x]$ determines a function $f : R \to R$, by setting

$$f(r) = a_0 + a_1 r + \cdots + a_d r^d$$

for all $r \in R$. Prove that if $R = k$ is an *infinite* field, then two polynomials $f(x), g(x) \in k[x]$ determine the same function $k \to k$ if and only if $f(x) = g(x)$ as polynomials. Find an example showing that the hypothesis that the field is infinite cannot be omitted.

Prove that if $f(x) \in k[x]$ and k is an infinite field, then $f(x)$ is constant as a polynomial if and only if it is constant as a function (cf. Exercise 7.4).

7.16   Explain how I constructed the field in Exercise 3.22.

7.17   Construct a field with 8 elements and a field with 9 elements. Is there a field with 10 elements?

7.18   ▷ Let $f(x) \in \mathbb{R}[x]$ be a polynomial with *real* coefficients. Prove that $f(\overline{z}) = \overline{f(z)}$ for all complex numbers z, where ⁻ denotes complex conjugation.

7.19   Let F be a field and let $\pi : \mathbb{C}[x] \to F$ be a surjective ring homomorphism. Prove that $F \cong \mathbb{C}$.

7.20   Let k be a field, and let $f(x)$ be an irreducible polynomial of degree 2 in $k[x]$. Let $F = k[t]/(f(t))$.

(i)   Prove that F is a field, and that there is an injective homomorphism $k \hookrightarrow F$. (We say that F is a 'field extension' of k.) Thus, we may view $f(x)$ as a polynomial in $F[x]$.

(ii)   Prove that $f(x)$ factors as a product of factors of degree 1 in $F[x]$.

7.21   ▷ Prove that if q is an irreducible integer, then q is irreducible as an element of $\mathbb{Z}[x]$.

7.22   Let $f(x) \in \mathbb{Z}[x]$ be a monic polynomial, and let $g(x), h(x) \in \mathbb{Q}[x]$ be monic polynomials such that $f(x) = g(x)h(x)$. Prove that $g(x)$ and $h(x)$ have integer coefficients.

7.23   ▷ Construct an alternative argument proving Theorem 7.22, by using the number of irreducible factors of an integer as a measure of its 'size' (cf. Remark 7.24).

7.24   ▷ Assuming the result mentioned in Remark 7.28 (to the effect that $R[x]$ is a UFD if R is a UFD), prove that $k[x_1, \ldots, x_n]$ is a UFD if k is a field.

7.25   Let $f(x) \in \mathbb{Z}[x]$ be a monic polynomial. Show that if $f(x)$ has a root in $\mathbb{Q}$, then in fact this root is an integer.

7.26   ▷ Formulate and prove a version of Proposition 7.33 for arbitrary integral domains R and prime ideals I.

7.27   Let $f(x) = a_3 x^3 + a_2 x^2 + a_1 x + a_0 \in \mathbb{Z}[x]$ be a polynomial of degree 3. Assume that a_0, $a_1 + a_2$, and a_3 are all odd. Prove that $f(x)$ is irreducible in $\mathbb{Q}[x]$.

7.28   Prove that $x^5 - 3x^4 - 6x^3 - 12x^2 + 9x + k$ is irreducible in $\mathbb{Q}[x]$ for infinitely many values of k.

7.29   Prove that $7x^3 + 6x^2 + 4x + 6$ is irreducible in $\mathbb{Q}[x]$, in (at least) three different ways.

7.30   Prove that $x^5 + x^4 + x - 1$ is irreducible in $\mathbb{Q}[x]$, by using Eisenstein's criterion. (Find a suitable 'shift' of x, as in Example 7.39.)

7.31   Let $f(x) \in \mathbb{Z}[x]$, and let p be a nonzero prime integer. Assume that p does not divide the constant term of $f(x)$, but it divides all the other coefficients; and further that p^2 does not divide the leading coefficient of $f(x)$. Prove that $f(x)$ is irreducible in $\mathbb{Q}[x]$.

7.32   ▷ Formulate and prove a version of Eisenstein's criterion (Theorem 7.35) for arbitrary integral domains and prime ideals.

7.33   ▷ *(Preview of coming attractions.)* Discuss the irreducibility of $x^2 + 1$ over the fields $\mathbb{Z}/p\mathbb{Z}$ (where p is prime). Find all the odd primes $p < 30$ for which this polynomial is *reducible*. See if you can conjecture a general statement based on this data, and try to prove that statement. (I don't believe the material we have covered so far suffices in order to obtain a reasonable proof of this statement, but I may be wrong. In Part II we will acquire tools that will make this problem very easy to solve—indeed, I will use its solution as a key lemma in a beautiful application down the road, in §10.4.)

Part II

Modules

8 Modules and Abelian Groups

8.1 Vector Spaces and Ideals, Revisited

Have you noticed that there is a remarkable similarity between *ideals* and *vector spaces?*

You likely have a level of familiarity with vector spaces, from previous encounters with linear algebra. A vector space over a field k is a set V endowed with an operation of addition, $+$, and a 'multiplication by scalars', denoted $\cdot$ or elegantly omitted altogether. Multiplication by scalars allows you to multiply a 'vector' $v \in V$ by a 'scalar' $c \in k$. If you visualize vectors as arrows, then, e.g., multiplying a vector v (in a vector space over $\mathbb{R}$) by 2 amounts to considering the vector $2v$ with the same direction as v, but twice its length.

Of course there are lots of vector spaces for which the 'arrow' interpretation is a rather poor visualization tool: for example, the polynomial ring $k[x]$ over a field k is a k-vector space (right? I can add polynomials, and multiply a polynomial by an element of k, and all properties I expect to hold in a vector space hold just fine), but I do not think of a polynomial as an 'arrow'.

We could be more precise and just spell out the axioms that define a vector space. You would agree that the usual axioms (i)–(iv) for addition from the list in Definition 3.1 hold in every vector space, and there are a few simple axioms satisfied by multiplication by scalars:

- $\forall v \in V$, we have $1_k v = v$;
- $\forall a, b \in k, \forall v \in V, (ab)v = a(bv)$;
- $\forall a, b \in k, \forall v \in V, (a + b)v = av + bv$;
- $\forall a \in k, \forall v, w \in V, a(v + w) = av + aw$.

These axioms look like axioms from the definition of a ring: the first one looks like the definition of a multiplicative identity, the second one looks like associativity, and the others smack of distributivity. If you are paying attention, you know that this is not quite right: multiplication by scalars is *not* an operation 'in' the vector space V, rather an operation which combines elements of V with elements of something else, the field k in this case.

So something else is going on, and there is a name for it: the axioms say that multiplication by scalars defines an *action* of k on V. In a not-too-distant future we will deal extensively with 'actions' in different contexts (see, e.g., §11.2), and they will become part of our active vocabulary. Right now it is not too important that we think deeply about what an 'action' really is; but work out Exercise 8.1 if you want to appreciate where the axioms listed above come from. We will come back to this more formally in §10.1, particularly Proposition 10.6 and Remark 10.7.

Right now I am more interested in the observation that if R is a ring and I is an ideal of R, then I *satisfies the same requirements* with respect to addition and 'multiplication by scalars', where now the scalars are taken from R and the action is given by multiplication in R. The (left) absorption property in Definition 5.8,

$$(\forall r \in R)(\forall a \in I): \quad ar \in I,$$

tells us that this 'multiplication by scalars' acts as expected, $R \times I \to I$; and the four axioms listed above hold just because ordinary multiplicative associativity, etc., hold in a ring.

Summarizing, we can think of an *ideal* of a ring R as some kind of 'vector space over R'. There is a name for that, too.

DEFINITION 8.1   Let R be a ring. A 'module over R' (or 'R-module') $(M, +, \cdot)$ consists of a set M along with two operations $+: M \times M \to M$ and $\cdot: R \times M \to M$ (called 'addition' and 'multiplication by scalars'), satisfying the following properties:

(i)	$\forall a, b, c \in M$	$(a + b) + c = a + (b + c)$
(ii)	$\exists 0_M \in M \; \forall a \in M$	$a + 0_M = 0_M + a = a$
(iii)	$\forall a \in M \; \exists z \in M$	$a + z = z + a = 0_M$
(iv)	$\forall a, b \in M$	$a + b = b + a$
(v)	$\forall a \in M$	$1_R \cdot a = a$
(vi)	$\forall r, s \in R \; \forall a \in M$	$(rs) \cdot a = r \cdot (s \cdot a)$
(vii)	$\forall r, s \in R \; \forall a \in M$	$(r + s) \cdot a = r \cdot a + s \cdot a$
(viii)	$\forall r \in R \; \forall a, b \in M$	$r \cdot (a + b) = r \cdot a + r \cdot b$

We usually refer to an R-module $(M, +, \cdot)$ by the name of the set M. The element 0_M is usually simply denoted 0. The $\cdot$ in the multiplication by scalars is usually omitted.

To be precise, Definition 8.1 tells us what a *left*-module is; I will leave you the pleasure of defining what a *right*-module should be. From this point of view, a *left*-ideal is already a module, since the multiplication by scalars in Definition 8.1 corresponds to absorption on the left, and absorption on the right plays no role in the axioms listed above.

Example 8.2   A singleton $\{*\}$ is an R-module for every ring: just define $r \cdot * = *$ for all $r \in R$. We call this 'the trivial R-module', and usually denote it by $\{0\}$ or even simply 0, since the single element in the set is in particular the 0 of this module.

Example 8.3 A k-vector space is nothing but an R-module in the particular case in which the ring $R = k$ happens to be a field.

Example 8.4 If R is a ring, then we may view R as an R-module: multiplication by scalars is simply multiplication in the ring.

Example 8.5 More generally, a (left-)ideal in a ring R is a subset $I \subseteq R$ such that $(I, +, \cdot)$ is an R-module, with the same operations $+, \cdot$ as in R.

Example 8.6 For a fancier-looking example, consider a ring R and a ring homomorphism $f : R \to S$. We can define an action of R on S by setting $\forall r \in R, \forall s \in S$,

$$r \cdot s = f(r)s.$$

(The notation here looks particularly confusing: the 'product' on the left is the multiplication by scalars that we are defining; the product on the right is the *bonafide* product in the ring S.) You can verify that this definition makes S into an R-module (Exercise 8.2).

Example 8.7 In particular, if I is an ideal of R, then R/I also is an R-module: indeed, we have a ring homomorphism $R \to R/I$, so we can view R/I as an R-module as in Example 8.6. Note that every $r + I$ may be written as $r \cdot (1 + I)$: that is, once we have $1 + I$ and multiplication by scalars, then we have all elements of R/I. We say that $1 + I$ 'generates' R/I. This makes R/I a *cyclic* R-module.

Example 8.8 Also keep in mind that R itself is part of the information of what an R-module is. Often 'R' may be omitted if it is understood in context, but the same set may be a module over different rings. For example, $\mathbb{Z}$ is of course a module over itself; but we can consider the ring homomorphism $\mathbb{Z}[x] \to \mathbb{Z}$ sending x to (for example) 0, and this makes $\mathbb{Z}$ into a $\mathbb{Z}[x]$-module as we have seen in Example 8.6. These two realizations of $\mathbb{Z}$ are *different* modules, in the sense that they are modules over different rings. (If we just write '$\mathbb{Z}$', we are usually referring to the first realization.) Similarly: $\mathbb{C}$ is a $\mathbb{C}$-vector space of dimension 1; but it is also an $\mathbb{R}$-vector space of dimension 2; and a $\mathbb{Q}$-vector space of *uncountable* dimension. Again, the 'ground' ring makes a difference.

There are a few simple consequences of the axioms listed in Definition 8.1, which should remind you of analogous facts we proved in the context of rings. For example, additive cancellation holds (the proof of Proposition 3.17 only uses axioms (i)–(iii) from Definition 3.1, and those same axioms hold for modules, so the same proof works for modules); the element 0_M is uniquely determined; and for a given $a \in M$, the 'additive inverse' z satisfying (iii) is also uniquely determined, and is denoted $-a$. The proofs of such facts are identical to the ones given in §3.2, so they will not be reproduced here. The point is that those proofs only used ring axioms (i)–(iv), and those axioms hold for modules as well. Similarly, the following observation closely mirrors Corollary 3.18.

PROPOSITION 8.9   *Let M be an R-module. Then $\forall m \in M$, $0_R \cdot m = 0_M$.*

Proof   We have $0_R \cdot m = (0_R + 0_R) \cdot m$, therefore $0_R \cdot m = 0_R \cdot m + 0_R \cdot m$, and $0_R \cdot m = 0_M$ follows by cancellation.   □

In particular we have $1 \cdot a + (-1) \cdot a = (1 - 1) \cdot a = 0 \cdot a = 0$, and therefore $(-1) \cdot a = -a$, as you would expect.

There is one important special case of Definition 8.1, which deserves its own name and on which we will focus later on in this text (Chapter 10 will be devoted to it): the case in which the ground ring R is $\mathbb{Z}$. $1_{\mathbb{Z}} \cdot a = a$ as prescribed by axiom (v), there really is no choice for what $n \cdot a$ should be, for any $n \in \mathbb{Z}$: if $n > 0$,

$$n \cdot a = \underbrace{(1 + \cdots + 1)}_{n \text{ times}} \cdot a = \underbrace{1 \cdot a + \cdots + 1 \cdot a}_{n \text{ times}} = \underbrace{a + \cdots + a}_{n \text{ times}}$$

by axioms (vii) and (v), and $(-n) \cdot a = (-a) + \cdots + (-a)$ by essentially the same token. You will recognize that this is nothing but the 'multiple' notation, also introduced in §3.2. We got thoroughly used to this when working with rings, and anything we proved about 'multiples' back then holds in this new context. In particular, we have the following convenient observation.

PROPOSITION 8.10   *If $(M, +)$ satisfies axioms (i)–(iv) in Definition 8.1, the (unique) $\mathbb{Z}$-action defined above satisfies axioms (v)–(viii). That is, M is then a $\mathbb{Z}$-module in a natural way.*

This is a straightforward consequence of the definition of 'multiple', so it does not seem necessary to give it a formal proof now. (Besides, you already tried your hand at that in Exercise 3.8.) In due time (Remark 10.7) we will understand the situation a little better, and we will have a more sophisticated point of view on Proposition 8.10. For now, just note that axiom (v) *is* the definition of $1_{\mathbb{Z}} \cdot a$; axioms (vi), (vii), (viii) just amount to the fact that we can move parentheses (because + is associative) and switch the order of summands (because + is commutative). For example, what (viii) says for $r = 3$ is that

$$(a + b) + (a + b) + (a + b) = (a + a + a) + (b + b + b),$$

which indeed holds because + is associative (axiom (i)) and commutative (axiom (iv)). Make up similar examples illustrating (vi) and (vii).

Simple as Proposition 8.10 is, it offers a simplified view on R-modules in the special case $R = \mathbb{Z}$: axioms (i)–(iv) are enough to define a $\mathbb{Z}$-module, since in this case the other axioms are automatically verified. A more common name for $\mathbb{Z}$-module is *abelian group*.

DEFINITION 8.11   An 'abelian group' $(A, +)$ consists of a set A along with a binary operation $+$ on A such that

(i)	$\forall a, b, c \in A$	$(a + b) + c = a + (b + c)$
(ii)	$\exists 0_A \in A \; \forall a \in A$	$a + 0_A = 0_A + a = a$
(iii)	$\forall a \in A \; \exists z \in A$	$a + z = z + a = 0_A$
(iv)	$\forall a, b \in A$	$a + b = b + a$

We usually refer to an abelian group $(A, +)$ by the name A of the set, understanding the operation $+$. Similarly, 0_A is usually simply denoted by 0. The adjective 'abelian' is derived from the name of Niels Abel, a remarkable Norwegian mathematician from the nineteenth century. Here,[1] 'abelian' is just another name for 'commutative'.

By Proposition 8.10, an abelian group is nothing but a $\mathbb{Z}$-module: axioms (v)–(viii) from Definition 8.1 are automatically satisfied if $R = \mathbb{Z}$, with the action defined by 'multiples' as discussed above. So we can view abelian groups as a particular case of R-modules: the case for which $R = \mathbb{Z}$. A little paradoxically, abelian groups are also 'more general' than modules, in the sense that every module is in particular an abelian group: you can view an R-module as an abelian group endowed with an action of a ring R.

Abelian groups are common in nature, and you are already familiar with a very large collection of examples of abelian groups.

Example 8.12   If $(R, +, \cdot)$ is a ring, then $(R, +)$ (same set, same addition) is an abelian group. Indeed, axioms (i)–(iv) in the definition of ring, Definition 3.1, coincide with the axioms listed in Definition 8.11. (Do you see what this has to do with Example 8.6?)

Thus, *every* ring you have encountered gives you an example of an abelian group: just forget the multiplication, and what is left is an abelian group.

Example 8.13   If M is any R-module (with any ring R), we may also view M as an abelian group: this simply amounts to forgetting R and the R-action altogether. For example, every vector space is in particular an abelian group. By the same token, ideals (in every ring) are in particular abelian groups.

In fact, it is not uncommon to cover this material in the 'opposite direction'. We could have started off by defining abelian groups, and then we could define rings as an abelian group (that is, a set with *one* operation $+$ satisfying the axioms in Definition 8.11) endowed with a *second* operation $\cdot$, satisfying the other axioms listed in Definition 3.1. We could also view R-modules as a variation on the definition of abelian groups, in which the natural $\mathbb{Z}$-action is replaced by the action of a more general ring as prescribed in Definition 8.1. So which should come first? Abelian groups or rings? or modules? This is a classic chicken vs. egg dilemma.

---

[1] There are also 'abelian categories', 'abelian integrals', 'abelian varieties', and more uses of the adjective 'abelian' where it does *not* mean 'commutative'.

Summarizing: There are interesting algebraic structures, called *R-modules,* over arbitrary rings *R*; if *k* is a field, *k*-modules are called *k-vector spaces;* $\mathbb{Z}$-modules are also called *abelian groups.* As it happens, the basic theory of *R*-modules is just as easy or as hard as the theory of abelian groups; we will study basic definitions and facts for *R*-modules, and you should keep in mind that these generalities will hold in particular for abelian groups. We will focus on the particular case of modules over integral domains, and especially Euclidean domains; again, you should keep in mind that everything we will say then will apply to abelian groups as well, since $\mathbb{Z}$ is a Euclidean domain. And then we will spend some time looking at abelian groups proper, before taking the next step in generalization and moving on to *groups.*

On the other hand—as we know, most rings, even most integral domains, are *not* Euclidean domains; as we have discussed at length in Part I, $\mathbb{Z}$ is a *very* special ring. Correspondingly, the case of abelian groups is also very special. To name one paramount example, every finitely generated $\mathbb{Z}$-module is a product of cyclic modules (as will will prove), and this is simply not the case for modules over an arbitrary ring. So our intuition about modules should not be based too firmly on the case of abelian groups.

8.2   The Category of *R*-Modules

Given our experience with rings, you should expect what comes next: we have now defined the *objects* in a 'category of *R*-modules', and the next natural task is to define the *morphisms* in this category,[2] that is, a sensible notion of 'function between *R*-modules'. Thinking back to rings, we defined 'ring homomorphisms' to be functions between rings which would preserve the information defining a ring, that is, the operations and the identity element. We will do just the same in this new case.

DEFINITION 8.14   Let *R* be a ring, and let *M, N* be *R*-modules. A function $f: M \to N$ is a 'homomorphism of *R*-modules' (or '*R*-module homomorphism') if $\forall r \in R, \forall a, b \in M$,

$$f(a + b) = f(a) + f(b),$$
$$f(r \cdot a) = r \cdot f(a).$$

We also say that *f* is an '*R*-linear map'.

This notion should look familiar: this is how you define 'linear maps' between vector spaces in linear algebra.

Example 8.15   If $f: R \to S$ is a ring homomorphism, then we have seen that we can view *S* as an *R*-module (Example 8.6); and of course *R* is itself an *R*-module. Then *f* is also a homomorphism *of R-modules.* Indeed, denoting multiplication by scalars by · and ordinary multiplication in *R* and *S* by juxtaposition, we have $\forall r, a, b \in R$:

$$f(a + b) = f(a) + f(b) \quad \text{and} \quad f(r \cdot a) = f(ra) = f(r)f(a) = r \cdot f(a).$$

---

[2] If you are uncertain about these terms, look through the brief 'Before we Begin' section preceding the main body of this text.

Note however that the requirement to be a *ring* homomorphism is much stronger than the requirement to be a *module* homomorphism. For example, there is only one ring homomorphism $\mathbb{Z} \to \mathbb{Z}$ (and in fact from $\mathbb{Z}$ to any ring: Exercise 4.14), while there are infinitely many different $\mathbb{Z}$-module homomorphisms, that is, 'homomorphisms of abelian groups'. To see this, let $n \in \mathbb{Z}$ be *any* integer, and define

$$f_n : \mathbb{Z} \to \mathbb{Z}, \quad f_n(a) = na\,.$$

Then only f_1 (that is, the identity) is a *ring* homomorphism, while *every* f_n is a $\mathbb{Z}$-module homomorphism, because $\forall a, b, r \in \mathbb{Z}$,

$$f_n(a + b) = n(a + b) = na + nb = f_n(a) + f_n(b)\,,$$
$$f_n(r \cdot a) = f_n(ra) = n(ra) = r(na) = r \cdot f_n(a)\,.$$

Example 8.16 More generally, if R is a *commutative* ring and M is an R-module, then 'multiplication by $r \in R$'

$$M \xrightarrow{\ r\cdot\ } M$$

defined by $m \mapsto r \cdot m$, is an R-module homomorphism. (Exercise 8.6.) In fact, this is the case even if R is not necessarily commutative, so long as r commutes with every element of R (i.e., r is in the *center* of R).

Example 8.17 In particular, 'multiplication by 0' is an R-module homomorphism. This is just the function $M \to M$ sending everything to 0_M. In fact, if M and N are any two R-modules, we can define an R-module homomorphism $M \to N$ by sending every element of M to 0_N. This is called the *trivial* homomorphism.

This is another feature of the category of modules which distinguishes it from the category of rings. If I give you two random rings R and S, in general there are *no* ring homomorphisms $R \to S$; while if M and N are any two R-modules, there always is at least one R-module homomorphism $M \to N$, that is, the trivial morphism. (You may recall that sending everything to 0 is *not* an option for a ring homomorphism, if the target is not a trivial ring; see Example 4.17.)

Example 8.18 In order to define an R-module homomorphism $f : R \to M$, it suffices to specify $f(1)$. Indeed, if $f(1) = m \in M$, then necessarily $\forall r \in R$, $f(r) = f(r1) = r\,f(1) = rm$, so this choice (and the fact that f is a homomorphism) determines f.

Conversely, for any fixed $m \in M$, the function $f(r) = rm$ is an R-module homomorphism. Indeed, $f(m + n) = r(m + n) = rm + rn$ (by axiom (viii)), and $\forall a \in R$, $f(ar) = (ar)m = a(rm) = a\,f(r)$ (by axiom (vi)).

Note that we have not defined a notion of homomorphism between an R-module and an S-module for different rings R and S. *Each* ring R gives rise to a new category, the category of R-modules. There are ways to have different categories interact, but this is not our focus here. Even if I talk about the 'category of modules', I will have implicitly chosen a ring R and I will be confined within the category of modules over that ring R.

When we propose a definition such as Definition 8.14 for the morphisms in a category, we have to verify a couple of properties. Here they are.

PROPOSITION 8.19   *Let R be a ring, and let L, M, N, P be R-modules.*

- *The identity function $\mathrm{id}_M \colon M \to M$ is an R-module homomorphism.*
- *Let $f \colon L \to M$, $g \colon M \to N$ be R-module homomorphisms. Then the composition $g \circ f \colon L \to N$ is an R-module homomorphism.*

Composition is associative: for all R-module homomorphisms $f \colon L \to M$, $g \colon M \to N$, $h \colon N \to P$, we have

$$h \circ (g \circ f) = (h \circ g) \circ f.$$

The proof is left to you—in case of difficulties, look back at Proposition 4.27, where we took care of the analogous statements for rings.

It is easy to verify that if $f \colon M \to N$ is an R-module homomorphism, then $f(0_M) = 0_N$. Again, the proof in the context of ring homomorphisms goes through without any change in the new context of modules.

The following definition should not be unexpected. (See Remark 4.34 for a discussion of isomorphisms in categories.)

DEFINITION 8.20   Let $f \colon M \to N$ be a homomorphism of R-modules. We say that f is an 'isomorphism' if it is a bijection. Equivalently, f is an isomorphism if and only if there exists an R-module homomorphism $g \colon N \to M$ such that $g \circ f = \mathrm{id}_M$ and $f \circ g = \mathrm{id}_N$.

There is a small proposition hidden in this statement, to the effect that the two stated conditions are equivalent: we should verify that if f is a bijective homomorphism, and g is its set-theoretic inverse, then g is also an R-module homomorphism. For rings, this was done in Proposition 4.30; the proof in the context of modules would be entirely similar. (Make sure you agree with me!)

Given that we have a notion of isomorphism of R-modules, we can introduce a corresponding relation, matching Definition 4.32.

DEFINITION 8.21   We say that two R-modules M, N are 'isomorphic', and write $M \cong N$, if there exists an isomorphism of R-modules $f \colon M \to N$.

Just as in the context of rings, this relation turns out to be an *equivalence* relation (cf. Corollary 4.33). Isomorphic R-modules are indistinguishable in terms of their R-module-theoretic properties.

Here is another feature in which module theory and ring theory differ. We have trained ourselves to think of *ideals* in a ring as rather different entities than rings. For example, $2\mathbb{Z}$ is a rng, not a ring (cf. Example 3.5); in ring theory, it is very different from $\mathbb{Z}$ itself.

However, the function $\mathbb{Z} \rightarrow 2\mathbb{Z}$ defined by multiplication by 2, $a \mapsto 2a$, is surjective by definition of $2\mathbb{Z}$, is evidently injective, and is a homomorphism of $\mathbb{Z}$-modules (Example 8.16). Therefore $2\mathbb{Z}$ (or, for that matter, $n\mathbb{Z}$ for every $n \neq 0$) is *isomorphic to $\mathbb{Z}$ as a $\mathbb{Z}$-module,* i.e., as an abelian group. You should verify that this observation extends to all PIDs (Exercise 8.10).

More differences: $\mathbb{Z}$ is isomorphic to $\mathbb{Z}$ as a ring *in only one way*—there is only one ring isomorphism $\mathbb{Z} \rightarrow \mathbb{Z}$, that is, the identity. (Right? Why?) On the other hand, $\mathbb{Z}$ is isomorphic to $\mathbb{Z}$ as a module *in two ways:* both the identity function $n \rightarrow n$ and the function $n \rightarrow -n$ are $\mathbb{Z}$-module isomorphisms. Modules and rings are very different environments.

8.3   Submodules, Direct Sums

Other examples of R-module homomorphisms arise in certain template situations.

DEFINITION 8.22   Let R be a ring, M an R-module, and N a subset of M. Assume that N is also an R-module. We say that N is a 'submodule' of M if the inclusion map $i \colon N \hookrightarrow M$ is an R-module homomorphism.

In other words, N is a submodule of M if the addition in N and the action of R on N are nothing but the restriction of the corresponding notions in M.

Verifying that a given subset of a module is a submodule can be done efficiently by using the following observation (which may remind you of Proposition 4.14).

PROPOSITION 8.23   *Let M be an R-module. Then a subset $M' \subseteq M$ is a submodule if and only if it is nonempty and closed with respect to addition and multiplication by scalars.*

Proof   The 'only if' implication is immediate. To verify the 'if' implication, we should check that the addition and multiplication by scalars in M' satisfy axioms (i)–(viii). Since (i) and (iv)–(viii) only depend on universal quantifiers and hold in M, they are automatically satisfied in M'. (Look back at the proof of Proposition 4.14 if this sounds mysterious.) Next, since $M' \neq \emptyset$, there is some $m \in M'$. Then $0 \in M'$ since $0 = 0 \cdot m$ and M' is closed with respect to multiplication by scalars. Further, for all $m \in M'$ we have $-m = (-1) \cdot m$, so $-m \in M'$, again since M' is closed under multiplication by scalars. This checks axioms (ii) and (iii), so we are done.   □

Using Proposition 8.23, it is straightforward to verify that the following give examples of submodules.

Example 8.24   The inclusion of an ideal I in a ring R makes I a sub-R-module of R. We could adopt this as a 'new' definition of *left*-ideals: they are the subsets of a ring which happen to be submodules. If R is commutative, then the distinction between left-

and right-ideals evaporates, so that the ideals of R are precisely its submodules in this case.

Example 8.25   If $m_1, \ldots, m_n$ are elements of an R-module M, then the set of linear combinations $r_1 m_1 + \cdots + r_n m_n$ with coefficients $r_i \in R$ determines a submodule of M. This is the smallest submodule of M containing $m_1, \ldots, m_n$.

Example 8.26   If N_1 and N_2 are submodules of an R-module M, then $N_1 + N_2$ denotes the smallest submodule of M containing both N_1 and N_2. This consists of all $m \in M$ such that there exist $n_1 \in N_1$, $n_2 \in N_2$ such that $m = n_1 + n_2$.

Example 8.27   In the same situation, $N_1 \cap N_2$ is also a submodule of M. More generally, if N_i, $i \in I$, is any family of submodules of M, then $\cap_{i \in I} N_i$ is a submodule of M.

The submodule in Example 8.25 deserves its own notation, and the particular case in which $n = 1$ is especially important.

DEFINITION 8.28   Let M be an R-module, and let $m_1, \ldots, m_n$. The submodule consisting of the linear combinations of $m_1, \ldots, m_n$ is denoted $\langle m_1, \ldots, m_n \rangle$, and it is called the submodule of M 'generated by' $m_1, \ldots, m_n$.

DEFINITION 8.29   An R-module M is 'finitely generated' if there exist $m_1, \ldots, m_n$ such that $M = \langle m_1, \ldots, m_n \rangle$. It is 'cyclic' if $M = \langle m \rangle$ for some $m \in M$.

Example 8.30   If I is any ideal of R, then R/I is a cyclic R-module: as observed in Example 8.7, R/I is generated by the coset $1 + I$. We will verify later (Example 8.45) that *every* cyclic module is of this type.

Example 8.31   If R is a commutative ring, then its ideals are the submodules of R viewed as R-modules (as seen in Example 8.24). Then an ideal of R is cyclic as an R-module if and only if it is principal as an ideal (as in Definition 5.17).

Homomorphisms preserve submodules, in both directions.

PROPOSITION 8.32   *Let $f : M \to N$ be a homomorphism of R-modules. Then:*

(i) If M' is a submodule of M, then $f(M')$ is a submodule of N.
(ii) If N' is a submodule of N, then $f^{-1}(N')$ is a submodule of M.

Proof   Both statements are straightforward.
(i) By Proposition 8.23, we just have to check that $f(M')$ is closed with respect to

addition and multiplication by scalars. If $n_1, n_2 \in f(M')$, then there exist $m_1, m_2 \in M'$ such that $m_1 = f(n_1), m_2 = f(n_2)$. Then $m_1 + m_2 \in M'$, therefore

$$n_1 + n_2 = f(m_1) + f(m_2) = f(m_1 + m_2)$$

is also an element of $f(M')$; this shows that $f(M')$ is closed with respect to $+$. Similarly, if $r \in R$ and $n \in f(M')$, then there exists $m \in M'$ such that $n = f(m)$; and then

$$r \cdot n = r \cdot f(m) = f(r \cdot m),$$

and this shows that $r \cdot n \in f(M')$ since $r \cdot m \in M'$. Therefore $f(M')$ is closed with respect to multiplication by scalars, and we are done.

(ii) Exercise 8.12. □

For example, the image $\operatorname{im} f = f(M)$ of an R-module homomorphism $f \colon M \to N$ is a submodule of N.

Inclusions of submodules are our prototype of *injective* homomorphisms of modules. What about *surjective* homomorphisms? The following construction gives us many examples.

DEFINITION 8.33 Let R be a ring, and let M, N be R-modules. The *direct sum* of M and N, denoted $M \oplus N$, is the R-module $(M \times N, +, \cdot)$, where $M \times N$ is the ordinary Cartesian product of sets and the operations $+, \cdot$ are defined *componentwise:*

$$(m_1, n_1) + (m_2, n_2) := (m_1 + m_2, n_1 + n_2),$$

$$r \cdot (m, n) := (rm, rn).$$

You may wonder why we do not use the notation $M \times N$ for this module. The reason is 'categorical': the direct sum of R-modules satisfies a broader universal property than the notation $M \times N$ would suggest. It would take us too far afield to discuss this in detail. You will get a sense of what is going on if you work out Exercise 8.16.

Definition 8.33 extends in the evident way to the direct sum of finitely many modules: $M_1 \oplus M_2 \oplus \cdots \oplus M_n$.

Example 8.34 You will recall from linear algebra that if k is a field (you may have studied more specifically the case $k = \mathbb{R}$, but the same holds more generally), then a 'vector space of dimension n' consists of the set of n-tuples of elements of k, with componentwise addition and multiplication by scalars. We have just learned a good notation for this: it is nothing but the direct sum $\underbrace{k \oplus \cdots \oplus k}_{n \text{ times}}$, also denoted $k^{\oplus n}$ or k^n.

Thus the ordinary real plane is $\mathbb{R}^2$, three-dimensional space is $\mathbb{R}^3$, and so on.

More generally, if R is a ring, then we can consider the module $R^{\oplus n}$. This is called the 'free' R-module of 'rank' n (cf. Definition 9.1).

It can be proved that if k is a field, then *every k-module is free;* in fact, we will recover this result in short order, in §9.4. The same statement does *not* hold for other rings, even for friendly rings like $\mathbb{Z}$: for example, $\mathbb{Z}/2\mathbb{Z}$ is not free as a $\mathbb{Z}$-module. We will come back to these definitions, and have much more to say about free modules in Chapter 9.

Example 8.35   If we have ring homomorphisms $f: R \to S$ and $g: R \to T$, then we can view S and T as R-modules (Example 8.6). The two homomorphisms determine a ring homomorphism $h: R \to S \times T$ making the diagram

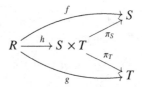

commute; concretely, $h(r) = (f(r), g(r))$. (Take this as a hint for Exercise 8.17.) This allows us to see $S \times T$ as an R-module, and you can verify that it then agrees with the definition of $S \oplus T$.

As in the case of products of rings, the direct sum comes endowed with canonical surjective R-module homomorphisms ('projections')

$$M \oplus N \xrightarrow{\pi_M} M \qquad M \oplus N \xrightarrow{\pi_N} N$$

given by $\pi_M((m, n)) := m$, $\pi_N((m, n)) := n$.

Well, here is another difference with the situation with rings. (The notation $\oplus$ has to do with this.) Not only do we have *projections* from $M \oplus N$ to M and N; we also have *injections* from M and N to $M \oplus N$. Define

$$M \xrightarrow{\iota_M} M \oplus N \qquad N \xrightarrow{\iota_N} M \oplus N$$

by $\iota_M(m) = (m, 0)$, $\iota_N(n) = (0, n)$. It is immediately verified that these are injective homomorphisms of R-modules. For rings, the analogous functions do *not* define ring homomorphisms, for example because they do not take 1 to 1. Thus, we may view, e.g., M as a submodule of $M \oplus N$ (since ι_M is an injective R-module homomorphism), while this identification does not work at the level of rings (cf. Example 4.10).

Taken along with these maps ι_M, ι_N, $M \oplus N$ is said to be a 'coproduct' in the category of R-modules. What this means is spelled out in Exercise 8.16. For now, it is not important to pay much attention to this—just file somewhere the notion that you have more flexibility in dealing with R-modules than you had in dealing with rings. The next considerations (dealing with quotients and such things) should only reinforce this conclusion. The category of R-modules is considerably friendlier than the category of rings; it satisfies a number of nice properties. My guess is that these were first noticed for the special case $R = \mathbb{Z}$, i.e., for *abelian groups*. And indeed, these properties are summarized by saying that the category of R-modules is an *abelian category*.

8.4   Canonical Decomposition and Quotients

In dealing with rings, one of our main general observations was that *every* ring homomorphism can be written as the composition of injective and surjective homomorphisms. Our next task is the verification that the same pleasant state of affairs holds for modules.

In fact, as pointed out in §5.1, the primary result here is the fact that this holds already for sets. So let M, N be R-modules, and let $f: M \to N$ be a homomorphism of R-modules. In particular, f is a function *of sets;* therefore we know already that we can decompose it as in the following diagram.

$$M \xrightarrow{\pi} M/\!\sim \xrightarrow[\tilde{f}]{\sim} f(M) \xrightarrow{\iota} N.$$

with f spanning from M to N.

Here $\sim$ is the equivalence relation that will be recalled below, π is a surjective function, ι is an injective function, and $\tilde{f}$, defined by

$$\tilde{f}([m]_\sim) = f(m),$$

is a bijective function. The challenge is to upgrade this whole diagram to R-modules, just as we upgraded it to rings in Chapter 5.

Now $f(M)$ stands for the image of f, and at this point we already know (by Proposition 8.32) that it is a submodule of N. Therefore: $f(M)$ is an R-module, and the inclusion $\iota: f(M) \hookrightarrow N$ is an R-module homomorphism. As in the case of rings, the heart of the upgrade is the matter of defining $M/\!\sim$ as an R-module. From the set-theory story, we know that the equivalence relation $\sim$ is defined by

$$m_1 \sim m_2 \iff f(m_1) = f(m_2),$$

and our task is to express this condition in some way having to do with R-modules.

You should be able to anticipate what is going to happen, because it follows closely what we did for rings. First, we observe that $f(m_1) = f(m_2) \iff f(m_1 - m_2) = 0$. This motivates the following definition, which will look very familiar.

DEFINITION 8.36 Let $f: M \to N$ be a homomorphism of R-module. The 'kernel' of f, denoted $\ker f$, is the submodule of M

$$\ker f = f^{-1}(0_N) = \{m \in M \mid f(m) = 0\}.$$

Thus, $r \in \ker f$ if and only if $f(r) = 0$.

Note that I am pointing out within the definition that $\ker f$ is a *submodule* of M. Indeed, I am defining it as the inverse image of 0_N, and $\{0_N\}$ is a submodule of N; therefore $f^{-1}(0_N)$ is a submodule of M by Proposition 8.32. This is a point of departure with ring theory: the kernel of a ring homomorphism is *not* a ring (if the target is nontrivial), while the kernel of a module homomorphism *is* a module. This is another 'nice' feature of R-modules compared with rings.

With Definition 5.3 in mind, we observe that

$$m_1 \sim m_2 \iff m_2 - m_1 \in \ker f.$$

Let's then consider *any* submodule K of an R-module M, and the equivalence relation $\sim_K$ on M defined by

$$m_1 \sim_K m_2 \iff m_2 - m_1 \in K.$$

(Make sure that you can verify that $\sim_K$ is indeed an equivalence relation!) In complete analogy with what we did for rings, we can make the following definition.

DEFINITION 8.37 Let R be a ring, M an R-module, K a submodule of M, and $\sim_K$ the relation introduced above. The equivalence class of $m \in M$ with respect to $\sim_K$ is the 'coset' of m (modulo K), denoted $m + K$. We will denote by M/K the set of cosets; M/K is the 'quotient of M modulo K'.

With the coset notation, we can write that $m + K = m' + K \iff m' - m \in K$; we should be used to this already from the ring story.

Thus, we know what M/K is as a set. I claim that we can make it into an R-module in a natural way. Indeed, we can define an addition on M/K, by setting $\forall m_1, m_2 \in M$,

$$(m_1 + K) + (m_2 + K) := (m_1 + m_2) + K;$$

and a multiplication by scalars from R by setting $\forall r \in R$, $\forall m \in M$,

$$r \cdot (m + K) := (r \cdot m) + K.$$

CLAIM 8.38 *These operations are well-defined.*

Please look back at §5.3! The same issue was discussed there in the process of defining quotient rings. If you have absorbed that construction, then you know that Claim 8.38 follows immediately if you can verify that the equivalence relation $\sim_K$ is suitably compatible with addition and multiplication by scalars. This means the following parallel of Lemma 5.25.

LEMMA 8.39 *Let K be a submodule of M. Then if $m_1' - m_1 \in K$ and $m_2' - m_2 \in K$,*

$$(m_1' + m_2') - (m_1 + m_2) \in K.$$

Further, if $r \in R$ and $m' - m \in K$, then

$$r \cdot m' - r \cdot m \in K.$$

Proof Both statements follow immediately from the fact that K is a submodule of M: submodules are closed with respect to addition and to multiplication by scalars. □

Otherwise put, if $m_1' + K = m_1 + K$ and $m_2' + K = m_2 + K$, then

$$(m_1' + m_2') + K = (m_1 + m_2) + K;$$

and further, if $r \in R$ and $m + K = m' + K$, then

$$r \cdot m' + K = r \cdot m + K.$$

This is precisely what Claim 8.38 says.

After all these preliminaries, the following result should not be unexpected.

> THEOREM 8.40 *Let K be a submodule of an R-module. Then the structure $(M/K, +, \cdot)$ is an R-module, with $0_{M/K} = 0 + K = K$.*

Proof This amounts to checking axioms (i)–(viii) from Definition 8.1. It is straightforward work, and a good exercise for you. □

Once again, note that the situation is very similar to the situation for rings, but in a sense even better. In ring theory, I cannot take the 'quotient of a ring by a ring': there is no such thing. We were able to define quotients, but we had to quotient rings out by *something else,* that is, ideals. With R-modules, there is no such silliness. I can take the quotient of a module by any submodule, without having to invent a whole new class of mathematical objects.

However, the end result is similar. Now we have a new R-module M/K for every R-module M and sub-R-module K. It comes endowed with a natural projection

$$\pi \colon M \longrightarrow M/K$$

defined by $\pi(m) = m + K$. We have the parallel of Proposition 5.35.

> PROPOSITION 8.41 *Let M be an R-module, and let K be a submodule of M. Then the natural projection $\pi \colon M \to M/K$ is a surjective homomorphism of R-modules, and its kernel is K.*

Proof You should not have any problem verifying this statement: all the work went into coming up with the right definitions. For example,

$$\ker \pi = \{m \in M \mid m + K = K\} = \{m \in M \mid m \in K\} = K$$

as needed. □

Remark 8.42 Note that this proves that *every* submodule of an R-module M is the kernel of some homomorphism of R-modules. Cf. Remark 5.36. ⌐

We are essentially done putting together our canonical decomposition of R-module homomorphisms. The definition of $\sim_K$ was concocted so that the equivalence relation $\sim$ from which we started would equal $\sim_{\ker f}$. Therefore, $M/\sim \, = M/\ker f$.

> THEOREM 8.43 (Canonical decomposition) *Let $f \colon M \to N$ be an R-module homomorphism. Then we have a commutative diagram*
>
> $$M \xrightarrow{\ \ \pi\ \ } (M/\ker f) \xrightarrow[\tilde{f}]{\ \sim\ } f(M) \overset{\iota}{\hookrightarrow} N,$$
>
> with f spanning over the top.

where π is the natural projection, ι is the inclusion homomorphism, and $\tilde{f}$ is the induced homomorphism. Further, $\tilde{f}$ is an isomorphism.

Proof Almost everything that needs to be proved here has been proved already. By Proposition 8.41, π is a surjective homomorphism of R-modules. The inclusion ι is also a homomorphism. The induced function $\tilde{f}$ is defined by

$$\tilde{f}(m + \ker f) = f(m)$$

for all $m \in M$. This is what set theory tells us, and it tells us that $\tilde{f}$ is automatically a bijection and that $f = \iota \circ \tilde{f} \circ \pi$, that is, that the diagram commutes. All that is left to verify is that $\tilde{f}$ is a homomorphism of R-modules, and this just takes a moment: $\forall m, m_1, m_2 \in M, \forall r \in R$:

$$\tilde{f}((m_1 + \ker f) + (m_2 + \ker f)) = \tilde{f}((m_1 + m_2) + \ker f) = f(m_1 + m_2) = f(m_1) + f(m_2)$$
$$= \tilde{f}(m_1 + \ker f) + \tilde{f}(m_2 + \ker f),$$
$$\tilde{f}(r \cdot (m + \ker f)) = \tilde{f}((r \cdot m) + \ker f) = f(r \cdot m) = r \cdot f(m)$$
$$= r \cdot \tilde{f}(m + \ker f)$$

as needed. □

8.5 Isomorphism Theorems

As in the case of rings, a particular case of the canonical decomposition gives a 'First Isomorphism Theorem' for R-modules, describing all homomorphic images of a given R-module M, that is, of all the modules N for which there is a *surjective* homomorphism $M \to N$. It shows that the true 'prototype' of a surjective R-module homomorphism is the canonical projection $M \to M/K$ for some submodule K of M.

THEOREM 8.44 (First Isomorphism Theorem) *Let $f : M \to N$ be a surjective homomorphism of R-modules. Then $N \cong M/\ker f$.*
In fact, f induces an isomorphism $\tilde{f} : M/\ker f \xrightarrow{\sim} N$, defined by $\tilde{f}(m + \ker f) = f(m)$.

Proof By the canonical decomposition, f induces an isomorphism $\tilde{f} : M/\ker f \cong f(M)$ defined by $\tilde{f}(m + \ker f) = f(m)$. Since f is surjective $f(M) = N$, and the statement follows. □

Example 8.45 Let R be a commutative ring. Then every cyclic R-module (Definition 8.29) is isomorphic to R/I for some ideal I of R (cf. Example 8.7).

Indeed, suppose $M = \langle m \rangle$ is generated by $m \in M$. Consider the homomorphism of R-modules $R \to M$ defined by $r \mapsto r \cdot m$. Definition 8.29 implies that this homomorphism is surjective if M is cyclic. Its kernel I is some submodule of R; by the First Isomorphism

Theorem, $M \cong R/I$. Submodules of a commutative ring are ideals (Example 8.24), so this is precisely the statement.

Example 8.46 More generally (and over any ring R), $M = \langle m_1, \ldots, m_n \rangle$ is generated by n elements if and only if there is a *surjective* homomorphism

$$R^{\oplus n} = \underbrace{R \oplus \cdots \oplus R}_{n \text{ times}} \longrightarrow M$$

defined by $(r_1, \ldots, r_n) \mapsto r_1 m_1 + \cdots + r_n m_n$; this is just another way of viewing Example 8.25 and Definition 8.28. By the First Isomorphism Theorem, $M \cong R^{\oplus n}/K$ for some submodule K of the direct sum $R^{\oplus n}$.

Therefore, M is generated by n elements if and only if it is a quotient of $R^{\oplus n}$. We say that M is 'finitely generated' if this is the case for some n (Definition 8.29).

What about the 'Third Isomorphism Theorem'? If K and L are submodules of an R-module M, and $K \subseteq L \subseteq M$, we may view K as a submodule *of L*, and consider the quotient L/K. As a subset of M/K, this consists of all cosets $\ell + K$ with $\ell \in L$; in other words, it equals $\pi(L)$, where $\pi: M \to M/K$ is the natural projection.

THEOREM 8.47 (Third Isomorphism Theorem) *Let M be an R-module, and let K be a submodule of M. Let L be a submodule of M containing K. Then L/K is a submodule of M/K, and all submodules of M/K are of this type. Further,*

$$(M/K)/(L/K) \cong M/L.$$

Proof We can adapt to modules the argument given for rings in Theorem 5.57.

Let $\pi: M \to M/K$ be the natural projection. If L is a submodule of M containing K, then $L/K = \pi(L)$ is a submodule of M/K by Proposition 8.32(i). Conversely, if $\overline{L}$ is a submodule of M/K, let $L = \pi^{-1}(\overline{L})$. Then L is a submodule of M by Proposition 8.32(ii), L contains $K = \pi^{-1}(0)$ since $\overline{L}$ contains 0, and $\overline{L} = \pi(L) = L/K$. This verifies that all submodules of M/K are of the form L/K as stated.

The last part follows from the First Isomorphism Theorem: composing the natural projections gives a surjective homomorphism $\rho: M \to M/K \to (M/K)/(L/K)$; concretely $\rho(m) = (m + K) + L/K$. The kernel of this homomorphism is

$$\rho^{-1}(0) = \{m \in M \mid (m + K) + L/K = L/K\}$$
$$= \{m \in M \mid (m + K) \in L/K\}$$
$$= \{m \in M \mid \pi(m) \in L/K\}$$
$$= \pi^{-1}(L/K)$$
$$= L.$$

By the First Isomorphism Theorem, $(M/K)/(L/K) \cong M/L$ as stated. $\qquad\square$

So far, so good: both theorems match their ring-theoretic counterparts. In fact, maybe the Third Isomorphism Theorem for modules clarifies a little the statement of the same for rings: in Theorem 5.57 we had to work a little to view J/I as an ideal in R/I. We thought of J/I as a subset of R/I first, and then we checked by hand that this would be an ideal. In the context of modules, this work is not necessary—L/K is an R-module because it is the quotient of an R-module by an R-module.

More is true. Now that we are in this new and improved context, we can state and prove the 'Second Isomorphism Theorem', which was missing in the ring story. This also has to do with two submodules K, L of an R-module M, but now we are not assuming that one is contained in the other. We can then look at several submodules of M:

$$K, \quad L, \quad K \cap L, \quad K + L$$

(cf. Examples 8.26 and 8.27).

> THEOREM 8.48 (Second Isomorphism Theorem)   *Let K, L be submodules of an R-module M. Then*
>
> $$L/(K \cap L) \cong (K + L)/K .$$
>
> *In fact, the function $L/(K \cap L) \to (K + L)/K$ defined by $\ell + (K \cap L) \mapsto \ell + K$ is a well-defined isomorphism of R-modules.*

A good mnemonic device to remember the isomorphism $L/(K \cap L) \cong (K + L)/K$ is that there are as many Ks as Ls in this formula.

Proof   Let $\phi \colon L \to (K + L)/K$ be the function defined by

$$\phi(\ell) = \ell + K .$$

Maybe a little surprisingly, ϕ turns out to be *surjective*. Indeed, an arbitrary element of $(K + L)/K$ is of the form $(k + \ell) + K$, where $k \in K$ and $\ell \in L$; I claim this element is in the image of ϕ. Indeed, $\phi(\ell) = \ell + K$ by definition, and

$$\ell + K = (k + \ell) + K$$

since $(k + \ell) - \ell = k \in K$.

Since ϕ is surjective, we can apply the First Isomorphism Theorem: ϕ induces an isomorphism

$$\varphi \colon \quad L/\ker\phi \overset{\sim}{\longrightarrow} (K + L)/K .$$

Now just note that

$$\ker\phi = \{\ell \in L \mid \ell + K = K\} = \{\ell \in L \mid \ell \in K\} = K \cap L :$$

this concludes the verification that $L/(K \cap L) \cong (K + L)/K$.

The isomorphism $\varphi \colon L/(K \cap L) \to (K + L)/K$ is induced by ϕ, therefore it is defined by $\varphi(\ell + (K \cap L)) = \phi(\ell) = \ell + K$ as stated.[3]   □

---

[3] Hidden in this argument is the verification that φ is well-defined. You can easily check this directly, if you wish.

Now you see why I did not bother stating the Second Isomorphism Theorem in the context of rings. If I, J are ideals of a ring R, then Theorem 8.48 tells us that

$$J/(I \cap J) \cong (I + J)/I$$

as R-modules; this is a useful statement, but it has to do with modules, not with rings.

Remark 8.49   Needless to say, everything we have done applies to *abelian groups:* that is just the particular case $R = \mathbb{Z}$, so it does not require any additional argument.   ⌐

Exercises

8.1   ▷ Recall (from Example 3.13) that the set of *linear transformations* $V \to V$ of a k-vector space V forms a ring; call this ring End(V). Prove that multiplication by scalars determines a ring homomorphism $k \to$ End(V).

8.2   ▷ Verify that if $f: R \to S$ is a ring homomorphism, then S is an R-module in a natural way. (See Example 8.5.)

8.3   ▷ Let K be a field, and let $k \subseteq K$ be a subfield (that is, a subring that is also a field). Show that K is a k-vector space in a natural way. (We say that K is a 'field extension' of k. The whole of Part IV of this text will be devoted to field extensions.)

8.4   Convince yourself of the truth of Proposition 8.10: either give a formal proof or at least construct examples clarifying which axioms relating to + enter in the verification of (v)–(viii) from Definition 8.1 when $R = \mathbb{Z}$.

8.5   We have seen (Proposition 8.10) that every abelian group, a.k.a. $\mathbb{Z}$-module, has a *unique* $\mathbb{Z}$-module structure. For example, $\mathbb{Z}$ itself can be realized as a $\mathbb{Z}$-module in only one way. Prove that $\mathbb{Z}$ may be realized as a $\mathbb{Z}[x]$-module in *infinitely many* different ways. (*Hint:* Think along the lines of Example 8.6.)

8.6   ▷ (Cf. Example 8.16.) Prove that if R is a ring, $r \in R$ is in the center of R (i.e., it commutes with every element of R), and M is an R-module, then the function $M \to M$ defined by $m \mapsto rm$ is an R-module homomorphism. Find an example showing that something may go wrong if r is not in the center of R.

8.7   Prove that the composition of two R-module homomorphisms is an R-module homomorphism.

8.8   Let $f: M \to N$ be a homomorphism of R-modules. Prove that $f(0_M) = 0_N$.

8.9   Does the category of R modules have initial objects? Final objects? (See Exercise 4.14 for the terminology.)

8.10   ▷ Let R be a PID. Prove that every nonzero ideal of R is isomorphic to R as an R-module. If R is a more general ring, what condition on $a \in R$ guarantees that (a) is isomorphic to R as an R-module?

8.11   Let M be an R-module, and let $A \subseteq M$ be a (possibly infinite) subset. Prove that the set $\langle A \rangle \subseteq M$ of R-linear combinations of elements of A is a submodule of M, and in fact it equals the smallest submodule of M containing A. (We say that this is the 'submodule of M generated by A'.)

8.12 ▷ Let $f: M \to N$ be a homomorphism of R-modules. Prove that if N' is a submodule of N, then $f^{-1}(N')$ is a submodule of M.

8.13 Let $f: M \to N$ be a homomorphism of R-modules. Prove that f is injective if and only if $\ker f = \{0\}$.

8.14 Let R be a ring. A nonzero R-module M is 'simple' if its only submodules are $\{0\}$ and M. Let M, N be simple modules, and let $\phi: M \to N$ be a homomorphism of R-modules. Prove that either $\phi = 0$ or ϕ is an isomorphism. (This statement is known as *Schur's lemma.*)

8.15 ▷ Let $f: M \to N$ be a homomorphism of R-modules. Prove that if $g: L \to M$ is any R-module homomorphism such that $f \circ g = 0$ (the trivial homomorphism), then $\operatorname{im} g \subseteq \ker f$: in particular, there exists a unique homomorphism $g': L \to \ker f$ making the following diagram commute:

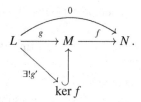

This statement is the 'universal property of kernels'.

8.16 ▷ Let R be a ring and let M, N, P be R-modules. Let $\mu: M \to P$, $v: N \to P$ be R-module homomorphisms. Prove that there is a unique R-module homomorphism $\mu \oplus v: M \oplus N \to P$ making the following diagram commute:

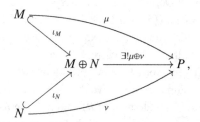

where ι_M, ι_N are the natural injections defined in §8.3. (This shows that the direct sum satisfies the 'universal property of coproducts' in the category of R-modules.)

8.17 ▷ Reverse all arrows in the property appearing in Exercise 8.16: this will tell you what the universal property of 'products' is. Verify that $M \oplus N$, along with the canonical projections to M and N, satisfies *also* this property.

As you may verify, Cartesian products of sets and rings satisfy this property (in their appropriate context, of course), but do *not* satisfy the property of coproducts.

8.18 Let $R = \mathbb{Z}[x]$. Prove that there is a ring homomorphism $\mathbb{Z}[x]/(x-1) \to \mathbb{Z}[x]$. Prove that there are no nonzero R-module homomorphisms $\mathbb{Z}[x]/(x-1) \to \mathbb{Z}[x]$.

8.19 ▷ Let M be an R-module. An element $m \in M$ is a 'torsion' element if there exists an $r \in R$, $r \neq 0$, such that $r \cdot m = 0$. Prove that if R is an integral domain, then the set of torsion elements is a submodule T of M. What goes wrong if R is not an integral domain?

8.20 We say that a module is 'torsion-free' if its only torsion element is 0. (See Exercise 8.19 for the definition of 'torsion'.)

Let R be an integral domain, and let M be an R-module. Prove that if M is a submodule of a free module, then it is torsion-free. What goes wrong if R has nonzero zero-divisors?

8.21 Let R be an integral domain; let M be an R-module, and let T be its torsion submodule. Prove that M/T is torsion-free. (See Exercises 8.19 and 8.20 for definitions.)

8.22 ▷ Let M_1, M_2 be R-modules, and let $L_1 \subseteq M_1$, $L_2 \subseteq M_2$ be submodules. Prove that $L_1 \oplus L_2$ is a submodule of $M_1 \oplus M_2$, and

$$(M_1 \oplus M_2)/(L_1 \oplus L_2) \cong (M_1/L_1) \oplus (M_2/L_2).$$

8.23 ▷ Let $f: M \to N$ be a homomorphism of R-modules, and assume that K is a submodule of M such that $K \subseteq \ker f$. Prove that f induces a unique homomorphism of R-modules $\tilde{f}: M/K \to N$ making the diagram

commute. (This shows that Proposition 5.39 holds for R-modules without any change; this is the universal property of quotients.)

9 Modules over Integral Domains

9.1 Free Modules

Let us fix a ring R. For convenience I will soon assume that R is an integral domain: these are the rings we are most familiar with, and the most sophisticated results we will prove will in fact hold for *Euclidean domains*. In fact, we will eventually focus on modules over $R = \mathbb{Z}$, i.e., the case of abelian groups. Integral domains are a vastly more general class of rings, and at the same time they form a comfortable enough environment that they do not present obstacles such as those exemplified by Exercise 8.19. Nevertheless, many facts we will review would hold over more general (say, commutative) rings, and further assumptions are not necessary, at least at the beginning. I will just write 'ring' for short; you are welcome to assume that R is an integral domain to begin with, or pay close attention to the proofs and spot when the presence of nonzero zero-divisors may cause difficulties.

At this point we have sketched some general features of the category of R-modules, but it is fair to say that we do not have many examples to play with—beyond the observation that many other things we may have encountered in the past, such as vector spaces, rings, ideals, are all examples of modules. It would be good to have some kind of description of what 'all' modules over a given ring R are.

This loose question does have an answer in some cases. For example, you may have seen in Linear Algebra that finite-dimensional vector spaces over a field are 'classified' by their *dimension:* up to isomorphism, there is exactly one vector space of any given dimension. (We will in fact recover this result below, in §9.4.) Therefore, the theory of R-modules is relatively simple when R is a field. What makes it simple is that all modules over a field are 'free'. This notion was mentioned in passing in Example 8.34; let's look at it more closely.

DEFINITION 9.1   Let $m \geq 0$ be an integer. An R-module F is 'free' of 'rank' m if $F \cong R^{\oplus m}$.

Remark 9.2   Free modules of infinite rank may be considered, and they are not a farfetched notion: for example, the polynomial ring $\mathbb{Z}[x]$ may be viewed as a $\mathbb{Z}$-module of (countably) infinite rank. However, we will mostly deal with free modules of finite rank, and I will occasionally neglect to state this finiteness condition explicitly in what follows. ⌐

There is a sense in which free modules are the 'easiest' to deal with, even when the ring R is not necessarily a field. Elements of a free module $R^{\oplus m}$ are simply m-tuples $(r_1, \ldots, r_m)$ of elements of R, with operations defined exactly as you do in linear algebra. It is convenient to give a name to the special tuples $(1, 0, \ldots, 0), \ldots, (0, \ldots, 0, 1)$: we let $\underline{e}_i = (0, \ldots, 0, 1, 0, \ldots, 0)$ be the element of $R^{\oplus m}$ with the ith entry equal to 1 and all others equal to 0. Note that the m tuples $\underline{e}_1, \ldots, \underline{e}_m$ generate $R^{\oplus m}$, in the sense that every $\underline{r} \in R^{\oplus m}$ may be written as a linear combination of the $\underline{e}_i$. Indeed, if $\underline{r} = (r_1, \ldots, r_m)$, then

$$\underline{r} = (r_1, 0, \ldots, 0) + (0, r_2, 0, \ldots, 0) + \cdots + (0, \ldots, 0, r_m)$$
$$= r_1 \underline{e}_1 + \cdots + r_m \underline{e}_m.$$

They are also *linearly independent* in the same sense you learn in ordinary linear algebra (but with coefficients in the ring R).

DEFINITION 9.3 The elements $x_1, \ldots, x_m$ of an R-module M are 'linearly independent' if $\forall r_1, \ldots, r_m \in R$,

$$r_1 x_1 + \cdots + r x_m = 0 \implies r_1 = \cdots = r_m = 0.$$

Remark 9.4 We also say that $x_1, \ldots, x_m$ form a 'linearly independent set'. This is a slight abuse of language, since sets do not allow for repetitions of the elements, while in principle we may want to consider the possibility that, e.g., $x_1 = x_2$. In this case the 'set' would not be linearly independent, since $x_1 - x_2$ would be a nontrivial linear combination giving 0. ⌐

The elements $\underline{e}_1, \ldots, \underline{e}_m$ are linearly independent because if $r_1 \underline{e}_1 + \cdots + r_m \underline{e}_m = 0$, then $(r_1, \ldots, r_m) = (0, \ldots, 0)$, and this says precisely that $r_1 = \cdots = r_m = 0$.

Thus, $\underline{e}_1, \ldots, \underline{e}_m$ form a 'linearly independent generating set'. There is a famous name for such a thing.

DEFINITION 9.5 An m-tuple $(x_1, \ldots, x_m)$ of elements of an R-module is a 'basis' if it is a linearly independent generating set.

PROPOSITION 9.6 *A module F is free of rank m if and only if it admits a basis $(x_1, \ldots, x_m)$.*

Proof Indeed, we have already observed that if $F \cong R^{\oplus m}$, then $(\underline{e}_1, \ldots, \underline{e}_m)$ is a linearly independent generating set. Conversely, if $(x_1, \ldots, x_m)$ is a basis of F, then we can define an R-module homomorphism $\varphi : R^{\oplus m} \to F$ by setting

$$\varphi(r_1, \ldots, r_m) := r_1 x_1 + \cdots + r_m x_m.$$

Then φ is surjective (because $x_1, \ldots, x_m$ generate F; cf. Example 8.46) and it is injective (because $x_1, \ldots, x_m$ are linearly independent), therefore it is an isomorphism. Thus F is isomorphic to $R^{\oplus m}$ in this case, and that means it is free of rank m according to Definition 9.1. □

Most modules are *not* free!

Example 9.7   The abelian group $\mathbb{Z}/2\mathbb{Z}$ is *not* free: indeed, a free abelian group (i.e., $\mathbb{Z}$-module) is necessarily infinite (Exercise 9.2). The one nonzero element $x = [1]_2 \in \mathbb{Z}/2\mathbb{Z}$ generates it, but it is not linearly independent *over* $\mathbb{Z}$, since $2x = 0$ even though $2 \neq 0$ in $\mathbb{Z}$.

On the other hand, $\mathbb{Z}/2\mathbb{Z}$ is a field, and in particular it is a vector space over itself; it is then free *as a* $(\mathbb{Z}/2\mathbb{Z})$-*module*. The element $x = [1]_2$ is a basis over $\mathbb{Z}/2\mathbb{Z}$.

Since free modules have bases, it is very easy to construct homomorphisms from a free module F: it suffices to assign (arbitrarily) the image of the basis elements. This observation generalizes Example 8.18.

> PROPOSITION 9.8   *Let R be a ring, and let $f: R^{\oplus m} \to M$ be an R-module homomorphism. Then there exist elements $a_1, \ldots, a_m \in M$ such that*
> $$f(r_1, \ldots, r_m) = r_1 a_1 + \cdots + r_m a_m. \tag{9.1}$$
> *With notation as above, $a_i = f(\underline{e}_i)$. Conversely, every choice $a_1, \ldots, a_m$ of m elements in M determines an R-module homomorphism $f: R^{\oplus m} \to M$ satisfying (9.1).*

Proof   If $f: R^{\oplus m} \to M$ is an R-module homomorphism, let $a_i := f(\underline{e}_i) \in M$ for $i = 1, \ldots, m$. By R-linearity,

$$f(r_1, \ldots, r_m) = f(r_1\underline{e}_1 + \cdots + r_m\underline{e}_m) = r_1 f(\underline{e}_1) + \cdots + r_m f(\underline{e}_m) = r_1 a_1 + \cdots + r_m a_m$$

as needed. Conversely, for any choice of $a_1, \ldots, a_m \in M$, (9.1) determines a well-defined function $f: R^{\oplus m} \to M$, and it is straightforward to verify that this function is an R-module homomorphism. ☐

Remark 9.9   Proposition 9.8 spells out yet another universal property, which characterizes free modules. It says that 'the free functor is left-adjoint to the forgetful functor'. (I am only telling you in order to amuse you. But if you persist in studying algebra, you will find out what this means and, if all goes well, this sentence will then seem distressingly obvious.) ⌐

Note that, in general, the situation is more complex: if a module is not free, then homomorphisms from that module are more constrained.[1]

Example 9.10   Again, the abelian group $\mathbb{Z}/2\mathbb{Z}$ is not free as an abelian group. It consists of two elements, 0 and 1 for short. If $f: \mathbb{Z}/2\mathbb{Z} \to M$ is a homomorphism to another abelian group M, necessarily $f(0) = 0$; so f is determined by $f(1) = m$. But unlike in the free case, the element $m \in M$ cannot be chosen arbitrarily! Indeed, necessarily

$$2 \cdot m = 2 \cdot f(1) = f(2 \cdot 1) = f(0) = 0.$$

---

[1] They are 'less free'!

By virtue of being the image of an element from $\mathbb{Z}/2\mathbb{Z}$, the element m must satisfy the relation $2 \cdot m = 0$.

The content of Proposition 9.8 is that no relations need be satisfied when considering homomorphisms from a free module. This is a huge simplification.

As a consequence of Proposition 9.8, we see that giving a homomorphism *from* a free module *to* a free module is the same thing as specifying a *matrix*. Indeed, let $f : R^{\oplus n} \to R^{\oplus m}$ be an R-module homomorphism. By Proposition 9.8, f is determined by the images of the basis elements $\underline{e}_j$, $j = 1, \ldots, n$. These images are elements of $R^{\oplus m}$, so they are m-tuples of elements of R. Let $f(\underline{e}_j) = (a_{1j}, \ldots, a_{mj})$ be these m-tuples. It is costumary to denote elements of a free module as column 'vectors': so we should actually write

$$f(\underline{e}_j) = \begin{pmatrix} a_{1j} \\ \vdots \\ a_{mj} \end{pmatrix}.$$

We assemble all these column vectors into a matrix:

$$A := \begin{pmatrix} a_{11} & a_{12} & \cdots & a_{1n} \\ a_{21} & a_{22} & \cdots & a_{2n} \\ \vdots & \vdots & \ddots & \vdots \\ a_{m1} & a_{m2} & \cdots & a_{mn} \end{pmatrix}.$$

With this notation, we can state the following consequence of Proposition 9.8.

COROLLARY 9.11   *Let $R^{\oplus m}$, $R^{\oplus n}$ be free R-modules, and let $f : R^{\oplus n} \to R^{\oplus m}$ be a homomorphism of R-modules. Then there exists a matrix A such that $\forall \underline{r} \in R^{\oplus n}$,*

$$f(\underline{r}) = A \cdot \underline{r}. \tag{9.2}$$

The jth column of the matrix A is $f(\underline{e}_j)$. Conversely, every $m \times n$ matrix A with entries in R determines an R-module homomorphism $f : R^{\oplus n} \to R^{\oplus m}$ satisfying (9.2).

Note the switch in the order of m and n: the homomorphism goes from $R^{\oplus n}$ to $R^{\oplus m}$ ('n first'); and the matrix is an $m \times n$ matrix ('n second'). The number of *columns* is n, i.e., the rank of the *source* of the homomorphism.

Remark 9.12   Let me recommend that you firmly remember the indication that *the jth column of the matrix A is $f(\underline{e}_j)$*. This is invaluable when you are performing concrete computations.   ⌋

Proof of Corollary 9.11   There is nothing to prove here that has not been proved already; but one has to get used to the column notation. The proof of Proposition 9.8

would run as follows in this particular case:

$$f(\underline{r}) = f\begin{pmatrix} r_1 \\ r_2 \\ \vdots \\ r_n \end{pmatrix} = f\begin{pmatrix} r_1 \\ 0 \\ \vdots \\ 0 \end{pmatrix} + f\begin{pmatrix} 0 \\ r_2 \\ \vdots \\ 0 \end{pmatrix} + \cdots + f\begin{pmatrix} 0 \\ 0 \\ \vdots \\ r_n \end{pmatrix}$$

$$= r_1 \cdot f\begin{pmatrix} 1 \\ 0 \\ \vdots \\ 0 \end{pmatrix} + r_2 \cdot f\begin{pmatrix} 0 \\ 1 \\ \vdots \\ 0 \end{pmatrix} + \cdots + r_n \cdot f\begin{pmatrix} 0 \\ 0 \\ \vdots \\ 1 \end{pmatrix}$$

$$= r_1 \begin{pmatrix} a_{11} \\ a_{21} \\ \vdots \\ a_{m1} \end{pmatrix} + r_2 \begin{pmatrix} a_{12} \\ a_{22} \\ \vdots \\ a_{m2} \end{pmatrix} + \cdots + r_n \begin{pmatrix} a_{1m} \\ a_{2n} \\ \vdots \\ a_{mn} \end{pmatrix}$$

$$= \begin{pmatrix} a_{11} & a_{12} & \cdots & a_{1n} \\ a_{21} & a_{22} & \cdots & a_{2n} \\ \vdots & \vdots & \ddots & \vdots \\ a_{m1} & a_{m2} & \cdots & a_{mn} \end{pmatrix} \cdot \begin{pmatrix} r_1 \\ r_2 \\ \vdots \\ r_n \end{pmatrix},$$

and this is precisely what (9.2) says. Conversely, for any matrix A, (9.2) defines a function $R^{\oplus n} \to R^{\oplus m}$ that is clearly a homomorphism of R-modules. □

Example 9.13 Thus, giving a homomorphism of abelian groups $\mathbb{Z}^{\oplus 3} \to \mathbb{Z}^{\oplus 2}$ (for instance) is the same as giving a 2×3 matrix of integers. The matrix

$$\begin{pmatrix} -4 & 1 & 7 \\ -2 & 0 & 4 \end{pmatrix}$$

is a 'matrix representation' of the homomorphism $f: \mathbb{Z}^{\oplus 3} \to \mathbb{Z}^{\oplus 2}$ for which

$$f\begin{pmatrix} 1 \\ 0 \\ 0 \end{pmatrix} = \begin{pmatrix} -4 \\ -2 \end{pmatrix}, \quad f\begin{pmatrix} 0 \\ 1 \\ 0 \end{pmatrix} = \begin{pmatrix} 1 \\ 0 \end{pmatrix}, \quad f\begin{pmatrix} 0 \\ 0 \\ 1 \end{pmatrix} = \begin{pmatrix} 7 \\ 4 \end{pmatrix}.$$

Reverting to row vectors, you could write this out as

$$f(r_1, r_2, r_3) = (-4r_1 + r_2 + 7r_3, -2r_1 + 4r_3);$$

maybe you will agree that the matrix representation is easier to parse.

A particularly good feature of matrix representations of homomorphisms of free modules is that *composition of homomorphisms corresponds to matrix multiplication.*

PROPOSITION 9.14   *Let $R^{\oplus n}$, $R^{\oplus m}$, $R^{\oplus \ell}$ be free modules, and let $f: R^{\oplus n} \to R^{\oplus m}$, $g: R^{\oplus m} \to R^{\oplus \ell}$ be R-module homomorphisms. Assume that f is represented by the matrix A and g by the matrix B. Then $g \circ f: R^{\oplus n} \to R^{\oplus \ell}$ is represented by $B \cdot A$.*

Proof   This follows right away from (9.2). Indeed, $\forall \underline{r} \in R^{\oplus n}$,

$$(g \circ f)(\underline{r}) = g(f(\underline{r})) = g(A \cdot \underline{r}) = B \cdot (A \cdot \underline{r}) = (B \cdot A) \cdot \underline{r}$$

as claimed. (I have used the property that matrix multiplication is *associative;* cf. Example 3.9 for the case of square matrices. The general case is analogous, and you probably saw it worked out in Linear Algebra.)   □

In particular, a homomorphism of free modules is an *isomorphism* precisely when the corresponding matrix A is *invertible*, i.e., when there exists another matrix B such that $A \cdot B = B \cdot A$ is the identity matrix. For matrices over a field, this is the same as saying that A is a square matrix with nonzero determinant; this is yet another fact that you have likely studied thoroughly in Linear Algebra, and you are encouraged to review it now. Over more general rings, invertible matrices are still necessarily square matrices, but determinant considerations are subtler. For example, the matrix

$$\begin{pmatrix} 1 & 0 \\ 0 & 2 \end{pmatrix} \tag{9.3}$$

has determinant $2 \neq 0$, but it is *not* invertible as a matrix over $\mathbb{Z}$, in the sense that its inverse

$$\begin{pmatrix} 1 & 0 \\ 0 & \frac{1}{2} \end{pmatrix}$$

is not a matrix with integer entries. Therefore, the homomorphism of free abelian groups $\mathbb{Z}^{\oplus 2} \to \mathbb{Z}^{\oplus 2}$ determined by the matrix (9.3) is *not* an isomorphism. (The homomorphism of $\mathbb{Q}$-vector spaces $\mathbb{Q}^{\oplus 2} \to \mathbb{Q}^{\oplus 2}$ determined by the same matrix *is* an isomorphism, since the matrix is invertible over $\mathbb{Q}$.)

We will be especially interested in isomorphisms corresponding to the classical 'elementary operations'. There are three basic types of such operations, and corresponding matrices.

- First, matrices obtained from the diagonal matrix by changing *one* diagonal entry to an arbitrary unit $u \in R$ are invertible:

$$\begin{pmatrix} 1 & 0 & 0 & \cdots & 0 \\ 0 & u & 0 & \cdots & 0 \\ 0 & 0 & 1 & \cdots & 0 \\ \vdots & \vdots & \vdots & \ddots & \vdots \\ 0 & 0 & 0 & \cdots & 1 \end{pmatrix} \cdot \begin{pmatrix} 1 & 0 & 0 & \cdots & 0 \\ 0 & u^{-1} & 0 & \cdots & 0 \\ 0 & 0 & 1 & \cdots & 0 \\ \vdots & \vdots & \vdots & \ddots & \vdots \\ 0 & 0 & 0 & \cdots & 1 \end{pmatrix} = \begin{pmatrix} 1 & 0 & 0 & \cdots & 0 \\ 0 & 1 & 0 & \cdots & 0 \\ 0 & 0 & 1 & \cdots & 0 \\ \vdots & \vdots & \vdots & \ddots & \vdots \\ 0 & 0 & 0 & \cdots & 1 \end{pmatrix}.$$

Multiplying a matrix A by such a matrix on the left amounts to multiplying all entries of one row by u; multiplying A on the right amounts to multiplying all entries of one column by u:

$$
\begin{pmatrix} 1 & 0 & 0 & 0 \\ 0 & u & 0 & 0 \\ 0 & 0 & 1 & 0 \\ 0 & 0 & 0 & 1 \end{pmatrix} \cdot \begin{pmatrix} a_{11} & a_{12} & a_{13} \\ a_{21} & a_{22} & a_{23} \\ a_{31} & a_{32} & a_{33} \\ a_{41} & a_{42} & a_{43} \end{pmatrix} = \begin{pmatrix} a_{11} & a_{12} & a_{13} \\ ua_{21} & ua_{22} & ua_{23} \\ a_{31} & a_{32} & a_{33} \\ a_{41} & a_{42} & a_{43} \end{pmatrix},
$$

$$
\begin{pmatrix} a_{11} & a_{12} & a_{13} & a_{14} \\ a_{21} & a_{22} & a_{23} & a_{24} \\ a_{31} & a_{32} & a_{33} & a_{34} \end{pmatrix} \cdot \begin{pmatrix} 1 & 0 & 0 & 0 \\ 0 & u & 0 & 0 \\ 0 & 0 & 1 & 0 \\ 0 & 0 & 0 & 1 \end{pmatrix} = \begin{pmatrix} a_{11} & ua_{12} & a_{13} & a_{14} \\ a_{21} & ua_{22} & a_{23} & a_{24} \\ a_{31} & ua_{32} & a_{33} & a_{34} \end{pmatrix}.
$$

- A second type consists of matrices obtained by adding a single nonzero off-diagonal entry $r \in R$ to the identity matrix: something like

$$
\begin{pmatrix} 1 & 0 & 3 \\ 0 & 1 & 0 \\ 0 & 0 & 1 \end{pmatrix}. \tag{9.4}
$$

These matrices are automatically invertible: for this example,

$$
\begin{pmatrix} 1 & 0 & 3 \\ 0 & 1 & 0 \\ 0 & 0 & 1 \end{pmatrix} \cdot \begin{pmatrix} 1 & 0 & -3 \\ 0 & 1 & 0 \\ 0 & 0 & 1 \end{pmatrix} = \begin{pmatrix} 1 & 0 & 0 \\ 0 & 1 & 0 \\ 0 & 0 & 1 \end{pmatrix}.
$$

Multiplying *on the left* by matrix (9.4) corresponds to adding to the first row of a matrix 3 times the third row:

$$
\begin{pmatrix} 1 & 0 & 3 \\ 0 & 1 & 0 \\ 0 & 0 & 1 \end{pmatrix} \cdot \begin{pmatrix} a_{11} & a_{12} & a_{13} & a_{14} \\ a_{21} & a_{22} & a_{23} & a_{24} \\ a_{31} & a_{32} & a_{33} & a_{34} \end{pmatrix} = \begin{pmatrix} a_{11} + 3a_{31} & a_{12} + 3a_{32} & a_{13} + 3a_{33} & a_{14} + 3a_{34} \\ a_{21} & a_{22} & a_{23} & a_{24} \\ a_{31} & a_{32} & a_{33} & a_{34} \end{pmatrix}
$$

while multiplying *on the right* by the same matrix corresponds to adding to the third column 3 times the first column:

$$
\begin{pmatrix} a_{11} & a_{12} & a_{13} \\ a_{21} & a_{22} & a_{23} \\ a_{31} & a_{32} & a_{33} \\ a_{41} & a_{42} & a_{43} \end{pmatrix} \cdot \begin{pmatrix} 1 & 0 & 3 \\ 0 & 1 & 0 \\ 0 & 0 & 1 \end{pmatrix} = \begin{pmatrix} a_{11} & a_{12} & 3a_{11} + a_{13} \\ a_{21} & a_{22} & 3a_{21} + a_{23} \\ a_{31} & a_{32} & 3a_{31} + a_{33} \\ a_{41} & a_{42} & 3a_{41} + a_{43} \end{pmatrix}.
$$

- Finally, we will be interested in matrices obtained by switching two rows (or, equivalently, columns) of the identity matrix. Such matrices are clearly invertible, since they are their own inverses. For example,

$$
\begin{pmatrix} 0 & 0 & 1 & 0 \\ 0 & 1 & 0 & 0 \\ 1 & 0 & 0 & 0 \\ 0 & 0 & 0 & 1 \end{pmatrix}
$$

is obtained by switching the first and third rows (columns) in the 4×4 identity matrix, and it is its own inverse. The effect of multiplying a matrix *on the left* by this matrix is to switch the first and third rows:

$$\begin{pmatrix} 0 & 0 & 1 & 0 \\ 0 & 1 & 0 & 0 \\ 1 & 0 & 0 & 0 \\ 0 & 0 & 0 & 1 \end{pmatrix} \cdot \begin{pmatrix} a_{11} & a_{12} & a_{13} \\ a_{21} & a_{22} & a_{23} \\ a_{31} & a_{32} & a_{33} \\ a_{41} & a_{42} & a_{43} \end{pmatrix} = \begin{pmatrix} a_{31} & a_{32} & a_{33} \\ a_{21} & a_{22} & a_{23} \\ a_{11} & a_{12} & a_{13} \\ a_{41} & a_{42} & a_{43} \end{pmatrix},$$

while multiplying on the *right* has the effect of switching the first and third columns:

$$\begin{pmatrix} a_{11} & a_{12} & a_{13} & a_{14} \\ a_{21} & a_{22} & a_{23} & a_{24} \\ a_{31} & a_{32} & a_{33} & a_{34} \end{pmatrix} \cdot \begin{pmatrix} 0 & 0 & 1 & 0 \\ 0 & 1 & 0 & 0 \\ 1 & 0 & 0 & 0 \\ 0 & 0 & 0 & 1 \end{pmatrix} = \begin{pmatrix} a_{13} & a_{12} & a_{11} & a_{14} \\ a_{23} & a_{22} & a_{21} & a_{24} \\ a_{33} & a_{32} & a_{31} & a_{34} \end{pmatrix}.$$

Summary: Performing these elementary operations on a matrix A corresponds to composing the homomorphism corresponding to A with isomorphisms, on the left or on the right as the case may be. A useful mnemonic: An elementary operation performed on the *rows* of the matrix is the result of multiplying on the *left* by the same operation performed on the identity matrix; and an elementary operation performed on the *columns* of the matrix is the result of multiplying on the *right* by the same operation performed on the identity matrix.

I will adopt the following somewhat nonstandard terminology.

DEFINITION 9.15   Two matrices A, A' with entries in a ring R are 'equivalent' (over R) if there exist matrices P, Q, invertible over R, such that $A' = PAQ$.

This *is* an equivalence relation, as you should verify. Of course there are many other possible equivalence relations we could define on the set of matrices; since we will focus on the notion specified in Definition 9.15, I am taking the liberty of just calling this specific one 'equivalence'. Don't take this as a standard, commonly adopted definition.

With this terminology, we have seen that if two matrices can be obtained one from the other by a sequence of elementary operations as above, then they are equivalent. These simple operations will soon lead us to a powerful classification theorem (Theorem 9.40).

Remark 9.16   In general, elementary operations are *not* enough to capture equivalence of matrices: there are rings, even PIDs, for which there are equivalent matrices which cannot be obtained one from the other by a sequence of elementary operations.

However, elementary operations will suffice for the applications we are going to explore.   ⌐

9.2   Modules from Matrices

In §9.1 we have discovered that homomorphisms between finitely generated free R-modules are 'the same as' matrices with entry in R. We will push this envelope a little further, and be able to use matrices to describe a large class of R-modules, going beyond the free case. The class I have in mind consists of *finitely presented* modules. In order to talk efficiently about this, we introduce a few new pieces of notation.

DEFINITION 9.17   A sequence of R-modules and R-module homomorphisms

$$\cdots \xrightarrow{f_{i+2}} M_{i+1} \xrightarrow{f_{i+1}} M_i \xrightarrow{f_i} M_{i-1} \xrightarrow{f_{i-1}} \cdots$$

is 'exact at M_i' if im $f_{i+1} = \ker f_i$, and it is 'exact' if it is exact at all those M_i that are both source and target of homomorphisms in the sequence.

Example 9.18   The sequence

$$M \xrightarrow{f} N \longrightarrow 0$$

is exact if and only if f is *surjective*. Indeed, the second homomorphism is necessarily trivial, so its kernel is the whole of N; and then the exactness condition is equivalent to im $f = N$, that is, to the surjectivity of f. Exactness at M or at 0 is not an issue, since there is no homomorphism pointing to M or from 0 in this sequence.

Example 9.19   Similarly,

$$0 \longrightarrow L \xrightarrow{g} M$$

is exact if and only if g is *injective* (Exercise 9.4).

Example 9.20   A *short exact sequence* is an exact sequence of the form

$$0 \longrightarrow L \xrightarrow{g} M \xrightarrow{f} N \longrightarrow 0.$$

You should develop the habit of immediately viewing N as an isomorphic copy of M/L when you run into such a sequence. Indeed, L may be identified with a submodule of M by g (since g is injective, as we have seen in Example 9.19); by exactness at M, this submodule $g(L) = $ im g equals $\ker f$; and $N \cong M/\ker f$ by the First Isomorphism Theorem (since f is surjective, as we have seen in Example 9.18). For instance, we have a short exact sequence of abelian groups

$$0 \longrightarrow \mathbb{Z} \xrightarrow{2\cdot} \mathbb{Z} \longrightarrow \mathbb{Z}/2\mathbb{Z} \longrightarrow 0,$$

where the injective homomorphism is multiplication by 2 (as indicated), identifying $\mathbb{Z}$ with its isomorphic copy $2\mathbb{Z}$ within $\mathbb{Z}$ itself.

A standard source of exact sequences of the type seen in Example 9.20 comes from surjective homomorphisms: if $f : M \to N$ is a surjective R-module homomorphism, then there is a natural short exact sequence

$$0 \longrightarrow \ker f \longrightarrow M \xrightarrow{f} N \longrightarrow 0.$$

What if f is not surjective? The 'failure' of surjectivity of f is measured by how far $\operatorname{im} f = f(M)$ is from N, and there is a precise way to quantify this.

DEFINITION 9.21 Let $f: M \to N$ be a homomorphism of R-modules. The 'cokernel' of f, denoted $\operatorname{coker} f$, is the quotient $N/\operatorname{im} f$.

Why the name 'cokernel'? By now you are thoroughly familiar with *kernels*, and if you have worked out Exercise 8.15, then you know what universal property they satisfy. If you *reverse* all the arrows in this property, you get another universal property, and as it happens *cokernels* satisfy this other requirement. This is what the prefix 'co' summarizes for the *cognoscenti*.

If $f: M \to N$ is a homomorphism of R-modules, then the sequence

$$0 \longrightarrow \ker f \longrightarrow M \xrightarrow{f} N \longrightarrow \operatorname{coker} f \longrightarrow 0$$

is exact, where the homomorphism $\ker f \to M$ is the inclusion and the homomorphism $N \to \operatorname{coker} f = N/\operatorname{im} f$ is the natural projection. Exactness should be clear at all modules in this sequence. For example, the sequence is exact at N because the kernel of the map from N to $\operatorname{coker} f = N/\operatorname{im} f$ is $\operatorname{im} f$, as required.

The notion of exact sequence (and more generally of 'complexes of modules') turns out to be very useful in a more general context. I am mostly introducing it because it is a convenient shorthand, and it pays off to get used to it. We can use it to recast and develop further a notion that I have already mentioned in Chapter §8.

DEFINITION 9.22 An R-module M is 'finitely generated' if there is an exact sequence

$$R^{\oplus m} \longrightarrow M \longrightarrow 0$$

with $R^{\oplus m}$ a free R-module of *finite* rank.

If this definition looks familiar, that's because we already discussed it in Example 8.46. Hopefully, the following definition will now look like a natural extension of that definition.

DEFINITION 9.23 An R-module M is 'finitely presented' if there is an exact sequence

$$R^{\oplus n} \longrightarrow R^{\oplus m} \longrightarrow M \longrightarrow 0$$

with $R^{\oplus m}$, $R^{\oplus n}$ free R-modules of finite rank.

In other words, a finitely presented module M is finitely generated, so that there is a surjective R-module homomorphism $f: R^{\oplus m} \twoheadrightarrow M$; further, if M is finitely presented, then the kernel of this homomorphism is *also* finitely generated, so there is another surjective homomorphism $g: R^{\oplus n} \twoheadrightarrow \ker f$. Viewing g as a homomorphism $R^{\oplus n} \to R^{\oplus m}$ gives rise to the exact sequence in the definition.

Equivalently, M is finitely presented if it is isomorphic to the cokernel of a homomorphism of finitely generated free modules. Now, we have seen in Corollary 9.11

that every such homomorphism is determined by a matrix. Therefore, a finitely presented module admits an amazingly concrete description, in terms of a *matrix* defining a corresponding homomorphism of free modules. In principle, any 'module-theoretic operation' you may want to perform on M can be translated into some kind of operation you can perform on a corresponding matrix. Computer implementations of module operations[2] work on this principle.

Example 9.24   The matrix in Example 9.13,

$$\begin{pmatrix} -4 & 1 & 7 \\ -2 & 0 & 4 \end{pmatrix},$$

defines a certain finitely presented $\mathbb{Z}$-module, i.e., the cokernel of the corresponding homomorphism $\mathbb{Z}^{\oplus 3} \to \mathbb{Z}^{\oplus 2}$. *Challenge:* Can you describe even more explicitly what this module is, up to isomorphism? That is, can you find a 'famous' abelian group that is isomorphic to this cokernel?

In order to answer the question posed in Example 9.24, we have to study a little more carefully the correspondence between matrices and their cokernels.

At the end of §9.1 we have observed that performing standard row/column operations on a matrix has the effect of composing the corresponding homomorphism of free modules by isomorphisms. The following result is straightforward (even if it may look somewhat 'abstract'), but yields a powerful tool: it implies that the cokernel does not change up to isomorphism as we perform such operations.

PROPOSITION 9.25   *Let $f : M \to N$ and $f' : M' \to N'$ be R-module homomorphisms. Assume $\phi : M' \to M$ and $\psi : N \to N'$ are isomorphisms such that the diagram*

$$
\begin{array}{ccc}
M & \xrightarrow{\ f\ } & N \\
{\scriptstyle \phi}\big\uparrow{\scriptstyle \wr} & & {\scriptstyle \wr}\big\downarrow{\scriptstyle \psi} \\
M' & \xrightarrow{\ f'\ } & N'
\end{array}
$$

commutes. (That is, such that $f' = \psi \circ f \circ \phi$.) Then $\operatorname{coker} f \cong \operatorname{coker} f'$.

Proof   Since ϕ is an isomorphism, it has an inverse ϕ^{-1}, and we can redraw the above diagram with ϕ^{-1} instead (and of course reversing the corresponding arrow). We can also extend the diagram to one including the cokernels:

$$
\begin{array}{ccccccc}
M & \xrightarrow{\ f\ } & N & \xrightarrow{\ \pi\ } & \operatorname{coker} f & \longrightarrow & 0 \\
{\scriptstyle \wr}\big\downarrow{\scriptstyle \phi^{-1}} & & {\scriptstyle \wr}\big\downarrow{\scriptstyle \psi} & & & & \\
M' & \xrightarrow{\ f'\ } & N' & \xrightarrow{\ \pi'\ } & \operatorname{coker} f' & \longrightarrow & 0
\end{array}
$$

---

[2] 'Macaulay2' is one such package, very powerful and available for free.

with exact rows. I claim that this diagram can be filled in with an isomorphism between the cokernels:

$$
\begin{array}{ccccccc}
M & \xrightarrow{\ f\ } & N & \xrightarrow{\ \pi\ } & \operatorname{coker} f & \longrightarrow & 0 \\
{\scriptstyle\wr}\downarrow{\scriptstyle\phi^{-1}} & & {\scriptstyle\wr}\downarrow{\scriptstyle\psi} & & {\scriptstyle\wr}\downarrow{\scriptstyle\tilde{\psi}} & & \\
M' & \xrightarrow{\ f'\ } & N' & \xrightarrow{\ \pi'\ } & \operatorname{coker} f' & \longrightarrow & 0
\end{array}
$$

in such a way that it remains commutative. To see this, note that we have a surjective homomorphism

$$
N \xrightarrow{\ \pi' \circ \psi\ } \operatorname{coker} f' \ .
$$

Its kernel is

$$
\ker(\pi' \circ \psi) = (\pi' \circ \psi)^{-1}(0) = \psi^{-1}(\pi'^{-1}(0)) = \psi^{-1}(\ker \pi') = \psi^{-1}(f'(M')) \ ;
$$

here I have used the exactness of the bottom row (which amounts to the definition of $\operatorname{coker} f'$). Since ϕ^{-1} is surjective, $M' = \phi^{-1}(M)$; therefore

$$
f'(M') = f'(\phi^{-1}(M)) = (f' \circ \phi^{-1})(M) = (\psi \circ f)(M) = \psi(f(M))
$$

by the commutativity of the diagram. Since $f(M) = \ker \pi$ by the exactness of the first row, this shows that

$$
\ker(\pi' \circ \psi) = \psi^{-1}(f'(M')) = \psi^{-1}(\psi(f(M))) = f(M) = \ker \pi
$$

(ψ is an isomorphism).

Summarizing, $\pi' \circ \psi \colon N \to \operatorname{coker} f'$ is a surjective homomorphism and we have found that its kernel is $\ker \pi$. Then the First Isomorphism Theorem (look at the fine print in Theorem 8.44) implies that there is an induced isomorphism

$$
\tilde{\psi} \colon N/\ker \pi \xrightarrow{\ \sim\ } \operatorname{coker} f'
$$

and we are done, since $N/\ker \pi = N/\operatorname{im} f = \operatorname{coker} f$. $\qquad \square$

This argument may sound fancy, but if you paid attention, then you recognize that it is just a dressed-up version of the First Isomorphism Theorem. It is no harder or easier than, e.g., Exercise 5.23 from way back. 'Diagram chases' like the one performed in the proof of Proposition 9.25 quickly become second nature; this proof is another example of an argument that could safely be omitted in a more advanced text. If it sounds complicated, it's just because sometimes it is complicated to fully describe simple things by using words. The way to go is to develop a 'calculus of diagrams' dealing with exact sequences, which encompasses this type of situation. Learning this calculus (which you will, if you study algebra at a more advanced level) makes it quite easy to prove statements such as Proposition 9.25. Perform an Internet search for the 'snake lemma' if you want to see an example of this calculus.

COROLLARY 9.26 *Let M be an R-module defined by a matrix A (as discussed above), and let A′ be a matrix equivalent to A in the sense of Definition 9.15. Then the module M′ defined by A′ is isomorphic to M.*

Make sure you understand why Corollary 9.26 follows from Proposition 9.25. The point is that if A and $A′$ are equivalent in the sense of Definition 9.15, then the corresponding homomorphisms of free R-modules differ by composition on the left or on the right by isomorphisms (as we discussed in §9.1) and then Proposition 9.25 tells us that the cokernels of these homomorphisms are isomorphic. This is precisely the statement of the corollary.

In particular, this is the case if $A′$ is obtained from A by a sequence of 'elementary operations', as described in §9.1: multiplying a row or column by a unit in R; or adding to a row an R-multiple of another row, or adding to a column an R-multiple of another column; or switching two rows or two columns.

Example 9.27   Let's go back to the matrix in Example 9.24:

$$\begin{pmatrix} -4 & 1 & 7 \\ -2 & 0 & 4 \end{pmatrix},$$

whose cokernel defines a certain abelian group G. Question: *what* abelian group? Let's perform several elementary operations on this matrix:
—Switch the first and the second columns:

$$\begin{pmatrix} 1 & -4 & 7 \\ 0 & -2 & 4 \end{pmatrix}.$$

—Add (-2) times the second row to the first row:

$$\begin{pmatrix} 1 & 0 & -1 \\ 0 & -2 & 4 \end{pmatrix}.$$

—Add 2 times the second column to the third column:

$$\begin{pmatrix} 1 & 0 & -1 \\ 0 & -2 & 0 \end{pmatrix}.$$

—Add the first column to the third column:

$$\begin{pmatrix} 1 & 0 & 0 \\ 0 & -2 & 0 \end{pmatrix}.$$

—Multiply the second column by -1 (a unit in $\mathbb{Z}$):

$$\begin{pmatrix} 1 & 0 & 0 \\ 0 & 2 & 0 \end{pmatrix}.$$

According to Corollary 9.26, G is isomorphic to the cokernel of the homomorphism $\varphi\colon \mathbb{Z}^{\oplus 3} \to \mathbb{Z}^{\oplus 2}$ defined by this new matrix. It is easy to 'compute' this cokernel! The image of φ is the submodule

$$\mathbb{Z} \oplus 2\mathbb{Z} \subseteq \mathbb{Z} \oplus \mathbb{Z},$$

therefore

$$\operatorname{coker} \varphi = (\mathbb{Z} \oplus \mathbb{Z})/(\mathbb{Z} \oplus 2\mathbb{Z}) \cong \mathbb{Z}/2\mathbb{Z}$$

(hopefully you have worked out Exercise 8.22). The conclusion is that $G \cong \mathbb{Z}/2\mathbb{Z}$: the given matrix is a complicated way to define $\mathbb{Z}/2\mathbb{Z}$. You can't get any more explicit than this.

9.3   Finitely Generated vs. Finitely Presented

I pointed out that (finite-rank) free modules are the 'easiest' modules with which we may work. The moral of §9.2 is that the more general *finitely presented* modules are also very easy to work with: one can manipulate finitely presented R-modules very concretely. Handling finitely presented modules amounts simply to handling matrices. As we have seen in Example 9.27, one can sometimes use this correspondence to identify given modules in terms of standard constructions. We will return to this point very soon, as it will allow us to *classify* finitely generated modules over Euclidean domains (§9.5).

The next natural question is whether finitely presented modules are common, or whether they are rarities. The bad news is that it is relatively easy to construct modules that are finitely generated (even cyclic, i.e., with a single generator) but not finitely presented. Even $\mathbb{Z}$ itself may be realized as a cyclic module over a somewhat nasty integral domain R, in such a way that it is not finitely presented as a module over this ring.[3] If you run into a module in a dark alley, chances are it is *not* finitely presented, and that could mean trouble. The good news is that *every finitely generated module over a Noetherian ring is finitely presented.* For example, *all* the finitely generated modules over rings appearing in 'classical algebraic geometry' are finitely presented. This is what we will figure out in this section.

Back in §6.3 we ran across the following notion.

DEFINITION 9.28   A ring R is 'Noetherian' if every ascending chain of ideals in R stabilizes.

You may want to glance back at Proposition 6.32 for a reminder of the terminology: there we proved that PIDs satisfy this condition. It is easy to verify (Exercise 9.10) that the Noetherian condition is in fact equivalent to the condition that every ideal of R is finitely generated.

THEOREM 9.29   *Every finitely generated module over a Noetherian ring is finitely presented.*

The proof of this theorem is not difficult, once one disposes of the following observations.

---

[3] If you want to construct such a realization yourself, rewind all the way back to the ring you encountered in Exercise 6.20 and keep Example 8.6 in mind.

LEMMA 9.30   *Let*

$$0 \longrightarrow L \longrightarrow M \longrightarrow N \longrightarrow 0$$

be an exact sequence of R modules, and assume that L and N are finitely generated. Then M is finitely generated.

Proof   Denote by $j: L \to M$ and $\pi: M \to N$ the two homomorphisms in the sequence, and let $\ell_1, \ldots, \ell_s$ be generators for L, and $n_1, \ldots, n_t$ be generators for N. Since j is injective, we may identify L with its image $j(L)$, a submodule of M. Since π is surjective, we may choose (arbitrarily) a preimage $\bar{n}_i \in M$ for each $n_i \in N$. Then I claim that the elements $\ell_1, \ldots, \ell_s, \bar{n}_1, \ldots, \bar{n}_t$ generate M.

To verify this, let $m \in M$. Since $n_1, \ldots, n_t$ generate N, there exist elements $r_i \in R$ such that

$$\pi(m) = r_1 n_1 + \cdots + r_t n_t.$$

Consider then the element $m - r_1 \bar{n}_1 - \cdots - r_t \bar{n}_t \in M$. Note that

$$\pi(m - r_1 \bar{n}_1 - \cdots - r_t \bar{n}_t) = \pi(m) - r_1 n_1 - \cdots - r_t n_t = 0:$$

this shows that $m - r_1 \bar{n}_1 - \cdots - r_t \bar{n}_t \in \ker \pi$. By exactness, $\ker \pi = \operatorname{im} j = L$ under our identification. Since $\ell_1, \ldots, \ell_s$ generate L, there exist elements $r'_i \in R$ such that

$$m - r_1 \bar{n}_1 - \cdots - r_t \bar{n}_t = r'_1 \ell_1 + \cdots + r'_s \ell_s.$$

Therefore,

$$m = r'_1 \ell_1 + \cdots + r'_s \ell_s + r_1 \bar{n}_1 + \cdots + r_t \bar{n}_t$$

is a linear combination of $\ell_1, \ldots, \ell_s, \bar{n}_1, \ldots, \bar{n}_t$, and this verifies my claim and concludes the proof of the lemma.   □

The full converse of Lemma 9.30 does *not* hold: if M is finitely generated and L is a submodule of M, in general L is not necessarily finitely generated. (The other 'half' of the converse does hold: see Exercise 9.12.) This is where the Noetherian hypothesis comes in. The following result highlights the one case we need to consider here.

LEMMA 9.31   *Let R be a Noetherian ring. Then for all $m \geq 0$, every submodule of $R^{\oplus m}$ is finitely generated.*

Proof   For $m = 0$ there is nothing to prove. Arguing by induction, in order to prove the lemma it suffices to verify that every submodule M of $R^{\oplus m}$ is finitely generated for a fixed $m > 0$, assuming that all submodules of $R^{\oplus k}$ are finitely generated for $k < m$.

View $R^{\oplus m}$ as $R^{\oplus(m-1)} \oplus R$. We have an injective homomorphism $j: R^{\oplus m-1} \to R^{\oplus m}$ and a surjective homomorphism $\pi: R^{\oplus m} \to R$, defined by

$$j(r_1, \ldots, r_{m-1}) = (r_1, \ldots, r_{m-1}, 0), \quad \pi(r_1, \ldots, r_m) = r_m;$$

we may use j to identify $R^{\oplus(m-1)}$ with a submodule $F \subseteq R^{\oplus m}$, and π may then be viewed as the projection $R^{\oplus m} \twoheadrightarrow R^{\oplus m}/F \cong R$.

If $M \subseteq R^{\oplus m}$ is a submodule, consider the intersection $M \cap F$ as a submodule of M. We have the short exact sequence

$$0 \longrightarrow M \cap F \longrightarrow M \longrightarrow M/(M \cap F) \longrightarrow 0. \qquad (9.5)$$

Now, $M \cap F$ is finitely generated because it is a submodule of $F \cong R^{\oplus(m-1)}$ and all submodules of $R^{\oplus(m-1)}$ are finitely generated by the induction hypothesis. On the other hand,

$$M/(M \cap F) \cong (M + F)/F$$

by the Second Isomorphism Theorem (Theorem 8.48); and we may view $(M + F)/F$ as a submodule of $R^{\oplus m}/F \cong R$, so this module is finitely generated since R is Noetherian by hypothesis. (The sub-R-modules of R are just the ideals of R, cf. Example 8.24.)

We can then apply Lemma 9.30 to the sequence at (9.5), and conclude that M is finitely generated, as was to be shown. □

Theorem 9.29 is a direct consequence of Lemma 9.31.

Proof of Theorem 9.29   Let M be a finitely generated module over a Noetherian ring R; therefore there exists a free module $R^{\oplus m}$ and a surjective homomorphism $f \colon R^{\oplus m} \to M$. Now $\ker f$ is a submodule of $R^{\oplus m}$, therefore it is finitely generated as R is Noetherian, by Lemma 9.31. Therefore there exists a free module $R^{\oplus n}$ and a surjective homomorphism $g \colon R^{\oplus n} \to \ker f$. Since $\ker f$ is a submodule of $R^{\oplus m}$, we can view g as a homomorphism $R^{\oplus n} \to R^{\oplus m}$ with $\operatorname{im} g = \ker f$. The sequence

$$R^{\oplus n} \xrightarrow{\ g\ } R^{\oplus m} \xrightarrow{\ f\ } M \longrightarrow 0$$

is then exact, showing that M is finitely presented. □

Example 9.32   Every finitely generated abelian group is finitely presented (as a $\mathbb{Z}$-module).

More generally, if R is any PID, then every finitely generated R-module is finitely presented. (Indeed, PIDs are Noetherian, by Proposition 6.32.) For instance, if k is a field, then every finitely generated $k[t]$-module is finitely presented.

Example 9.33   Finite-dimensional vector spaces are finitely presented. (Indeed, fields are Noetherian.) In fact, they are free—as you may have seen in Linear Algebra, and as we will soon be able to prove from scratch (Theorem 9.35).

Example 9.34   The rings appearing in classical algebraic geometry are all quotients of polynomial rings $\mathbb{C}[x_1, \ldots, x_n]$. This polynomial ring is Noetherian, by Hilbert's basis theorem (mentioned, but not proved, in §7.2). Quotients of $\mathbb{C}[x_1, \ldots, x_n]$ are therefore Noetherian, by Exercise 9.11. Therefore, as advertised earlier, finitely generated modules over these rings are finitely presented.

9.4   Vector Spaces are Free Modules

Let $R = k$ be a field. I have stated several times that the theory of k-modules, i.e., k-vector spaces, is simplified by the fact that *vector spaces are free.* If you have seen something like this in Linear Algebra, it was probably in the form of a theorem of 'existence of bases': every vector space has a *basis,* i.e., a linearly independent generating set (and this is the same as being free, by Proposition 9.6). From the point of view we have taken here, we can recover the same result, at least in the finite-dimensional case, essentially as a consequence of Theorem 9.29.

> THEOREM 9.35   *Every finitely generated vector space is free: if k is a field, and V is a finitely generated k-vector space, then $V \cong k^{\oplus d}$ for some $d \geq 0$.*

In the context of vector spaces, the integer d, that is, the rank, is called the *dimension* of V. Theorem 9.35 also holds for *infinite-dimensional* vector spaces, but we will only deal with the finite-dimensional case here.

How do we prove Theorem 9.35? The first observation is that every field is Noetherian. By Theorem 9.29, every finitely generated k-vector space V is a cokernel of a homomorphism of free k-vector spaces, determined by a matrix A:

$$k^{\oplus n} \xrightarrow{\ A\ } k^{\oplus m} \longrightarrow V \longrightarrow 0$$

(Example 9.33). By Corollary 9.26, we can determine a vector space isomorphic to V by performing elementary operations to A.

> LEMMA 9.36   *Let $A = (a_{ij})$ be an $m \times n$ matrix with entries in a field k. Then there is a sequence of elementary operations transforming A into an $m \times n$ matrix of the form*
>
> $$\begin{bmatrix} \begin{array}{cccc|ccc} 1 & 0 & \cdots & 0 & 0 & \cdots & 0 \\ 0 & 1 & \cdots & 0 & 0 & \cdots & 0 \\ \vdots & \vdots & \ddots & \vdots & \vdots & \ddots & \vdots \\ 0 & 0 & \cdots & 1 & 0 & \cdots & 0 \\ \hline 0 & 0 & \cdots & 0 & 0 & \cdots & 0 \\ \vdots & \vdots & \ddots & \vdots & \vdots & \ddots & \vdots \\ 0 & 0 & \cdots & 0 & 0 & \cdots & 0 \end{array} \end{bmatrix} \tag{9.6}$$
>
> *with r diagonal entries equal to 1, for some $r \geq 0$.*

(*Note:* Despite appearances, the matrix is not necessarily a square matrix.)

The integer r is called the *rank* of the matrix.

Proof   We can argue by induction on $\min(m, n)$. If $\min(m, n) = 1$, that is, A is a single row or column vector, then either it is 0 (and then the statement is true) or some entry of A is not zero. By a row/column switch we may then assume that a_{11} is nonzero; as such, a_{11} is a unit (since k is a field), so by multiplying the vector by a_{11}^{-1} we may assume $a_{11} = 1$. And then we can use row/column operations to ensure that the other entries are 0, and this produces the form (9.6) in this case.

For $\min(m, n) > 1$, again either $A = 0$ (and we are done already), or else we may assume that $a_{11} \neq 0$ after performing row and/or column switches. By multiplying the first row by a_{11}^{-1} (again, we can do this because k is a field), we may assume that $a_{11} = 1$. By adding multiples of the first row to the other rows, we may ensure that $a_{1j} = 0$ for $j > 0$; and then by adding multiples of the first column to the other columns, we may ensure that $a_{i1} = 0$ for $i > 0$. At this stage the matrix A has been put into the form

$$\begin{pmatrix} 1 & 0 & \cdots & 0 \\ 0 & a_{22} & \cdots & a_{2n} \\ \vdots & \vdots & \ddots & \vdots \\ 0 & a_{m2} & \cdots & a_{mn} \end{pmatrix}.$$

By induction, the 'inside' matrix

$$\begin{pmatrix} a_{22} & \cdots & a_{2n} \\ \vdots & \ddots & \vdots \\ a_{m2} & \cdots & a_{mn} \end{pmatrix}$$

may be put into the form (9.6) by a sequence of elementary operations (since $\min(m - 1, n - 1) = \min(m, n) - 1$), and this shows that A may be reduced to the same form, concluding the proof. □

Lemma 9.36 is all we need in order to prove Theorem 9.35.

Proof of Theorem 9.35 By Theorem 9.29, Lemma 9.36, and Corollary 9.26, V is isomorphic to the cokernel of the homomorphism of free k-vector spaces

$$k^{\oplus n} \xrightarrow{\ I_r\ } k^{\oplus m} \longrightarrow 0,$$

where I_r is the matrix at (9.6). The image of this homomorphism consists of the first r factors of $k^{\oplus m}$:

$$\underbrace{k \oplus \cdots \oplus k}_{r} \oplus \underbrace{0 \oplus \cdots \oplus 0}_{m-r} \subseteq \underbrace{k \oplus \cdots \oplus k}_{m}.$$

Therefore the cokernel is isomorphic to

$$\underbrace{k/k \oplus \cdots \oplus k/k}_{r} \oplus \underbrace{k/0 \oplus \cdots \oplus k/0}_{m-r} \cong \underbrace{0 \oplus \cdots \oplus 0}_{r} \oplus \underbrace{k \oplus \cdots \oplus k}_{m-r} \cong k^{\oplus(m-r)},$$

and this proves the statement, with $d = m - r$. □

The process described in the proof of Lemma 9.36 is a form of *Gaussian elimination*, with which you are likely to be familiar from Linear Algebra. Note that we have crucially used the fact that we are working over a field: the process involves multiplying by the inverses of certain entries of the matrix, based only on the fact that we know those entries are not 0. For this, we need the base ring to be a field.

If you are like me, your eyes may glaze over the argument given to prove Lemma 9.36. There is no better way to process such an argument than working it out in a specific example.

Example 9.37 Let's recycle again the matrix in Example 9.24:

$$\begin{pmatrix} -4 & 1 & 7 \\ -2 & 0 & 4 \end{pmatrix}.$$

In Example 9.27 we have reduced this matrix to the form

$$\begin{pmatrix} 1 & 0 & 0 \\ 0 & 2 & 0 \end{pmatrix}$$

by working *over the integers*. If we see the original matrix as a matrix with $\mathbb{Q}$ entries, therefore defining a homomorphism $\mathbb{Q}^{\oplus 3} \to \mathbb{Q}^{\oplus 2}$, then the elimination process described in the proof of Lemma 9.36 produces a different result.

- First, note that the $(1, 1)$ entry is not zero, so we do not need to swap rows or columns.
- We multiply the top row by the inverse of the $(1, 1)$ entry, that is, by $\frac{1}{4}$, to get

$$\begin{pmatrix} 1 & -\frac{1}{4} & -\frac{7}{4} \\ -2 & 0 & 4 \end{pmatrix}.$$

- We add to the second row 2 times the first row, in order to clear the $(2, 1)$ entry:

$$\begin{pmatrix} 1 & -\frac{1}{4} & -\frac{7}{4} \\ 0 & -\frac{1}{2} & \frac{1}{2} \end{pmatrix}.$$

- We add to the second and third column multiples of the first column, thereby clearing entries $(1, 2)$ and $(1, 3)$:

$$\begin{pmatrix} 1 & 0 & 0 \\ 0 & -\frac{1}{2} & \frac{1}{2} \end{pmatrix}.$$

- In the proof we now invoke induction, which tells us that the smaller matrix (just $\begin{pmatrix} -\frac{1}{2} & \frac{1}{2} \end{pmatrix}$ in this example) can be reduced as stated, and we are done. In this small example we can simply multiply the bottom row by -2 (that is, the inverse of $= -\frac{1}{2}$):

$$\begin{pmatrix} 1 & 0 & 0 \\ 0 & 1 & -1 \end{pmatrix}$$

and then add to the third column the second column, and get the form advertised in the statement:

$$\begin{pmatrix} 1 & 0 & 0 \\ 0 & 1 & 0 \end{pmatrix}.$$

The conclusion is that the $\mathbb{Q}$-vector space defined by this matrix is actually trivial: $m = r = 2$, so $d = 0$ with notation as in the statements of Theorem 9.35 and Lemma 9.36. (We have seen that the $\mathbb{Z}$-module defined by the same matrix is *not* trivial.)

9.5   Finitely Generated Modules over Euclidean Domains

The process used in §9.4 can be 'upgraded' to Euclidean domains. The analogue to Lemma 9.36 in this context is the following result.

LEMMA 9.38   *Let $A = (a_{ij})$ be an $m \times n$ matrix with entries in a Euclidean domain R. Then there is a sequence of elementary operations transforming A into a matrix of the form*

$$
\begin{pmatrix}
d_1 & 0 & \cdots & 0 & 0 & \cdots & 0 \\
0 & d_2 & \cdots & 0 & 0 & \cdots & 0 \\
\vdots & \vdots & \ddots & \vdots & \vdots & \ddots & \vdots \\
0 & 0 & \cdots & d_s & 0 & \cdots & 0 \\
0 & 0 & \cdots & 0 & 0 & \cdots & 0 \\
\vdots & \vdots & \ddots & \vdots & \vdots & \ddots & \vdots \\
0 & 0 & \cdots & 0 & 0 & \cdots & 0
\end{pmatrix}, \tag{9.7}
$$

where $d_1 \mid d_2 \mid \cdots \mid d_s$. (This matrix is 0 if $s = 0$.)

A formal proof of this lemma goes along the lines of the proof of Lemma 9.36, by induction on $\min(m, n)$, but is even more awkward to write down; I'll do my best in what follows. Looking at the case $m = 1$, i.e., when A is a single row vector $\underline{a}$, already shows the essential steps (and the essential awkwardness). Consider

$$ \underline{a} = (a_1, \ldots, a_n). $$

Each a_i is an element of R, and we are assuming that R is a Euclidean domain (Definition 6.24); let v be the Euclidean valuation.

If all entries are equal to 0, then $\underline{a}$ is already in the desired form. Otherwise, we can switch columns and assume that $a_1 \neq 0$. In the proof of Lemma 9.36 we could then multiply the whole vector by a_1^{-1}, and use column operations to ensure that all other entries were equal to 0. We cannot operate directly in the same way now, because a_1 is not necessarily a unit in R.

However, we can apply the following operations to the vector $(a_1, \ldots, a_n)$.

- Input: $\underline{a} = (a_1, \ldots, a_n)$, with $a_1 \neq 0$.
- For $i = 2, \ldots, n$, perform division with remainder of a_i by a_1: there exist $q_i, r_i \in R$ such that

$$ a_i = a_1 q_i + r_i $$

  with either $r_i = 0$ or $v(r_i) < a_1$.
- For $i = 2, \ldots, n$, add to the ith entry a_i the multiple $-q_i a_1$ of the first entry; this has the effect of replacing the nonzero entries a_i, $i \geq 2$, by the remainders r_i.
- (i) If the resulting vector is not of the form $(d, 0, \ldots, 0)$, switch columns to replace the first entry with any other nonzero entry, and start from the beginning with this new vector $\underline{a}'$.

- (ii) If the resulting vector is of the form $(d, 0, \ldots, 0)$, EXIT.

I claim that this procedure *terminates*. Indeed, whenever the procedure goes through (i), the valuation of the first entry of $\underline{a}'$ is *strictly less* than the valuation of a_1. Since the valuation takes values in $\mathbb{Z}^{\geq 0}$, it cannot be decreased infinitely many times. Therefore the procedure must reach (ii) and exit after finitely many steps. When it does so, its output is a vector $(d, 0, \ldots, 0)$ as required by the statement. Let me encourage you to try it out in one example, to see the cogs turn (Exercise 9.14).

This procedure has the effect of cleaning up a row: it applies elementary transformations to bring a given row to the form $(d, 0, \ldots, 0)$; and keep in mind that nontrivial applications of this 'row step' *decrease* the valuation of the first entry. This argument proves[4] the statement of Lemma 9.38 in the case $m = 1$. An analogous 'column step' takes care of the case $n = 1$.

For $\min(m, n) > 1$ and $A \neq 0$, we can switch rows/columns to ensure that $a_{11} \neq 0$, then apply a row step as above to set the first-row entries $a_{12}, \ldots, a_{1n}$ to 0; and then a column step to set the first-column entries $a_{21}, \ldots, a_{m1}$ to 0. However, in doing so the first row may again acquire nonzero entries. So the row step may have to be applied again, possibly followed by another application of the column step, and so on. Again the process must terminate, since the valuation of the $(1, 1)$ entry of the matrix cannot decrease indefinitely. This process brings the given matrix to the form

$$\begin{pmatrix} d_1 & 0 & \cdots & 0 \\ 0 & a_{22} & \cdots & a_{2n} \\ \vdots & \vdots & \ddots & \vdots \\ 0 & a_{m2} & \cdots & a_{mn} \end{pmatrix}.$$

This looks like progress. By induction, the 'inside' matrix may be brought into the form specified by the statement of the lemma, and this will produce an equivalent matrix to the original matrix A, and of the form

$$\begin{pmatrix} d_1 & 0 & \cdots & 0 & 0 & \cdots & 0 \\ 0 & d_2 & \cdots & 0 & 0 & \cdots & 0 \\ \vdots & \vdots & \ddots & \vdots & \vdots & \ddots & \vdots \\ 0 & 0 & \cdots & d_s & 0 & \cdots & 0 \\ 0 & 0 & \cdots & 0 & 0 & \cdots & 0 \\ \vdots & \vdots & \ddots & \vdots & \vdots & \ddots & \vdots \\ 0 & 0 & \cdots & 0 & 0 & \cdots & 0 \end{pmatrix}, \tag{9.8}$$

where $d_2 \mid d_3 \mid \cdots \mid d_s$.

Are we done? No! The statement of the lemma demands that $d_1 \mid d_2$, and this condition has not been verified. But notice that the process so far has produced a matrix of the form (9.8), *and in the process the valuation of the top $(1, 1)$ entry has decreased*. If

---

[4] Somewhat informally—the 'row step' described above hides an induction argument.

d_1 does *not* divide d_2, perform a row operation to obtain the matrix

$$\begin{pmatrix} d_1 & d_2 & \cdots & 0 & 0 & \cdots & 0 \\ 0 & d_2 & \cdots & 0 & 0 & \cdots & 0 \\ \vdots & \vdots & \ddots & \vdots & \vdots & \ddots & \vdots \\ 0 & 0 & \cdots & d_s & 0 & \cdots & 0 \\ 0 & 0 & \cdots & 0 & 0 & \cdots & 0 \\ \vdots & \vdots & \ddots & \vdots & \vdots & \ddots & \vdots \\ 0 & 0 & \cdots & 0 & 0 & \cdots & 0 \end{pmatrix}$$

(note the new $(1, 2)$ entry) and *start all over*. This will again produce a matrix of the form (9.8), maybe with different $d_2 \mid d_3 \mid \cdots \mid d_s$, and with lower valuation $v(d_1)$. Once more this valuation cannot decrease indefinitely, so after finitely many iterations of this process one must reach the form (9.8) *and* satisfy the further requirement $d_1 \mid d_2$. This concludes the proof of Lemma 9.38.   □

I will grant you that this argument is dreadful, but it has one redeeming aspect: it is an actual procedure which can be conceivably carried out in concrete cases. There may be substantially more elegant ways to achieve the same result, but probably they will not be as concrete.

Example 9.39   Running the algorithm described above on the matrix

$$\begin{pmatrix} -2 & 0 & 4 \\ -4 & 1 & 7 \end{pmatrix}$$

results in the following steps:

$$\begin{pmatrix} -2 & 0 & 4 \\ -4 & 1 & 7 \end{pmatrix} \rightarrow \begin{pmatrix} -2 & 0 & 0 \\ -4 & 1 & -1 \end{pmatrix} \rightarrow \begin{pmatrix} -2 & 0 & 0 \\ 0 & 1 & -1 \end{pmatrix} \rightarrow \begin{pmatrix} -2 & 0 & 0 \\ 0 & 1 & 0 \end{pmatrix}$$

$$\xrightarrow{*} \begin{pmatrix} -2 & 1 & 0 \\ 0 & 1 & 0 \end{pmatrix} \rightarrow \begin{pmatrix} 1 & -2 & 0 \\ 1 & 0 & 0 \end{pmatrix} \rightarrow \begin{pmatrix} 1 & 0 & 0 \\ 1 & 2 & 0 \end{pmatrix} \rightarrow \begin{pmatrix} 1 & 0 & 0 \\ 0 & 2 & 0 \end{pmatrix}.$$

The first three steps produce a matrix of the form (9.8), but with $d_1 \nmid d_2$; the row operation marked $*$ produces a new $(1, 2)$ entry, and the remaining steps (a new 'row step' followed by a 'column step') run the reduction again and produce a matrix in the form prescribed by the statement of Lemma 9.38.

Lemma 9.38 implies the classification theorem that follows, and this is probably one of the most important theorems in this text. Recall that a *cyclic R-module* is a module generated by a single element (Definition 8.29). Every cyclic module is isomorphic to R/I for an ideal I (Example 8.45).

| THEOREM 9.40 (Classification of finitely generated modules over Euclidean domains)

> Let M be a finitely generated module over a Euclidean domain R. Then M is isomorphic to a direct sum of cyclic modules. More precisely, there exist elements $d_1, \ldots, d_s$ in R, with $d_1 \mid d_2 \mid \cdots \mid d_s$ and d_1 not a unit, and an integer $r \geq 0$, such that
>
> $$M \cong R^{\oplus r} \oplus (R/(d_1)) \oplus (R/(d_2)) \oplus \cdots \oplus (R/(d_s)).$$

Proof   Euclidean domains are PIDs (Theorem 6.26), and PIDs are Noetherian (Proposition 6.32), so M is finitely *presented* by Theorem 9.29. Therefore there is an exact sequence

$$R^{\oplus n} \xrightarrow{\;\varphi\;} R^{\oplus m} \longrightarrow M \longrightarrow 0,$$

with φ given by an $m \times n$ matrix with entries in R. By Lemma 9.38 and Corollary 9.26, M is isomorphic to the cokernel of the homomorphism $\varphi \colon R^{\oplus n} \to R^{\oplus m}$ defined by a matrix of the form (9.7), with $d_1 \mid d_2 \mid \cdots \mid d_s$. The image of this homomorphism is

$$\operatorname{im}\varphi = (d_1) \oplus (d_2) \oplus \cdots \oplus (d_s) \oplus 0 \oplus \cdots \oplus 0 \subseteq R^{\oplus m},$$

and it follows that $\operatorname{coker}\varphi = R^{\oplus m}/\operatorname{im}\varphi$ has the form given in the statement. Note that we can eliminate factors $R/(d_i)$ where d_i is a unit, since those factors are trivial. Thus we may assume that d_1 is not a unit, as stated.   □

Remark 9.41   The form obtained in Theorem 9.40 is a true 'classification', in the sense that the integer r (the 'rank' of M) and the factors $R/(d_i)$ are actually *uniquely determined* by M.

The proof of this fact is not difficult, but it requires some care, and I will not get into the details. First one may argue that the 'free' part $F = R^{\oplus r}$ and the 'torsion' part

$$T = R/(d_1) \oplus R/(d_2) \oplus \cdots \oplus R/(d_s)$$

are uniquely determined by M; this is a good exercise for you (Exercise 9.18). Therefore, it suffices to prove that (i) if $R^{\oplus r_1} \cong R^{\oplus r_2}$, then $r_1 = r_2$; and (ii) if

$$R/(d_1) \oplus \cdots \oplus R/(d_s) \cong R/(e_1) \oplus \cdots \oplus R/(e_t)$$

with $d_1 \mid \cdots \mid d_s$, $e_1 \mid \cdots \mid e_t$ and d_1, e_1 not units, then it follows that $s = t$ and $R/(d_i) \cong R/(e_i)$ for all i.

Assertion (i) boils down to the fact that invertible matrices are necessarily square. Assertion (ii) may be proved by an inductive procedure, extracting one factor at a time. One can, for example, prove that necessarily $(d_s) = (e_t)$, since this ideal is the *annihilator* of M. Again, this is a good exercise for you (Exercise 9.17, where you will find the definition of 'annihilator').   ⌐

Remark 9.42   In fact, the conclusion of Theorem 9.40 holds over every PID, including the uniqueness considerations in Remark 9.41. The proof is more involved in this more general case.   ⌐

Remark 9.43   There is a very remarkable observation hidden within the proof of Theorem 9.40: *every submodule of a finitely generated free module over a Euclidean domain is free* (cf. Exercise 9.19). This in fact holds for every PID, but not beyond: if an integral domain R is not a PID, then some submodule of R itself (i.e., some ideal) is *not* free.   ⌐

There is a useful alternative formulation of Theorem 9.40. The elements $d_i \in R$ (defined up to a unit) appearing in the statement are called the 'invariant factors' of the module. Since the ring R is assumed to be a Euclidean domain, it is in particular a UFD; therefore, the invariant factors have factorizations into irreducibles. By the Chinese Remainder Theorem (Theorem 5.52), if $d = q_1^{a_1} \cdots q_r^{a_r}$ is a factorization into relatively prime factors, then

$$R/(d) \cong R/(q_1^{a_1}) \oplus \cdots \oplus R/(q_r^{a_r}). \tag{9.9}$$

Indeed, the Chinese Remainder Theorem tells us that $R/(d)$ is isomorphic *as a ring* to the product of the terms on the right-hand side; and we know that this means that $R/(d)$ is isomorphic *as a module* to the direct sum of the same terms, viewed as R-modules (Example 8.35). Therefore, we can draw the following consequence.

> COROLLARY 9.44   *Every finitely generated R-module over a Euclidean domain is isomorphic to a direct sum of cyclic modules of the form $R/(q^a)$, where $q \in R$ is irreducible and $a \geq 0$.*

Further, this decomposition is unique (Remark 9.41). The q^a appearing in the decomposition are called the 'elementary divisors' of the module.

Summary: If R is a Euclidean domain, the modules $R/(q^a)$ are the basic bricks out of which every finitely generated module is constructed, in the simplest possible way (i.e., as a direct sum). As I mentioned in passing, this in fact holds over PIDs. The situation over more general rings is quite a bit more complex.

Applications of Theorem 9.40 abound. I will quickly review a famous example in §9.6; we will use Theorem 9.40 to great effect in the case $R = \mathbb{Z}$ of abelian groups in Chapter 10.

9.6 Linear Transformations and Modules over $k[t]$

A brilliant example of a Euclidean domain is the ring of polynomials $k[t]$ over a field k (Theorem 7.2). There is an application of the classification theorem from §9.5 to linear algebra which uses this ring. I will digress a moment to describe this application in general terms; but for details you will have to look into more advanced texts.

Let k be a field, and let V be a k-vector space. Recall that a 'linear transformation' of V is a linear map $\alpha \colon V \to V$, that is, a homomorphism of vector spaces $V \to V$. Since vector spaces are free as k-modules (as we have proved in §9.4, in the finite dimensional case), if V has dimension d, then α may be viewed as a $d \times d$ matrix with entries in k.

The following remark is simple but powerful (and may sound mysterious the first time you come across it): *giving a linear transformation $\alpha \colon V \to V$ is the same as giving V a structure of a $k[t]$-module extending the action of k.*

Indeed:

- Given a linear transformation $\alpha \colon V \to V$, we may define an action $k[t] \times V \to V$ by setting $t \cdot v = \alpha(v)$ and extending the operation in the natural way. That is: for $f(t) \in k[t]$, $f(t) = a_0 + a_1 t + a_2 t^2 + \cdots + a_n t^n$, set

$$f(t) \cdot v = a_0 v + a_1 \alpha(v) + a_2 \alpha^2(v) + \cdots + a_n \alpha^n(v),$$

where α^k stands for the composition $\underbrace{\alpha \circ \cdots \circ \alpha}_{k \text{ times}}$. It is easy to verify that with this action, and its own addition operation, V is a module over $k[t]$, and the definition is concocted so that if $\lambda \in k$ is viewed as a constant polynomial, then its action agrees with multiplication by the scalar λ in V.

- Conversely, if an action of $k[t]$ on V extends the k-action, then 'multiplication by t' defines a k-linear transformation $V \to V$. Indeed, for $\lambda_1, \lambda_2 \in k$ and $v_1, v_2 \in V$,

$$t \cdot (\lambda_1 v_1 + \lambda_2 v_2) = \lambda_1 t \cdot v_1 + \lambda_2 t \cdot v_2,$$

by axiom (viii) in Definition 8.1 and the commutativity of the operation in $k[t]$.

With this understood, we summarize the situation by simply saying that α determines a $k[t]$-module structure on V where 'α acts as multiplication by t'.

Now assume that V is finite-dimensional. Its generators as a k-vector space will generate it as a $k[t]$ module, so V will be *finitely generated* as a $k[t]$-module. And now we have a classification theorem for finitely generated modules over $k[t]$, because $k[t]$ is a Euclidean domain! That is, we can describe *all* possible finitely generated $k[t]$-modules! We apply this to V.

By Theorem 9.40, there exist nonconstant monic polynomials $f_1(t), \ldots, f_s(t)$ with $f_1(t) \mid \cdots \mid f_s(t)$, such that

$$V \cong k[t]/(f_1(t)) \oplus \cdots \oplus k[t]/(f_s(t)). \tag{9.10}$$

(The free part vanishes, because it has infinite dimension over k, while we are assuming that V be finite-dimensional.)

The isomorphism (9.10) is extremely useful. The right-hand side may be viewed as a k-vector space, and as such it must be isomorphic to V. So, (9.10) tells us that in the presence of a linear transformation $\alpha: V \to V$, we may find an isomorphic copy of V over which α acts in a way ('multiplication by t') which is straightforward to analyze. For example, this leads to possible choices of bases for V with respect to which the matrix representation of α is particularly nice—representations obtained in this fashion are called 'canonical forms'. Precise statements and their proofs may be found in more advanced texts.

Also, the polynomials $f_1(t), \ldots, f_s(t)$ are refinements of notions from 'ordinary' linear algebra. By chasing definitions, it is easy to verify that

- The *minimal polynomial* of α equals the polynomial $f_s(t)$.
- The *characteristic polynomial* of α equals $f_1(t) \cdots f_s(t)$.

These polynomials are famous quantities, and you may have encountered them in Linear Algebra. For example, the characteristic polynomial of a matrix A is $\det(tI - A)$; it is the polynomial whose roots define the *eigenvalues* of A. From the considerations sketched above, it is *evident* that *the minimal polynomial divides the characteristic polynomial.* This implies a famous result.

THEOREM 9.45 (Cayley–Hamilton) *Let α be a linear transformation of V, and let $P(t)$ be its characteristic polynomial. Then $P(\alpha) = 0$.*

This is highly non-obvious[5] in itself, but may be viewed as a straightforward consequence of the classification theorem for finitely generated modules over $k[t]$, a very particular case of Theorem 9.40. Once again, a progression of comparatively simple steps has advanced our understanding to the point that objectively 'difficult' results become relatively straightforward. This is as it should be.

Exercises

9.1 Let $F = R^{\oplus n}$ be a free module, and let $(x_1, \ldots, x_n)$ be a basis of F. Let $\varphi : F \to F$ be an isomorphism. Prove that $(\varphi(x_1), \ldots, \varphi(x_n))$ is also a basis of F.

9.2 ▷ 'Most' modules are not free. Verify that this is the case for the following examples.

• Let A be an abelian group. Prove that if A is finite and nontrivial, then it is not free as a $\mathbb{Z}$-module.

• Prove that $\mathbb{Q}$ is not free as a $\mathbb{Z}$-module. (*Hint:* Let $a = (a_1, \ldots, a_n) \in \mathbb{Z}^{\oplus n}$. Can you 'divide' a by *every* nonzero integer?)

• Let $V \neq 0$ be a $k[t]$-module, where k is a field. Prove that if V is finite-dimensional as a k-vector space, then it is not free as a $k[t]$-module.

• Prove that the ideal (x, y) of $\mathbb{C}[x, y]$ is not free as a $\mathbb{C}[x, y]$-module. (Since the ideal is not principal, it needs at least two generators—for example, x and y. Why can't two polynomials in $\mathbb{C}[x, y]$ form a *basis* over $\mathbb{C}[x, y]$?)

9.3 Prove that 'equivalence' of matrices (as in Definition 9.15) is an equivalence relation.

9.4 ▷ Prove that a sequence of R-modules

$$0 \longrightarrow M \overset{f}{\longrightarrow} N$$

is exact if and only if f is *injective*.

9.5 Assume that there is an exact sequence of R-modules

$$0 \longrightarrow M \longrightarrow N \longrightarrow 0 .$$

Prove that $M \cong N$.

---

[5] You may think that it *is* obvious: it sounds like it is claiming that $\det(A \cdot I - A) = \det(A - A) = \det(0) = 0$, and isn't this evident? Think again—take any 3×3 matrix A with real entries, write out the matrix $tI - A$, then replace t by A. What you get is a strange matrix in which some of the entries are matrices and some are numbers. In any case, you do not get just '0'.

9.6   Prove that there is *no* short exact sequence

$$0 \longrightarrow \mathbb{Z}/2\mathbb{Z} \longrightarrow \mathbb{Z} \longrightarrow \mathbb{Z} \longrightarrow 0 \ .$$

9.7   Let M, N be R-modules. Recall that the direct sum $M \oplus N$ comes endowed with certain injective and surjective homomorphisms. Use them to construct exact sequences

$$0 \longrightarrow M \longrightarrow M \oplus N \longrightarrow N \longrightarrow 0,$$

$$0 \longrightarrow N \longrightarrow M \oplus N \longrightarrow M \longrightarrow 0.$$

9.8   Assume that two abelian groups G_1, G_2 have matrix representations

$$\begin{pmatrix} 3 & 7 & -2 \\ 1 & 4 & 5 \end{pmatrix}, \quad \begin{pmatrix} -1 & 4 \\ 2 & 9 \\ 0 & 5 \end{pmatrix}$$

(i.e., they are cokernels of the corresponding homomorphisms of free modules). What is a matrix representation for $G_1 \oplus G_2$?

9.9   Suppose that the ith row or the jth column of a matrix A consists of 0s, with the single (i, j)-entry equal to 1. Prove that the module corresponding to A is isomorphic to the module corresponding to the matrix obtained from A by erasing both the ith row and the jth column.

9.10   ▷ Let R be a commutative ring.

• Assume that every ideal of R is finitely generated. Prove that every ascending chain of ideals in R stabilizes. (Upgrade the proof of Proposition 6.32.)

• Conversely, assume that every ascending chain of ideals in R stabilizes. Prove that every ideal of R is finitely generated. (Let I be an ideal of R, and inductively construct a chain of ideals as follows: $I_0 = (0)$, and $I_n = I_{n-1} + (r_n)$ for $n > 0$, where r_n is any element in $I \setminus I_{n-1}$. What happens now?)

A ring is 'Noetherian' if it satisfies these equivalent conditions.

9.11   ▷ Let R be a Noetherian ring, and let $I \subseteq R$ be an ideal. Prove that R/I is a Noetherian ring.

9.12   ▷ Let $M \longrightarrow N \longrightarrow 0$ be an exact sequence of R-modules, and assume that M is finitely generated. Prove that N is finitely generated. (This is half of a converse to Lemma 9.30. As mentioned in the notes, the other half does not hold in general.)

9.13   Let R be a Noetherian ring, and let M be a finitely generated R-module. Prove that every submodule of M is finitely generated.

9.14   ▷ Run the 'row step' described in the proof of Lemma 9.38 on the row vector $(0, -12, 9, -3, -6)$, over $\mathbb{Z}$.

9.15   Determine (as a direct sum of cyclic groups) the cokernel of the homomorphism $\mathbb{Z}^{\oplus 3} \rightarrow \mathbb{Z}^{\oplus 3}$ defined by the matrix

$$\begin{pmatrix} 0 & -2 & 0 \\ 6 & 4 & -2 \\ 2 & 4 & 2 \end{pmatrix}.$$

9.16 Determine the cokernel of the homomorphism $\mathbb{R}[t]^{\oplus 2} \to \mathbb{R}[t]^{\oplus 2}$ defined by the matrix

$$\begin{pmatrix} t-1 & 2 \\ -1 & t+1 \end{pmatrix}.$$

9.17 ▷ Let R be a commutative ring, and let M be an R-module. The 'annihilator' of M is the ideal $\mathrm{Ann}(M) := \{r \in R \mid \forall m \in M, rm = 0\}$. If $d_1 \mid \cdots \mid d_s$, prove that

$$\mathrm{Ann}\,(R/(d_1) \oplus \cdots \oplus R/(d_s)) = (d_s).$$

9.18 ▷ Let M be a finitely generated module over a Euclidean domain, and let T be the *torsion submodule* of M (Exercise 8.19). Prove that M/T is *free*. (*Hint:* Use the classification Theorem 9.40.) Over more general integral domains, the quotient M/T is in general only *torsion-free*, cf. Exercise 8.21.

In particular, a finitely generated torsion-free module over a Euclidean domain is free. Give an example of an integral domain R and a torsion-free R-module M that is not free.

9.19 ▷ Let R be a Euclidean domain, and let N be a submodule of a free module $F \cong R^{\oplus m}$; so N is finitely generated (Euclidean domains are Noetherian) and free by Exercise 9.18.

• Explain why Exercise 9.18 implies that N is free.

• Prove the following more precise result: there exists a basis $(x_1, \ldots, x_m)$ of F and a basis $(y_1, \ldots, y_n)$ of N, with $s \le m$, and elements $d_1, \ldots, d_n \in R$ with $d_1 \mid \cdots \mid d_n$, such that $y_i = d_i x_i$ for $i = 1, \ldots, n$. (Use Lemma 9.38.)

This result may be viewed as a generalization of the fact that Euclidean domains are PIDs: indeed, this is what the result says for $m = 1$.

9.20 Let k be a field, and consider the k-vector space

$$V = k[t]/(t+1) \oplus k[t]/(t^2 - 1).$$

Then V is also a $k[t]$-module; as discussed in §9.6, multiplication by t defines a linear transformation $\alpha: V \to V$.

• Find a basis for V as a k-vector space.

• Obtain a matrix A representing α with respect to this basis.

• Verify that the characteristic polynomial $P(t) = \det(tI - A)$ equals $(t+1)(t^2-1)$. (As mentioned in §9.6, this is a general feature of (9.10).)

• Verify that $P(A) = 0$. (This example illustrates Cayley–Hamilton.)

10 Abelian Groups

10.1 The Category of Abelian Groups

Abelian groups occupy a special place in algebra: on the one hand, they may be viewed as generalizations of modules and of rings, since they are defined by a subset of the axioms defining these structures; on the other hand, they are also a particular case of *R*-modules, i.e., the case of modules over the ring $R = \mathbb{Z}$. It is useful to study abelian groups, for the first reason: we are eventually going to generalize even further (to the case of *groups,* relying on even fewer axioms), and abelian groups are a natural stop on the way. On the other hand, we have already done most of the work, because of the second fact mentioned above.

Thus, we already know what the category of abelian groups is, and we know about sub(-abelian)groups, direct sums, quotients, kernels, and cokernels. We have a canonical decomposition of morphisms of abelian groups, and First, Second, and Third Isomorphism Theorems. We can take a quotient of an abelian group by any subgroup: in the context of abelian groups, kernel $\iff$ subgroup. We do not need to do any new work to establish all these things: we have already done that work over arbitrary rings R in Chapter 8, and the case of abelian groups is simply the special case obtained by taking $R = \mathbb{Z}$.

Further, since $\mathbb{Z}$ is a Euclidean domain, we already know that every finitely generated abelian group is a direct sum of *cyclic* groups (Theorem 9.40 for $R = \mathbb{Z}$). By all accounts, we are in good shape. We will just pause to record a few further observations on the category of abelian groups, and then spend some time looking more carefully at cyclic groups (again, these are the bricks of the theory, at least for what concerns finitely generated abelian groups); and then we will review some applications of the classification theorem. In fact, we will be mostly interested in *finite* abelian groups, that is, abelian groups whose *order* is finite.

DEFINITION 10.1 The 'order' of an abelian group A, denoted $|A|$, is the cardinality of its underlying set.

Caveat on notation. The operation endowing a set A with the structure of an abelian group has been denoted + and called *addition*. The main reason to do this was to distinguish this operation from the 'multiplication' in a ring and from the 'multiplication by scalar' in a module, both denoted by $\cdot$. If we move squarely into the context of abelian groups, then there will really only be *one* operation at work, and it will be useful to allow for other notational styles. A popular alternative to '+' is '·': we often denote by ·, or simply juxtaposition, the operation in an abelian group. This will not lead to confusion in practice, precisely because we will only consider one operation at a time. However, this choice has ripple effects. For example, while using the 'additive' notation + we denote multiples as, well, multiples:

$$a + a + a + a + a = 5a \, ;$$

the same concept is expressed by powers if we are using the 'multiplicative' notation:

$$ggggg = g^5 \, .$$

In multiplicative notation, the identity element is written e (a popular alternative is 1). In the general considerations which follow, I will stick with +, particularly since we are still thinking of abelian groups as $\mathbb{Z}$-modules. In any case, using + tells the reader that the operation is *commutative,* that is, that axiom (iv) in the usual list is satisfied. The notation · does not have this automatic connotation.

Example 10.2 Of course, examples of abelian groups abound. Rings, ideals, and modules give 'additive' examples right away (cf. Examples 8.12 and 8.13). A good 'multiplicative' example to keep in mind is $(R^*, \cdot)$, where R is a commutative ring, R^* denotes its set of units (Definition 3.27), and · is multiplication in R. Multiplication is associative (in any ring) and commutative if R is commutative, 1 is an identity element, and units have inverses by definition. In particular, examples of this type arise when $R = k$ is a field, so $R^* = k^{\neq 0}$ consists of its nonzero elements. For instance, $\mathbb{Q}^{\neq 0}$ is an abelian group under multiplication. As another example, the set of nonzero classes in $\mathbb{Z}/5\mathbb{Z}$ is a group under multiplication, of order 4. We can write the operation table as follows:

·	1	2	3	4
1	1	2	3	4
2	2	4	1	3
3	3	1	4	2
4	4	2	2	1

modulo the usual abuse of notation ($1 = [1]_5$, etc.). Convince yourself that $((\mathbb{Z}/5\mathbb{Z})^*, \cdot)$ satisfies the axioms defining abelian groups. By the same token, $((\mathbb{Z}/p\mathbb{Z})^*, \cdot)$ is an abelian group of order $p - 1$, for every prime p. We will soon have something remarkable and quite nontrivial to say about these abelian groups (Theorem 10.24).

One simplification occurs in the definition of homomorphism of abelian groups. Let $(A, +), (B, +)$ be abelian groups, and let $f : A \to B$ be a function. We know that f is a homomorphism of abelian groups (i.e., of $\mathbb{Z}$-modules) if $\forall n \in \mathbb{Z}, \forall a, a' \in A$,

$$f(a + a') = f(a) + f(a'),$$
$$f(n \cdot a) = n \cdot f(a)$$

(Definition 8.14).

PROPOSITION 10.3   *Let A, B be abelian groups. A function $f: A \to B$ is a homomorphism of abelian groups if and only if $\forall a, a' \in A$,*

$$f(a + a') = f(a) + f(a'), \tag{10.1}$$

In other words, the second requirement for homomorphisms of $\mathbb{Z}$-modules is implied by the first.

You should try to give a formal proof (by induction) of this statement: Exercise 10.1. The point is that, e.g.,

$$f(3a) = f(a + a + a) \overset{*}{=} f(a) + f(a) + f(a) = 3f(a),$$

where $\overset{*}{=}$ holds by (10.1). The $\mathbb{Z}$-action is by 'multiples', and a function preserving addition already preserves multiples.

Similarly, the verification that a subset S of an abelian group A is a subgroup[1] is streamlined with respect to the analogous verification for submodules.

PROPOSITION 10.4   *Let A be an abelian group. Then a subset $S \subseteq A$ is a subgroup if and only if it is nonempty and closed with respect to subtraction.*

Proof   The 'only if' implication is immediate. For the other implication, assume that S is a nonempty subset of A and that it is closed under subtraction. Since S is nonempty, it contains an element b; and then it must contain $0 = b - b$, since it is closed with respect to subtraction; and therefore it must contain $-b = 0 - b$, by the same token. For all $a, b \in S$, since $-b \in S$ and S is closed under subtraction, we see that $a + b = a - (-b) \in S$. Therefore S is closed under addition, and under multiplication by -1. It follows that S is closed under multiples by any $n \in \mathbb{Z}$, and then S is a sub-$\mathbb{Z}$-module by Proposition 8.23. □

In multiplicative notation, all such statements have to be adapted. If $(A, +)$ and $(G, \cdot)$ are abelian groups, a homomorphism $\varphi: A \to G$ is a function such that $\varphi(a_1 + a_2) = \varphi(a_1)\varphi(a_2)$, and so on. For example, the second requirement in the definition of homomorphism (which, as we have seen, is implied by the first) would state that $\varphi(ka) = \varphi(a)^k$ for this function.

Another interesting feature of the category of abelian groups also has to do with homomorphisms. I am really only mentioning it to tickle your curiosity; taking off in this direction would take us too far afield. (So don't worry if the rest of this subsection sounds somewhat mysterious; it will not haunt you in the material which we will

---

[1] For subgroups, the adjective 'abelian' is usually dropped.

develop in the rest of the text.) In general, 'morphisms' between two objects A, B in a category $\mathbf{C}$ form a *set*,[2] denoted $\text{Hom}_{\mathbf{C}}(A, B)$; the subscript $\mathbf{C}$ may be omitted if the category is understood. But of course we are not dealing with categories in general here; so I will reproduce this definition for the category at hand.

DEFINITION 10.5 Let A, B be abelian groups. Then $\text{Hom}(A, B)$ denotes the set of homomorphisms of abelian groups $f : A \to B$.

The feature I would like to advertise is that $\text{Hom}(A, B)$ is in fact *itself* an abelian group. Indeed, I can define an operation $+$ on $\text{Hom}(A, B)$ by defining, $\forall f, g \in \text{Hom}(A, B)$, a new homomorphism $f + g \in \text{Hom}(A, B)$:

$$\forall a \in A \qquad (f + g)(a) = f(a) + g(a).$$

This should not strike you as very unusual; for instance, look back at Example 3.13. It is perhaps worth checking that the 'sum' of two homomorphisms is a homomorphism, and that amounts to the following verification: $\forall a, a' \in A$,

$$(f + g)(a + a') = f(a + a') + g(a + a') = f(a) + f(a') + g(a) + g(a')$$

$$\overset{!}{=} f(a) + g(a) + f(a') + g(a') = (f + g)(a) + (f + g)(a'),$$

where $\overset{!}{=}$ holds by the commutativity of the operation in B.

It is immediate that $(\text{Hom}(A, B), +)$ satisfies the axioms (i)–(iv) defining abelian groups (as in Definition 8.11). Thus, the Hom sets in the category of abelian groups may be viewed as objects of the same category. Not all categories satisfy this property! (Cf. Remark 9.1.)

Going one step further, let A be an abelian group and consider the abelian group $\text{End}(A) := \text{Hom}(A, A)$. (This is the abelian group of 'endomorphisms' of A, i.e., homomorphisms $A \to A$.) Then $\text{End}(A)$ has a *second* operation, that is, composition: if $f : A \to A$ and $g : A \to A$ are homomorphisms, then $f \circ g : A \to A$ is also a homomorphism.

It will take you a moment to verify that $(\text{End}(A), +, \circ)$ is a *ring* (Exercise 10.6). The 0 in this ring is the trivial homomorphism, while the multiplicative identity is the identity *function* $\text{id}_A : A \to A$.

With this understood, start with an *abelian group M* and consider the information needed to specify a *ring homomorphism* $R \to \text{End}(M)$, where R is any ring.

PROPOSITION 10.6 *Let M be an abelian group, and let R be a ring. Every ring homomorphism $f : R \to \text{End}(M)$ determines an R-module structure on M. Conversely, every R-module structure on M determines a ring homomorphism $R \to \text{End}(M)$. These correspondences are inverses of each other.*

---

[2] This is actually one of the axioms defining categories.

Proof If $f\colon R \to \mathrm{End}(M)$ is a ring homomorphism, define a 'multiplication by scalar' $\cdot\colon R \times M \to M$ by setting, $\forall r \in R$, $\forall a \in M$,

$$r \cdot a := f(r)(a). \tag{10.2}$$

(That is: $f(r) \in \mathrm{End}(M)$, so $f(r)$ is a homomorphism $M \to M$; $r \cdot a$ is the result of applying this homomorphism to a.) I claim that this multiplication, together with the addition in M, satisfies the module axioms in Definition 8.1.

Indeed, axioms (i)–(iv) are satisfied because $(M, +)$ is an abelian group; (v):

$$1_R \cdot a = f(1_R)(a) = \mathrm{id}_M(a) = a$$

because ring homomorphisms send 1 to 1; (vi):

$$(rs) \cdot a = f(rs)(a) = (f(r) \circ f(s))(a) = f(r)(f(s)(a)) = f(r)(s \cdot a) = r \cdot (s \cdot a)$$

because f preserves multiplication; (vii):

$$(r + s) \cdot a = f(r + s)(a) = (f(r) + f(s))(a) = f(r)(a) + f(s)(a) = r \cdot a + s \cdot a$$

because f preserves addition; (viii):

$$r \cdot (a + b) = f(r)(a + b) = f(r)(a) + f(r)(b) = r \cdot a + r \cdot b$$

because elements of $\mathrm{End}(M)$ preserve addition.

This concludes the proof of the first assertion. Conversely, if $R \times M \to M$ is the multiplication by scalar making M into an R-module, define $f\colon R \to \mathrm{End}(M)$ by the same formula (10.2), viewed as a definition: $\forall r \in R$, the element $f(r) \in \mathrm{End}(M)$ is defined by setting, $\forall a \in M$,

$$f(r)(a) := r \cdot a.$$

The fact that $f(r)$ is a homomorphism $M \to M$ follows from axiom (viii) in Definition 8.1; f preserves addition by (vii), multiplication by (vi), and the multiplicative identity by (v). (The verifications are entirely analogous to the ones carried out a moment ago.)

It is clear that the assignments are inverses of each other, since they are defined by the same prescription (10.2). $\qquad\square$

Remark 10.7 The proof of Proposition 10.6 is straightforward—once you come up with (10.2), all the verifications are immediate. Still, the result is interesting. First of all, it highlights what a 'ring action' really is: if the set of morphisms $\mathrm{End}_{\mathbf{C}}(X)$ from an object X of a category $\mathbf{C}$ to itself happens to be a ring, then 'acting' on X with a ring R means specifying a ring homomorphism $R \to \mathrm{End}_{\mathbf{C}}(X)$. An R-module M may be viewed as an abelian group M endowed with an action of the ring R on M. We will have good reason to consider other examples of actions, based on *groups* rather than rings, later on in these notes (cf. especially §11.2).

Second, we again see why we should identify the notion of abelian group and the notion of $\mathbb{Z}$-module: given an abelian group M, there is *only one* ring homomorphism $\mathbb{Z} \to \mathrm{End}(M)$, because there is only one homomorphism from $\mathbb{Z}$ to *any* ring (Exercise 4.14: $\mathbb{Z}$ is 'initial' in the category of rings). Therefore, there is one and only one way to make an abelian group into a $\mathbb{Z}$-module. We add nothing to an abelian group by viewing it as a $\mathbb{Z}$-module. This was something we noticed as far back as Proposition 8.10, but that we (possibly) understand better now.

Third, the fact that the End of an abelian group A is a *ring* may be viewed as motivation for the ring axioms: a ring is nothing but the abstraction of the general properties that End(A) satisfies, regardless of what abelian group A may be. In this sense, the ring axioms are *not* an arbitrary invention; they are something one inevitably stumbles upon when studying a more basic type of algebraic structure, that is, abelian groups.

Incidentally, here is another good reason why the 'right' definition of ring *should* include the multiplicative identity: it does in this prototypical case of End(A).   ⌙

10.2   Cyclic Groups, and Orders of Elements

Since $\mathbb{Z}$ is a Euclidean domain and abelian groups are $\mathbb{Z}$-modules, Theorem 9.40 implies that every finitely generated abelian group is a direct sum of *cyclic $\mathbb{Z}$-modules*. These are simply called 'cyclic groups'. Here is Definition 8.29 again, for the case of abelian groups.

DEFINITION 10.8   Let G be an abelian group. We say that G is a 'cyclic group' if $\exists g \in G$ such that $G = \langle g \rangle$.

Quite simply, a cyclic group consists of all the multiples of its 'generator' g. We already know (Example 8.45) that all cyclic R-modules are isomorphic to R/I for some ideal I of R; since $\mathbb{Z}$ is a PID, all cyclic groups are isomorphic to $\mathbb{Z}/n\mathbb{Z}$ for some $n \geq 0$. Thus there is exactly one *infinite* cyclic group up to isomorphism, i.e., $\mathbb{Z}$; and for every positive integer $n > 0$ there is exactly one cyclic group of order (that is, cardinality) n up to isomorphism, i.e., $\mathbb{Z}/n\mathbb{Z}$. A cyclic group of order $n > 0$ is often denoted C_n, particularly (but not exclusively) when we use multiplicative notation: so if $C_n = \langle g \rangle$ and we are adopting multiplicative notation, then every element of C_n may be written as g^k for some $k \in \mathbb{Z}^{\geq 0}$.

We have studied the structures $\mathbb{Z}/n\mathbb{Z}$ thoroughly from the ring-theoretic point of view. The emphasis shifts a little when we look at them again as abelian groups; for example, properties such as 'being a unit' only make sense in the setting of rings, since they have to do with multiplication, and this operation is not available in the context of abelian groups. Still, much of what we know can be reinterpreted from the point of view of abelian groups.

PROPOSITION 10.9   *The infinite cyclic group $\mathbb{Z}$ has two generators: -1 and $+1$.*
   The cyclic group $\mathbb{Z}/n\mathbb{Z}$ has $\phi(n)$ generators. In fact, $[a]_n$ generates $\mathbb{Z}/n\mathbb{Z}$ if and only if $\gcd(a, n) = 1$, if and only if $[a]_n \in (\mathbb{Z}/n\mathbb{Z})^*$.

(Recall that $\phi(n)$ denotes Euler's 'totient' function, i.e., the number of positive integers k, $1 \leq k \leq n$, that are relatively prime to n; see §5.6.)

Proof   The first assertion is clear. For the second, note that $[a]_n$ generates $\mathbb{Z}/n\mathbb{Z}$ if and only if every class $[b]_n$ can be written as a multiple of $[a]_n$. In particular, if $[a]_n$ is a generator, then there exists a $k \in \mathbb{Z}$ such that $k[a]_n = [1]_n$, i.e., $[k]_n[a]_n = [1]_n$. This implies that $[a]_n$ is a unit in $\mathbb{Z}/n\mathbb{Z}$. Conversely, if $[a]_n$ is a unit, then there exists a class $[k]_n$ such that $[ka]_n = [1]_n$; and then every class $[b]_n$ can be written as a multiple of $[a]_n$, since $[b]_n = b[1]_n = bk[a]_n$.

This shows that $[a]_n$ is a generator of the *abelian group* $\mathbb{Z}/n\mathbb{Z}$ if and only if it is a unit in the *ring* $\mathbb{Z}/n\mathbb{Z}$, i.e., $[a]_n \in (\mathbb{Z}/n\mathbb{Z})^*$. The result then follows from Proposition 3.28.   □

The proof shows an interplay with ring theory, but note that while, e.g., $1 \in \mathbb{Z}$ is a very special element from the point of view of ring theory, from the point of view of abelian groups it is just one of two possible generators of the infinite cyclic group $\mathbb{Z}$; it is exactly as special as -1 may be. The function

$$i \colon \mathbb{Z} \to \mathbb{Z} \quad \text{defined by } i(n) = -n$$

is an *isomorphism* of abelian groups, while it is not even a homomorphism of rings.

The next result is also borrowed from ring theory. We know all subgroups of $\mathbb{Z}$: a subgroup is a sub-$\mathbb{Z}$-module, and the sub-R-modules of an R-module are its ideals; therefore, since $\mathbb{Z}$ is a PID, its subgroups must be all of the form $d\mathbb{Z}$, $d \geq 0$. The situation for finite cyclic groups is similar. Further, we can also classify all possible homomorphic images of a cyclic group.

PROPOSITION 10.10   *(i) Every subgroup of a cyclic group is cyclic. In fact, the cyclic group $C_n = \langle g \rangle$ has exactly one cyclic subgroup of order m for every positive $m \mid n$; this subgroup is generated by $g^{\frac{n}{m}}$.*

(ii) Every homomorphic image of a cyclic group is cyclic. In fact, there is a surjective homomorphism of abelian groups $C_n \to C_m$ if and only if $m \mid n$, and every homomorphic image of C_n is isomorphic to C_m for some $m \mid n$.

Proof   (i) Since $(C_n, \cdot) \cong (\mathbb{Z}/n\mathbb{Z}, +)$, we deal with the latter. We may choose the isomorphism so that g corresponds to $[1]_n$. By the Third Isomorphism Theorem (Theorem 8.47), the subgroups of $\mathbb{Z}/n\mathbb{Z}$ correspond to the subgroups of $\mathbb{Z}$ containing $n\mathbb{Z}$. We know every subgroup of $\mathbb{Z}$ is of the form $d\mathbb{Z}$ for some $d \in \mathbb{Z}$; and $n\mathbb{Z} \subseteq d\mathbb{Z}$ implies that $d \mid n$. Therefore every subgroup of $\mathbb{Z}/n\mathbb{Z}$ is of the form $d\mathbb{Z}/n\mathbb{Z}$, and there is exactly one such subgroup for every positive $d \mid n$. Now note that

$$d\mathbb{Z}/n\mathbb{Z} \cong \mathbb{Z}/(\tfrac{n}{d})\mathbb{Z}.$$

Indeed, this follows from the First Isomorphism Theorem, since we have a surjective homomorphism

$$\mathbb{Z} \xrightarrow{\;d\;} d\mathbb{Z} \longrightarrow d\mathbb{Z}/n\mathbb{Z},$$

whose kernel is

$$\{a \in \mathbb{Z} \mid da \in n\mathbb{Z}\} = \tfrac{n}{d}\mathbb{Z}$$

(keep in mind that $\frac{n}{d} \in \mathbb{Z}$, since $d \mid n$). The first assertion follows, with $m = \frac{n}{d}$. Further, the subgroup $d\mathbb{Z}/n\mathbb{Z}$ is generated by $[d]_n = [\frac{n}{m}]_n$, which corresponds to $g^{\frac{n}{m}}$ in multiplicative notation, and this concludes the proof.

(ii) Again switching to additive notation, assume that A is a homomorphic image of $\mathbb{Z}/n\mathbb{Z}$: that is, assume that there is a surjective homomorphism $f : \mathbb{Z}/n\mathbb{Z} \to A$. Then I claim that $A = \langle f([1]_n) \rangle$. Indeed: if $a \in A$, then $\exists k \in \mathbb{Z}^{\geq 0}$ such that $a = f([k]_n)$; but then $a = f([k]_n) = f(k[1]_n) = kf([1]_n) \in \langle f([1]_n) \rangle$.

This shows that A is cyclic: $A \cong \mathbb{Z}/m\mathbb{Z}$ for some m, and we may assume $f([1]_n) = [1]_m$. Since $[n]_n = [0]_n$, we see that

$$[n]_m = n[1]_m = n f([1]_n) = f(n[1]_n) = f([n]_n) = f([0]_n) = [0]_m .$$

This shows that $n \equiv 0 \bmod m$, that is, $m \mid n$.

All that is left to show is that if $m \mid n$, then there is a surjective homomorphism $\mathbb{Z}/n\mathbb{Z} \twoheadrightarrow \mathbb{Z}/m\mathbb{Z}$. If $m \mid n$, then $n\mathbb{Z} \subseteq m\mathbb{Z}$, and by the Third Isomorphism Theorem

$$(\mathbb{Z}/n\mathbb{Z})/(m\mathbb{Z}/n\mathbb{Z}) \cong \mathbb{Z}/m\mathbb{Z} .$$

Composing this isomorphism with the natural projection $\mathbb{Z}/n\mathbb{Z} \twoheadrightarrow (\mathbb{Z}/n\mathbb{Z})/(m\mathbb{Z}/n\mathbb{Z})$ gives the sought-for surjective homomorphism $\mathbb{Z}/n\mathbb{Z} \twoheadrightarrow \mathbb{Z}/m\mathbb{Z}$. □

The argument proving (i) should look familiar—it is simply recovering the set of ideals in the *ring* $\mathbb{Z}/n\mathbb{Z}$, cf Example 5.59. The analogue of (ii) in ring theory also holds, with the same proof.

Example 10.11 What (i) says is that the 'lattice of subgroups' of $C_n \cong \mathbb{Z}/n\mathbb{Z}$ matches the structure of positive divisors of n. For instance, for $n = 12$ we have the following situation:

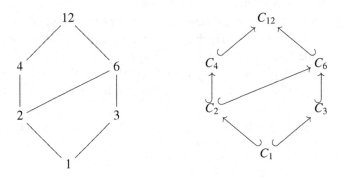

where I am indicating the divisibility relation by connecting numbers with a —. The subgroups are as follows, in terms of generators or listing elements (and writing a for $[a]_{12}$):

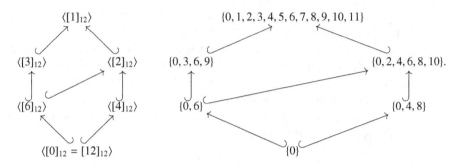

The lattice is very simple for powers of a prime p, because so is the divisibility relation:

$$C_1 \hookrightarrow C_p \hookrightarrow C_{p^2} \hookrightarrow C_{p^3} \hookrightarrow C_{p^4} \hookrightarrow C_{p^5} \hookrightarrow \cdots .$$

For $C_{2^4} \cong \mathbb{Z}/16\mathbb{Z}$, this sequence is

$$\{0\} \subseteq \{0, 8\} \subseteq \{0, 4, 8, 12\} \subseteq \{0, 2, 4, 6, 8, 10, 12, 14\}$$
$$\subseteq \{0, 1, 2, 3, 4, 5, 6, 7, 8, 9, 10, 11, 12, 13, 14, 15\} .$$

By the way, note that $g \in C_{p^k}$ is a generator for C_{p^k} if and only if it is not contained in the largest proper subgroup $C_{p^{k-1}}$. (This should be evident; and work out Exercise 10.7 for good measure.) It follows that $\phi(p^k) = p^k - p^{k-1}$. This recovers the key point of Exercise 5.33.

Intimately related with cyclic groups is the following useful definition.

DEFINITION 10.12   Let G be an abelian group, and let $g \in G$. The 'order' of g, denoted $|g|$, is the order of the cyclic group $\langle g \rangle \subseteq G$ generated by g.

Thus, $|g| = \infty$ if $\langle g \rangle \cong \mathbb{Z}$, and $|g| = n$ if $\langle g \rangle \cong \mathbb{Z}/n\mathbb{Z}$. As g corresponds to a generator of $\mathbb{Z}/n\mathbb{Z}$ in the latter case, we see that $|g| = n > 0$ if (i) $ng = 0$ and (ii) $kg \neq 0$ for all positive integers $k < n$. Therefore, $|g|$ is the *smallest positive integer n* such that $ng = 0$, or $|g| = \infty$ if there is no such positive integer. This is a useful interpretation of the notion of 'order', and may be adopted as an alternative definition.

For example, order considerations clarify the following version of the 'Chinese Remainder Theorem', already recalled in (9.9) in the more general context of modules.

PROPOSITION 10.13   *Let $m, n > 0$ be integers. Then $C_m \times C_n \cong C_{mn}$ if and only if m and n are relatively prime.*

Proof   If $\gcd(m, n) = 1$, then $C_{mn} \cong C_m \times C_n$ by (9.9).
If $\gcd(m, n) = d \neq 1$, then $\forall a \in C_m \cong \mathbb{Z}/m\mathbb{Z}$ and $\forall b \in C_n \cong \mathbb{Z}/n\mathbb{Z}$ we have

$$\frac{mn}{d}(a, b) = \left(\frac{n}{d} \cdot ma, \frac{m}{d} \cdot nb \right) = (0, 0) .$$

In particular, *all* elements have order $\leq \frac{mn}{d} < mn$. Since no element has order mn, $C_m \times C_n \not\cong C_{mn}$, as claimed. □

Definition 10.12, combined with the previous results, makes it very easy to count elements of given order in a cyclic group.

Example 10.14 The cyclic group C_n contains an element of order 4 if and only if 4 divides n. Indeed, such an element generates a subgroup of C_n isomorphic to C_4, and C_n contains an isomorphic copy of C_4 if and only if $4 \mid n$ by Proposition 10.10. If $4 \mid n$, then C_n contains exactly *two* elements of order 4. Indeed, they are the generators of this isomorphic copy of C_4, and C_4 has $\phi(4) = 2$ generators by Proposition 10.9.

More generally, C_n contains elements of order p^k, where p is prime, if and only if p^k divides n. If p^k divides n, then C_n contains exactly $\phi(p^k) = p^k - p^{k-1}$ elements of order p^k.

Example 10.15 Further, if $p^k \mid C_n$, then C_n contains exactly p^k elements whose order divides p^k. Indeed, the subgroup of C_n isomorphic to C_{p^k} consists precisely of these elements. (If $\ell \leq k$ and $g \in C_n$ generates a subgroup isomorphic to C_{p^ℓ}, then this subgroup must be contained in $C_{p^k} \subseteq C_n$: indeed, C_{p^k} contains a subgroup isomorphic to C_{p^ℓ}, and there is only one such subgroup in C_n. In particular, $g \in C_{p^k}$.)

More generally, C_n contains $\min(p^k, p^\ell)$ elements whose order divides p^k, where p^ℓ is the largest power of p dividing n.

The following result is very simple, but occasionally very useful.

PROPOSITION 10.16 *Let A be an abelian group, and let $a \in A$, $N \in \mathbb{Z}$. Then $Na = 0$ if and only if $|a|$ divides N.*

(In multiplicative notation: $g^N = e$ if and only if $|g|$ divides N.)

Proof If $|a| = \infty$, then $Na \neq 0$ for all $N > 0$, and there is nothing to prove.

If $|a| = n$, then by definition there is an isomorphism $\alpha: \langle a \rangle \to \mathbb{Z}/n\mathbb{Z}$, and we may assume that $\alpha(a) = [1]_n$. Therefore, the statement is equivalent to the assertion that $N[1]_n = [0]_n$ if and only if $n \mid N$. This is clear: $N[1]_n = [N]_n$, and $[N]_n = [0]_n$ if and only if $n \mid N$ by definition of congruence modulo n. □

With this understood, we can refine Proposition 10.10(ii) and describe all homomorphisms from a cyclic group to another abelian group A. The case of $\mathbb{Z}$ is already covered by Proposition 9.8, since $\mathbb{Z}$ is free as a $\mathbb{Z}$-module: *every* choice of an element $a \in A$ determines a unique homomorphism of abelian groups $f: \mathbb{Z} \to A$ such that $f(1) = a$. The following result takes care of all finite cyclic groups.

PROPOSITION 10.17 *Let $C_n = \langle g \rangle$ be a cyclic group of order n generated by an element g, and let A be an abelian group.*

- *Let $f: C_n \to A$ be a homomorphism of abelian groups. Then the order of $f(g)$ divides n.*
- *Conversely, let $a \in A$ be an element whose order divides n. Then there exists a unique homomorphism $f: C_n \to A$ such that $f(g) = a$.*

Proof If $f: C_n \to A$ is a homomorphism of abelian groups, then $f(C_n) = \langle f(g) \rangle$ is a homomorphic image of C_n. It is cyclic of order m for some $m \mid n$, by Proposition 10.10(ii). This proves the first part.

For the second part, let $a \in A$ be an element whose order divides n. Since $\mathbb{Z}$ is free, we have a homomorphism $\overline{f}: \mathbb{Z} \to A$, determined by setting $\overline{f}(1) = a$ (Proposition 9.8). We have

$$\overline{f}(n) = n \cdot \overline{f}(1) = n \cdot a = 0$$

since $|a|$ divides n by hypothesis (and using Proposition 10.16). This shows that $n \in \ker \overline{f}$; therefore $n\mathbb{Z} \subseteq \ker \overline{f}$, and we get an induced homomorphism $f: \mathbb{Z}/n\mathbb{Z} \to A$ making the diagram

commute. (Here I am using Exercise 8.23, which is nothing but a rehash of Proposition 5.39.) In fact, $f([1]_n) = \overline{f}(1) = a$. This gives the statement (modulo the isomorphism $C_n \cong \mathbb{Z}/n\mathbb{Z}$, matching g with $[1]_n$). □

10.3 The Classification Theorem

In §9.5 we proved a 'classification theorem' for all finitely generated modules over a Euclidean domain, Theorem 9.40. As $\mathbb{Z}$ is a Euclidean domain, and abelian groups are $\mathbb{Z}$-modules, the classification theorem holds for finitely generated *abelian groups*. Thus, we already know that every finitely generated abelian group is a direct sum of cyclic groups. Restricting further to the case of finite abelian groups, we can state this particular case of Theorem 9.40 as follows.

THEOREM 10.18 (Classification of finite abelian groups) *Let A be a finite abelian group. Then there exist integers $1 < d_1, \ldots, d_s$, with $d_1 \mid d_2 \mid \cdots \mid d_s$, such that*

$$A \cong C_{d_1} \oplus \cdots \oplus C_{d_s}.$$

Further, this decomposition is unique.

Equivalently, there exist prime integers $p_1, \ldots, p_r$ and exponents $\alpha_{ij} > 0$, uniquely determined by A, such that

$$A \cong \bigoplus_{i,j} (\mathbb{Z}/p_i^{\alpha_{ij}}\mathbb{Z}).$$

The equivalence between the two forms of the statement follows from the Chinese Remainder Theorem, in the form given in Proposition 10.13. (Also cf. Corollary 9.44, giving the corresponding statement for modules over more general Euclidean domains.)

The existence of the decompositions as direct sums in Theorem 10.18 was fully proved in §9.5. The uniqueness was discussed, but not proved in detail, in Remark 9.41. In the case of finite abelian groups, our work on cyclic groups in §10.2 simplifies the argument. Since the two formulations in Theorem 10.18 are equivalent, it suffices to show the uniqueness of the 'elementary divisors' $p_i^{\alpha_{ij}}$.

> PROPOSITION 10.19   *Let $p_1, \ldots, p_r, q_1, \ldots, q_s$ be positive prime integers, and let $\alpha_{ij} > 0, \beta_{k\ell} > 0$ be integers. Assume that*
>
> $$A \cong \bigoplus_{i,j}(\mathbb{Z}/p_i^{\alpha_{ij}}\mathbb{Z}) \cong \bigoplus_{k,\ell}(\mathbb{Z}/q_k^{\beta_{k\ell}}\mathbb{Z}).$$
>
> *Then $r = s$, and up to a reordering we have $p_i = q_i$, $\alpha_{ij} = \beta_{ij}$ for all i and j.*

Proof   For a prime p, we can extract the part of the two decompositions for which $p_i = q_k = p$:

$$(\mathbb{Z}/p^{\alpha_1}\mathbb{Z}) \oplus \cdots \oplus (\mathbb{Z}/p^{\alpha_m}\mathbb{Z}) \cong (\mathbb{Z}/p^{\beta_1}\mathbb{Z}) \oplus \cdots \oplus (\mathbb{Z}/p^{\beta_n}\mathbb{Z}) \qquad (10.3)$$

with $1 \le \alpha_1 \le \cdots \le \alpha_m$, $1 \le \beta_1 \le \cdots \le \beta_n$ (for example, these direct sums equal 0 if p is *not* a prime appearing in the decompositions). These two groups are isomorphic because they are both isomorphic to the subgroup of A consisting of elements whose order is a power of p (cf. Exercise 10.11). We have to prove that $m = n$ and the exponents in these two decompositions agree. To verify this, consider the number of elements of order $\le p^d$ in both, for all d; since the direct sums are isomorphic, this number is the same for both. An element

$$(a_1, \ldots, a_m) \in (\mathbb{Z}/p^{\alpha_1}\mathbb{Z}) \oplus \cdots \oplus (\mathbb{Z}/p^{\alpha_m}\mathbb{Z})$$

has order $\le p^d$ if and only if each a_i has order $\le p^d$, and it follows that this number equals

$$\prod_{i=1}^{m} p^{\min(\alpha_i, d)}$$

for the left-hand side of (10.3), and

$$\prod_{k=1}^{n} p^{\min(\beta_k, d)}$$

for the right-hand side (cf. Example 10.15). This shows that

$$\prod_{i=1}^{m} p^{\min(\alpha_i, d)} = \prod_{k=1}^{n} p^{\min(\beta_k, d)}$$

for all $d \ge 0$, and it follows easily that the lists (α_i), (β_k) agree (Exercise 10.15).   □

Theorem 10.18 allows us to 'list' all abelian groups of a given size, up to isomorphism.

Example 10.20   How many abelian groups of order 360 are there, up to isomorphism? Answer: 6. Indeed, by Theorem 10.18 there is one such group for every sequence of integers $1 < d_1 < d_2 < \cdots$ such that $d_1 \mid d_2 \mid \cdots$ and whose product is $360 = 2^3 3^2 5$. There are six such lists:

$$360$$
$$2 \mid 180$$
$$2 \mid 2 \mid 90$$
$$3 \mid 120$$
$$6 \mid 60$$
$$2 \mid 6 \mid 30$$

How do I know? List columns with decreasing powers of the primes 2, 3, 5, totaling $2^3 = 8$, $3^2 = 9$, $5^1 = 5$, then take the products of the numbers in each row. For example, the second-to-last list comes from this table:

$60 =$	4	3	5
$6 =$	2	3	

Listing all possible such tables is straightforward.

The classification theorem tells us that the six distinct isomorphism classes of abelian groups of order 360 are

$$C_{360}, \quad C_2 \oplus C_{180}, \quad C_2 \oplus C_2 \oplus C_{90}, \quad C_3 \oplus C_{120}, \quad C_6 \oplus C_{60}, \quad C_2 \oplus C_6 \oplus C_{30}.$$

Every abelian group of order 360 you will ever run across is isomorphic to one and only one in this list.

It is hard to overemphasize how useful a classification theorem is. Armed with Theorem 10.18, we can reduce questions about *arbitrary* finite abelian groups to questions about *a specific type* of group, that is, direct sums of cyclic groups. Often this allows us to dispose of the question quickly and systematically, removing the need for clever constructions, induction arguments, and whatnot.

Here is a simple example.

PROPOSITION 10.21   *Let A be a finite abelian group, and let $a \in A$. Then $|a|$ divides $|A|$.*

Proof   By the classification theorem,

$$A \cong (\mathbb{Z}/d_1\mathbb{Z}) \oplus \cdots \oplus (\mathbb{Z}/d_s\mathbb{Z})$$

for some positive integers $1 < d_1 \mid \cdots \mid d_s$. If $\underline{a} = ([a_1]_{d_1}, \ldots, [a_s]_{d_s})$ is an element of the direct sum, then

$$d_s \underline{a} = d_s([a_1]_{d_1}, \ldots, [a_s]_{d_s}) = ([d_s a_1]_{d_1}, \ldots, [d_s a_s]_{d_s}) = 0 \,.$$

Indeed, $d_s a_i$ is a multiple of d_i since d_s is a multiple of d_i, for all i. Since $|A| = d_1 \cdots d_s$, we see that $|A|a = 0$; and this implies that $|a|$ divides $|A|$, by Proposition 10.16. □

We will recover this result in the more general setting of *groups* from *Lagrange's theorem* (Theorem 12.3), which is actually a much simpler deal than the classification theorem. Perhaps a better illustration of the usefulness of the classification theorem is the following result, which we will need in due time. (It will be an ingredient in the proof of the *first Sylow's first theorem,* in §12.5.)

> THEOREM 10.22 (Cauchy's Theorem, abelian version) *Let p be a prime divisor of the order of an abelian group A. Then A contains an element of order p.*

Proof By Theorem 10.18, A is a direct sum of cyclic groups: $A \cong C_{d_1} \oplus \cdots \oplus C_{d_s}$. Since p divides $|A| = d_1 \cdots d_s$ and $d_1 \mid \cdots \mid d_s$, p divides d_s. By Proposition 10.10(i), there exists a cyclic subgroup of order p in C_{d_s}. As C_{d_s} may be viewed as a subgroup of $C_{d_1} \oplus \cdots \oplus C_{d_s} \cong A$, this shows that A contains a cyclic subgroup of order p. If a is a generator of this subgroup, we get that $|a| = p$, concluding the proof. □

It is not difficult to prove Cauchy's theorem by an induction argument on $|A|$; you are welcome to try if you want. If you do, you will agree that the proof given above is simpler, once you can use the classification theorem.

The following result is also straightforward from the classification theorem, and it will have a nice consequence.

> LEMMA 10.23 *Let A be a finite abelian group, and assume that for every integer $n > 0$ the number of elements $a \in A$ such that $na = 0$ is at most n. Then A is cyclic.*

Proof By the classification theorem,

$$A \cong (\mathbb{Z}/d_1\mathbb{Z}) \oplus \cdots \oplus (\mathbb{Z}/d_s\mathbb{Z})$$

for some positive integers $1 < d_1 \mid \cdots \mid d_s$. In the proof of Proposition 10.16 we have verified that $d_s a = 0$ for all $a \in A$. Therefore, the number of elements $a \in A$ such that $d_s a = 0$ is $|A|$. The hypothesis implies $|A| \le d_s$. Since $|A| = d_1 \cdots d_s \ge d_s$, we conclude that necessarily $|A| = d_s$, that is, $s = 1$: $A \cong \mathbb{Z}/d_s\mathbb{Z}$ is cyclic. □

For example, this lemma is the key ingredient in the proof of the following theorem, which will be important in applications, for example to the classification of finite fields (§14.5). We have already observed that if k is a field, then $(k^*, \cdot)$ is an abelian group (Example 10.2).

THEOREM 10.24   *Let k be a field, and let G be a finite subgroup of the multiplicative abelian group $(k^*, \cdot)$. Then G is a cyclic group.*

Proof   For every n, the equation $x^n - 1$ has at most n solutions: this is an easy consequence of the fact that unique factorization holds in $k[x]$ if k is a field (Corollary 7.8); in fact, you worked this out in Exercise 7.14.

In particular, we see that there are at most n elements $g \in G$ for which $g^n = 1$, for all n. Since G is a finite abelian group, Lemma 10.23 implies that G must be cyclic. (The condition $g^n = 1$ is the version in multiplicative notation of the condition $na = 0$ appearing in Lemma 10.23.)   □

Example 10.25   Look way back at Exercise 2.9: there you were looking for a 'congruence class $[a]_7$ in $\mathbb{Z}/7\mathbb{Z}$ such that every class $[b]_7$ except $[0]_7$ equals a power of $[a]_7$'. You now realize that this awkward sentence is asking for a generator of the multiplicative group $((\mathbb{Z}/7\mathbb{Z})^*, \cdot)$, and at this point we know that *this group admits a generator,* since it is cyclic by Theorem 10.24.

Example 10.26   More generally, $((\mathbb{Z}/p\mathbb{Z})^*, \cdot)$ is a cyclic group of order $p - 1$ for every positive prime integer p. This has always struck me as most remarkable. The integer $p = 87178291199$ is allegedly prime; I find it amazing that we know without any further work whatsoever that the abelian group

$$((\mathbb{Z}/87178291199\,\mathbb{Z})^*, \cdot)$$

is a cyclic group (of order 87178291198). Thus, there must be an integer a between 1 and 87178291198 such that *every* integer in the same range is congruent to a power of a modulo 87178291199. (In fact, there are *a lot* of integers satisfying this property, cf. Exercise 10.27.)

Example 10.27   It is easy to check that $x^3 + x + 1$ is irreducible over $\mathbb{Z}/5\mathbb{Z}$ (Proposition 7.15), and it follows that

$$(\mathbb{Z}/5\mathbb{Z})[x]/(x^3 + x + 1)$$

is a field (irreducible elements generate maximal ideals in a PID, by Proposition 6.33). I have no special intimate knowledge of this field with 125 elements, and yet I know (since I know Theorem 10.24) that the corresponding multiplicative group

$$(((\mathbb{Z}/5\mathbb{Z})[x]/(x^3 + x + 1))^*, \cdot)$$

is a *cyclic* group of order 124. I know this even if I have no idea of what a generator of this group may be; finding one would require some actual work. Isn't this remarkable?

10.4   Fermat's Theorem on Sums of Squares

The concepts we have explored in the last several sections are objectively important for the development of the standard material; but we can take a little detour on the way and see that what we have learned allows us to do some concrete things that would be quite hard without these notions. Around 1640, Fermat stated the following theorem[3] (without providing a proof).

> THEOREM 10.28   *Let $p > 2$ be a prime integer. Then p may be written as a sum of two squares if and only if $p \equiv 1$ mod 4.*

For example, the 10,000th positive prime is $104729 \equiv 1$ mod 4. According to Theorem 10.28, it must be possible to find positive integers a, b, such that $104729 = a^2 + b^2$. And indeed, $104729 = 20^2 + 323^2$. The 10,001st prime is $104743 \equiv 3$ mod 4. Therefore, this prime can*not* be written as a sum of two squares, according to Theorem 10.28.

It took *one hundred* years before a proof of this theorem was published (by Euler), and that proof was in a sense trickier (even if 'elementary') than the proof we are going to go through. Take this as an example of the power of the abstract machinery: by knowing a little about Euclidean domains, cyclic groups, and so on, you are in the position of understanding in a few minutes a result that at one time was perceived as quite challenging.[4]

How do we go about proving Theorem 10.28? The theorem actually consists of two statements: if $p > 0$ is prime, then

- If $p \equiv 1$ mod 4, then p *can* be written as the sum of two squares, and
- If $p \equiv 3$ mod 4, then p can*not* be written as the sum of two squares.

The second statement is actually nearly immediate: you disposed of its contrapositive as soon as you learned modular arithmetic, in Exercise 2.10. In two words, if $p = a^2 + b^2$, then $[p]_4 = [a]_4^2 + [b]_4^2$; but the class of a square modulo 4 is necessarily $[0]_4$ or $[1]_4$, so adding two of them cannot get you $[3]_4$.

The challenging statement is the first one: to show that for an odd prime p, if $p \equiv 1$ *mod* 4, then $p = a^2 + b^2$ *for some integers a and b.*

Let's set the stage to study this question. On the one hand, we have the condition $p \equiv 1$ mod 4; this looks straightforward enough, but we will have to find a way to encode this condition so that it has something to do with the other condition, $p = a^2 + b^2$, which may look more mysterious. How can we process the condition of being 'the sum of two squares'?

If you stare at it long enough, you may hit upon the idea of viewing $a^2 + b^2$ as a product $(a + ib)(a - ib)$, an operation performed in the ring $\mathbb{Z}[i]$ of *Gaussian integers* we

---

[3] A version of the statement was actually published earlier by Albert Girard.
[4] There are modern clever proofs of this result which use different tools and can be compressed into as little as *one sentence*. See Don Zagier's 1990 paper with the title 'A One-Sentence Proof That Every Prime $p \equiv 1$ mod 4 Is a Sum of Two Squares', in the *American Mathematical Monthly*.

encountered briefly in Example 4.12. This is

$$\mathbb{Z}[i] := \{a + bi \in \mathbb{C} \mid a \in \mathbb{Z}, b \in \mathbb{Z}\} :$$

the subring of $\mathbb{C}$ consisting of complex numbers whose real and imaginary parts are integers. A simple application of the First Isomorphism Theorem for rings shows that

$$\mathbb{Z}[i] \cong \mathbb{Z}[x]/(x^2 + 1)$$

(Example 5.42); and since $\mathbb{Z}[i]$ is a subring of a field (that is, $\mathbb{C}$), it is an integral domain.

We need to become a little more familiar with this ring. Here is a picture of some of its elements, drawn in the ordinary complex plane:

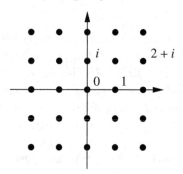

The complex plane has an interesting symmetry, that is, *conjugation* (see Example 4.23). It maps $z = a + bi$ to $\bar{z} = a - bi$, and in particular it sends $\mathbb{Z}[i]$ to itself: complex conjugation induces a function $\mathbb{Z}[i] \to \mathbb{Z}[i]$. Since this function clearly preserves 1, addition, and multiplication, it is a ring homomorphism; and it is its own inverse, so it is in fact an *isomorphism*. This observation will be useful in a moment.

It is easy to visualize principal ideals in this ring. For example, the ideal $(2+i)$ consists of a tilted, enlarged square lattice superimposed on the lattice $\mathbb{Z}[i]$:

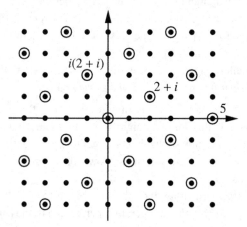

This will be clear to you if you have any familiarity whatsoever with complex numbers. The integer multiples of $2 + i$ are the lattice points on the line through 0 and $2 + i$; the integer multiples of the complex number $i(2+i)$ are the lattice points on the line through the origin and *perpendicular* to the same line; and all elements of the ideal $(2 + i)$, that is, all the multiples $z(2 + i)$ with $z = a + bi \in \mathbb{Z}[i]$, are combinations of these.

Next, recall that the 'squared norm' of a complex number $z = x+iy$ is the nonnegative real number

$$N(z) = |z|^2 = z\bar{z} = (x + iy)(x - iy) = x^2 + y^2 .$$

Geometrically, this is the square of the distance from 0 to z in the usual representation of $\mathbb{C}$ as a plane. Thus, $N(w - z)$ is the square of the distance between the complex numbers z and w. This function restricts to a function $N : \mathbb{Z}[i] \to \mathbb{Z}^{\geq 0}$. Its very existence gives us useful information. For example, note that N preserves multiplication:

$$N(zw) = |zw|^2 = |z^2||w^2| = N(z)N(w).$$

The following observation is a simple consequence.

LEMMA 10.29   *The units in $\mathbb{Z}[i]$ are $\pm 1, \pm i$.*

Proof If $z \in \mathbb{Z}[i]$ is a unit, then there exists a w in $\mathbb{Z}[i]$ such that $zw = 1$. Since N is multiplicative, $N(z)N(w) = N(zw) = N(1) = 1$. This shows that $N(z)$ is a unit *in* $\mathbb{Z}$, that is, $N(z) = 1$ (since $N(z)$ is a nonnegative integer). The only elements $z \in \mathbb{Z}[i]$ with $N(z) = 1$ are $z = \pm 1, z = \pm i$, and this gives the statement.   □

But much more is true. By our good fortune, the function N turns out to be a *Euclidean valuation!* (Definition 6.24.) That is, we have the following result.

PROPOSITION 10.30   *The ring $\mathbb{Z}[i]$ is a Euclidean domain.*

Proof All the King's horses and all the King's notation could be used here, but my feeling is that the following picture clarifies why N is a Euclidean valuation in a matter of seconds and just as convincingly.

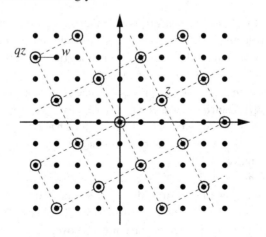

Given $z, w \in \mathbb{Z}[i]$, with $z \neq 0$, w will fall within one of the 'boxes' of the larger lattice determined by the ideal (z). Let $qz \in (z)$ be the nearest corner of this box to w. Then the distance between qz and w is less than the length of the side of the box. This says that, setting $r = w - qz$, we have

$$w = qz + r \quad \text{and} \quad N(r) < N(z).$$

(This allows for the possibility that $w = qz$, in which case $r = 0$.) Therefore, 'division with remainder' holds in $\mathbb{Z}[i]$, with N as a Euclidean valuation, as prescribed by Definition 6.24. This proves that $\mathbb{Z}[i]$ is a Euclidean domain.   □

Once we know that $\mathbb{Z}[i]$ is a Euclidean domain, a number of sophisticated statements become available: we explored this theme at length in §6.3 and §6.4. In particular, this tells us that $\mathbb{Z}[i]$ is a PID and a UFD, and hence that *irreducible elements in $\mathbb{Z}[i]$ are prime* (by Proposition 6.33). Let's record this fact, in the form that will be useful for the task at hand.

> COROLLARY 10.31   *Let $z \in \mathbb{Z}[i]$. If z is not prime in $\mathbb{Z}[i]$, then it is not irreducible in $\mathbb{Z}[i]$.*

If you think that a prime *integer* $p \in \mathbb{Z}$ is naturally going to be prime *in* $\mathbb{Z}[i]$, think again: for example, 5 is prime in $\mathbb{Z}$, but it has a nontrivial factorization in $\mathbb{Z}[i]$: $5 = (2 + i)(2 - i)$. In the pictures drawn above, 5 is a fat point in the lattice since it is an element of the ideal $(2 + i)$. On the other hand, you could verify that 3 *does* remain prime in $\mathbb{Z}[i]$. (In fact, if you work out Exercise 10.29 right away, probably what follows will make even better sense.) There must be some special property of a prime integer $p \in \mathbb{Z}$ which determines whether p remains prime in $\mathbb{Z}[i]$ or does not. And here is that special property.

> COROLLARY 10.32   *Let p be a positive prime integer. Then if p is* not *prime in $\mathbb{Z}[i]$, then it is the sum of two perfect squares in $\mathbb{Z}$: $\exists a, b \in \mathbb{Z}$ such that $p = a^2 + b^2$.*

Proof   Keep in mind that $\mathbb{Z}[i]$ is a UFD, as we have shown. Therefore every nonzero, nonunit element of $\mathbb{Z}[i]$ has a unique factorization into irreducibles in $\mathbb{Z}[i]$.

The statement holds for $p = 2$: 2 is not prime in $\mathbb{Z}[i]$ since $2 = (1 + i)(1 - i)$, and $2 = 1^2 + 1^2$ is the sum of two squares.

Next, let $p > 2$ be an odd prime in $\mathbb{Z}$, and let q be a factor in its irreducible factorization in $\mathbb{Z}[i]$. Then q is not an integer, since the only integer factors of p are $\pm 1, \pm p$, while ± 1 are units, so they are not irreducible, and $\pm p$ are not prime in $\mathbb{Z}[i]$ by hypothesis, and therefore are not irreducible by Corollary 10.31. By the same token, q is also not an integer times i.

Since $q \notin \mathbb{Z}$ and $q \notin \mathbb{Z}i$, then $q = \alpha + \beta i$ with $\alpha, \beta \in \mathbb{Z}$ and both α, β not 0. Now note that if q is a factor of p, then $\bar{q} = \alpha - \beta i$ also is a factor: indeed, if $p = qr$, then $p = \bar{p} = \bar{q}\bar{r}$. Further, note that if q is irreducible, then so is $\bar{q}$: indeed, conjugation is a ring isomorphism as we have noted above. Further still, q and $\bar{q}$ do *not* differ by a multiplicative unit. Indeed, the units in $\mathbb{Z}[i]$ are $\pm 1, \pm i$ by Lemma 10.29. (We have

$\alpha - i\beta \neq \pm(\alpha + i\beta)$ since $\alpha \neq 0$ and $\beta \neq 0$. If, e.g., we had $\alpha - i\beta = i(\alpha + i\beta)$, this would imply $\beta = -\alpha$, so $\alpha + i\beta = \alpha(1 - i)$. But $(1 - i)$ is a factor of 2 in $\mathbb{Z}[i]$, so this would imply that $\gcd(2, p) \neq 1$, which is not the case since p is odd.)

The conclusion is that, *by uniqueness of factorizations, both q and $\bar{q}$ must appear in the factorization of p in $\mathbb{Z}[i]$.* This tells us that the irreducible factors of q come in pairs $q, \bar{q}$:

$$p = q_1 \cdots q_r \bar{q}_1 \cdots \bar{q}_r.$$

We are done! Let $q_1 \cdots q_r = a + bi \in \mathbb{Z}[i]$; then $\bar{q}_1 \cdots \bar{q}_r = a - bi$ and

$$p = (a + bi)(a - bi) = a^2 + b^2$$

as needed. □

A little more work would allow us to describe the situation in more detail. For example, it turns out that if an integer prime p is not prime in $\mathbb{Z}[i]$, then it has exactly two factors in $\mathbb{Z}[i]$ (that is, $r = 1$ in the proof of the corollary); we could completely describe all the irreducible elements of $\mathbb{Z}[i]$; and we could upgrade Corollary 10.32 to an if-and-only-if statement. You can try your hand at some of this in the exercises at the end of the chapter. For the purpose of proving Theorem 10.28, Corollary 10.32 is just what we need.

Recall that we have to show that if $p > 2$ is a prime integer and $p \equiv 1 \bmod 4$, then p may be expressed as the sum of two perfect squares. Notice that everything we have done so far only used ring theory, and especially what we learned when studying Euclidean domains. But we still have to process the condition that $p \equiv 1 \bmod 4$, and *this* is where knowing about abelian groups helps.

LEMMA 10.33 *Let $p > 2$ be a prime integer. The polynomial $x^2 + 1$ is reducible over $\mathbb{Z}/p\mathbb{Z}$ if and only if $p \equiv 1 \bmod 4$.*

You likely conjectured this statement when you worked on Exercise 7.33.

Proof The polynomial $x^2 + 1$ is reducible over a field k if and only if it has roots in k (Proposition 7.15). Therefore, the statement is that there exists an element $g \in k = \mathbb{Z}/p\mathbb{Z}$ such that $g^2 = -1$ if and only if $p \equiv 1 \bmod 4$. We know this already! Indeed, $g^2 = -1$ if and only if $g^4 = 1$ and $g^k \neq 1$ for $1 \leq k < 4$; that is, if and only if g has order 4 in the group $((\mathbb{Z}/p\mathbb{Z})^*, \cdot)$. This group of order $p - 1$ is cyclic by Theorem 10.24, therefore it has an element of order 4 if and only if $4 \mid (p - 1)$, as we saw in Example 10.14. The condition $4 \mid (p - 1)$ is equivalent to $p \equiv 1 \bmod 4$, so we are done. □

Simple as this argument sounds, it relies crucially on Theorem 10.24, which we proved as a consequence of Theorem 10.18, the classification theorem for finite abelian groups.

Now all the pieces of the puzzle are in place. We have proved that, for an integer prime $p > 2$:

- If p is *not* prime in $\mathbb{Z}[i]$, then p is the sum of two squares (Corollary 10.32), and

- If $p \equiv 1 \bmod 4$, then $x^2 + 1$ is reducible over $\mathbb{Z}/p\mathbb{Z}$ (Lemma 10.33).

All that remains is a bridge connecting these two conditions, that is, the following result.

> LEMMA 10.34   *Let p be a positive prime integer. Then p is prime in $\mathbb{Z}[i]$ if and only if $x^2 + 1$ is irreducible in $(\mathbb{Z}/p\mathbb{Z})[x]$.*

Proof   Since $(\mathbb{Z}/p\mathbb{Z})[x]$ is a PID (by Corollary 7.7, since $\mathbb{Z}/p\mathbb{Z}$ is a field), $x^2 + 1$ is irreducible if and only if it is prime (Theorem 6.23, Proposition 6.33). We have the ring isomorphisms

$$\mathbb{Z}[i]/(p) \cong (\mathbb{Z}[x]/(x^2 + 1))/(p) \overset{!}{\cong} \mathbb{Z}[x]/(p, x^2 + 1)$$

$$\overset{!}{\cong} (\mathbb{Z}[x]/(p))/(x^2 + 1) \cong (\mathbb{Z}/p\mathbb{Z})[x]/(x^2 + 1)$$

due to various applications of known isomorphisms, particularly the Third Isomorphism Theorem for rings (Theorem 5.57) at the places marked by !.

Therefore, $\mathbb{Z}[i]/(p)$ is an integral domain if and only if $(\mathbb{Z}/p\mathbb{Z})[x]/(x^2 + 1)$ is an integral domain, and this is the statement.   □

With all this background out of the way, the proof of Fermat's theorem on sums of squares is now exceedingly straightforward.

Proof of Theorem 10.28   A prime $p > 2$ can be written as a sum of two squares if and only if p is not prime in $\mathbb{Z}[i]$ (Corollary 10.32), if and only if $x^2 + 1$ is not irreducible in $(\mathbb{Z}/p\mathbb{Z})[x]$ (Lemma 10.34), if and only if $p \equiv 1 \bmod 4$ (Lemma 10.33).   □

Remark 10.35   The representation of integers by quadratic forms (of which the expression $n = a^2 + b^2$ is an example) is a beautiful topic in classical number theory, hiding many gems.

For example, let ρ_n be the number of ways a positive integer n can be represented as a sum of two squares of nonnegative integers. For example, $\rho_3 = 0$; $\rho_5 = 2$ since $5 = 1^2 + 2^2$ and $5 = 2^2 + 1^2$; while $\rho_{65} = 4$ since $65 = 1^2 + 8^2 = 4^2 + 7^2 = 7^2 + 4^2 = 8^2 + 1^2$. Refining the analysis that led us to the proof of Theorem 10.28, one could express ρ_n in terms of the irreducible factorization of n, particularly distinguishing between factors which are $\equiv 1 \bmod 4$ and those which are $\equiv 3 \bmod 4$. It would seem that ρ_n should be privy to some deep 'algebraic' information concerning the integer n.

On the other hand, let $R(N) = \sum_{n=1}^{N} \rho_n$. Then I claim that

$$\lim_{N \to \infty} \frac{R(N)}{N} = \frac{\pi}{4}. \tag{10.4}$$

That is, the average number of representations of n as sums of squares of nonnegative integers, from 1 to N, converges to $\frac{\pi}{4}$ as $N \to \infty$.

Isn't this beautiful? Surprisingly, a proof of this fact boils down to *calculus,* not to sophisticated algebraic arguments. Draw on a Cartesian plane the pairs (a, b) counted by ρ_n, i.e., such that $a^2 + b^2 = n$, for, say, $n = 1, \ldots, 100$. Chances are that in the process you will 'see' why (10.4) must be true.

But we are digressing, and it is time to come back to 'algebra' proper.

Exercises

10.1 ▷ Let A, B be abelian groups, and let $f: A \to B$ be a function. Assume that $\forall a, a' \in A$, $f(a + a') = f(a) + f(a')$. Prove that $\forall n \in \mathbb{Z}$, $\forall a \in A$, $f(n \cdot a) = n \cdot f(a)$.

(In other words, the second requirement for a homomorphism of $\mathbb{Z}$-modules is implied by the first, as pointed out in §10.1.)

10.2 Let A, B, C be abelian groups, and let $f: A \to B$ be a homomorphism of abelian groups. We can define a function $\mathrm{Hom}(C, A) \to \mathrm{Hom}(C, B)$ by sending $\alpha: C \to A$ to $f \circ \alpha: C \to B$; and a function $\mathrm{Hom}(B, C) \to \mathrm{Hom}(A, C)$ by sending $\beta: B \to C$ to $\beta \circ f: A \to C$. (I find it helpful to look at these two diagrams:

$$C \xrightarrow{\ \alpha\ } A \xrightarrow{\ f\ } B \qquad\qquad A \xrightarrow{\ f\ } B \xrightarrow{\ \beta\ } C$$
$$\underbrace{}_{f \circ \alpha} \qquad\qquad \underbrace{}_{\beta \circ f}$$

when thinking about these functions.)

Prove that these two functions are homomorphisms of abelian groups. (These prescriptions define two famous 'functors' associated with Hom.)

10.3 Using the definitions introduced in Exercise 10.2, prove that if $f: A \to B$ is injective, then the corresponding homomorphism $\mathrm{Hom}(C, A) \to \mathrm{Hom}(C, B)$ is injective, and if $f: A \to B$ is surjective, then the corresponding homomorphism $\mathrm{Hom}(B, C) \to \mathrm{Hom}(A, C)$ is again injective

10.4 Again using the definitions introduced in Exercise 10.2: if $f: A \to B$ is surjective, is the corresponding homomorphism $\mathrm{Hom}(C, A) \to \mathrm{Hom}(C, B)$ necessarily surjective? Let $A = \mathbb{Z}$, $B = \mathbb{Z}/2\mathbb{Z}$, and let f be the natural projection $\mathbb{Z} \to \mathbb{Z}/2\mathbb{Z}$. For $C = \mathbb{Z}/2\mathbb{Z}$, prove that the homomorphism $\mathrm{Hom}(\mathbb{Z}/2\mathbb{Z}, \mathbb{Z}) \to \mathrm{Hom}(\mathbb{Z}/2\mathbb{Z}, \mathbb{Z}/2\mathbb{Z})$ is *not* surjective. (*Note:* There is method in this madness, but to learn about it you will have to study *homological algebra*.[5])

10.5 Let A be an abelian group. Prove that $\mathrm{Hom}(\mathbb{Z}^{\oplus n}, A) \cong A^{\oplus n}$ as abelian groups. (This is essentially just a restatement of Proposition 9.8 for $R = \mathbb{Z}$. Your task is to understand why.)

10.6 ▷ Verify that if A is an abelian group, then $(\mathrm{End}(A), +, \circ)$ is a ring. (See §10.1 for the definition of $\mathrm{End}(A)$.)

10.7 ▷ Let U be the union of all *proper* subgroups of C_n. Prove that g is a generator of C_n if and only if $g \notin U$.

10.8 For integers $n \geq 1$, let $\phi(n)$ denote Euler's totient function. Prove that the sum $\sum_{1 \leq d \mid n} \phi(d)$ equals n. (For example: $\phi(1) = 1$, $\phi(2) = 1$, $\phi(3) = 2$, $\phi(4) = 2$, $\phi(6) = 2$, $\phi(12) = 4$, and indeed $1 + 1 + 2 + 2 + 2 + 4 = 12$. *Hint:* Think in terms of subgroups of C_n.)

---

[5] Item 18G in the Mathematics Subject Classification.

10.9   Let k, n be positive integers. Recall that the cyclic group C_n contains (an isomorphic copy of) C_k if and only if $k \mid n$. Determine C_n/C_k as a better known group.

10.10   Let A be an abelian group, and let $n > 0$ be an integer. Consider the set

$$O_n = \{a \in A \mid \text{the order } |a| \text{ of } a \text{ divides } n\}.$$

Prove that O_n is a subgroup of A.

10.11   ▷ Let A be an abelian group, and let $p > 0$ be a prime integer. Consider the set

$$O_p = \{a \in A \mid \text{the order } |a| \text{ of } a \text{ is a power of } p\}.$$

Prove that O_p is a subgroup of A.

10.12   What is $\mathrm{Hom}(\mathbb{Z}/n\mathbb{Z}, A)$, as an abelian group? (*Hint:* Exercise 10.10.)

10.13   Let $C_n = \langle g \rangle$ be a cyclic group of order n, generated by g. Prove that $|mg| = n/\gcd(m, n)$.

10.14   Let $p > 0$ be a prime integer. Prove that if $p^k \mid n$, then there are exactly p^k homomorphisms $C_{p^k} \to C_n$. (*Hint:* Example 10.15.)

10.15   ▷ Let $1 \le \alpha_1 \le \cdots \le \alpha_m$, $1 \le \beta_1 \le \cdots \le \beta_n$ be two lists of integers. Let p be a positive prime integer, and assume that

$$\prod_{i=1}^{m} p^{\min(\alpha_i, d)} = \prod_{j=1}^{n} p^{\min(\beta_j, d)} \tag{10.5}$$

for all $d \ge 0$. Prove that $m = n$ and $\alpha_i = \beta_i$ for all i. (*Hint:* What does (10.5) say for $d = 1$? What about $d = 2$?)

10.16   Choose a random largish positive integer N and list all abelian groups of order N up to isomorphism.

10.17   List all distinct abelian groups of order ≤ 12, up to isomorphism.

10.18   ▷ Prove that there are 42 abelian groups of order 1024, up to isomorphism. (*Hint:* Look up the number of 'partitions' of 10.)

10.19   Let A be a finite abelian group, and assume $|A| = 2n$, with n odd. Use the classification theorem to prove that A contains exactly one element of order 2.

10.20   Let A be a finite abelian group. Prove that A is *not* cyclic if and only if there is a prime p and a subgroup of A isomorphic to $C_p \times C_p$.

10.21   Let A be a finite abelian group, whose order is a power of a prime p. Assume that A has only one subgroup of order p. Prove that A is cyclic.

10.22   ▷ Let A be a finite abelian group. For any prime integer $p > 0$, let p^r be the largest power of p dividing the order $|A|$ of A. Prove that A has a subgroup of order p^k for all k, $0 \le k \le r$, as an application of the classification theorem for finite abelian groups, Theorem 10.18.

Like Cauchy's theorem, this result will hold more generally for noncommutative groups: it will be one of the *Sylow theorems;* see Theorem 12.29.

10.23   ▷ Let A be a finite abelian group, and let $|A| = p_1^{r_1} \cdots p_s^{r_s}$ be the prime factorization of its order. Prove that for $i = 1, \ldots, s$, A contains a *unique* subgroup of order $p_i^{r_i}$.

10.24 ▷ Generalize the result of Exercise 10.22 as follows.

Let A be a finite abelian group, and let b be a positive integer that divides the order of A. Prove that there exists a subgroup $B \subseteq A$ such that $|B| = b$. (As in the case of Exercise 10.22, this may be obtained as a direct application of the classification theorem. If you want an argument with a different feel, you can argue as follows. Let p be a prime dividing b; use Cauchy's theorem to find an element g of order p, mod out by $\langle g \rangle$, and use induction.)

Note: While Cauchy's theorem and Exercise 10.22 will generalize to noncommutative groups, the result of this exercise will *not* (cf. Example 12.6).

10.25 A theorem of Dirichlet states that if a and b are relatively prime positive integers, then there are infinitely many primes of the form $a + nb$, with $n > 0$. Assuming this theorem, prove that every finite abelian group G is isomorphic to a quotient of $(\mathbb{Z}/n\mathbb{Z})^*$ for some positive integer n, as follows.

- Assuming Dirichlet's theorem, prove that for every integer $N > 0$ there are infinitely many primes $p > 0$ such that $p \equiv 1 \bmod N$.
- Prove that for any r positive integers $n_1, \ldots, n_r$ there exist distinct positive primes p_i, $i = 1, \ldots, r$, such that $p_i \equiv 1 \bmod n_i$ for all i.
- Prove that for any r positive integers $n_1, \ldots, n_r$ there exist distinct positive primes p_i, $i = 1, \ldots, r$, such that $\mathbb{Z}/n_i\mathbb{Z}$ is isomorphic to a quotient of $(\mathbb{Z}/p_i\mathbb{Z})^*$.
- Let $n = p_1 \cdots p_r$. Prove that $(\mathbb{Z}/n\mathbb{Z})^* \cong (\mathbb{Z}/p_1\mathbb{Z})^* \times \cdots \times (\mathbb{Z}/p_r\mathbb{Z})^*$. (Use the Chinese Remainder Theorem.)
- Now put everything together and use the classification theorem and the First Isomorphism Theorem to prove that if G is a finite abelian group, then there exists a positive integer n and a subgroup K of $(\mathbb{Z}/n\mathbb{Z})^*$ such that G is isomorphic to the quotient $(\mathbb{Z}/n\mathbb{Z})^*/K$.

10.26 Reprove Fermat's little theorem (Theorem 2.18) as a consequence of Theorem 10.24.

10.27 ▷ Prove that there are 43157569800 integers a between 1 and 87178291198 with the property that every integer in the same range is congruent to a power of a modulo 87178291199.

10.28 Let m, n be integers. Prove that $\gcd(m, n) = 1$ in $\mathbb{Z}$ if and only if $\gcd(m, n) = 1$ in $\mathbb{Z}[i]$.

10.29 ▷ Verify 'directly' that 3 is prime when viewed as an element of $\mathbb{Z}[i]$: prove that $\mathbb{Z}[i]/(3)$ is an integral domain. (Prove that $\mathbb{Z}[i]/(3)$ is isomorphic to $(\mathbb{Z}/3\mathbb{Z})[x]/(x^2 + 1)$, and then verify that this ring is an integral domain.)

10.30 Prove the converse to Corollary 10.32: for a positive prime integer p, if $p = a^2 + b^2$, then p is not prime in $\mathbb{Z}[i]$. Conclude that a prime integer $p > 2$ is prime in $\mathbb{Z}[i]$ if and only if $p \equiv 3 \bmod 4$.

10.31 Let $p > 2$ be a prime integer, and let a be any integer. We say that an integer a that is not a multiple of p is a 'quadratic residue' mod p if $[a]_p$ is a square in $\mathbb{Z}/p\mathbb{Z}$. Prove that a is a quadratic residue mod p if and only if $a^{\frac{p-1}{2}} \equiv 1 \bmod p$. (Use Theorem 10.24 to write $[a]_p = g^k$, for a generator g of $(\mathbb{Z}/p\mathbb{Z})^*$. What can you say about this, if a is a quadratic residue?)

This result is known as *Euler's criterion* (1748). Explain in what sense it generalizes Lemma 10.33.

10.32 Let $q \in \mathbb{Z}[i]$. Prove that if $N(q)$ is a prime integer, then q is irreducible.

10.33 Let p be a positive prime integer, and assume that p is *not* prime in $\mathbb{Z}[i]$. Prove that $p = q\bar{q}$ for an irreducible $q \in \mathbb{Z}[i]$. (Use Exercise 10.32.)

10.34 Prove that an integer $a \neq 0$ is a sum of two perfect squares if and only if every prime $p \equiv 3 \mod 4$ appears in the factorization of a with an *even* power, as follows.

- Prove that the product of two integers that are sums of squares is also a sum of squares (this is particularly straightforward in terms of complex norms).
- Deduce the 'if' part of the statement.
- For the 'only if' part, assume that $n = a^2 + b^2 = N(a + bi)$. Use unique factorization to relate the factorization of $a + bi$ in $\mathbb{Z}[i]$ with the factorization of n in $\mathbb{Z}$.
- Use the result of Exercise 10.30 to conclude that primes $p \equiv 3 \mod 4$ appear an even number of times in the factorization of n in $\mathbb{Z}$.

Part III

Groups

11 Groups—Preliminaries

11.1 Groups and their Category

We can think of *rings* and *modules* as structures built out of abelian groups by introducing further information. In this sense, abelian groups are 'more general' than either rings or modules—quite simply, they are defined by a subset of the axioms defining these other structures.

What would be more general than abelian groups? Can we consider some useful structure defined by even fewer axioms? Yes! It turns out that removing the axiom prescribing the commutativity of the operation yields a tremendously useful general notion.

DEFINITION 11.1 A 'group' $(G, \cdot)$ consists of a set G along with a binary operation $\cdot$ on G such that

(i) $\forall g, h, k \in G$ $(g \cdot h) \cdot k = g \cdot (h \cdot k)$

(ii) $\exists e_G \in G \; \forall g \in G$ $g \cdot e_G = e_G \cdot g = g$

(iii) $\forall g \in G \; \exists h \in G$ $g \cdot h = h \cdot g = e_G$

We usually refer to a group $(G, \cdot)$ by the name G of the underlying set. The 'order' of a group G is its cardinality, denoted $|G|$ (thus, we adopt for groups the terminology introduced in Definition 10.1 for abelian groups).

Just as with multiplication in a ring, the operation $\cdot$ is often just replaced by juxtaposition: we write gh for $g \cdot h$. We also usually write e for e_G. We could also denote this identity element by 1, but e is more commonly used for it. Let's also dispose right away of basic but useful observations.

- The identity element e is *unique:* if axiom (ii) is satisfied by taking e_G to be e_1 or e_2, then $e_1 = e_2$.
- The inverse of a given element $g \in G$ is also unique: if h_1 and h_2 both satisfy the requirement for h in (iii), then $h_1 = h_2$. This element is denoted g^{-1}.

It is hopefully clear that $(g^{-1})^{-1} = g$. The proofs of these assertions are identical to the proofs we have seen for other operations (in particular, Propositions 3.14 and 3.15), and can be safely left to you (Exercise 11.1). Just as simple is the observation that *cancellation* holds in groups: if $g_1 h = g_2 h$, then

$$g_1 = g_1(hh^{-1}) = (g_1 h)h^{-1} = (g_2 h)h^{-1} = g_2(hh^{-1}) = g_2 .$$

(Cancellation 'on the left' works just as well.) In particular, if $g \neq e$, then $gh \neq h$ for every $h \in G$.

We adopt the usual power notation: $g^5 = ggggg$, and $g^{-3} = g^{-1}g^{-1}g^{-1}$. Clearly, $g^{m+n} = g^m g^n$, for all $m, n \in \mathbb{Z}$.

The fact that the operation in a group is not necessarily commutative forces us to be careful about the order in which we write expressions. We are already used to this from our experience with rings. Keep in mind that expressions such as $(gh)^2 = ghgh$ cannot be simplified further, and $(gh)^{-1} = h^{-1}g^{-1}$—the apparent switching is necessary for this identity to hold.

Two elements g, h in a group G 'commute' if $gh = hg$. If all elements in G commute, then we say that the group is 'commutative'. A commutative group is of course precisely the same thing as an *abelian* group. One might think that the commutativity axiom is a small enhancement of the three axioms defining a group, but this is far from being the case: the theory of *groups* is substantially more complex than the theory of *abelian groups*. For example, we were able to establish with relative ease a classification theorem for all finite abelian groups (Theorem 10.18); a corresponding theorem for finite 'simple' groups is available, but it is a behemoth filling many thousand pages in articles written by hundreds of authors over the span of 50 years—it is not stuff for mere mortals. If you want a specific example, recall that there are 42 *abelian groups* of order 1024, up to isomorphism: you found this out when you solved Exercise 10.18, and you could write all of them down if need be, by using the classification theorem. Well, I am told that there are 49,487,365,402 distinct *groups* of order 1024, up to isomorphism.

Maybe surprisingly, the 'category of groups' is substantially less friendly than the category of abelian groups. Its definition is natural enough: the objects of this category are groups, and its morphisms are defined as you might expect.

DEFINITION 11.2   Let G, H be groups. A function $\varphi: G \to H$ is a 'homomorphism of groups' (or 'group homomorphism') if, $\forall g_1, g_2 \in G$,

$$\varphi(g_1 g_2) = \varphi(g_1)\varphi(g_2).$$

A group homomorphism maps the identity to the identity, and inverses to inverses (Exercise 11.2). A little work shows that Definition 11.2 satisfies our basic requirements for morphisms of a category.

PROPOSITION 11.3   *Let G, H, and K be groups.*

- *The identity function $\mathrm{id}_G: G \to G$ is a group homomorphism.*
- *Let $\varphi: G \to H$, $\psi: H \to K$ be group homomorphisms. Then the composition $\psi \circ \varphi: G \to K$ is a group homomorphism.*

The operation of composition of group homomorphisms is associative.

Proof   Exercise 11.3.   □

You already know many examples of group homomorphisms. If G and H are groups, the function that sends every element of G to e_H is a group homomorphism; this is the 'trivial' homomorphism. More interestingly, every homomorphism of abelian groups is in particular a group homomorphism. In fact, if A and B are abelian groups, then the set of homomorphisms of abelian groups $f: A \to B$, denoted $\mathrm{Hom}(A, B)$ in §10.1, is the same as the set of homomorphism of 'ordinary' groups. (This makes the category of abelian groups a 'full subcategory' of the category of groups.)

Remark 11.4   While we have seen that $\mathrm{Hom}(A, B)$ is itself an abelian group in a natural way, the set $\mathrm{Hom}(G, H)$ of group homomorphisms between two groups is in general *not* a group in a natural way. This is a little disappointing: while we have not pursued this line of thought, the fact that $\mathrm{Hom}(A, B)$ is an abelian group when A and B are abelian groups leads to interesting constructions. These developments are not available (at least in a straightforward fashion) in group theory.

It is actually instructive to see what goes wrong in the noncommutative setting. If φ and ψ are two group homomorphisms $G \to H$, we could try to define a new group homomorphism $\varphi\psi$ by the same strategy we used following Definition 10.5 in the abelian case, setting $\forall g \in G$,

$$(\varphi\psi)(g) = \varphi(g)\psi(g).$$

This prescription does define a function $\varphi\psi: G \to H$. Is it again a group homomorphism? For $g_1, g_2 \in G$ we have

$$(\varphi\psi)(g_1 g_2) = \varphi(g_1 g_2)\,\psi(g_1 g_2) = \varphi(g_1)\,\varphi(g_2)\,\psi(g_1)\,\psi(g_2),$$
$$(\varphi\psi)(g_1)(\varphi\psi)(g_2) = \varphi(g_1)\psi(g_1)\,\varphi(g_2)\psi(g_2) = \varphi(g_1)\,\psi(g_1)\,\varphi(g_2)\,\psi(g_2),$$

and here we are stuck: these two expressions are in general *not* equal, since in general there is no reason why $\varphi(g_2)\,\psi(g_1)$ should equal $\psi(g_1)\,\varphi(g_2)$. We are missing the commutativity axiom (iv) that made this work in the abelian setting. Thus, in general $\varphi\psi$ is not a group homomorphism. This says that $\mathrm{Hom}(G, H)$ is 'nothing more' than a set.[1] ⌟

Isomorphisms of groups are also defined as you probably expect by now.

DEFINITION 11.5   Let $\varphi: G \to H$ be a homomorphism of groups. We say that φ is an 'isomorphism' if there exists a group homomorphism $\psi: H \to G$ such that $\psi \circ \varphi = \mathrm{id}_G$ and $\varphi \circ \psi = \mathrm{id}_H$. Equivalently, φ is an isomorphism if and only if it is a bijection.

We say that two groups G, H are 'isomorphic', and write $G \cong H$, if there exists an isomorphism of groups $\varphi: G \xrightarrow{\sim} H$.

As a review of things we have already worked out together in different settings, prove that the two conditions appearing in Definition 11.5 are indeed equivalent as stated (Exercise 11.4). Also verify that 'isomorphic' is an equivalence relation on any set of groups (cf. Corollary 4.33).

In order to define *subgroups* I will adopt the same strategy used for submodules.

---

[1] Actually this is not quite true: $\mathrm{Hom}(G, H)$ always has a distinguished element, that is, the trivial homomorphism. This makes it a 'pointed set'.

DEFINITION 11.6   Let G be a group and let H be a subset of G. Assume that H is also a group. We say that H is a 'subgroup' of G if the inclusion map $i: H \hookrightarrow G$ is a group homomorphism.

This really just says that the operation making H into a group is the restriction of the operation on G. As for the previous cases we have encountered, it is convenient to streamline the verification that a given subset is a subgroup.

PROPOSITION 11.7   *A nonempty set H of a group G is a subgroup if and only if for all $g, h \in H$, $gh^{-1} \in H$.*

Proof   This is the analogue of Proposition 10.4, and there is very little to check. The 'only if' implication is clear. For the 'if' implication, we have to show that a nonempty subset H satisfying the stated condition is necessarily a group with the induced operation.

Since H is nonempty, we can pick $h \in H$. Taking $g = h$, the given condition tells us that $e = hh^{-1} \in H$. Taking $g = e$, it tells us that $h^{-1} = eh^{-1} \in H$. Finally, if $g, h \in H$, we know $h^{-1} \in H$, and the condition tells us that $gh = g(h^{-1})^{-1} \in H$. Therefore H is closed with respect to the operation (which is then necessarily associative), contains the identity, and contains the inverse of any of its elements. This means that H is a group with the operation induced from G, as needed.   □

The following examples are simple applications of Proposition 11.7, and could be verified by even more direct arguments; please work them out to make sure you agree with me.

Example 11.8   If $H_1, \ldots, H_r$ are subgroups of a group G, then $H_1 \cap \cdots \cap H_r$ is a subgroup of G. More generally, if $\{H_\alpha\}$ is any (not necessarily finite) family of subgroups of a group G, then the intersection $\cap_\alpha H_\alpha$ is a subgroup of G.

Example 11.9   If $\varphi: G \to G'$ is a group homomorphism, and H is a subgroup of G, then $\varphi(H)$ is a subgroup of G'. In particular, the image of a group homomorphism, $\operatorname{im} \varphi = \varphi(G)$, is a subgroup of G'.

Example 11.10   If $\varphi: G \to G'$ is a group homomorphism, and H' is a subgroup of G', then $\varphi^{-1}(H')$ is a subgroup of G. In particular, $\ker \varphi = \varphi^{-1}(e_{G'})$ is a subgroup of G. We will come back to kernels in due time (§11.4).

DEFINITION 11.11   Let G be a group, and let $S \subseteq G$ be a subset of G. The subgroup of G 'generated by S', denoted $\langle S \rangle$, is the smallest subgroup of G containing S.

It is clear that this subgroup exists: the intersection of *all* subgroups containing S is a subgroup, cf. Example 11.8, and it is clearly the smallest one containing S.

Remark 11.12   There is a slight complication here, in that the subgroup $\langle g_1, \ldots, g_n \rangle$ generated by finitely many elements $g_1, \ldots, g_n$ is a little harder to describe than in the abelian case (or more generally the case of modules, see Example 8.25, Definition 8.28). For an abelian group A, the subgroup generated by $a_1, \ldots, a_n$ consists of all sums of multiples $m_1 a_1 + \cdots + m_n a_n$ of the generators a_i. The corresponding 'multiplicative' notion would be the set of all products $g_1^{m_1} \cdots g_n^{m_n}$, but these do *not* form a group in general: unless g_1 and g_2 commute, there is no reason why (for example) a product $g_1 g_2 g_1 g_2$ should be expressible as $g_1^{m_1} g_2^{m_2}$ for any exponents m_1, m_2. As you can check, the subgroup generated by $g_1, \ldots, g_n$ consists of all products of the generators g_i and of their inverses g_i^{-1}, in any order: a typical such element may look like

$$g_3 g_2^{-1} g_1 g_2^2 g_3^{-1} g_1^{-3} \; ;$$

but in general there is no way to simplify such a 'word' any further.

Of course this problem does not arise in the case of subgroups generated by a single element g: $\langle g \rangle$ does consist of all powers g^i, $i \in \mathbb{Z}$, in direct analogy with the abelian case (Definition 10.8).   ⌐

Products are also defined as in previous encounters.

DEFINITION 11.13   Let G, H be rings. The *Cartesian product* of G and H is the group $(G \times H, \cdot)$, where $\cdot$ is defined componentwise:

$$(g_1, h_1) \cdot (g_2, h_2) := (g_1 g_2, h_1 h_2)$$

for all $g_1, g_2 \in G$, $h_1, h_2 \in H$.

We gave a similar definition for modules (and hence for abelian groups), but we opted to call the result 'direct sum', and we noted that in that context it satisfied a property making it into a *coproduct* as well as a *product* (Exercise 8.16). The situation for general groups is a little different. Just as in the case of abelian groups, there are surjective group homomorphisms $\pi_G \colon G \times H \to G$, $\pi_H \colon G \times H \to H$, defined as usual: $(g, h) \mapsto g$, $(g, h) \mapsto h$. With respect to these, $G \times H$ satisfies the 'universal property of products' you should have spelled out in Exercise 8.17. However, unlike in the abelian case, $G \times H$ does *not* satisfy the universal property of *coproducts*. Particularly if this sounds a little mysterious, you should take the time to try and see what goes wrong in attempting to verify for groups the property stated for modules in Exercise 8.16 (Exercise 11.7). You will see that again the problem is simply that a key step requires the commutativity of the operation, and that is no longer available since we have given up axiom (iv).

Incidentally, it is possible to define a 'coproduct' in the category of groups. However, this new group (denoted $G * H$ for two groups G, H) is not as easy to handle as $G \times H$. We will not bother looking further into it.

Summarizing: The usual main constructions may be carried out for groups in essentially the same fashion as for the other structures we have examined so far. However, some features of the theory that were true and easily established for abelian groups, simply do not hold for groups. This makes it harder to work with groups rather than abelian groups. On the other hand, groups are considerably more general, and we will have a lot to say about them.

Example 11.14   The simplest group is of course the *trivial* group, consisting of a singleton: $\{*\}$, with operation $* \cdot * = *$. Since the notation we usually adopt for groups is multiplicative, we do not use 0 to denote this group. The standard notation for it is 1.

Example 11.15   *All* the structures we have studied in detail—rings, ideals, modules— are groups, simply because they are abelian groups. Thus, we already have many, many examples of groups.

Example 11.16   However, these examples are all very special: they are commutative. In this sense they are not really representative of what a random group looks like. A 'better' example, from this point of view, consists of the set R^* of invertible elements in a (possibly noncommutative) ring R; this is an upgrade of Example 10.2, where we were dealing with commutative rings. For instance, we can consider the ring of $n \times n$ matrices with entries in a (say, commutative) ring, and restrict attention to *invertible* matrices, with the operation of multiplication. For $n = 2$, and entries in $\mathbb{R}$, this is the group

$$\mathrm{GL}_2(\mathbb{R}) := \left\{ \begin{pmatrix} a & b \\ c & d \end{pmatrix} \mid a, b, c, d \in \mathbb{R}, ad - bc \neq 0 \right\}.$$

The quantity $ad - bc$ appearing in this definition is the *determinant* $\det(M)$ of the matrix M, and you learn in Linear Algebra that an $n \times n$ matrix with entries in $\mathbb{R}$ (or, more generally, a field) is invertible if and only if its determinant is nonzero. The group of invertible $n \times n$ matrices with real entries is denoted $\mathrm{GL}_n(\mathbb{R})$, and it is called the 'general linear group'.

In fact, in Linear Algebra you learn that for $n \times n$ matrices A, B,

$$\det(AB) = \det(A)\det(B); \tag{11.1}$$

I already mentioned this in Example 4.26. This identity shows that $\det \colon GL_n(\mathbb{R}) \to \mathbb{R}^*$ is a group homomorphism. The real numbers are not special here: we can consider a general linear group over any field, and even over rings; the determinant is a group homomorphism in all these contexts. We can also consider more special conditions: for example, the 'special linear' group consists of those matrices with determinant equal to 1; by (11.1), the product of two such matrices is also a matrix of the same kind, and it follows easily that the special linear group is a subgroup of the general linear group. Various other conditions may be considered, and lead to groups of fundamental importance in, e.g., physics. Further, the entries can be taken in more general (say, commutative) rings, and again some of these examples are very important. For instance,

$$\mathrm{SL}_2(\mathbb{Z}) := \left\{ \begin{pmatrix} a & b \\ c & d \end{pmatrix} \mid a, b, c, d \in \mathbb{Z}, ad - bc = 1 \right\}$$

is a group of fundamental importance in number theory (as is its closed relative $\mathrm{PSL}_2(\mathbb{Z})$, called the 'modular group').

As you likely learned in Linear Algebra (and as we also reviewed in §9.1), matrices represent linear transformations. In this sense, all the examples I mentioned in Example 11.16 are examples of groups whose elements may be interpreted as certain types of functions. This is not a coincidence, as we will see next.

11.2 Why Groups? Actions of a Group

The main reason why groups are important is that they *act* on other structures. We talked about the notion of 'action' when we defined modules, in §8.1: with the terminology we have acquired along the way, we can say that an R-module is an abelian group A taken along with an *action* of the ring R, by which we mean a ring homomorphism $R \to \text{End}(A)$. See Proposition 10.6 and Remark 10.7: in fact, the ring axioms could be viewed as an abstraction of the axioms naturally satisfied by $\text{End}(A) = \text{Hom}(A, A)$, the set of homomorphisms of an abelian group to itself.

Similarly, the group axioms may be viewed as an abstraction of the axioms satisfied by the set of *isomorphisms* of an object A of *any* category to itself. There is a name for this set: $\text{Aut}_C(A)$ denotes the set of 'automorphisms' of A in **C**, that is, the set of isomorphisms $A \to A$ in **C**. To focus on one very concrete example, consider the category of sets. If A is a set, then an 'automorphism' of A in the category of sets is nothing but a bijection $\alpha: A \to A$. Let's call S_A the set of bijections $A \to A$. What axioms does S_A satisfy, independently of A? To begin with, the composition of two bijections is a bijection. That is, we have a binary operation $\circ$ defined on S_A.

- This operation is associative: if α, β, γ are bijections $A \to A$, then we have the equality $(\alpha \circ \beta) \circ \gamma = \alpha \circ (\beta \circ \gamma)$.
- It has an identity element: the identity function $\text{id}_A: A \to A$ is a bijection.
- And every bijection has an inverse: if $\alpha: A \to A$ is a bijection, $\alpha^{-1}: A \to A$ is also a bijection, and by definition $\alpha \circ \alpha^{-1} = \alpha^{-1} \circ \alpha = \text{id}_A$.

The astute reader will recognize that these are *nothing but the group axioms listed in Definition 11.1:* the set S_A is a group! Further, you could observe that the same axioms are satisfied by the set of automorphisms of a ring (that is, ring isomorphisms from that ring to itself), or of an abelian group, or of an R-module for every ring R: each of these automorphism sets is a group in a natural way. The general linear group encountered in Example 11.16 is a concrete realization of the group of isomorphisms of a vector space.

Bottom line: the group axioms capture a very fundamental feature of *every* object of *every* category! At the risk of oversimplifying, I could state that *this is the reason why* we study groups.

The case of sets is of primary importance, so the group S_A has an official name.

DEFINITION 11.17   The 'symmetric' (or 'permutation') group on a set A, denoted S_A, is the group of bijections $A \to A$.

Why 'permutation'? If A is a finite set, then a bijection $A \to A$ is just a scrambling of the elements of A; technically this is called a 'permutation' of the elements of A.

Example 11.18   If $A = \{1, 2, 3\}$ is a set with three elements, then S_A consists of the $6 = 3!$ permutations of three elements. We could list them all:

$$
\begin{cases} 1 \mapsto 1 \\ 2 \mapsto 2, \\ 3 \mapsto 3 \end{cases}
\begin{cases} 1 \mapsto 1 \\ 2 \mapsto 3, \\ 3 \mapsto 2 \end{cases}
\begin{cases} 1 \mapsto 2 \\ 2 \mapsto 1, \\ 3 \mapsto 3 \end{cases}
\begin{cases} 1 \mapsto 2 \\ 2 \mapsto 3, \\ 3 \mapsto 1 \end{cases}
\begin{cases} 1 \mapsto 3 \\ 2 \mapsto 1, \\ 3 \mapsto 2 \end{cases}
\begin{cases} 1 \mapsto 3 \\ 2 \mapsto 2. \\ 3 \mapsto 1 \end{cases}
$$

More generally, if $A = \{1, \ldots, n\}$, then S_A is a certain group with $n!$ elements. The permutation group on $A = \{1, \ldots, n\}$ is in fact usually denoted S_n.

If you feel that we could use better notation for the elements of S_n, you are right—we will improve the situation a little when we explore this example further.

And what is 'symmetric' about this group? Your guess is actually as good as mine. *Symmetries* are transformations of an object which preserve some salient feature. For example, take an equilateral triangle T; flipping it about an axis realizes a 'symmetry' of T.

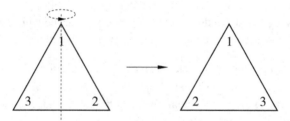

As sets are featureless (any element of a set is like any other element), *every* bijection of a set A counts as a symmetry of A; so it is perhaps not unreasonable to call S_A the 'symmetric group' of A.

Symmetric groups already provide us with a rich source of examples of groups—there is one for every set A!—and these examples are much more representative of what groups really look like than, say, abelian groups. In §11.3 we will study this group S_A in more detail, particularly in the case where A is a finite set. There is also a technical sense in which, if we understood everything about the groups S_A, then we would understand everything about *every* group; this should become clear in a moment (cf. Theorem 11.29).

According to the principle stated in Remark 10.7, a *group action* of a group G on an object A of a category $\mathbf{C}$ should be a group homomorphism from G to the group $\mathrm{Aut}_\mathbf{C}(A)$. For sets, an *action of a group G on a set A* should be nothing but a group homomorphism $G \to S_A$. We can spell out what this means.

DEFINITION 11.19   Let A be a set, and let G be a group. An 'action' of G on A is a function $\rho\colon G \times A \to A$ satisfying the following properties: for all $a \in A$,

- $\rho(e_G, a) = a$;
- $\rho(gh, a) = \rho(g, \rho(h, a))$.

It is common to omit ρ from the notation, and write ga for $\rho(g, a)$. Then the defining properties are simply stating that $e_G a = a$ and $(gh)a = g(ha)$.

> PROPOSITION 11.20   *Let A be a set, and let G be a group. Every action of G on A as in Definition 11.19 determines a group homomorphism $\sigma\colon G \to S_A$, defined by*
>
> $$\forall g \in G,\ \forall a \in A\colon \quad \sigma(g)(a) = \rho(g, a). \tag{11.2}$$
>
> *Conversely, every action of G on A is determined by a homomorphism of groups $\sigma\colon G \to S_A$, with ρ defined by the same identity (11.2).*

The identity (11.2) may be hard to parse at first. If $\sigma\colon G \to S_A$ is a function, then $\sigma(g)$ is a bijection $A \to A$, for every $g \in G$. Therefore it makes sense to apply $\sigma(g)$ to an element a of A: this is what $\sigma(g)(a)$ stands for.

Proof   We have to verify that if $\rho\colon G \times A \to A$ is a function as in Definition 11.19, then (11.2) defines a function $\sigma\colon G \to S_A$, and this function is a group homomorphism.

For every $g \in G$, identity (11.2) determines $\sigma(g)$ as a function $A \to A$. Note that for all $a \in A$,

$$\sigma(e_G)(a) = \rho(e_G, a) = a$$

by the first property listed in Definition 11.19: this shows that $\sigma(e_G) = \mathrm{id}_A$. Next, for all $g, h \in G$ and all $a \in A$,

$$\sigma(gh)(a) = \rho(gh, a) \overset{!}{=} \rho(g, \rho(h(a))) = \sigma(g)(\rho(h(a))) = \sigma(g)(\sigma(h)(a))$$
$$= (\sigma(g) \circ \sigma(h))(a),$$

where the equality marked with ! holds because of the second property listed in Definition 11.19. This shows that $\sigma(gh) = \sigma(g) \circ \sigma(h)$ for all $g, h \in G$, that is, σ preserves the operation. In particular, for all $g \in G$,

$$\sigma(g) \circ \sigma(g^{-1}) = \sigma(gg^{-1}) = \sigma(e_G) = \mathrm{id}_A :$$

therefore $\sigma(g)$ is in fact a *bijection* $A \to A$. So indeed $\sigma(g) \in S_A$ for all $g \in G$. Summarizing, σ is a function $G \to S_A$ preserving the group operation, that is, a group homomorphism. This verifies the first assertion.

Conversely, given a group homomorphism $\sigma\colon G \to S_A$, we can view (11.2) as the definition of a function $\rho\colon G \times A \to A$:

$$\rho(g, a) := \sigma(g)(a).$$

You should take a moment to verify that this function ρ satisfies the properties listed in Definition 11.19 (Exercise 11.14), and this will conclude the proof. $\square$

When a group G acts on a set A, every element $g \in G$ determines a specific transformation of A, it 'moves' the elements $a \in A$ in a specified manner.

Example 11.21   View the set $A = \{1, 2, 3\}$ as the set of vertices of an equilateral triangle—this is just in order to visualize what an action on this set may do.

We can act with the cyclic group C_3 on A by rotating the triangle, for example clockwise. Let g be a generator of C_3, so $C_3 = \{e, g, g^2\}$. View g as a $120°$ rotation:

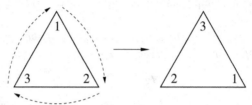

Under the action, e will leave the triangle alone; g will rotate it $120°$ clockwise; and g^2 will rotate it $240°$. The effect on the vertices is as follows:

e : 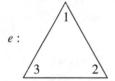 g : 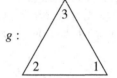 g^2 :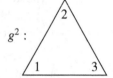

and this action corresponds to the group homomorphism $C_3 \to S_3$ given by

$$e \mapsto \begin{cases} 1 \mapsto 1 \\ 2 \mapsto 2, \\ 3 \mapsto 3 \end{cases} \qquad g \mapsto \begin{cases} 1 \mapsto 2 \\ 2 \mapsto 3, \\ 3 \mapsto 1 \end{cases} \qquad g^2 \mapsto \begin{cases} 1 \mapsto 3 \\ 2 \mapsto 1. \\ 3 \mapsto 2 \end{cases}$$

Similarly, the flip about the axis through the center and vertex 1, depicted earlier, corresponds to the action of $C_2 = \{e, f\}$ on A given by

$$e \mapsto \begin{cases} 1 \mapsto 1 \\ 2 \mapsto 2, \\ 3 \mapsto 3 \end{cases} \qquad f \mapsto \begin{cases} 1 \mapsto 1 \\ 2 \mapsto 3, \\ 3 \mapsto 2 \end{cases}$$

where f 'sends 2 to 3 and 3 to 2'.

There is some useful terminology associated with actions.

DEFINITION 11.22   Let A be a set, and let G be a group acting on A. The 'orbit' of an element $a \in A$ is the subset $\{ga \mid g \in G\}$ of A.

That is, the orbit of a is the subset of A consisting of the elements to which a is sent by elements of G. We say that the orbit of a is 'trivial' if it consists of a itself, that is, if a is a 'fixed point' of the action.

The orbits of an action clearly form a partition of the set A. This simple observation will lead to powerful counting principles down the road, such as the 'class equation' (Theorem 12.21).

Example 11.23   The orbits of the action of C_2 on $A = \{1, 2, 3\}$ in Example 11.21 are $\{1\}$ and $\{2, 3\}$.

It may happen that A consists of a single orbit, and there is a name for this situation.

DEFINITION 11.24   An action of a group G on a set A is 'transitive' if A consists of a single orbit under the action. That is, the action is transitive if for all a and b in A there exists a $g \in G$ such that $b = ga$.

Example 11.25   The action of C_3 on $A = \{1, 2, 3\}$ in Example 11.21 is transitive. What this says is that every vertex of the triangle can be reached by any other vertex by suitably rotating the triangle.

The definition of 'action' does not prescribe that a nonidentity element $g \in G$ should necessarily move any element of A. As an extreme example, we can act 'trivially' with any group G on any set A by simply setting $ga = a$ for all $g \in G$ and $a \in A$. You do not expect to learn much from such an action. There is a term for actions which, on the contrary, act nontrivially by every nonidentity element.

DEFINITION 11.26   An action of a group G on a set A is 'faithful' if the corresponding group homomorphism $\sigma \colon G \to S_A$ is injective. That is, if for all $g \neq e$ there exists some $a \in A$ such that $ga \neq a$.

An action is 'free' if nonidentity elements move *every* element of A.

Example 11.27   Both actions in Example 11.21 are faithful. The C_3 action is free, while the C_2 action is not (since vertex 1 is not moved by the nonidentity element $f \in C_2$).

Whenever a group G acts on a set A (say, nontrivially) we learn something about both G and A. In fact, most of the higher-level results we will prove on groups (such as the *Sylow theorems,* see §12.4 and §12.5) will be obtained by analyzing suitable actions. In fact, a lot of information will be gleaned by letting G act onto *itself.* This can be done in different ways, and we can already appreciate one of these ways and a remarkable (even if essentially trivial!) consequence.

The operation $\cdot$ in a group G is really a function $G \times G \to G$, denoted by juxtaposition:

$(g, h) \mapsto gh$. The requirements of Definition 11.19 are immediately verified for this function:

- $eg = g$, since e is the identity in G; and
- $(gh)k = g(hk)$, since the operation is associative.

We say that 'G acts on itself by left-multiplication'.

> LEMMA 11.28 *Let G be a group. Then the action of G on itself by left-multiplication is free (and in particular it is faithful).*

Proof If g is an element of G other than e, then $gh \neq h$ for every $h \in G$. □

Believe it or not, these very general, simple considerations already suffice to prove a 'named' theorem.[2]

> THEOREM 11.29 (Cayley's theorem) *Every group G is isomorphic to a subgroup of a symmetric group S_A, for some set A. (In fact, we may take A = G.)*

Proof This is an immediate consequence of Lemma 11.28: since G acts faithfully on itself (by left-multiplication), we have an injective group homomorphism $G \to S_G$. □

Example 11.30 Relabel the set A (i.e., the vertices of the triangle) in the C_3 action given in Example 11.21: $1 \leftrightarrow e$, $2 \leftrightarrow g$, $3 \leftrightarrow g^2$. Here is the picture again, after this change is implemented:

$e:$ 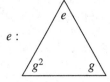 $g:$ 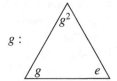 $g^2:$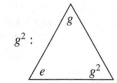

We see that this is nothing but the left-multiplication action:[3]

$$e \mapsto \begin{cases} e \mapsto e \\ g \mapsto g, \\ g^2 \mapsto g^2 \end{cases} \qquad g \mapsto \begin{cases} e \mapsto g \\ g \mapsto g^2, \\ g^2 \mapsto e \end{cases} \qquad g^2 \mapsto \begin{cases} e \mapsto g^2 \\ g \mapsto e. \\ g^2 \mapsto g \end{cases}$$

This can be interpreted as a group homomorphism $C_3 \hookrightarrow S_{\{1,g,g^2\}}$; we see that it is injective and identifies C_3 with a subgroup of $S_{\{1,g,g^2\}}$, as prescribed by Cayley's theorem.

---

[2] Incidentally, Cayley may have been the first mathematician to define groups in the 'modern' way, as in Definition 11.1.

[3] Here I am interpreting the second picture as telling us that, e.g., e is sent to the place that was occupied by g in the leftmost picture, g to g^2, g^2 to e.

Cayley's theorem is conceptually important: it is the formal statement I alluded to earlier, when I mentioned that complete knowledge of symmetric groups would yield complete knowledge about *all* groups. In practice, symmetric groups are too large. After all, S_{70} has $70! > 10^{100}$ elements, substantially more than the estimated *number of elementary particles in the universe*. Any attempt at classifying groups must go beyond a brute-force analysis of all subgroups of all symmetric groups.

Also, symmetric groups—that is, groups of automorphisms in the category of sets—are not the only game in town. One can use very successfully the category of *vector spaces* over a field: every homomorphism of a group G to Aut(V) for a vector space V determines a 'linear' action of G on V. If V is finite-dimensional, then Aut(V) is isomorphic to the 'general linear group' mentioned in Example 11.16, so these actions can be expressed very concretely in terms of matrices. Such an action is called a 'representation' of G: when the action is faithful, it is a way to 'represent' G as a group of matrices. *Representation theory*[4] is a whole field of mathematics in its own right.

A group G acts on itself in at least another way, and this action will be very useful in the next chapter. If $a \in G$ and $g \in G$, we can let

$$\rho(g, a) := gag^{-1}.$$

Such an element is called a 'conjugate' of a. I claim that this prescription defines an action, according to Definition 11.19. Indeed, $\forall a \in A$, $\forall g \in G$:

- $\rho(e, a) = eae^{-1} = a$;
- $\rho(gh, a) = (gh)a(gh)^{-1} = ghah^{-1}g^{-1} = g(hah^{-1})g^{-1} = \rho(g, hah^{-1}) = \rho(g, \rho(h, a))$.

This action is in general *not* faithful: in fact, g acts trivially by conjugation if and only if g commutes with every element of G. If G is commutative, then the conjugation action is trivial.

DEFINITION 11.31   The 'conjugacy class' of $a \in G$ is its orbit under the conjugation action, that is, the set of all the conjugates gag^{-1} of a.

With this terminology, a group G is abelian if and only if all its conjugacy classes consist of single elements. Conjugation will be one of our main tools in the more sophisticated results on groups that we will prove in Chapter 12.

11.3   Cyclic, Dihedral, Symmetric, Free Groups

The next natural topics are canonical decomposition, quotients, and isomorphism theorems. However, generalities are of little use in the absence of concrete examples to which we may apply them. Therefore, I want to first take a moment and look more carefully at a few actual (and important) classes of groups.

We already know quite a bit about *cyclic groups:* we studied these in the context of

---

[4] Item 20C in the Mathematics Subject Classification.

abelian groups, in §10.2. In the context of *groups,* I will mostly use the notation C_n for the cyclic group of order n, rather than its 'additive' alternative $\mathbb{Z}/n\mathbb{Z}$; this is an inessential distinction, and it may occasionally be convenient to revert to the abelian notation.

We used cyclic groups to define the notion of *order* of an element of an abelian group (Definition 10.12). That same definition can be adopted in the more general context of groups. Thus, $g \in G$ has 'order n', denoted $|g| = n$, if the subgroup $\langle g \rangle$ generated by g in G is isomorphic to C_n, and $|g| = \infty$ if $\langle g \rangle \cong \mathbb{Z}$. Equivalently, $|g|$ is either n if n is the smallest positive integer such that $g^n = e$, or ∞ if $g^n \neq e$ for all $n > 0$.

Propositions 10.16 and 10.17 did not rely on the commutativity of the relevant groups, so they hold without any change. For example, $g^N = e$ if and only if N is a multiple of $|g|$. One item whose proof *did* rely on the commutativity hypothesis is Cauchy's theorem (Theorem 10.22), stating that if $p > 0$ is prime and p divides the order of an abelian group A, then A contains an element of order p. This statement does upgrade to the setting of groups, but the proof given in §10.3 will not work, since it relied on the classification theorem for finite *abelian* groups. The same applies to Proposition 10.21: we will be able to verify (quite easily, actually) that if G is finite and $g \in G$, then $|g|$ divides $|G|$, but by a different argument than the one used in the abelian setting.

A concrete realization of C_n is available, as the group of (say, clockwise) rotations of an n-sided regular polygon; we looked at the case of C_3 in Example 11.21. The group C_n acts on the set of vertices of the polygon, so we may realize it as a group of symmetries of a regular polygon.

Dihedral groups extend the C_n action to *all* symmetries of a regular polygon. In §11.2 we have seen at work both ways in which we can rigidly move a polygon so that its 'shadow' does not change: we can flip it about a symmetry axis through the center, or rotate it by multiples of $\frac{360}{n}$ degrees, where n is the number of sides. We call the results of these motions the 'symmetries' of the polygon.

DEFINITION 11.32   Let $n > 2$ be an integer. The 'dihedral group' D_{2n} is the group of symmetries of a regular n-gon.

Example 11.33   The group D_6 consists of the symmetries of a triangle.

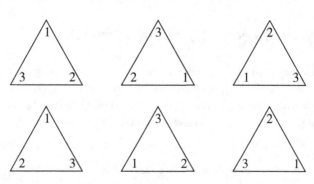

The top row of this picture consists of rotations, and reproduces the action of C_3. The bottom row represents the three flips.

Example 11.34 The group D_8 consists of the symmetries of a square.

Here, the top row corresponds to the action of C_4, and the bottom row represents the flips about the 13-axis, a vertical axis, the 24-axis, and a horizontal axis.

In general, D_{2n} consists of $2n$ elements: there are n possible rotations (including the identity), and n symmetry axes (through the center and vertices or midpoints of the sides). The rotations reproduce the action of C_n: this means that C_n may be viewed as a subgroup of D_{2n}. Each flip generates a copy of C_2 within D_{2n}; thus, D_{2n} also contains n subgroups isomorphic to C_2.

Performing 'computations' with elements of D_{2n} from their geometric description is somewhat cumbersome. One way to streamline computations is to extract more information from the fact that D_{2n} acts on the set $\{1, 2, \ldots, n\}$ of vertices of the regular n-gon: as we know (Proposition 11.20), this action determines a group homomorphism $D_{2n} \to S_n$, and since the action is faithful (every element of D_{2n} other than the identity moves some vertex) we can in fact identify D_{2n} with its image in S_n. We still have not developed a good notation to deal with permutations (we will soon do so), but this gets us going. For example, the eight elements of D_8 depicted in Example 11.34 map to the following bijections of $\{1, 2, 3, 4\}$:

$$
\begin{cases} 1 \mapsto 1 \\ 2 \mapsto 2 \\ 3 \mapsto 3 \\ 4 \mapsto 4 \end{cases}
\begin{cases} 1 \mapsto 2 \\ 2 \mapsto 3 \\ 3 \mapsto 4 \\ 4 \mapsto 1 \end{cases}
\begin{cases} 1 \mapsto 3 \\ 2 \mapsto 4 \\ 3 \mapsto 1 \\ 4 \mapsto 2 \end{cases}
\begin{cases} 1 \mapsto 4 \\ 2 \mapsto 1 \\ 3 \mapsto 2 \\ 4 \mapsto 3 \end{cases}
$$

$$
\begin{cases} 1 \mapsto 1 \\ 2 \mapsto 4 \\ 3 \mapsto 3 \\ 4 \mapsto 2 \end{cases}
\begin{cases} 1 \mapsto 4 \\ 2 \mapsto 3 \\ 3 \mapsto 2 \\ 4 \mapsto 1 \end{cases}
\begin{cases} 1 \mapsto 3 \\ 2 \mapsto 2 \\ 3 \mapsto 1 \\ 4 \mapsto 4 \end{cases}
\begin{cases} 1 \mapsto 2 \\ 2 \mapsto 1 \\ 3 \mapsto 4 \\ 4 \mapsto 3 \end{cases}.
$$

(11.3)

There are actually different ways to encode this information, and it is not clear that this is preferable to others. To clarify how I am doing this, let me redraw the last diagram on the first row of the picture in Example 11.34, and let me also mark on the outside the numbers of the vertices in their 'initial' position:

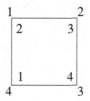

Vertex 1 has moved to occupy the place that was occupied by 4, 2 has replaced 1, etc. I am encoding this as the bijection $1 \mapsto 4, 2 \mapsto 1, 3 \mapsto 2, 4 \mapsto 3$, as indicated at the end of the first row in (11.3). This is three clockwise 90° rotations, or equivalently one counterclockwise rotation.

For all D_{2n}, denote by x the flip about the axis through the center and vertex 1, and by y the basic clockwise rotation by $\frac{360}{n}$ degrees. For the $n = 4$ case we are looking at, x is the flip about the NW–SE axis, and y is the 90° clockwise rotation. The corresponding bijections are

$$x \mapsto \begin{cases} 1 \mapsto 1 \\ 2 \mapsto 4 \\ 3 \mapsto 3 \\ 4 \mapsto 2 \end{cases} \qquad y \mapsto \begin{cases} 1 \mapsto 2 \\ 2 \mapsto 3 \\ 3 \mapsto 4 \\ 4 \mapsto 1. \end{cases}$$

Since we are viewing elements of D_{2n} as bijections, and in particular as functions, the element xy corresponds to *first* doing y, *then* x (just as when we write $g \circ f$ we mean first apply f, then g). If you want to perform a bit of mental gymnastics, figure out what xy is: where do vertices go, if you first rotate an n-gon clockwise by $\frac{360}{n}$ degrees, then flip about the axis through the center and the position where vertex 1 was? And then double down and figure out what $xyxy$ is.

In case your geometric inner eye fails you (most times it fails me), or in case you are a computer, you can still perform these operations with 100% accuracy by manipulating the corresponding bijections. For example, for $n = 4$:

$$\begin{cases} 1 \overset{y}{\mapsto} 2 \overset{x}{\mapsto} 4 \overset{y}{\mapsto} 1 \overset{x}{\mapsto} 1 \\ 2 \mapsto 3 \mapsto 3 \mapsto 4 \mapsto 2 \\ 3 \mapsto 4 \mapsto 2 \mapsto 3 \mapsto 3 \\ 4 \mapsto 1 \mapsto 1 \mapsto 2 \mapsto 4. \end{cases}$$

Aha: xy turns out to act by $1 \mapsto 4, 2 \mapsto 3, 3 \mapsto 2, 4 \mapsto 1$. It is a flip about a horizontal axis, the second diagram in the second row as depicted in Example 11.34. Perhaps remarkably, $xyxy$ turns out to equal *the identity*.

Is this last observation a coincidence? Of course not.

PROPOSITION 11.35 *For any $n > 2$, let $x \in D_{2n}$ be the flip about the axis through the center and vertex 1, and let $y \in D_{2n}$ be the clockwise rotation by $\frac{360}{n}$ degrees. Then $D_{2n} = \langle x, y \rangle$, and x, y satisfy the following relations:*

$$x^2 = e, \quad y^n = e, \quad xy = y^{-1}x.$$

Every element of D_{2n} may be written uniquely as $x^i y^j$ for some $i = 0, 1$ and $j = 0, \ldots, n - 1$.

Proof It is clear that $x^2 = e$ (flipping twice returns the polygon to its original position) and $y^n = e$ (this corresponds to an $n \cdot \frac{360}{n} = 360°$ rotation). Note that in particular $x^{-1} = x$. The other relation, $xy = y^{-1}x$, is equivalent to $xyx^{-1}y = e$, that is, $xyxy = e$: it is the relation we verified above for $n = 4$. We have to verify it for arbitrary n.

There are many ways to do this; here is an argument which requires no 'visualization'. We know that D_{2n} consists of the subgroup $C_n = \langle y \rangle$ and of n elements of order 2 (the 'flips'); x is one of these elements. Consider the element xy. Can xy belong to C_n? No: if $xy = y^i$ for some i, then $x = y^{i-1} \in \langle y \rangle = C_n$, a contradiction. Therefore xy must be an element of order 2, and this says precisely that $xyxy = e$.

It remains to prove that D_{2n} is generated by x and y, and that in fact every element of D_{2n} may be written as stated. For this, I claim that the elements $e, y, \ldots, y^{n-1}$, $x, xy, \ldots, xy^{n-1}$ of $\langle x, y \rangle$ are all distinct. Indeed, assume that $x^i y^j = x^k y^\ell$, with i, k between 0 and 1 and j, ℓ between 0 and $n - 1$. Then

$$x^{i-k} = y^{\ell - j}$$

and since $x \notin \langle y \rangle$, it follows that $x^{i-k} = y^{\ell - j} = e$. Therefore $i \equiv k \bmod 2$ and $j \equiv \ell \bmod n$, and this implies $i = k$, $j = \ell$ because of the bounds on these exponents.

This shows that $\langle x, y \rangle$ consists of at least $2n$ elements, and then it must exhaust D_{2n} as this group consists of exactly $2n$ elements. Therefore $\langle x, y \rangle = D_{2n}$. □

As we have just shown, the elements of D_{2n} are $e, y, \ldots, y^{n-1}, x, xy, \ldots, xy^{n-1}$. For example, the elements of D_8 appear in the following array in Example 11.34:

$$
\begin{array}{cccc}
e & y & y^2 & y^3 \\
x & xy & xy^2 & xy^3
\end{array}.
$$

You will now agree that while mental gymnastics is fun, and while the identification of D_8 with a subgroup of S_4 is useful, this last description of D_{2n} as generated by two elements x, y satisfying certain relations is probably the most efficient for actual computations.

Example 11.36 Consider the elements xy^3 and xy^5 of D_{14}. We can determine their product very quickly by using the relations $x^2 = e$, $y^7 = e$, and $xy = y^{-1}x$ (or equivalently $yx = xy^{-1}$):

$$(xy^3)(xy^5) = xyyyxy^5 = xyyxy^{-1}y^5 = xyxy^{-1}y^{-1}y^5 = x^2y^{-1}y^{-1}y^{-1}y^5 = y^2.$$

Therefore, the element $(xy^3)(xy^5)$ is a clockwise rotation by $2 \cdot \frac{360}{7}$ degrees. Visualize *that,* if you can.

Let's move on to *symmetric groups*. The action of D_{2n} on $\{1, \ldots, n\}$ is faithful, since every nonidentity symmetry moves some vertex, therefore the corresponding group homomorphism $D_{2n} \rightarrow S_n$ is injective. For $n = 3$ both D_6 and S_3 have six elements; therefore the homomorphism $D_6 \rightarrow S_3$ *must be an isomorphism.* What this says is that every permutation of $\{1, 2, 3\}$ may be realized by a rigid symmetry of an equilateral triangle; maybe that is not so surprising. Since we now understand dihedral groups rather well, we can claim a good understanding of the symmetric group S_3: this is a noncommutative group of order 6, with a single cyclic subgroup of order 3 and three cyclic subgroups of order 2. It is generated by two elements, subject to relations as we have seen above for D_6.

One obstacle in dealing with larger symmetric groups is notation. So far I have written elements of S_n explicitly as bijections from $\{1, \ldots, n\}$ to itself. For example, an element of S_7 may be denoted

$$\begin{cases} 1 \mapsto 4 \\ 2 \mapsto 3 \\ 3 \mapsto 5 \\ 4 \mapsto 1 \\ 5 \mapsto 7 \\ 6 \mapsto 2 \\ 7 \mapsto 6. \end{cases} \qquad (11.4)$$

This is clearly not very efficient. A minor advantage is obtained by stacking the same information horizontally, in a matrix; this permutation would be displayed as

$$\begin{pmatrix} 1 & 2 & 3 & 4 & 5 & 6 & 7 \\ 4 & 3 & 5 & 1 & 7 & 2 & 6 \end{pmatrix}.$$

For example, the elements of S_3 could be listed as

$$\begin{pmatrix} 1 & 2 & 3 \\ 1 & 2 & 3 \end{pmatrix} \begin{pmatrix} 1 & 2 & 3 \\ 2 & 3 & 1 \end{pmatrix} \begin{pmatrix} 1 & 2 & 3 \\ 3 & 1 & 2 \end{pmatrix} \begin{pmatrix} 1 & 2 & 3 \\ 1 & 3 & 2 \end{pmatrix} \begin{pmatrix} 1 & 2 & 3 \\ 3 & 2 & 1 \end{pmatrix} \begin{pmatrix} 1 & 2 & 3 \\ 2 & 1 & 3 \end{pmatrix}:$$

the first three matrices represent the subgroup C_3 of S_3, and the last three are the three elements of order 2.

Multiplying elements of S_n amounts to composing the corresponding bijections. For example,

$$\begin{pmatrix} 1 & 2 & 3 & 4 & 5 \\ 2 & 1 & 3 & 5 & 4 \end{pmatrix} \begin{pmatrix} 1 & 2 & 3 & 4 & 5 \\ 5 & 1 & 3 & 4 & 2 \end{pmatrix} = \begin{pmatrix} 1 & 2 & 3 & 4 & 5 \\ 4 & 2 & 3 & 5 & 1 \end{pmatrix} \qquad (11.5)$$

because

$$\begin{pmatrix} 1 & 2 & 3 & 4 & 5 \\ 2 & 1 & 3 & 5 & 4 \end{pmatrix} \circ \begin{pmatrix} 1 & 2 & 3 & 4 & 5 \\ 5 & 1 & 3 & 4 & 2 \end{pmatrix} (1) = \begin{pmatrix} 1 & 2 & 3 & 4 & 5 \\ 2 & 1 & 3 & 5 & 4 \end{pmatrix} (5) = 4,$$

$$\begin{pmatrix} 1 & 2 & 3 & 4 & 5 \\ 2 & 1 & 3 & 5 & 4 \end{pmatrix} \circ \begin{pmatrix} 1 & 2 & 3 & 4 & 5 \\ 5 & 1 & 3 & 4 & 2 \end{pmatrix} (2) = \begin{pmatrix} 1 & 2 & 3 & 4 & 5 \\ 2 & 1 & 3 & 5 & 4 \end{pmatrix} (1) = 2,$$

and similarly for the other entries.

The matrices appearing here are just notational devices. An alternative notation for elements of S_n uses 'actual' $n \times n$ matrices: for this, consider the action of S_n on an n-dimensional $\mathbb{R}$-vector space obtained by permuting vectors of the standard basis. This action is faithful, so we obtain an injective group homomorphism $S_n \to GL_n(\mathbb{R})$ identifying S_n with the set of 'permutation' matrices. These are all and only the matrices with only 0 and 1 entries, and exactly one 1 in each row, exactly one 1 in each column. For example, the element of S_7 shown at (11.4) corresponds to the matrix

$$M = \begin{pmatrix} 0 & 0 & 0 & 1 & 0 & 0 & 0 \\ 0 & 0 & 0 & 0 & 0 & 1 & 0 \\ 0 & 1 & 0 & 0 & 0 & 0 & 0 \\ 1 & 0 & 0 & 0 & 0 & 0 & 0 \\ 0 & 0 & 1 & 0 & 0 & 0 & 0 \\ 0 & 0 & 0 & 0 & 0 & 0 & 1 \\ 0 & 0 & 0 & 0 & 1 & 0 & 0 \end{pmatrix}.$$

Indeed, denoting by $\underline{e}_1, \underline{e}_2, \dots$ the vectors of the standard basis (as we did in §9.1), we have

$$M \cdot \underline{e}_1 = M \cdot \begin{pmatrix} 1 \\ 0 \\ 0 \\ 0 \\ 0 \\ 0 \\ 0 \end{pmatrix} = \begin{pmatrix} 0 \\ 0 \\ 0 \\ 1 \\ 0 \\ 0 \\ 0 \end{pmatrix} = \underline{e}_4, \quad M \cdot \underline{e}_2 = \underline{e}_3, \quad M \cdot \underline{e}_3 = \underline{e}_5, \quad \text{etc.,}$$

telling us that $1 \mapsto 4$, $2 \mapsto 3$, $3 \mapsto 5$, etc.: the same information we have at (11.4). The six elements of S_3 correspond to the matrices

$$\begin{pmatrix} 1 & 0 & 0 \\ 0 & 1 & 0 \\ 0 & 0 & 1 \end{pmatrix} \begin{pmatrix} 0 & 0 & 1 \\ 1 & 0 & 0 \\ 0 & 1 & 0 \end{pmatrix} \begin{pmatrix} 0 & 1 & 0 \\ 0 & 0 & 1 \\ 1 & 0 & 0 \end{pmatrix} \begin{pmatrix} 1 & 0 & 0 \\ 0 & 0 & 1 \\ 0 & 1 & 0 \end{pmatrix} \begin{pmatrix} 0 & 0 & 1 \\ 0 & 1 & 0 \\ 1 & 0 & 0 \end{pmatrix} \begin{pmatrix} 0 & 1 & 0 \\ 1 & 0 & 0 \\ 0 & 0 & 1 \end{pmatrix}$$

in the same order as listed above. Operations in S_n correspond to ordinary products of matrices. For example, the product in (11.5) can be carried out as

$$\begin{pmatrix} 0 & 1 & 0 & 0 & 0 \\ 1 & 0 & 0 & 0 & 0 \\ 0 & 0 & 1 & 0 & 0 \\ 0 & 0 & 0 & 0 & 1 \\ 0 & 0 & 0 & 1 & 0 \end{pmatrix} \cdot \begin{pmatrix} 0 & 1 & 0 & 0 & 0 \\ 0 & 0 & 0 & 0 & 1 \\ 0 & 0 & 1 & 0 & 0 \\ 0 & 0 & 0 & 1 & 0 \\ 1 & 0 & 0 & 0 & 0 \end{pmatrix} = \begin{pmatrix} 0 & 0 & 0 & 0 & 1 \\ 0 & 1 & 0 & 0 & 0 \\ 0 & 0 & 1 & 0 & 0 \\ 1 & 0 & 0 & 0 & 0 \\ 0 & 0 & 0 & 1 & 0 \end{pmatrix}.$$

On the whole it may seem we have not gained much, but we do get something almost for free from this representation. Since matrices have determinants, we may associate a number with every permutation, that is, the determinant of its permutation matrix.

LEMMA 11.37   *The determinant of a permutation matrix is* ±1.

For the proof, recall that the determinant of an $n \times n$ matrix can be computed by 'expanding' with respect to any row or column. For example, using the first row one may compute the determinant of $A = (a_{ij})$ as

$$\det A = \sum_{j=1}^{n} (-1)^{1+j} a_{1j} \det A^{1j},$$

where A^{1j} denotes the $(n-1) \times (n-1)$ matrix obtained by removing the first row and the jth column from A.

Example 11.38   You have likely practiced this way of computing determinants in Linear Algebra, at least for small values of n. For $n = 3$, the expansion gives

$$a_{11} \det \begin{pmatrix} a_{11} & a_{12} & a_{13} \\ a_{21} & a_{22} & a_{23} \\ a_{31} & a_{32} & a_{33} \end{pmatrix} - a_{12} \det \begin{pmatrix} a_{11} & a_{12} & a_{13} \\ a_{21} & a_{22} & a_{23} \\ a_{31} & a_{32} & a_{33} \end{pmatrix} + a_{13} \det \begin{pmatrix} a_{11} & a_{12} & a_{13} \\ a_{21} & a_{22} & a_{23} \\ a_{31} & a_{32} & a_{33} \end{pmatrix},$$

that is,

$$a_{11}(a_{22}a_{33} - a_{23}a_{32}) - a_{12}(a_{21}a_{33} - a_{23}a_{31}) + a_{13}(a_{21}a_{32} - a_{22}a_{31}),$$

matching every other way you may have to compute $\det A$. The same procedure works for square matrices of any size.

Proof of Lemma 11.37   Argue by induction on n. If $n = 1$, the only permutation matrix is (1), and there is nothing to prove. For larger n, assume that A is an $n \times n$ permutation matrix; compute $\det A$ by expanding (for example) along the first row. Since the first row of A only contains one nonzero entry, equal to 1, $\det A$ equals $\pm$ the determinant of an $(n-1) \times (n-1)$ permutation matrix, which equals ±1 by the induction hypothesis. It follows that $\det A = \pm1$ as stated.   □

DEFINITION 11.39   The 'sign' of a permutation $\sigma \in S_n$, denoted $(-1)^\sigma$, is the determinant of the permutation matrix corresponding to σ.
We say that σ is 'even' if $(-1)^\sigma = 1$, and 'odd' if $(-1)^\sigma = -1$.

Since the determinant is a group homomorphism (as recalled in Example 11.16), so is the function

$$S_n \to \{-1, 1\} \cong C_2$$

sending σ to $(-1)^\sigma$. This will be quite important later on. It implies that the product of two even permutations is an even permutation, and it follows that the set of *even* permutations is a subgroup of S_n. This famous subgroup of S_n has a name.

DEFINITION 11.40   The 'alternating group' $\mathcal{A}_n$ is the subgroup of S_n consisting of even permutations.

For example, A_3 coincides with the isomorphic copy of C_3 within S_3. The alternating group will take center stage in our more sophisticated considerations on groups, particularly in §12.6.

Taking a little more seriously the fact that S_n acts on $\{1, \ldots, n\}$ leads to what may be the most efficient way to denote its elements. For an element $\sigma \in S_n$, consider the action of $\langle \sigma \rangle$ on $\{1, \ldots, n\}$. A given $a \in \{1, \ldots, n\}$ will visit several others as powers of σ act on it, then eventually return to itself:

$$a \xmapsto{\sigma} \sigma(a) \xmapsto{\sigma} \sigma^2(a) \xmapsto{\sigma} \sigma^3(a) \xmapsto{\sigma} \cdots \xmapsto{\sigma} \sigma^{r-1}(a) \xmapsto{\sigma} \sigma^r(a) = a \,,$$

and from there cycle back through the same elements. The action of $\langle \sigma \rangle$ partitions $\{1, \ldots, n\}$ into orbits, and σ acts within each orbit by cycling through it. Our running example in S_7,

$$\begin{pmatrix} 1 & 2 & 3 & 4 & 5 & 6 & 7 \\ 4 & 3 & 5 & 1 & 7 & 2 & 6 \end{pmatrix},$$

does this as follows:

$$1 \mapsto 4 \mapsto 1, \quad 2 \mapsto 3 \mapsto 5 \mapsto 7 \mapsto 6 \mapsto 2 \,. \tag{11.6}$$

There are two orbits $\{1, 4\}$ and $\{2, 3, 5, 6, 7\}$, and σ cycles through them as displayed in the following picture.

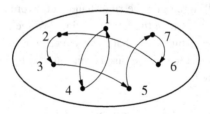

We can define permutations that *cycle* through one orbit and leave the other elements of $\{1, \ldots, n\}$ alone. Such permutations are called, rather unimaginatively, 'cycles'.

DEFINITION 11.41   An 'r-cycle' $(a_1 \ldots a_r)$ is a permutation cyclically permuting r elements $a_1, a_2, \ldots, a_r$, i.e., mapping a_i to a_{i+1} for $i = 1, \ldots, r-1$ and a_r to a_1, and acting trivially on all other elements. A 'transposition' is a 2-cycle.

For example, the two cycles shown in (11.6) are

$$(1\,4) \quad \text{and} \quad (2\,3\,5\,7\,6).$$

They are the bijections

$$
\begin{cases}
①\mapsto④\\
2\mapsto 2\\
3\mapsto 3\\
④\mapsto①\\
5\mapsto 5\\
6\mapsto 6\\
7\mapsto 7
\end{cases}
\quad \text{and} \quad
\begin{cases}
1\mapsto 1\\
②\mapsto③\\
③\mapsto⑤\\
4\mapsto 4\\
⑤\mapsto⑦\\
⑥\mapsto②\\
⑦\mapsto⑥
\end{cases}
$$

where I have circled the elements in the nontrivial orbit of each.

Remark 11.42 The notation I have just introduced is a little quirky. First of all, the notation does not mention 'what n is': $(2\,5)$ could be an element of S_5 or of S_{1000}. Second, since we are cycling through a set of elements, it does not matter which we start from. Therefore,

$$(2\,3\,5\,7\,6) = (3\,5\,7\,6\,2) = (5\,7\,6\,2\,3) = (7\,6\,2\,3\,5) = (6\,2\,3\,5\,7).$$

Third, a '1-cycle' (a) is just the identity: it maps a to a, and all other elements to themselves as well. Therefore $(1) = (2) = \cdots = (n) = e_{S_n}$. ⌐

Remark 11.43 We say that two cycles are 'disjoint' if their nontrivial orbits are disjoint. It should be clear that *disjoint cycles commute,* since each acts just as the identity on the elements appearing in the other. ⌐

Cycles offer the alternative notation I was looking for. We can write

$$\begin{pmatrix} 1 & 2 & 3 & 4 & 5 & 6 & 7 \\ 4 & 3 & 5 & 1 & 7 & 2 & 6 \end{pmatrix} = (1\,4)(2\,3\,5\,7\,6),$$

and you will agree that the right-hand side expresses much more concisely what the permutation does than the left-hand side. We see right away that this permutation has two nontrivial orbits, and what it does in each orbit.

We formalize the above discussion into the following statement, which should not require a proof.

> PROPOSITION 11.44 *Every permutation $\sigma \ne e_{S_n}$ may be written as a product of disjoint nontrivial cycles.*

'Trivial' cycles, i.e., 1-cycles (a), may be omitted since they just represent the identity.

One defect of the cycle notation is that it is not unique. Our running example could also be written in a variety of different ways:

$$(1\,4)(2\,3\,5\,7\,6) = (5\,7\,6\,2\,3)(1\,4) = (4\,1)(7\,6\,2\,3\,5) = \ldots.$$

In terms of cycle notation, the elements of S_3 may be written

$$e = (1), \quad (1\,2\,3), \quad (1\,3\,2), \quad (2\,3), \quad (1\,3), \quad (2\,3)$$

in the same order as listed earlier. This notation conveys more readily what the permutations do than the alternatives we have seen.

Another advantage of the cycle notation is that it makes it easy to compute standard information associated with a permutation. For example, the order of $\sigma = (1\,4)(2\,3\,5\,7\,6)$ is 10: indeed,

$$\sigma^{10} = (1\,4)^{10}(2\,3\,5\,7\,6)^{10} = e_{S_7} \cdot e_{S_7} = e_{S_7},$$

and $\sigma^i \neq e_{S_7}$ for $i = 1, \ldots, 9$. (The order of a cycle of length r is clearly r.)

It is occasionally useful to write a permutation as a product of not necessarily disjoint cycles. For example, we can write every cycle as a product of (nondisjoint!) transpositions, i.e., of 2-cycles.

LEMMA 11.45

$$(a_1 \ldots a_r) = (a_1 a_r) \cdots (a_1 a_3)(a_1 a_2). \tag{11.7}$$

Proof It suffices to show that the image of a_i under application of the right-hand side is a_{i+1} for $i = 1, \ldots, r - 1$, and a_1 for $i = r$, and this is straightforward. For example, $a_2 \mapsto a_3$:

$$((a_1 a_r) \cdots (a_1 a_3)(a_1 a_2))(a_2) = ((a_1 a_r) \cdots (a_1 a_3))(a_1) = a_3$$

since all transpositions $(a_1 a_4), \ldots, (a_1 a_r)$ act trivially on a_3. □

This simple observation has useful consequences.

LEMMA 11.46 *An r-cycle is even or odd according to whether r is odd or even. That is: if σ is an r-cycle, then $(-1)^\sigma = (-1)^{r-1}$.*

Proof Transpositions are odd, since the matrix of a transposition is obtained from the identity matrix by switching two rows. Since an r-cycle is a product of $r - 1$ transpositions by Lemma 11.45, the statement follows from the fact that the sign function is a group homomorphism. □

For example, $\sigma = (1\,4)(2\,3\,5\,7\,6)$ is an *odd* permutation, since it is the product of an odd cycle (that is, $(1\,4)$) and an even one (that is, $(2\,3\,5\,7\,6)$).

PROPOSITION 11.47 *The group S_n is generated by transpositions.*

Proof Every permutation is a product of cycles, and every cycle is a product of transpositions. □

What Proposition 11.47 says is that you can perform any permutation by switching two elements at a time. This can be used to give an alternative proof of Lemma 11.37: indeed, it implies that every permutation matrix may be obtained from the identity matrix (which has determinant equal to 1) by a sequence of column switches, and the determinant switches sign whenever you switch two columns.

You will use Proposition 11.47 to prove that S_n can be generated very economically: $(1\,2)$ and $(1\,\ldots\,n)$ suffice to generate S_n (Exercise 11.28). For $n = 3$, this says that S_3 is generated by $(1\,2)$ and $(1\,2\,3)$, and a moment's thought will show that this matches something we know already, since $S_3 \cong D_6$ (cf. Proposition 11.35).

As a last class of general examples, I will mention *free groups*. Free *modules,* and in particular free *abelian groups,* played an important role in Part II. The salient feature of free modules is that there are no constraints on the homomorphisms from a free R-module F to another R-module M: every homomorphism determines and is determined by the images in M of the elements in a basis of F, and these images may be chosen arbitrarily. This ('universal') property is stated precisely and proved in Proposition 9.8. It does not hold for all modules: in general, the images of a set of generators must satisfy suitable relations, see Example 9.10. *Free groups* satisfy an analogous property: in order to specify a group homomorphism from the 'free group $F(A)$ on a set A' to an arbitrary group G, all we need to specify is the images in G of the elements of the set A; and every choice in G of images of elements in A determines a unique group homomorphism $F(A) \to G$.

We will not use free groups in what follows, so I will refrain from giving a complete, rigorous construction of such groups. The gist of the construction is, however, easy to appreciate. Treat the elements of the set A as 'letters' of an alphabet. You can write 'words' by stringing any finite number of letters of this alphabet (including no letter at all if you so choose) in any order you wish, with any integer exponent you wish. For example, if $A = \{a, b\}$ consists of two letters a and b, two 'words' may be

$$b^{-3}a^2a^3b^4a^{-1}ab\,, \quad b^{-2}b^{-3}a^2b^2b^{-3}\,.$$

Now pretend that this is happening in a group: then you can carry out the implicit operations to remove redundancies; these two words would simplify to the 'reduced' forms

$$b^{-3}a^5b^5\,, \quad b^{-5}a^2b^{-1}\,.$$

We get an operation by juxtaposing two words and reducing the resulting word. The product of these two words would be

$$b^{-3}a^5b^5b^{-5}a^2b^{-1} \mapsto b^{-3}a^5a^2b^{-1} \mapsto b^{-3}a^7b^{-1}\,.$$

It is not hard to verify that the set $F(A)$ of reduced words, together with this operation, is a group. The identity e of this group is the 'empty' word. Inverses are obtained by flipping words and replacing each exponent by its opposite: for example, the inverse of the word we just computed would be $ba^{-7}b^3$, since

$$(ba^{-7}b^3)(b^{-3}a^7b^{-1}) \mapsto ba^{-7}a^7b^{-1} \mapsto bb^{-1} \mapsto e$$

reduces to the empty word, and so does $(b^{-3}a^7b^{-1})(ba^{-7}b^3)$.

In a sense, this is the 'purest' way to make a group out of the information of a set A: the elements of A are just strung together to get elements of $F(A)$, and no extra information is added in the process. (On the other hand, the reader is warned that $F(A)$ is a rather unwieldy group if A contains two or more elements.) In order to define a group homomorphism $\varphi\colon F(A) \to G$, it suffices to prescribe $\varphi(a) \in G$ for every letter $a \in A$: imposing the homomorphism property determines the value of $\varphi(g)$ for every word g. Since the letters of A satisfy no relation in $F(A)$, we are *free* to choose their images $\varphi(a)$ as we wish, without any constraint. So $F(A)$ satisfies a similar requirement to the property of free *modules* recalled above. This requirement is the 'universal property of free groups'.

Example 11.48 If A is a set of generators of a group G, then we have a natural *surjective* homomorphism $F(A) \to G$: we get a homomorphism as mentioned a moment ago, and this is surjective since every element of G may be written as a product of powers of the generators. For example, we can interpret Proposition 11.35 as stating that for every $n > 2$ there is an onto homomorphism from $F(\{x, y\})$ to D_{2n}, and the relations satisfied by x and y in D_{2n}, listed in that statement, amount to a description of the kernel of this homomorphism.

We will not have the opportunity to use free groups, but they play a fundamental role in group theory, and they occur spontaneously in nature, for example in topology. The 'fundamental group' of the figure 8, i.e., two circles joined at a point, is isomorphic to the free group $F(\{a, b\})$ on a set with two elements.

11.4 Canonical Decomposition, Normality, and Quotients

Let $\varphi\colon G \to H$ be a group homomorphism. As we have done in the case of rings and modules (and, in particular, abelian groups), we are going to prove that φ admits a canonical decomposition as a surjective homomorphism, followed by an isomorphism, followed by an injective homomorphism. The diagram is going to be

$$G \xrightarrow[\pi]{} G' \xrightarrow[\tilde\varphi]{\sim} H' \overset{\iota}{\hookrightarrow} H \qquad (11.8)$$

and you should review §5.1 and §8.4 if this does not look very familiar. In fact, we should know exactly what to do, based on our previous encounters.

- We are going to let $H' = \operatorname{im}\varphi = \varphi(G)$; this is a subgroup of H (Example 11.9), so the inclusion function $\iota\colon H' \to H$ is an injective group homomorphism.
- We know that, as a set, $G' = G/\sim$, where $\sim$ is the equivalence relation defined by

$$a \sim b \iff \varphi(a) = \varphi(b).$$

The crux of the matter is to endow G' with a group structure, so that all the maps in (11.8) are homomorphisms of groups. Set theory already tells us that π is surjective and $\tilde\varphi$ is a bijection, where $\tilde\varphi$ is defined by

$$\tilde\varphi([a]_\sim) = \varphi(a). \qquad (11.9)$$

Everything will work out just fine, but there is an interesting phenomenon that is 'new' to groups, and we need to come to terms with it. The discussion will be quite a bit more involved than the analogous discussion for modules (and in particular *abelian* groups) in §8.4.

Let's proceed using our previous experience as a guide. As usual, the first observation is that

$$a \sim b \iff \varphi(a) = \varphi(b) \iff \varphi(a^{-1}b) = e_{G'} \iff a^{-1}b \in \ker \varphi, \qquad (11.10)$$

where $\ker \varphi$ is the *kernel* of φ.

DEFINITION 11.49   Let $\varphi: G \to H$ be a group homomorphism. The 'kernel' of φ, denoted $\ker \varphi$, is the subset of G

$$\ker \varphi = \{g \in G \,|\, \varphi(G) = e_H\}.$$

Thus, $g \in \ker \varphi$ if and only if $\varphi(g) = e_H$.

We already know (Example 11.10) that $\ker \varphi$ is a subgroup of G. It should be clear that φ is injective if and only if $\ker \varphi = \{e_G\}$; if this is not immediately transparent to you, verify it now.

On the basis of (11.10), we may hit upon the idea of defining a relation 'like $\sim$' for every subgroup of G: if $K \subseteq G$ is a subgroup, we could define

$$a \sim'_K b \iff a^{-1}b \in K.$$

If $K = \ker \varphi$, then this new relation agrees with $\sim$ according to (11.10). For every subgroup K, the relation $\sim'_K$ is an equivalence relation (Exercise 11.34), so we can define a 'quotient of G mod K', at least as a set: G/K could denote the set of equivalence classes with respect to the relation $\sim'_K$. These classes are easy to describe:

$$[a]_{\sim'_K} = \{b \in G \,|\, a^{-1}b \in K\} = \{b \in G \,|\, \exists k \in K, a^{-1}b = k\} = \{b \in G \,|\, \exists k \in K, b = ak\}$$
$$= aK,$$

where aK consists of all products ak for $k \in K$. This is the 'coset' of a modulo K. (Since in rings and modules the operation was $+$, there we used the notation $a + I$, etc.; but the math is just the same.) In fact, we call aK a *left*-coset, since we are multiplying K by a on the left.

DEFINITION 11.50   Let G be a group, and let K be a subgroup of G. The notation G/K denotes the set of left-cosets of K in G.

Now, if you feel optimistic, you may hope that this set G/K of left-cosets is a group in a natural way, and we would be home free. After all, this is what happens with abelian groups, modulo writing things additively rather than multiplicatively.

Well, the matter is a little more complex. Rewind just a little, and try this other way to think about the original equivalence relation $\sim$:

$$a \sim b \iff \varphi(a) = \varphi(b) \iff \varphi(ba^{-1}) = e_{G'} \iff ba^{-1} \in \ker \varphi. \qquad (11.11)$$

This is just as true as (11.10), but it leads you in a different direction. Given a subgroup K, we could define *yet another* equivalence relation, $\sim''_K$, by setting

$$a \sim''_K b \iff ba^{-1} \in K;$$

this would just be $\sim$ again if $K = \ker \varphi$. However, the equivalence classes with respect to *this* relation look somewhat different:

$$[a]_{\sim''_K} = \{b \in G \mid ba^{-1} \in K\} = \{b \in G \mid \exists k \in K, ba^{-1} = k\} = \{b \in G \mid \exists k \in K, b = ka\}$$
$$= Ka.$$

Instead of *left*-cosets, we get *right*-cosets. If $K = \ker \varphi$, then all these equivalence relations must agree, so the equivalence classes must be the same: that is, left-cosets and right-cosets must be equal. But is this going to be the case for *every* subgroup K of *every* group G?

Example 11.51 Let $G = D_6$, and $K = \langle x \rangle = \{e, x\}$ (using the notation introduced in §11.3). Choose $a = y$. Then

$$aK = \{y, yx\}, \quad Ka = \{y, xy\}.$$

Since (by Proposition 11.35) $xy = y^{-1}x \neq yx$ in D_6, we see that $aK \neq Ka$ in this example.

This is a problem. *If* we manage to make G/K into a group, in such a way that the natural projection $\pi : G \to G/K$ is a group homomorphism, then necessarily $K = \ker \pi$ is a kernel. But as we have just seen, for kernels it so happens that left- and right-cosets coincide, while for an arbitrary subgroup K it may well happen that left- and right-cosets do *not* coincide.

The conclusion is that we can*not* define a 'quotient group G/K' for *every* subgroup K. This is a point of departure with respect to the theory for *abelian* groups: for abelian groups, we *were* able to define a quotient by an arbitrary subgroup. As we have just observed, for general groups the subgroups for which a quotient group can be defined must satisfy a special property.

DEFINITION 11.52 A subgroup K of a group G is 'normal' if $gK = Kg$ for all $g \in G$ or, equivalently, $gKg^{-1} = K$ for all $g \in G$.

Remark 11.53 In verifying that a subgroup K is normal, it suffices to show that $\forall g \in G$, $gKg^{-1} \subseteq K$. Indeed, if this is true for all $g \in G$, then we also have $g^{-1}Kg \subseteq K$, and this latter condition is equivalent to $K \subseteq gKg^{-1}$. Since both inclusions hold, this verifies $gKg^{-1} = K$ as needed.   ⌐

Remark 11.54 Caveat. The statement $gK = Kg$ does *not* mean that $gk = kg$ for all $k \in K$. What it means is that for all $k \in K$ there exists $k' \in K$ such that $gk = k'g$. A subgroup K is normal if and only if $\forall k \in K, \forall g \in G, \exists k' \in K$ such that $gkg^{-1} = k'$.

In general, $k' \neq k$. Again, $gKg^{-1} = K$ does *not* mean that every element of K commutes with every element of G.   ⌐

Remark 11.55   The requirement $gKg^{-1} = K$ is equivalent to $gK = Kg$, so indeed we are prescribing that left- and right-cosets should agree. If K is normal, then the partition G/K of G into left-cosets and the partition into right-cosets (which may be denoted $K\backslash G$) coincide.

Conversely, if left-cosets and right-cosets agree, then K satisfies the requirement in Definition 11.52, so K is normal.   ⌐

Remark 11.56   Regardless of normality, the set of (say) left-cosets of a subgroup K is interesting, because G *acts* on it, by 'left-multiplication': if $g \in G$ and $aK \in G/K$, then we can define a new coset $g \cdot aK := (ga)K$. This action is defined regardless of whether K is normal in G, and it will be very useful in, e.g., the proof of the Sylow theorems in §12.5.   ⌐

Note that if G is commutative, then $\forall k \in K$ we have $gkg^{-1} = gg^{-1}k = k \in K$: this shows that every subgroup of a commutative group is normal, and this is why we did not have to face this issue when we dealt with abelian groups. One way to think about this is through the following observation, which is often useful.

> PROPOSITION 11.57   *A subgroup K of a group G is normal if and only if it is a union of conjugacy classes.*

(Conjugacy classes are defined in Definition 11.31.)

Proof   If K is a subgroup of G, let $\widehat{K}$ be the union of all the conjugacy classes of elements of K:

$$\widehat{K} := \{h \in G \mid \exists k \in K, \exists g \in G, h = gkg^{-1}\}.$$

Since every element is conjugate to itself, $K \subseteq \widehat{K}$. It suffices to show that a subgroup K is normal if and only if $K = \widehat{K}$, that is, if and only if $\widehat{K} \subseteq K$.

($\Rightarrow$) Assume that K is normal, and let $h \in \widehat{K}$. Then there exist $k \in K$ and $g \in G$ such that $h = gkg^{-1}$; and therefore $h \in gKg^{-1} \subseteq K$, as needed.

($\Leftarrow$) Conversely, assume that $\widehat{K} \subseteq K$, and let $g \in G$. For all $k \in K$, $gkg^{-1} \in \widehat{K}$ by definition of $\widehat{K}$. This shows that for all $g \in G$, $gKg^{-1} \subseteq \widehat{K}$; since $\widehat{K} \subseteq K$, we have that $gKg^{-1} \subseteq K$ for all $g \in G$. As observed in Remark 11.53, this suffices to prove that K is normal.   □

In an abelian group, the conjugacy class of an element k is k itself; therefore every subgroup is a 'union of conjugacy classes'. In general, Proposition 11.57 gives a constraint on the possible size of a normal subgroup in a group G, if we have information on the sizes of the conjugacy classes in G.

Let's formalize the observation that kernels are normal.

> PROPOSITION 11.58   *Let $\varphi: G \to H$ be a group homomorphism. Then $\ker \varphi$ is a normal subgroup of G.*

If you followed me in the analysis that got us here, then you know this is true. Anyway, it is easy to provide a direct argument.

Proof Let $g \in G$ and let $k \in \ker \varphi$. Then

$$\varphi(gkg^{-1}) = \varphi(g)\varphi(k)\varphi(g^{-1}) = \varphi(g)e_H\varphi(g)^{-1} = e_H :$$

therefore $gkg^{-1} \in \ker \varphi$. This shows that $\forall g \in G$, $g(\ker \varphi)g^{-1} \subseteq \ker \varphi$, which suffices in order to prove normality as observed in Remark 11.53. □

We already know lots of examples of normal subgroups, since subgroups of abelian groups are automatically normal. The situation is of course more interesting for non-commutative groups. Often, the simplest way to verify that a subgroup K of a group G is normal is to realize K as the kernel of a homomorphism from G to some other group (which implies it is normal, by Proposition 11.58).

Example 11.59 Suppose there are exactly two left-cosets: $G/K = \{K, aK\}$, for some $a \in G$, $a \notin K$. Then I claim that K is necessarily normal.

Indeed, as we have observed (Remark 11.56), G acts on this set, by left-multiplication; so we get a group homomorphism $\varphi : G \to S_2$. The kernel of this homomorphism is

$$\ker \varphi = \{g \in G \,|\, gK = K \text{ and } gaK = aK\}.$$

Note that $gK = K$ implies $g = g \cdot e \in K$; so $\ker \varphi \subseteq K$. On the other hand, if $g \in K$, then $gaK \neq K$: indeed, otherwise we would have $ga = ga \cdot e = k \in K$, and this would imply $a = g^{-1}k \in K$, a contradiction. Therefore, $g \in K \implies gaK = aK$ (as well as $gK = K$), and hence $K \subseteq \ker \varphi$.

The conclusion is that $K = \ker \varphi$, confirming that K is normal.

We will have a simpler way to think about this once we learn about Lagrange's theorem in §12.1.

Example 11.60 The *center* of a group G is the set $Z(g)$ of elements $g \in G$ which commute with all other elements: $Z(G) = \{z \in G \,|\, \forall g \in G, zg = gz\}$. It should just take you a moment to check that $Z(G)$ is a subgroup of G, and it takes a second moment to realize it is normal. Indeed, $\forall g \in G$, $\forall z \in Z(G)$,

$$gzg^{-1} = zgg^{-1} = z$$

since z commutes with every element. Therefore $\forall g \in G$, $gZ(G)g^{-1} \subseteq Z(G)$ as needed for normality.

Example 11.61 Let D_{2n}, $n \geq 3$ be the dihedral group studied in §11.3. As we have seen, it contains one subgroup isomorphic to C_n and n subgroups isomorphic to C_2. The latter are *not* normal. Indeed, given one such subgroup F, we may assume it is generated by the element we called x in the discussion in §11.3, so $F = \{e, x\}$; since

$$yxy^{-1} = xy^{-1}y^{-1} = xy^{-2} \notin F,$$

we see that $yFy^{-1} \neq F$, so F is not normal. By contrast, $K = \{e, y, \ldots, y^{n-1}\}$ *is* normal.

Once we have Lagrange's theorem (Theorem 12.3), this will be evident from Example 11.59: the fact that $|K| = n$ and $|D_{2n}| = 2n$ implies that there are exactly two cosets. In any case, it is instructive to check normality directly. Every element $g \in D_{2n}$ may be written as $g = x^i y^j$ for $0 \le i \le 1, 0 \le j \le n - 1$ (Proposition 11.35); if $i = 0$, then $g \in K$ and therefore $gKg^{-1} = K$; if $i = 1$, gKg^{-1} consists of elements

$$gy^k g^{-1} = xy^j y^k y^{n-j} x = xy^k x = y^{-k} xx = y^{-k} \in K.$$

This shows $gKg^{-1} \subseteq K$ and confirms that K is normal.

Alternatively, we can define a group homomorphism $\varphi: D_{2n} \to C_2 = \{e, f\}$ by letting $\varphi(g) = e$ if g does not flip the regular n-gon, and $\varphi(g) = f$ if it does. Spend a moment to realize that this is a homomorphism, and then it will follow that the group $K \cong C_n$ of rotations is normal, since it is the kernel of this homomorphism.

Example 11.62   The alternating group $\mathcal{A}_n$ (Definition 11.40), consisting of even permutations, is normal in $\mathcal{S}_n$. Indeed, it is by definition the kernel of the 'sign' homomorphism $\mathcal{S}_n \to \{+1, -1\}$. For $n = 3$, it corresponds to $C_3 \subseteq D_6 \cong S_3$; for $n \ge 4$, it is an interesting noncommutative group. The group $\mathcal{A}_4$ is isomorphic to the group of rotational symmetries of a regular tetrahedron (Exercise 11.23); the group $\mathcal{A}_5$ is isomorphic to the group of rotational symmetries of a regular icosahedron, or equivalently of a soccer ball.

Example 11.63   Within $\mathcal{A}_4$ we find the subgroup

$$V = \{e, (1\,2)(3\,4), (1\,3)(2\,4), (1\,4)(2\,3)\}$$

(using cycle notation). It is isomorphic to $C_2 \times C_2$, as you can check in a moment; it is called the *Klein Vierergruppe* ('Vier' just means 'four'). You can verify that V is normal in $\mathcal{A}_4$: for example, $(123)(12)(34)(321)$ maps 1 to 4, 2 to 3, 3 to 2, and 4 to 1, therefore

$$(1\,2\,3)(1\,2)(3\,4)(3\,2\,1) = (1\,4)(2\,3) \in V.$$

(You can bypass all computations if you make judicious use of Exercise 11.30.)

Challenge: Does $\mathcal{A}_n$ have nontrivial *normal* subgroups, for $n \ge 5$? We will figure this out: it will in fact be our main focus in §12.6. It is a more important question than it may seem, with a very significant application that we will explore in depth in Chapter 15.

Let's go back to the motivation that led us to the definition of normal subgroups: we had discovered that if we can take a quotient of a group G by a subgroup K, then K is necessarily normal; this is a consequence of Proposition 11.58. The good news is that normality is the *only* requirement on a subgroup K needed in order to define a quotient.

THEOREM 11.64   *Let G be a group, and let K be a* normal *subgroup of G. Then*

- *For all $a \in G$, $aK = Ka$.*
- *The two equivalence relations $\sim'_K$, $\sim''_K$ defined earlier coincide.*

- *The set G/K of (left-)cosets of K is a group, with operation defined by*

$$(aK) \cdot (bK) := (ab)K.$$

Proof The first point is just a restatement of the normality of K.

The second point follows from the first. Indeed, at the beginning of this discussion we had verified that the left-cosets are the equivalence classes with respect to $\sim'_K$ and the right-cosets are the equivalence classes with respect to $\sim''_K$. Since an equivalence relation is determined by its equivalence classes, $\sim'_K$ and $\sim''_K$ must be the same relation.

As for the third point, I claim that the given prescription for the product of two cosets,

$$(aK) \cdot (bK) := (ab)K, \tag{11.12}$$

is well-defined. For this, assume $aK = a'K$ and $bK = b'K$; then $a^{-1}a' \in K$, $b^{-1}b' \in K$. It follows that

$$(ab)^{-1}(a'b') = b^{-1}a^{-1}a'b' = b^{-1}a^{-1}a'(bb^{-1})b' = (b^{-1}a^{-1}a'b)(b^{-1}b')$$
$$\in (b^{-1}Kb)(b^{-1}b') \in K.$$

Indeed, $b^{-1}b' \in K$, and $b^{-1}Kb = K$ since K is normal. This proves that $(ab)K = (a'b')K$ and establishes that (11.12) is well-defined.

It follows that G/K is a group with this operation. Indeed, the operation is associative since the operation in G is associative, $eK = K$ satisfies the requirement for the identity element, and

$$(aK) \cdot (a^{-1}K) = (aa^{-1})K = K, \quad (a^{-1}K) \cdot (aK) = (a^{-1}a)K = K,$$

showing that $a^{-1}K$ satisfies the requirement for the inverse of aK. $\square$

DEFINITION 11.65 Let G be a group, and let K be a normal subgroup of G. The 'quotient' of G modulo K, denoted G/K, is the group of cosets of K, with the operation defined by $(aK)(bK) := (ab)K$.

(I have written the cosets as left-cosets. Of course, left-cosets equal right-cosets, since the subgroup is normal.)

The natural projection $\pi \colon G \to G/K$ is a surjective group homomorphism, by virtue of the definition of product in G/K:

$$\pi(ab) = (ab)K = (aK)(bK) = \pi(a)\pi(b);$$

and

$$\ker \pi = \{a \in G \mid aK = K\} = \{a \in G \mid a \in K\} = K.$$

Remark 11.66 We can now conclude that *the notion of kernel and of normal subgroup coincide:* kernels are normal by Proposition 11.58, and we have just verified that if K is normal, then K is the kernel of the natural projection $G \to G/K$.

This is the parallel for groups of the observation that the notions of kernel and ideal coincide in ring theory (Remark 5.36), and that kernels and submodules coincide in the theory of modules (Remark 8.42). If the group is commutative, then every subgroup is normal, so again we can observe that, for *abelian* groups, the notions of 'kernel' and 'subgroup' coincide.   ⌐

We are finally ready to go back to our canonical decomposition (11.8). We had observed that $G' = G/\sim$ as a set, and we have realized the relation $\sim$ as $\sim'_K$ for $K = \ker \varphi$. Therefore $G' = G/\ker \varphi$. Putting things together:

> THEOREM 11.67 (Canonical decomposition)   *Let $\varphi: G \to H$ be a group homomorphism. Then we have a commutative diagram*
>
> $$G \xrightarrow[\pi]{\quad\quad} (G/\ker \varphi) \xrightarrow[\tilde{\varphi}]{\sim} \varphi(G) \xrightarrow[\iota]{} H$$
>
> *where π is the natural projection, ι is the inclusion homomorphism, and $\tilde{\varphi}$ is the induced group homomorphism. Further, $\tilde{\varphi}$ is an* isomorphism.

Along the way we have understood that π is a surjective group homomorphism. The bijection $\tilde{\varphi}$ is defined by (11.9), which with the new notation reads

$$\tilde{\varphi}(a \ker \varphi) := \varphi(a).$$

The last item to prove is that this function is a homomorphism of groups, and this follows from the definitions: for all cosets $a \ker \varphi$ and $b \ker \varphi$ in $G/\ker \varphi$,

$$\tilde{\varphi}((a \ker \varphi)(b \ker \varphi)) = \tilde{\varphi}((ab) \ker \varphi) = \varphi(ab) = \varphi(a)\varphi(b) = \tilde{\varphi}(a \ker \varphi)\,\tilde{\varphi}(b \ker \varphi)$$

as needed.   □

11.5   Isomorphism Theorems

The *First Isomorphism Theorem* is, as we know, just a particular case of the canonical decomposition.

> THEOREM 11.68 (First Isomorphism Theorem)   *Let $\varphi: G \to H$ be a* surjective *group homomorphism. Then $H \cong G/\ker \varphi$.*
>   *In fact, φ induces an isomorphism $\tilde{\varphi}: G/\ker \varphi \xrightarrow{\sim} H$, defined by $\tilde{\varphi}(gK) = \varphi(g)$.*

Proof   Since φ is surjective, $\varphi(G) = H$ in Theorem 11.67.   □

We already know this is going to be useful—we have used the corresponding result for rings and modules many times in previous chapters.

Example 11.69   The quotient $S_n/\mathcal{A}_n$ is isomorphic to the cyclic group C_2 for $n \geq 2$. Indeed, $\mathcal{A}_n$ is the kernel of the sign morphism $S_n \to \{+1, -1\} \cong C_2$, which is surjective for $n \geq 2$. Similarly, for $n \geq 3$, $D_{2n}/C_n \cong C_2 = \{e, f\}$, where here C_n is identified with the subgroup of clockwise rotations in D_{2n}. (See Examples 11.61 and 11.62.)

Also, consider the group homomorphism $C_{2n} \to C_2 \cong \{[0]_{2n}, [n]_{2n}\}$ given by multiplication by $[n]_{2n}$. The kernel is the subgroup $\{[0]_{2n}, [2]_{2n}, \ldots, [2n-2]_{2n}\} \cong C_n$. In this sense, $C_{2n}/C_n \cong C_2$.

Example 11.70 Let G_1 and G_2 be groups, and let $K_1 \subseteq G_1$, $K_2 \subseteq G_2$ be normal subgroups. Then $K_1 \times K_2$ is a normal subgroup of $G_1 \times G_2$, and we have an isomorphism $(G_1 \times G_2)/(K_1 \times K_2) \cong (G_1/K_1) \times (G_2/K_2)$.

Indeed, there is a surjective group homomorphism $G_1 \times G_2 \to (G_1/K_1) \times (G_2/K_2)$, defined by sending (g_1, g_2) to $(g_1 K_1, g_2 K_2)$. Its kernel is

$$\{(g_1, g_2) \mid (g_1 K_1, g_2 K_2) = (K_1, K_2)\} = \{(g_1, g_2) \mid g_1 K_1 = K_1, g_2 K_2 = K_2\}$$
$$= \{(g_1, g_2) \mid g_1 \in K_1, g_2 \in K_2\} = K_1 \times K_2,$$

and the result follows directly from the First Isomorphism Theorem.

Example 11.71 We have briefly encountered free groups in §11.3; recall that $F(A)$ denotes the free group on the set A. In Example 11.48 I have pointed out that if A is a set of generators for a group G, then there is a natural onto homomorphism $\varphi \colon F(A) \to G$, determined by sending $a \in A$ to 'itself' in G, and extending this prescription to all *words* by using the homomorphism property.

As a consequence of this observation and of the First Isomorphism Theorem, $G \cong F(A)/\ker\varphi$: in particular, *every* group is isomorphic to a quotient of a free group. (This result should be compared with Theorem 11.29, which states that every group is isomorphic to a subgroup of a symmetric group.)

For a concrete example, recall that we have verified that D_{2n} is generated by two elements x, y, subject to the relations

$$x^2 = e, \quad y^n = e, \quad xy = y^{-1}x.$$

(Proposition 11.35). It follows that D_{2n} is isomorphic to the quotient $F(\{x, y\})/K$, where K is the smallest normal subgroup of $F(\{x, y\})$ containing x^2, y^n, and $xyxy$.

In general, a 'presentation' of a group G is a pair $(R \mid A)$ where A is a set and R (the 'relations') are a set of words in the alphabet A, in the sense we have introduced in our discussion of free groups in §11.3, such that

$$G \cong F(A)/K,$$

where K is the smallest normal subgroup of $F(A)$ containing R. For example, a presentation of D_{2n} may be written as $(x^2, y^n, xyxy \mid x, y)$.

Presentations are an efficient way to define groups, and they are used in software packages (such as 'GAP') dedicated to group theory. One obstacle is the so-called 'word problem', that is, the problem of detecting whether two words represent the same element of G. This is known to be undecidable in general, a result of Pyotr Novikov. However, the problem is solvable for many classes of groups.

In fact, the other isomorphism theorems are examples of applications of the first one. The 'Second Isomorphism Theorem' requires a preliminary observation. If H and K are subsets of a group G, HK denotes the set of products of elements of H and elements of K:

$$HK = \{hk \mid h \in H, k \in K\}.$$

If H and K are subgroups, HK is *not* necessarily a subgroup.

Example 11.72   Take $H = \{e, x\}$ and $K = \{e, xy\}$ in D_6 (cf. §11.3). Then

$$HK = \{e, x, xy, y\}$$

and this is not a subgroup of D_6, since (for example) it contains the elements xy and y but not $xy \cdot y = xy^2$.

However, if one of the groups is normal, then HK *is* a subgroup!

> PROPOSITION 11.73   *Let H, K be subgroups of a group G, and assume that K is normal in G. Then HK is a subgroup of G, and K is normal in HK.*
> *Further, $H \cap K$ is a normal subgroup of H.*

Proof   The set HK is certainly nonempty, since it contains the identity. By Proposition 11.7, in order to verify that it is a subgroup we have to verify that if $a, b \in HK$, then $ab^{-1} \in HK$. If $a, b \in HK$, then there exist $h_1, h_2 \in H$, $k_1, k_2 \in K$ such that $a = h_1 k_1$ and $b = h_2 k_2$. Then

$$ab^{-1} = h_1 k_1 (h_2 k_2)^{-1} = h_1 k_1 k_2^{-1} h_2^{-1} = (h_1 h_2^{-1})(h_2 k_1 k_2^{-1} h_2^{-1}),$$

and this shows that $ab^{-1} \in HK$ since $h_1 h_2^{-1} \in H$ (since H is a subgroup) and $h_2(k_1 k_2^{-1})h_2^{-1}$ is an element of K since it is a conjugate of an element of K *and K is normal.*

It is clear that $K = eK$ is a subgroup of HK. It is normal because $aKa^{-1} = K$ for all $a \in G$, and hence in particular for all $a \in HK$.

Finally, $H \cap K$ is clearly a subgroup of H. To see that it is normal, we have to verify that if $k \in H \cap K$ and $a \in H$, then $aka^{-1} \in H \cap K$. But this element is in H because it is a product of elements of H, and it is in K since K is normal, so it indeed is in $H \cap K$.   □

With this understood, we can state the Second Isomorphism Theorem for groups: the quotients appearing in the statement are defined as groups, since $H \cap K$ is normal in H and K is normal in HK, as we have just verified in Proposition 11.73.

> THEOREM 11.74 (Second Isomorphism Theorem)   *Let H, K be subgroups of a group G, and assume that K is normal in G. Then*
>
> $$H/(H \cap K) \cong HK/K.$$
>
> *In fact, the function $H/(H \cap K) \to HK/K$ defined by $h(H \cap K) \mapsto hK$ is a well-defined isomorphism of groups.*

Proof The argument is, *mutatis mutandis,* the same as the one proving the Second Isomorphism Theorem for modules (Theorem 8.48). We can define a function $\phi\colon H \to HK/K$, by setting

$$\phi(h) = hK.$$

Verify that ϕ is surjective, by adapting the proof of Theorem 8.48. Then apply the First Isomorphism Theorem to conclude that ϕ induces an isomorphism

$$\tilde{\phi}\colon \quad H/\ker\phi \xrightarrow{\sim} HK/K,$$

defined by mapping the coset of h to hK (as indicated in the statement). Since

$$\ker\phi = \{h \in H \mid hK = K\} = \{h \in H \mid h \in K\} = H \cap K,$$

we are done. $\qquad\square$

The Third Isomorphism Theorem is also similar in content and in proof to previous editions of the same theorem.

> THEOREM 11.75 (Third Isomorphism Theorem) *Let G be a group, and let K be a normal subgroup of G. Let H be a subgroup of G containing K. Then H/K is a subgroup of G/K, and all subgroups of G/K may be realized in this way. Further, H/K is normal in G/K if and only if H is normal in G, and in this case*
>
> $$(G/K)/(H/K) \cong G/H.$$

Proof Let $\pi\colon G \to G/K$ be the natural projection. It is clear that if H is a subgroup of G containing K, then H/K is a subgroup of G/K; in fact, $H/K = \pi(H)$ (cf. Example 11.9). Conversely, if $\overline{H}$ is a subgroup of G/K, then $H = \pi^{-1}(\overline{H})$ is a subgroup of H containing K (cf. Example 11.10), and $\overline{H} = \pi(H) = H/K$. This takes care of the first part of the statement.

For the second part, assume first that H is normal in G and contains K. Then the projection $G \to G/H$ induces a surjective group homomorphism $\rho\colon G/K \to G/H$ (Exercise 11.39). Since $H/K = \ker\rho$, it follows that this subgroup of G/K is normal, and the stated isomorphism follows from the First Isomorphism Theorem.

Conversely, if H/K is normal in G/K, then we can consider the surjective group homomorphism

$$G \longrightarrow (G/K)/(H/K)$$

obtained by sending g to the coset of gK modulo H/K. It takes a moment to verify that the kernel of this group homomorphism is H, and it follows that H is normal. Once again, the First Isomorphism Theorem yields the isomorphism given in the statement and concludes the proof. $\qquad\square$

Exercises

11.1 ▷ Prove that the identity element e in a group G is unique, and that the inverse of a given element $g \in G$ is also unique.

11.2 ▷ Let $\varphi \colon G \to H$ be a group homomorphism. Prove that $\varphi(e_G) = e_H$. Prove that $\forall g \in G$, $\varphi(g^{-1}) = \varphi(g)^{-1}$.

11.3 ▷ Prove that the identity function is a group homomorphism, and that the composition of two group homomorphisms is a group homomorphism.

11.4 ▷ Prove that if a group homomorphism $\varphi \colon G \to H$ is a bijection, then its inverse $\varphi^{-1} \colon H \to G$ is also a group homomorphism.

11.5 What is an initial object in the category of groups? A final object?

11.6 Let G be a group, and let $g_1, \dots, g_n \in G$. Prove that the subgroup generated by G (in the sense of Definition 11.11) consists of all products $g_{i_1}^{e_1} \cdots g_{i_k}^{e_k}$, where $e_j \in \mathbb{Z}$.

11.7 ▷ Discuss the 'universal property of coproducts' (Exercise 8.16) for $G \times H$. You will run into a problem—$G \times H$ is *not* a coproduct of G and H in the category of groups. On the other hand, revisit Exercise 8.17: prove that $G \times H$, along with the projections π_G, π_H, does satisfy the universal property of products in the category of groups.

11.8 Let H, K be subgroups of a finite group G, and assume $H \cap K = \{e\}$. Prove that the set HK consists of exactly $|H||K|$ elements. (Define an injective function $H \times K \to G$.)

11.9 Note that the set HK in Exercise 11.8 need not be a subgroup of G. Now assume that H and K are *normal* subgroups of G; so HK is a subgroup of G by Proposition 11.73. Assume that $H \cap K = \{e\}$. Prove that the function $H \times K \to HK$ sending (h, k) to hk is an isomorphism of groups. (First prove that $hk = kh$ for all $h \in H$, $k \in K$; and to see this, what can you say about the element $hkh^{-1}k^{-1}$?)

Do both subgroups need to be normal, for the conclusion to hold?

11.10 Prove that a group of order 4 is necessarily commutative. (*Hint:* Show that there are two nonidentity elements $g \neq h$ such that $gh \neq e$. What can you say about hg?)

11.11 List all elements of $GL_2(\mathbb{Z}/2\mathbb{Z})$: they form a noncommutative group of order 6. Prove that this group is isomorphic to S_3.

11.12 Let $p > 0$ be a prime integer. Prove that $GL_2(\mathbb{Z}/p\mathbb{Z})$ is a noncommutative group of order $p^4 - p^3 - p^2 + p$. (*Hint:* A square matrix with entries in a field is invertible if and only if its rows are linearly independent.)

11.13 Let $n > 0$ be an integer. Prove that the group of automorphisms of the abelian group $\mathbb{Z}/n\mathbb{Z}$ is isomorphic to $(\mathbb{Z}/n\mathbb{Z})^*$, therefore has $\phi(n)$ elements. What can you say about $\operatorname{Aut}(\mathbb{Z}/p\mathbb{Z})$ if $p > 0$ is prime?

11.14 ▷ Let G be a group. Prove that a group homomorphism $G \to S_A$ determines an action of G on A (cf. Proposition 11.20).

11.15 If you have successfully worked out Exercise 11.11, then you have obtained a group homomorphism (and in fact an isomorphism) $S_3 \to GL_2(\mathbb{Z}/2\mathbb{Z})$. Determine the corresponding representation of S_3. (Note that the 'plane' over $\mathbb{Z}/2\mathbb{Z}$ consists of four points with coordinates (x, y), $x, y = [0]_2$ or $[1]_2$.)

11.16 Prove that the conjugation relation (i.e., $a \sim b \iff \exists g \in G : b = gag^{-1}$) is an equivalence relation.

11.17 Let G be a group, and let $g \in G$. Prove that conjugation by g, i.e., the function $G \to G$ defined by $a \mapsto gag^{-1}$, is an automorphism of G. (These are called 'inner' automorphisms.)

11.18 Let g, h be conjugate elements in a group. Prove that $|g| = |h|$.

11.19 Let G be a finite group, and assume that every element of G is its own inverse. Prove that $G \cong C_2 \times \cdots \times C_2$.

11.20 Construct a multiplication table for D_8.

11.21 Prove that if σ is an r-cycle, then $|\sigma| = r$.

11.22 List all elements of S_4 in cycle notation. Which of these are in $\mathcal{A}_4$?

11.23 ▷ Label the vertices of a regular tetrahedron by the numbers $1, 2, 3, 4$. A rigid symmetry of the tetrahedron induces a permutation of the vertices: this gives a faithful action of the group G of symmetries on the set $\{1, 2, 3, 4\}$, identifying G with a subgroup of S_4. Prove that this subgroup is the alternating group $\mathcal{A}_4$.

11.24 Prove that every element of S_n may be written as the product of *at most* $n - 1$ transpositions.

11.25 Let $\sigma \in S_n$, and assume σ is written in any way as a product of transpositions: $\sigma = \tau_1 \cdots \tau_m$. Prove that the *parity* of m is determined by σ: if σ can be written as a product of an *even* number of transpositions, then it cannot be written as a product of an *odd* number of transpositions, and conversely.

11.26 Let $0 < a < b < n$ be integers. Prove that the product $(1\, 2\, \ldots\, n)(a\, b)(n\, \ldots\, 2\, 1)$ in S_n equals $(a + 1\ b + 1)$.

11.27 Let $1 \leq a < b \leq n$ be integers, and let $\sigma \in S_n$. Prove that $(a\, b)\sigma(a\, b)$ is the permutation with the same decomposition of σ as a product of disjoint cycles, with a replaced by b and b by a.

(For example, let $\sigma = (1\, 4)(2\, 3\, 5\, 7\, 6)$. Then $(4\, 7)\sigma(4\, 7) = (1\, 7)(2\, 3\, 5\, 4\, 6)$.)

11.28 ▷ Prove that S_n is generated by $(1\, 2\, \ldots\, n)$ and $(1\, 2)$. (*Hint:* Use Exercise 11.26 to show that $\langle (1\, 2), (1\, 2\, \ldots\, n) \rangle$ contains all transpositions $(a\ a + 1)$, and then Exercise 11.27 to prove it contains *all* transpositions. Therefore...)

11.29 If $p > 0$ is prime, prove that S_p is generated by $(1\, 2\, \ldots\, p)$ and *any* transposition.

11.30 ▷ The 'type' of a permutation $\sigma \in S_n$ is the collection of the lengths of its orbits, i.e., the lengths of the cycles in its cycle decomposition (including the omitted 1s).

Prove that for all $\sigma, \tau \in S_n$, $\tau\sigma\tau^{-1}$ has the same type as σ. That is, conjugation preserves the type of a permutation. (Use Exercise 11.27.) Conversely, prove that if σ and σ' have the same type, then they are conjugate in S_n.

11.31 Show that S_4 is partitioned into five conjugacy classes, of sizes $1, 3, 6, 6$, and 8. (*Hint:* Use Exercise 11.30.)

11.32 Prove that S_4 has no normal subgroups of size 6 or 8. (Use Proposition 11.57 and Exercise 11.31.)

11.33 ▷ Show that $\mathcal{A}_4$ is partitioned into four conjugacy classes, of sizes $1, 3, 4$, and 4.

11.34 ▷ Let K be a subgroup of a group G, and define relations $\sim'_K$, $\sim'_K$ on G by setting $a \sim'_K b \iff a^{-1}b \in K$, $a \sim''_K b \iff ba^{-1} \in K$. Prove that $\sim'_K$ and $\sim''_K$ are both equivalence relations.

11.35 Use the description of free groups given in §11.3 to prove that the free group over a singleton is isomorphic to $\mathbb{Z}$. Discuss the corresponding universal property in this case.

11.36 ▷ Let G_1 and G_2 be groups, both containing a normal subgroup isomorphic to a given group H. If $G_1/H \cong G_2/H$, does it follow that $G_1 \cong G_2$? (*Hint:* No.)

11.37 Let G be a group, and $K \subseteq G$ a normal subgroup. Prove that G/K is commutative if and only if $\forall g, h \in G$: $ghg^{-1}h^{-1} \in K$. (This element is called the 'commutator' of g and h.)

11.38 Let G be a group, and let K be the subgroup of G generated by all commutators $ghg^{-1}h^{-1}$, as g and h range in G. Prove that K is normal in G. This subgroup is called the 'commutator subgroup' of G.

11.39 ▷ Let $\varphi: G \to G'$ be a group homomorphism, and let K be a normal subgroup of G such that $K \subseteq \ker \varphi$. Prove that φ induces a unique group homomorphism $\tilde{\varphi}: G/K \to G'$ making the diagram

commute. This is the group-theoretic version of the universal property for quotients; see Proposition 5.39 and Exercise 8.23 for the versions for rings and for modules.

In particular, if $K \subseteq H$ are normal subgroups of G, then there is a natural surjective group homomorphism $G/K \to G/H$.

11.40 Make sense of the statement '$(G_1 \times G_2)/G_1 \cong G_2$', and prove it.

11.41 Recall from Example 11.16 that $GL_n(\mathbb{R})$ denotes the 'general linear group' of invertible matrices with entries in $\mathbb{R}$. We also denote by $SL_n(\mathbb{R})$ the 'special linear group', of matrices with determinant equal to 1. Prove that $SL_n(\mathbb{R})$ is normal in $GL_n(\mathbb{R})$, and $GL_n(\mathbb{R})/SL_n(\mathbb{R}) \cong \mathbb{R}^*$.

11.42 Let H, K be subgroups of a group G, and assume that K is normal. Prove that $HK = KH$.

12 Basic Results on Finite Groups

12.1 The Index of a Subgroup, and Lagrange's Theorem

There is an impressively straightforward and equally impressively powerful 'numerical' result concerning cosets of subgroups. I could have mentioned this much earlier in this text, e.g., while discussing cosets of ideals in rings or of subgroups in abelian groups; I have waited until now because its applications really shine in the context of the theory of *finite groups*. Here is a useful preliminary definition.

DEFINITION 12.1   Let G be a group, and let H be a subgroup of G. The 'index' of H in G is the number of left-cosets of H in G (if this number is finite). We denote the index of H in G by $[G : H]$.

The index is also the number of *right*-cosets (Exercise 12.1).

As you will recall, for every subgroup H, the notation G/H stands for the set of left-cosets of H in G (Definition 11.50). The index $[G : H]$ is the number of elements in this set (if finite). If H is normal, then G/H is in fact a *group*, and $[G : H]$ is its order; but the number makes sense in any case.

Lagrange's theorem relates the numbers $|G|$, $|H|$, $[G : H]$ (if G is finite). It boils down to the following lemma.

LEMMA 12.2   *Let H be a subgroup of a group G, and let $g \in G$. Then the function $H \to gH, h \mapsto gh$, is a bijection.*

Proof   The function is surjective by definition of left-coset gH, and it is injective by cancellation: if $gh_1 = gh_2$, then $h_1 = g^{-1}gh_1 = g^{-1}gh_2 = h_2$.   □

Of course an analogous statement holds for right-cosets. If G is finite, Lemma 12.2 tells us that the cosets gH (and Hg) all have the same number of elements, for all g. The following famous theorem is then an immediate consequence.

THEOREM 12.3 (Lagrange's theorem)   *Let G be a finite group, and let H be a subgroup of G. Then $|H|$ divides $|G|$; in fact,*

$$|G| = [G : H] \cdot |H|.$$ (12.1)

Proof   The group G is the union of $[G : H]$ disjoint cosets gH, each consisting of $|H|$ elements by Lemma 12.2.   □

Lagrange's theorem has a large number of useful consequences, and we will use it frequently. In fact, we could have made good use of it already. For example, from the point of view of Lagrange's theorem it is perhaps easier to verify that if $[G : K] = 2$, then K is normal (cf. Example 11.59): in this situation $|K|$ is half the size of $|G|$ by Lagrange's theorem, and $G \setminus K$ must be the other *left*-coset as well as the other *right*-coset. Left-cosets and right-cosets coincide, so K must be normal. Here is my mental picture for this situation (where $a \notin K$):

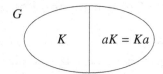

For instance, the subgroup K of D_{2n} isomorphic to C_n (cf. Example 11.61) is clearly normal: C_n is half the size of D_{2n}, therefore $[D_{2n} : C_n] = 2$ by Lagrange's theorem.

Using Lagrange's theorem, it is also immediate that, with notation as in Proposition 11.35, x and y must generate D_{2n}. Indeed, if a subgroup H of D_{2n} contains x and y, then it properly contains $K \cong C_n$, so $1 \le [D_{2n} : H] < [D_{2n} : K] = 2$: the only option is $[D_{2n} : H] = 1$, that is, $H = D_{2n}$.

Also, since $S_n/\mathcal{A}_n \cong C_2$ (Example 11.62), we have $[S_n : \mathcal{A}_n] = 2$, and hence $|\mathcal{A}_n| = \frac{1}{2}|S_n| = \frac{1}{2}n!$ by Lagrange's theorem. This is actually otherwise clear, as it says that there are as many odd permutations as there are even ones—and indeed, multiplying all even permutations by a fixed transposition gives all odd permutations. But having Lagrange's theorem, we are dispensed from the need to come up with such *ad hoc* arguments.

The following application is a surprisingly strong statement, and it is an immediate consequence of Lagrange's theorem.

PROPOSITION 12.4   *Let G be a finite group, and let $g \in G$. Then $|g|$ divides $|G|$, that is, $g^{|G|} = e$.*

Proof   Indeed, $|g|$ is the order of the subgroup $\langle g \rangle$ of G. (The second statement follows from the multiplicative version of Proposition 10.16.)   □

For abelian groups, we had proved Proposition 12.4 as an example of application of the classification theorem (Proposition 10.21). We now see that it was a much simpler business. And at this stage Fermat's 'little theorem' (Theorem 2.18), whose proof required some cleverness back in §2.5, should appear *evident* to you(!). If a, p are integers, $p > 0$ is prime, and $p \nmid a$, then $[a]_p$ is an element of the group $(\mathbb{Z}/p\mathbb{Z})^*$, whose order is $p - 1$. *Of course* $[a]_p^{p-1} = [1]_p$: this is an extremely special case of Proposition 12.4. Once you know Lagrange's theorem, no cleverness is needed in order to prove Fermat's little theorem. You will even be able to come up with and prove very quickly Euler's generalization of Fermat's little theorem (Exercise 12.2).

Here is another immediate but useful consequence.

PROPOSITION 12.5   *If $|G| = p$ is prime, then G is cyclic: $G \cong C_p$.*

Proof   Let $g \in G$, $g \neq e$. Then $|g| > 1$ and $|g|$ divides $p = |G|$ (Proposition 12.4), therefore $|g| = p$ since p is prime. This implies $\langle g \rangle = G$, proving that G is cyclic.   □

It is tempting to hope that Lagrange's theorem may have some kind of 'converse'. The theorem tells us that *if H is a subgroup of G and $|H| = m$, then m* divides $|G|$. Might it be the case that *if m* divides $|G|$, then there must necessarily be a subgroup H of G of order m? This actually happens to be the case if G is commutative, as you know if you worked out Exercise 10.24. But this converse does not hold for arbitrary noncommutative groups.

Example 12.6   The group $\mathcal{A}_4$ has $\frac{1}{2}4! = 12$ elements. Therefore, if $H \subseteq \mathcal{A}_4$ is a subgroup, then (by Lagrange's theorem) $|H| = 1, 2, 3, 4, 6$, or 12.

Now suppose that there were a subgroup H of order 6 in $\mathcal{A}_4$. Then H would have index 2, hence it would be normal. Therefore, H would be a union of conjugacy classes (Proposition 11.57). However, $\mathcal{A}_4$ has exactly four conjugacy classes, of size 1, 3, 4, 4 (Exercise 11.33). Since 6 cannot be obtained as a sum of these numbers, we reach a contradiction.

Even aside from Lagrange's theorem, the index is a rather well-behaved quantity, and the following observations will be quite useful.

PROPOSITION 12.7   *Let G be a finite group, and let $H \subseteq K \subseteq G$ be subgroups. Then*

$$[G : H] = [G : K][K : H].$$

Note that this relation is particularly transparent if H and K are normal, by the Third Isomorphism Theorem and Lagrange's theorem. But this hypothesis is not needed.

Proof   The subgroup K is a disjoint union of left-cosets $k_i H$. It is straightforward to verify that the coset gK is then a disjoint union of the cosets $gk_i H$.

Therefore each coset of K in G splits as a union of $[K : H]$ cosets of H in G, and the statement follows.   □

12.2   Stabilizers and the Class Equation

Let's consider again the notion of *action* of a group G on a set A (Definition 11.19). If G acts on A, and $a \in A$, then we can consider an interesting subgroup of G.

DEFINITION 12.8   Let a group G act on a set A, and let $a \in A$. The 'stabilizer' of a is the subgroup

$$\text{Stab}_G(a) = \{g \in G \,|\, ga = a\}$$

of G consisting of elements which fix a.

It should be clear that $\text{Stab}_G(a)$ is indeed a subgroup of G: indeed, $e \in \text{Stab}_G(a)$ (so the stabilizer is not empty) and if g, h both stabilize a, then $ha = a$, and hence $a = h^{-1}a$ and $gh^{-1}a = ga = a$. (So Proposition 11.7 confirms that $\text{Stab}_G(a)$ is a subgroup.)

The reason we care about stabilizers is that they give us essential information about the *orbits* of the action.

LEMMA 12.9   *Let a group G act on a set A, and let $a \in A$. Then there is a bijection between the orbit of a and the set $G/\text{Stab}_G(a)$ of left-cosets of the stabilizers of a.*

Proof   Let $O_a =$ be the orbit of a, and define a function $\alpha : G/\text{Stab}_G(a) \to O_a$:

$$\alpha(g \,\text{Stab}_G(a)) := ga\,.$$

Of course we have to verify that α is well-defined. For this, assume that

$$g_1 \,\text{Stab}_G(a) = g_2 \,\text{Stab}_G(a) :$$

then $g_1^{-1}g_2 \in \text{Stab}_G(a)$, therefore $g_1^{-1}g_2a = a$, and this implies $g_1a = g_2a$ as needed. So α is well-defined.

The function α is clearly surjective: every $b \in O_a$ is of the form $b = ga$ for some g, and then $\alpha(g \,\text{Stab}_G(a)) = ga = b$.

Is α injective? Assume $\alpha(g_1 \,\text{Stab}_G(a)) = \alpha(g_2 \,\text{Stab}_G(a))$; then we have

$$g_1a = g_2a \implies g_1^{-1}g_2a = a \implies g_1^{-1}g_2 \in \text{Stab}_G(a)\,.$$

Therefore, $g_1 \,\text{Stab}_G(a) = g_2 \,\text{Stab}_G(a)$. This verifies that α is injective, and completes the proof.   □

If G is *finite*, Lemma 12.9 implies a powerful numerical tool, which will be at the root of some of our most sophisticated results.

THEOREM 12.10   *Let G be a finite group acting on a set A, and let O_a be the orbit of an element $a \in A$. Then $|O_a| = [G : \text{Stab}_G(a)]$. In particular, the cardinality of every orbit divides the order of the group.*

Proof   The index is by definition the number of (left-)cosets (Definition 12.1), so this follows from Lemma 12.9.   □

One reason that makes Theorem 12.10 powerful is that it greatly restricts the possible actions of a group on a set. Recall that an action of a group G on a set A is 'transitive' if for all $a, b \in A$, there is a $g \in G$ such that $b = ga$ (Definition 11.24). Also, a is a 'fixed point' of an action if $ga = a$ for all $g \in G$, that is, if the orbit of that point is trivial.

Example 12.11   There are *no* transitive actions of A_4 on a set with two elements: indeed, the stabilizer of one of the two elements would be a subgroup of A_4 of index 2, and there is no such subgroup (Example 12.6).

Example 12.12   Suppose C_{35} acts on a set A with $|A| = 4$. Then every point of A must be fixed by the action. Indeed, orbits of a C_{35} action have either size 1 (i.e., they correspond to fixed points), 5, 7, or 35 by Theorem 12.10, and there is not enough room in A for the latter types.

Example 12.13   Every action of D_8 on a set A with an odd number of elements has fixed points. Indeed, each of the nontrivial orbits has an even number of elements by Theorem 12.10, so A must contain some *trivial* orbits.

The last example has a useful generalization.

DEFINITION 12.14   Let $p > 0$ be a prime integer. A 'p-group' is a group G such that $|G| = p^r$ for some $r \geq 0$.

COROLLARY 12.15   *Let $p > 0$ be a prime integer, and let G be a p-group acting on a finite set A. Let F be the set of fixed points of the action. Then*

$$|F| \equiv |A| \mod p.$$

In particular, if p does not divide $|A|$, then the action necessarily has fixed points.

Proof   The complement $A \setminus F$ consists of nontrivial orbits, and by Theorem 12.10 the cardinality of each nontrivial orbit is a power of p. Therefore $|A \setminus F|$ is a multiple of p.   □

Example 12.16   Even the particular case of Corollary 12.15 where $|G| = 2$ is useful. A C_2-action on a set A is determined by the action of the nonidentity element; that gives a function $i: A \to A$ such that $i^2 = \mathrm{id}_A$. We say that such a function is an 'involution'. Now suppose that you have an involution on a set A, with exactly one fixed point. By Corollary 12.15, that implies that $|A|$ is odd, and (by Corollary 12.15 again) that *every* involution on A will have fixed points. Don Zagier's 'one-sentence proof' of Fermat's theorem on sums of squares (mentioned in footnote 4 of §10.4) consists of a very clever setup which reduces the theorem to this observation about involutions.

Another reason Theorem 12.10 is very powerful is that some objects of interest within G itself may be realized as orbits under an action of G.

COROLLARY 12.17   *Let G be a finite group, and let a be an element of G. Then the number of distinct conjugates gag^{-1} of a divides $|G|$.*

Proof   The conjugates of a form the orbit of a under the conjugation action (Definition 11.31).   □

Example 12.18   Look back at Exercise 11.33: there you found that the sizes of the conjugacy classes in $\mathcal{A}_4$ are 1, 3, 4, and 4; these are all divisors of $|\mathcal{A}_4| = 12$, as Corollary 12.17 mandates. Knowing that, e.g., 8 is not a possible size of a conjugacy class in $\mathcal{A}_4$ (because $8 \nmid 12$) would streamline some of the work you carried out in that exercise.

In fact, the stabilizer of a under conjugation has a name.

DEFINITION 12.19   Let G be a group, and let $a \in G$. The 'centralizer' of a, denoted $Z_G(a)$, is its stabilizer under the conjugation action. That is,

$$Z_G(a) = \{g \in G \mid gag^{-1} = a\} :$$

the subgroup $Z_G(a)$ consists of the elements of G which commute with a.

(Note that $gag^{-1} = a \iff ga = ag$.) A more precise version of Corollary 12.17 states that *the number of conjugates of a equals the index of its centralizer.* This is what Theorem 12.10 implies, and it is a useful slogan to remember.

The intersection of all centralizers is called the *center;* I mentioned this subgroup in passing already in Example 11.60.

DEFINITION 12.20   The 'center' of a group G is the subgroup

$$Z(G) = \{g \in G \mid \forall a \in G, ag = ga\}.$$

That is, $Z(G)$ consists of the elements of G which commute with every other element of G.

Thus, G is commutative if and only if $G = Z(G)$; in general, $Z(G)$ is the union of the trivial orbits of the conjugation action. With this in mind, we have the following famous consequence of Corollary 12.17.

THEOREM 12.21 (Class equation)   *Let G be a finite group. Then*

$$|G| = |Z(G)| + \sum_{a \in \Gamma} [G : Z_G(a)],$$

where the sum is over a set Γ of distinct representatives for the nontrivial conjugacy classes in G.

Proof The group G is the union of the orbits under conjugation. The trivial orbits are collected in $Z(G)$. The nontrivial ones are the nontrivial conjugacy classes, and the size of the conjugacy class of a is the index of its centralizer (cf. the comments following Corollary 12.17). □

This 'class equation' has spectacular applications.

COROLLARY 12.22   *Let $p > 0$ be a prime integer, and let G be a nontrivial p-group. Then the center $Z(G)$ is not trivial.*

Proof If G is a p-group, then all terms $[G : Z_G(a)]$ in the sum are multiples of p. The class equation then says that

$$p^r = |Z(G)| + \text{a multiple of } p,$$

with $r \geq 1$. Since $Z(G)$ is not empty ($e \in Z(G)$!), it follows that $|Z(G)|$ is a nonzero multiple of p, and in particular $|Z(G)| > 1$, which is the statement. □

For an alternative and maybe more economical point of view on the same result: G acts on itself by conjugation, and $Z(G)$ consists of the fixed points under this action, therefore (by Corollary 12.15) $|Z(G)| \cong |G| \bmod p$, with the same conclusion.

COROLLARY 12.23   *Let $p > 0$ be a prime integer, and let G be a group of order p^2. Then G is commutative.*

Proof By Corollary 12.22, $Z(G)$ is not trivial. By Lagrange's theorem, $|Z(G)| = p$ or p^2, and in the latter case $Z(G) = G$ and G is commutative. We have to exclude the possibility that $|Z(G)| = p$. In this case $Z(G) \neq G$, so we can pick $a \notin Z(G)$. Since $Z_G(a)$ contains $Z(G)$ and a, it *properly* contains $Z(G)$; so $|Z_G(a)| > p$ and $|Z_G(a)|$ divides p^2 (by Lagrange again). It follows that $Z_G(a) = G$. But this would imply $a \in Z(G)$, a contradiction. □

Note that D_8 is a noncommutative group of order $8 = 2^3$, so Corollary 12.23 is sharp.[1] Corollary 12.23 is a *very* strong statement: since we have already classified finite *abelian* groups (Theorem 10.18), we can conclude that a group of order p^2, with p a positive prime, is necessarily isomorphic to either C_{p^2} or $C_p \times C_p$.

The slogan 'the number of conjugates of a equals the index of its centralizer' extends to arbitrary subsets of a group G. A group acts on the set of subsets $S \subseteq G$ by conjugation, with g sending S to gSg^{-1}; the number of conjugates of S will be the index of its stabilizer, i.e., of the subgroup fixing S, by Theorem 12.10. If $S = H$ is a subgroup of G, this stabilizer has a name.

---

[1] We say that a bound is 'sharp' if it cannot be improved. Here we found that if $p > 0$ is prime, then groups of order p^2 *are* necessarily commutative while groups of order p^3 are *not* necessarily commutative. In this sense the bound p^2 is 'sharp'.

DEFINITION 12.24   Let G be a group, and let $H \subseteq G$ be a subgroup. The 'normalizer' of H, denoted $N_G(H)$, is its stabilizer under the conjugation action:

$$N_G(H) = \{g \in G \mid gHg^{-1} = H\}.$$

That is, $N_G(H)$ is the largest subgroup of G containing H as a normal subgroup.

Theorem 12.10 implies immediately that *the number of conjugates of H equals the index of its normalizer,* another useful slogan that you are well advised to remember.

12.3   Classification and Simplicity

Where are we heading? The goal of many fields in mathematics is to obtain a *classification* of the objects under study: a complete 'list' of all possible such objects, up to isomorphism. This is what we accomplished for finite *abelian* groups: armed with Theorem 10.18, we can list all isomorphism classes of abelian groups of any given order. If we have any question about finite abelian groups, in principle we can answer that question by a case-by-case analysis, testing it on *all* abelian groups provided by the classification theorem. We have used this principle several times, and with striking results—for instance, that is how we proved that a finite subgroup of the multiplicative group of a field is necessarily cyclic.

What is the situation for (not necessarily commutative) *groups?* We know enough to seemingly get a good start on the project of listing 'all' groups of given order n, for small n. In fact, here is what we know at this point:

- $|G| = 1 \implies G \cong C_1$ is trivial.
- $|G| = 2 \implies G \cong C_2$.
- $|G| = 3 \implies G \cong C_3$.
- $|G| = 4 \implies G \cong C_4$ or $C_2 \times C_2$.
- $|G| = 5 \implies G \cong C_5$.
- $|G| = 6 \implies G \cong C_6$ or S_3 or ?
- $|G| = 7 \implies G \cong C_7$.
- $|G| = 8 \implies G \cong C_8$ or $C_4 \times C_2$ or $C_2 \times C_2 \times C_2$ or D_8 or ?
- $|G| = 9 \implies G \cong C_9$ or $C_3 \times C_3$.
- $|G| = 10 \implies G \cong C_{10}$ or D_{10} or ?
- $|G| = 11 \implies G \cong C_{11}$.
- $|G| = 12 \implies G \cong C_{12}$ or $C_2 \times C_6$ or D_{12} or $\mathcal{A}_4$ or ?
- $|G| = 13 \implies G \cong C_{13}$.
- $\cdots$

The tools used here are:

- If $|G|$ is prime, then G is cyclic (Proposition 12.5).
- If $|G| = p^2$ for p prime, then G is abelian (Corollary 12.23), hence classified by Theorem 10.18: $G \cong C_{p^2}$ or $G \cong C_p \times C_p$.

- We also know that $C_m \times C_n \cong C_{mn}$ if and only if m and n are relatively prime (Proposition 10.13).

I am not including D_6 in the list, since we know it is isomorphic to S_3. I am including both D_{12} and $\mathcal{A}_4$, since these are *non-isomorphic* noncommutative groups of order 12 (Exercise 12.17).

Still, there are several question marks for mysteries we have not cleared up yet. For example, are there any groups of order 6, 8, 10, 12 other than the ones we have already encountered? Each of these cases can be tackled (see for example Exercise 12.18, where you establish that indeed there is no other group of order 6 up to isomorphism), but it seems that we are missing some essential tool to work this out systematically. This should not surprise us: the classification problem is *hard*.

One way to simplify the question is the following. Suppose G is a group, and H is a normal subgroup of G. Mimicking the notation for 'exact sequences' introduced in §9.2, we could represent this situation by a short exact sequence

$$1 \longrightarrow H \longrightarrow G \longrightarrow G/H \longrightarrow 1$$

(recall that the trivial group is denoted 1). The 'extension problem' seeks to reconstruct the object at the center of such a sequence from the sides H and G/H. You should know already that this is not too straightforward, since there may be several possibilities for the object in the middle for given objects on the sides (Exercise 11.36); and indeed, it is in itself a very challenging problem. Assuming that it could be approached successfully, the classification of finite groups would be reduced to the classification of finite groups *without proper nontrivial normal subgroups*.

There is a name for such groups.

DEFINITION 12.25   A nontrivial group G is 'simple' if it has no nontrivial proper normal subgroups.

The trivial group is *not* taken to be simple.

Example 12.26   A nontrivial *abelian* group G is simple if and only if $G \cong C_p$, with $p > 0$ prime. Indeed, if $|G|$ is *not* prime, let p be a prime factor of $|G|$. By Cauchy's theorem (Theorem 10.22), G contains an element g of order p, and hence a proper nontrivial subgroup $\langle g \rangle$. Since subgroups of abelian groups are automatically normal, this shows that G is not simple.

A simplified version of the classification problem would then amount to the task of classifying *simple* groups. Sadly, even this problem is many orders of magnitude harder than the classification of abelian groups. As I mentioned in §11.1, while a *classification of finite simple groups* has been achieved, it should be counted among the most complex single achievements in the whole of mathematics.

Even with this caveat, we can appreciate some further results aimed at deciding which simple groups exist. We will focus on the following question: *given a positive integer n, can we decide whether there can exist a simple group of order n?*

Example 12.27   Let p be a prime, and let G be a group of order p^r, $r > 1$. Then G is not simple. Indeed, we have proved that the center of such a group is not trivial (Corollary 12.22), and $Z(G)$ is normal in G.

If we go down the same list we obtained above, we see that we can answer this question for several orders n: there are simple groups of order $2, 3, 5, 7, 11, 13$ (Example 12.26); while groups of order $4, 8, 9$ cannot be simple (Example 12.27). For $n = 6, 10, 12$, some more serious work seems to be needed.

Example 12.28   Let G be a group of order $2p$, where $p > 2$ is prime. By Proposition 12.4, the order of a nonidentity element of G can only be 2, p, or $2p$. Can *all* nonidentity elements have order 2? If x, y, and xy all have order ≤ 2, then $xy = (xy)^{-1} = y^{-1}x^{-1} = yx$. Therefore, if all elements of G had order ≤ 2, then G would be commutative. However, by the classification theorem for abelian groups, G would then be cyclic, and in particular not all elements of G would have order 2.

This contradiction shows that G must have elements of order p or $2p$. If G has an element of order $2p$, then $G \cong C_{2p}$, therefore it is not simple. If it has an element g of order p, then the subgroup $\langle g \rangle$ has index 2, therefore it is normal in G (Example 11.59), hence again G is not simple.

The conclusion is that there are *no* simple groups of order $2p$, for p prime. In particular, there are no simple groups of order 6 or 10.

This argument should strike you as rather *ad hoc,* and require a certain amount of cleverness. The tools we are going to develop will streamline this type of argument considerably and remove the need for any cleverness whatsoever, at least when dealing with groups of very low order.

12.4   Sylow's Theorems: Statements, Applications

Peter Ludwig Mejdell Sylow is credited with three very sharp results dealing with the structure of p-subgroups of a given finite group G. Maybe surprisingly, the structure of these subgroups is strongly constrained. In fact, G must contain p-groups of the largest size allowed by Lagrange's theorem (Sylow I); any two such maximal subgroups of G are conjugate (Sylow II); and the number of these subgroups divides $|G|$ and is $\equiv 1 \bmod p$ (Sylow III).

Here are the precise statements of the theorems.

THEOREM 12.29 (Sylow's first theorem) *Let G be a finite group, and let p be a prime integer. Assume that p^k divides $|G|$. Then G contains a subgroup of order p^k.*

In particular, if $|G| = p^r m$ with $p \nmid m$, then G must contain subgroups of order p^r.

Remark 12.30   For *abelian* groups, this result is an easy consequence of the classification theorem, as you hopefully verified when you worked out Exercise 10.22. Lacking a handy 'classification theorem' for arbitrary groups, we will have to come up with a standalone argument to prove Theorem 12.29.   ⌋

DEFINITION 12.31   Let $p > 0$ be a prime integer. A 'p-Sylow subgroup' (or 'Sylow p-subgroup') of a finite group G is a subgroup of order p^r, where p^r is the maximum power of p dividing the order of G.

THEOREM 12.32 (Sylow's second theorem) *Let G be a finite group, and let p be a prime integer. Let P be a p-Sylow subgroup of G, and let H be a p-group contained in G. Then there exists $g \in G$ such that $H \subseteq gPg^{-1}$.*

In particular, if P_1 and P_2 are two p-Sylow subgroups, then P_1 and P_2 are conjugate: $\exists g \in G$ such that $P_2 = gP_1g^{-1}$.

THEOREM 12.33 (Sylow's third theorem) *Let G be a finite group, and let $p > 0$ be a prime integer. Assume $|G| = p^r m$, where $p \nmid m$. Then the number of p-Sylow subgroups of G divides m and is $\equiv 1 \bmod p$.*

Consequences of these theorems abound. First of all, we obtain an overdue upgrade of the version of Cauchy's theorem we proved in §10.3 (Theorem 10.22).

THEOREM 12.34 (Cauchy's theorem) *Let p be a prime divisor of the order of a group G. Then G contains an element of order p.*

Proof   By Sylow I, G contains a subgroup H of order p. Necessarily, H is cyclic (Proposition 12.5): $H = \langle g \rangle$ for some $g \in G$. Then g has order p, as required.   □

We also recover directly a consequence of the classification theorem of finite abelian groups (cf. Exercise 10.23, reproduced here word-for-word).

COROLLARY 12.35   *Let A be a finite abelian group, and let $|A| = p_1^{r_1} \cdots p_s^{r_s}$ be the prime factorization of its order. Then for $i = 1, \ldots, s$, A contains a* unique *subgroup of order $p_i^{r_i}$.*

Proof   This follows from Sylow II: the p_i-Sylow subgroups of A are all conjugate, and therefore they must coincide since A is commutative (hence $P = gPg^{-1}$ for all g).   □

For our purposes, the main consequence of Sylow's theorems is the following basic obstruction to the simplicity of a group.

COROLLARY 12.36 *Let $p > 0$ be a prime integer, and let G be a finite group of order $|G| = p^r m$, with $r \geq 1$, $m > 1$, and $p \nmid m$. Assume that the only divisor of m that is $\equiv 1$ mod p is 1. Then G is not simple, as it contains a (unique) normal p-Sylow subgroup.*

Proof By Sylow III, G contains a *unique* p-Sylow subgroup P, and P is nontrivial (since $r \geq 1$) and proper (since $m > 1$). On the other hand, if P is a p-Sylow subgroup, then so is gPg^{-1}, for all $g \in G$. It follows that $\forall g \in G$, $gPg^{-1} = P$, proving that P is normal in G.

Since G has a nontrivial proper normal subgroup, G is not simple. □

Example 12.37 For instance, let's recover very quickly from Corollary 12.36 the result painfully worked out in Example 12.28. Assume $|G| = 2p$, where p is an odd prime. The only divisor of 2 which is $\equiv 1$ mod p is 1. Corollary 12.36 implies immediately that G is not simple.

In fact, G contains a unique normal subgroup of order p.

Hopefully this impresses you a little. The following statement captures a useful circumstance when this result may be applied.

PROPOSITION 12.38 *Let G be a finite group of order $|G| = p^r m$, where p is prime, $r \geq 1$, and $1 < m \leq p$. Then G is not simple.*

Proof If $m = p$, then $|G|$ is a p-group of order $\geq p^2$, and hence it is not simple (Example 12.27). If $1 < m < p$, the only divisor of m which is $\equiv 1$ mod p is 1. The hypotheses of Corollary 12.36 are satisfied, hence G is not simple. □

Let's apply this proposition to groups of all orders ≤ 60. In the following table I am marking with • the orders n for which we know there is a simple group of order n (for now, that means that n is prime); I am marking with ○ the orders > 1 for which Proposition 12.38 tells us that a group of that order *cannot* be simple.

1	○	7	•	13	•	19	•	25	○	31	•	37	•	43	•	49	○	55	○
2	•	8	○	14	○	20	○	26	○	32	○	38	○	44	○	50	○	56	
3	•	9	○	15	○	21	○	27	○	33	○	39	○	45		51	○	57	○
4	○	10	○	16	○	22	○	28	○	34	○	40		46	○	52	○	58	○
5	•	11	•	17	•	23	•	29	•	35	○	41	•	47	•	53	•	59	•
6	○	12		18	○	24		30		36		42	○	48		54	○	60	

Remarkably, Proposition 12.38 rules out most orders in this range: the holdouts are

$$12 = 2^2 \cdot 3 \qquad 24 = 2^3 \cdot 3 \qquad 30 = 2 \cdot 3 \cdot 5$$
$$36 = 2^2 \cdot 3^2 \qquad 40 = 2^3 \cdot 5 \qquad 45 = 3^2 \cdot 5$$
$$48 = 2^4 \cdot 3 \qquad 56 = 2^3 \cdot 7 \qquad 60 = 2^2 \cdot 3 \cdot 5$$

In fact, just sharpening a little our use of Sylow III also disposes of most of these remaining cases; for example, Corollary 12.36 implies right away that there are no simple groups of order 40 or 45. (Do you agree?) The methods used for the remaining cases are a good illustration of the type of tricks one may have to deploy in order to get the most out of these theorems. One basic observation in these arguments is that two or more subgroups of order p can only intersect at the identity: indeed, if H is such a subgroup, then $H \cong C_p$, and H is generated by any nonidentity element; therefore a nonidentity element cannot be on more than one such subgroup.

Example 12.39   There are no simple groups of order 12.

Indeed, say G is simple and $|G| = 12$. By Sylow III, the number of 3-Sylow subgroups of G divides 4 and is $\equiv 1 \bmod 3$: that is, it can be 1 or 4. If it is 1, then the 3-Sylow subgroup is normal, and G is not simple. If it is 4, then these subgroups account for $4 \cdot 2 = 8$ elements of order 3. These elements and the identity account for nine elements of G; the three remaining elements and the identity are just enough for *one* 2-Sylow subgroup. Therefore if the 3-Sylow subgroup is not normal, then necessarily the 2-Sylow subgroup is normal. Either way, G is not simple.

Example 12.40   There are no simple groups of order 24.

To see this, let G be a group of order 24. By Sylow III, there are either *one* or *three* 2-Sylow subgroups. If there is only one 2-Sylow subgroup, then that is a normal subgroup of order 8, and G is not simple. Otherwise, G acts (nontrivially) by conjugation on the set of three 2-Sylow subgroups. This determines a nontrivial homomorphism $G \to S_3$, whose kernel is then a proper normal subgroup of G—hence G is not simple.

Example 12.41   There are no simple groups of order 30.

According to Sylow III, if $|G| = 30$ and G *is* simple, then G has at least ten 3-Sylow subgroups and six 5-Sylow subgroups. However, this would account for 20 elements of order 3 and 24 elements of order 5, and there is not enough room in G for all these elements. Therefore G cannot be simple.

Armed with these ideas, you should have no serious difficulty disposing of most of the other orders: you will be able to show that there are no simple groups of order 36, 40, 45, 48, and 56 (Exercise 12.23). Putting together what we know at this point, you will have proved the following fact.

> PROPOSITION 12.42   *There are no noncommutative simple groups of order < 60.*

(Hard as you may try, however, you will *not* be able to use Sylow's theorems to show that there is no simple group of order 60. Why do you think that may be?)

Occasionally, using Sylow's theorems you may solve the classification problem for a given order. The following is the standard example.

> PROPOSITION 12.43   *Let $p < q$ be positive prime integers, and assume $q \not\equiv 1$ mod p. Let G be a group of order pq. Then G is cyclic.*

Proof   Let N_p, resp. N_q, be the number of p-Sylow, resp. q-Sylow, subgroups of G. By Sylow III, N_q divides p and $N_q \equiv 1$ mod q. Since p is prime and $p < q$, this implies $N_q = 1$. Similarly, $N_p = 1$: indeed, $N_p = 1$ or q as $N_p \mid q$, and by hypothesis $q \not\equiv 1$ mod p.

Therefore G has a unique subgroup H of order p and a unique subgroup K of order q. By Lagrange's theorem, the order of an element of G divides pq (Proposition 12.4), so it can only be $1, p, q$ or pq. The identity is the only element of order 1. Every element of G of order p must be contained in H, and every element of G of order q must be contained in K (by Sylow II). That leaves $pq - (p-1) - (q-1) - 1 = (p-1)(q-1) > 0$ elements of G of order pq. Since G has elements of order pq, $G \cong C_{pq}$ is cyclic as stated.   □

For example, there is only one group of order 15 up to isomorphism.

One can of course also examine groups of order pq with $p < q$ prime and $q \equiv 1$ mod p. It turns out there are *only two* such groups up to isomorphism: the cyclic group C_{pq} and a noncommutative group. Sylow III is a key ingredient in the proof of this fact. In the particular case where $|G| = 2q$ with q odd and prime, then the only options are $G \cong C_{2q}$ and $G \cong D_{2q}$. Proving this latter fact would not involve anything you do not know at this point, but is a little involved. Look it up if you are curious, or (better) prove it yourself.

12.5   Sylow's Theorems: Proofs

The proofs of Sylow's theorems are all very good examples of the power of judicious use of group actions, by way of the class equation or by direct exploitation of a suitable action.

Proof of Sylow I   If the theorem is not true, there must be a group G, a prime p, and an integer k such that p^k divides $|G|$ but G does not contain a subgroup of order p^k. (Note that $k \geq 1$, since the identity is a subgroup of order p^0 contained in G!) We may assume that $|G|$ is as small as possible with this property: that is, every group of order smaller than $|G|$ satisfies the statement of Sylow I. We are seeking a contradiction.

Assume first that there is a proper subgroup H of G such that p does not divide $[G : H]$. In this case, since p^k divides $|G|$, it must divide $|H|$. But $|H| < |G|$, so the statement of Sylow I holds for H; therefore H contains a subgroup of order p^k. Then so does G, a contradiction.

Therefore, p divides the index of *all* proper subgroups of G. The class equation (Theorem 12.21)

$$|G| = |Z(G)| + \sum_{a \in \Gamma} [G : Z_G(a)]$$

then implies that p divides $|Z(G)|$, since it divides $|G|$ and the other summands in the formula. By Cauchy's theorem for abelian groups (Theorem 10.22), there is an element g of order p in $|Z(G)|$. Since g is in the center, the subgroup $\langle g \rangle$ is *normal* in G. Consider then the quotient $G' = G/\langle g \rangle$. By Lagrange's theorem, $|G'| = |G|/p$; and since p^k divides $|G|$, p^{k-1} divides $|G'|$. Since the order of G' is smaller than the order of G, the statement of Sylow I holds for G': therefore, $G' = G/\langle g \rangle$ contains a subgroup H' of order p^{k-1}. This subgroup corresponds to a subgroup H of G containing $\langle g \rangle$ (Theorem 11.75), and in fact such that $H' = H/\langle g \rangle$. We are done, since then $|H| = p|H'| = p \cdot p^{k-1} = p^k$, so G *does* contain a subgroup of order p^k, contradicting the assumption on G. □

Proof of Sylow II   Let $H \subseteq G$ be a subgroup of G as in the statement: so H is a p-group, for a prime $p > 0$. We can act with H by left-multiplication on the set G/P of left-cosets of a given p-Sylow subgroup P: $h \in H$ acts on gP by sending it to the coset hgP. Since G/P has order $[G : P]$, which is relatively prime to p, this action must have fixed points by Corollary 12.15. A fixed point gP is a coset for which $\forall h \in H$, $hgP = gP$. Now

$$(\forall h \in H) \; hgP = gP \implies (\forall h \in H) \; g^{-1}hgP = P \implies (\forall h \in H) \; g^{-1}hg \in P$$
$$\implies g^{-1}Hg \subseteq P \implies H \subseteq gPg^{-1},$$

and this is the statement. □

Proof of Sylow III   Let N_p be the number of p-Sylow subgroups of G; we are assuming that $|G| = p^r m$ with $p > 0$ prime and $p \nmid m$. By Sylow I, G has at least one p-Sylow subgroup P, and by Sylow II the set of p-Sylow subgroups is the set of conjugates of P. Therefore N_p equals the index of the normalizer of P:

$$N_p = [G : N_G(P)]$$

(remember the slogan advertised after Definition 12.24!). The fact that N_p divides m is an easy consequence, by the multiplicativity of indices (Proposition 12.7):

$$m = [G : P] = [G : N_G(P)] \cdot [N_G(P) : P] = N_p \cdot [N_G(P) : P]. \tag{12.2}$$

In order to prove that $N_p \equiv 1 \bmod p$, we need a different interpretation for the index $[N_G(P) : P]$. For this, we use yet another action: act with P on G/P, by left-multiplication.

CLAIM 12.44   *A coset $gP \in G/P$ is fixed under this action if and only if it is contained in $N_G(P)$. Therefore, $[N_G(P) : P]$ equals the number of fixed points of the action of P on G/P by left-multiplication.*

To see this, note that a coset gP is fixed if and only if $\forall h \in P$, $hgP = gP$. Arguing as we just did in the proof of Sylow II, this is equivalent to $gPg^{-1} \subseteq P$, or simply $gPg^{-1} = P$ since the two sets have the same number of elements. But this says precisely that $g \in N_G(P)$, proving the claim.

Summarizing, P is acting on a set with $[G : P] = m$ elements (i.e., G/P), and this action has $[N_G(P) : P]$ fixed points. Since P is a p-group, Corollary 12.15 tells us something very interesting about these numbers:

$$[N_G(P) : P] \equiv m \bmod p.$$

Now we can go back to (12.2) with this added knowledge, and we see that

$$m = N_p \cdot [N_G(P) : P] \equiv N_p \cdot m \bmod p.$$

Since $p \nmid m$, it follows that $1 \equiv N_p \bmod p$ (Remark 2.17!), and we are done.   □

12.6   Simplicity of $\mathcal{A}_n$

After successfully ruling out the existence of a noncommutative simple group of order less than 60 in §12.4, maybe you have attempted to refine your skills in the use of Sylow's theorems by trying to convince yourself that there is no simple group of order 60. Hopefully you have not managed to do that, because as it happens *the alternating group $\mathcal{A}_5$ is simple.* (And as we know, $|\mathcal{A}_5| = \frac{5!}{2} = 60$.)

The group $\mathcal{A}_2$ is trivial, and $\mathcal{A}_3$ is isomorphic to C_3, hence it is cyclic and simple. As for $\mathcal{A}_4$, we have run across a nontrivial proper subgroup of $\mathcal{A}_4$ in Example 11.63, so we know that $\mathcal{A}_4$ is *not* simple. Studying the simplicity of $\mathcal{A}_n$ for $n \geq 5$ requires studying these groups a little more thoroughly.

LEMMA 12.45   *For all $n \geq 2$, the alternating group $\mathcal{A}_n$ is generated by 3-cycles.*

Proof   We know that S_n is generated by transpositions (Proposition 11.47). Since transpositions are *odd,* $\mathcal{A}_n$ consists of those permutations which may be written as a product of an *even* number of transpositions.

If σ is a product of an even number of transpositions, pair these off to write σ as a product of permutations of the following type:

$$(i\,j)(k\,\ell).$$

It suffices to show that *these* can be written as products of 3-cycles. And indeed:

- If $(i\,j) = (k\,\ell)$, then the product is the identity.
- If $(i\,j), (k\,\ell)$ have exactly one element in common, say $j = \ell$, then

$$(i\,j)(k\,j) = (i\,j\,k)$$

is a 3-cycle.

- If $(i\,j), (k\,\ell)$ are disjoint, then

$$(i\,j)(k\,\ell) = (i\,j\,k)(j\,k\,\ell)$$

as you should verify.   □

Conjugation in S_n is actually a fairly simple business. You (should) have worked it out in Exercise 11.30; I will discuss it in detail here, since it is necessary for our study of the simplicity of $\mathcal{A}_n$.

LEMMA 12.46 *Let $\sigma, \tau \in S_n$, and assume that*

$$\sigma = (a_1 \ \ldots \ a_r)(b_1 \ \ldots \ b_s) \cdots (c_1 \ \ldots \ c_t);$$

then

$$\tau\sigma\tau^{-1} = (\tau(a_1) \ \ldots \ \tau(a_r))(\tau(b_1) \ \ldots \ \tau(b_s)) \cdots (\tau(c_1) \ \ldots \ \tau(c_t)).$$

Proof Since the conjugate of a product is the product of the conjugates, it suffices to verify this for a cycle, for which it is straightforward.   □

In the statement of Lemma 12.46 it is immaterial whether the cycles making up σ are or are not disjoint. The case in which they *are* disjoint gives the following important consequence, for which we introduce a convenient piece of terminology.

DEFINITION 12.47 The 'type' of a permutation $\sigma \in S_n$ is the partition[2] of n determined by the length of its orbits; that is, by the lengths of the cycles in the cycle decomposition of σ (including the omitted 1s).

For example, the type of $\sigma = (1\,4)(2\,3\,5\,7\,6) \in S_7$ is '2 + 5'.

COROLLARY 12.48 *Two permutations σ, σ' are conjugate in S_n if and only if they have the same type.*

Proof If $\sigma' = \tau\sigma\tau^{-1}$, then σ' has the same type as σ by Lemma 12.46. Conversely, if σ and σ' have the same type $r + s + \cdots + t$, say

$$\sigma = (a_1 \ \ldots \ a_r)(b_1 \ \ldots \ b_s) \cdots (c_1 \ \ldots \ c_t),$$
$$\sigma' = (a_1' \ \ldots \ a_r')(b_1' \ \ldots \ b_s') \cdots (c_1' \ \ldots \ c_t'),$$

then for any $\tau \in S_n$ such that $a_1' = \tau(a_1)$, $a_2' = \tau(a_2)$, ..., $\tau(c_t) = \tau(c_t')$ we will have $\sigma' = \tau\sigma\tau^{-1}$, again by Lemma 12.46. Thus σ and σ' are conjugate in this case.   □

Example 12.49 Let's return to our running example from §11.3:

$$\sigma = \begin{pmatrix} 1 & 2 & 3 & 4 & 5 & 6 & 7 \\ 4 & 3 & 5 & 1 & 7 & 2 & 6 \end{pmatrix} = (1\,4)(2\,3\,5\,7\,6).$$

The claim is that every permutation with the same type $2 + 5$ must be conjugate to σ. For example, let's choose (randomly)

$$\sigma' = (6\,5)(7\,2\,1\,3\,4).$$

The permutation

$$\tau = \begin{pmatrix} 1 & 2 & 3 & 4 & 5 & 6 & 7 \\ 6 & 7 & 2 & 5 & 1 & 4 & 3 \end{pmatrix}$$

satisfies

$$(\tau(1)\,\tau(4))\,(\tau(2)\,\tau(3)\,\tau(5)\,\tau(7)\,\tau(6)) = (6\,5)\,(7\,2\,1\,3\,4);$$

according to Lemma 12.46, we must have $\sigma' = \tau\sigma\tau^{-1}$. And indeed, $\tau\sigma\tau^{-1}$ equals

$$\begin{pmatrix} 1 & 2 & 3 & 4 & 5 & 6 & 7 \\ 6 & 7 & 2 & 5 & 1 & 4 & 3 \end{pmatrix}\begin{pmatrix} 1 & 2 & 3 & 4 & 5 & 6 & 7 \\ 4 & 3 & 5 & 1 & 7 & 2 & 6 \end{pmatrix}\begin{pmatrix} 1 & 2 & 3 & 4 & 5 & 6 & 7 \\ 5 & 3 & 7 & 6 & 4 & 1 & 2 \end{pmatrix}$$

$$= \begin{pmatrix} 1 & 2 & 3 & 4 & 5 & 6 & 7 \\ 3 & 1 & 4 & 7 & 6 & 5 & 2 \end{pmatrix}$$

and it takes a moment to verify that the cycle decomposition of this permutation is indeed $(6\,5)(7\,2\,1\,3\,4)$.

By Corollary 12.48, it is clear that *all 3-cycles are conjugate in S_n.* The same holds in $\mathcal{A}_n$, provided that $n \geq 5$.

LEMMA 12.50   *If $n \geq 5$, then all 3-cycles are conjugate in $\mathcal{A}_n$.*

Proof   It is enough to show that every 3 cycle $(i\,j\,k)$ is conjugate to $(1\,2\,3)$ in $\mathcal{A}_n$. Let τ be any permutation such that $\tau(i) = 1$, $\tau(j) = 2$, $\tau(k) = 3$; then

$$\tau(i\,j\,k)\tau^{-1} = (1\,2\,3)$$

as observed above. If $\tau \in \mathcal{A}_n$, we are done. If τ is odd, then

$$(4\,5)\tau(i\,j\,k)\tau^{-1}(4\,5) = (4\,5)(1\,2\,3)(4\,5) = (1\,2\,3);$$

note that we can do this since $n \geq 5$. Since τ is odd, $(4\,5)\tau$ is even, and we are done.   □

What if $n = 3$ or 4? (For $n = 1$ and 2 there are no 3-cycles, so the question is moot.) Since $\mathcal{A}_3 \cong C_3$ is commutative, conjugacy classes are trivial; so the two 3-cycles in $\mathcal{A}_3$ are *not* conjugate. In $\mathcal{A}_4$ there are *eight* 3-cycles, so they cannot all be conjugate: the size of a conjugacy class must divide the order of the group (Corollary 12.17) and 8 does not divide $|\mathcal{A}_4| = 12$. In fact, there are exactly two conjugacy classes of 3-cycles in $\mathcal{A}_4$, as you found out when you solved Exercise 11.33. The assumption that $n \geq 5$ in Lemma 12.50 is necessary.

We are almost ready to settle the simplicity of $\mathcal{A}_n$ for $n \geq 5$. The proofs I know of this fact are all somewhat technical; the argument I will present relies on the following lemma, which seems hard to motivate independently and whose proof is a somewhat uninspiring case-by-case analysis.

LEMMA 12.51   *Let $n \geq 5$ and let $\sigma \in \mathcal{A}_n$, $\sigma \neq (1)$. If σ is not a 3-cycle, then there exists an even permutation τ such that $\tau\sigma\tau^{-1}\sigma^{-1}$ is not the identity permutation and has more fixed points than σ.*

Proof   If σ is a product of disjoint 2-cycles, then it has at least two such cycles, since it is even. Up to a relabeling of the elements of $\{1, \ldots, n\}$, we may assume

$$\sigma = (1\,2)(3\,4)\rho\,,$$

where ρ stands for a product of (possibly) more disjoint 2-cycles. Since $n \geq 5$, we can let $\tau = (3\,4\,5)$. If σ has fixed points f, necessarily $f \geq 5$; for all fixed points $f \geq 6$ we have

$$\tau\sigma\tau^{-1}\sigma^{-1}(f) = \tau\sigma\tau^{-1}(f) = \tau\sigma(f) = \tau(f) = f\,,$$

so these points are also fixed by $\tau\sigma\tau^{-1}\sigma^{-1}$. Also,

$$\tau\sigma\tau^{-1}\sigma^{-1}(1) = \tau\sigma\tau^{-1}(2) = \tau\sigma(2) = \tau(1) = 1\,,$$
$$\tau\sigma\tau^{-1}\sigma^{-1}(2) = \tau\sigma\tau^{-1}(1) = \tau\sigma(1) = \tau(2) = 2\,.$$

It follows that $\tau\sigma\tau^{-1}\sigma^{-1}$ has more fixed points than σ. On the other hand,

$$\tau\sigma\tau^{-1}\sigma^{-1}(3) = \tau\sigma\tau^{-1}(4) = \tau\sigma(3) = \tau(4) = 5\,,$$

so $\tau\sigma\tau^{-1}\sigma^{-1}$ is not the identity. Therefore the statement holds in this case.

If σ is *not* a product of disjoint 2-cycles, then some cycle in its decomposition has length ≥ 3, that is, σ has an orbit consisting of at least three elements. Again up to a relabeling, we may assume that

$$\sigma = (1\,2\,3 \cdots)\rho\,,$$

with ρ acting on a subset of $\{4, \ldots, n\}$. We cannot have $\sigma = (1\,2\,3)$, by hypothesis. Then σ must move *at least two other elements,* which we may assume to be 4 and 5. Indeed, else $\sigma = (1\,2\,3\,r)$ for some r; but this is not an option, since 4-cycles are odd. In particular, if f is a fixed point of σ, then necessarily $f \geq 6$.

Again take $\tau = (3\,4\,5)$. For all fixed points f of σ we have

$$\tau\sigma\tau^{-1}\sigma^{-1}(f) = \tau\sigma\tau^{-1}(f) = \tau\sigma(f) = \tau(f) = f$$

since $f \geq 6$; so these points are also fixed by $\tau\sigma\tau^{-1}\sigma^{-1}$. Further,

$$\tau\sigma\tau^{-1}\sigma^{-1}(2) = \tau\sigma\tau^{-1}(1) = \tau\sigma(1) = \tau(2) = 2\,,$$

so 2 is an additional fixed point of $\tau\sigma\tau^{-1}\sigma^{-1}$; and

$$\tau\sigma\tau^{-1}\sigma^{-1}(3) = \tau\sigma\tau^{-1}(2) = \tau\sigma(2) = \tau(3) = 4\,,$$

veryfing that $\tau\sigma\tau^{-1}\sigma^{-1}$ is not the identity, and completing the proof of the lemma.   □

Lemma 12.51 makes the proof of the simplicity of $\mathcal{A}_n$ for $n \geq 5$ quite straightforward.

THEOREM 12.52   *The alternating group $\mathcal{A}_n$ is simple for $n \geq 5$.*

(Again, we already know that $\mathcal{A}_3 \cong C_3$ is simple and that $\mathcal{A}_4$ is not simple.)

Proof   It is enough to prove that if $n \geq 5$, then every nontrivial normal subgroup N of $\mathcal{A}_n$ contains a 3-cycle. Indeed, by Lemma 12.50 all 3-cycles are conjugate in $\mathcal{A}_n$, therefore if N is normal and contains a 3-cycle, then it contains *all* 3-cycles; and then N contains a set of generators for $\mathcal{A}_n$, by Lemma 12.45, hence $N = \mathcal{A}_n$. This will show that the only nontrivial normal subgroup of $\mathcal{A}_n$ is $\mathcal{A}_n$ itself, implying that $\mathcal{A}_n$ is simple.

Therefore, all we have to prove is that if N is a nontrivial normal subgroup of $\mathcal{A}_n$, with $n \geq 5$, then N contains a 3-cycle. For this, it makes sense to look at nonidentity elements $\sigma \in N$ with the maximum possible number of fixed points: if N contains a 3-cycle, this will indeed be such a permutation.

Let then $\sigma \in N$ be a nonidentity permutation with the maximum number of fixed points. By Lemma 12.51, if σ is *not* a 3-cycle, then there exists a $\tau \in \mathcal{A}_n$ such that $\tau\sigma\tau^{-1}\sigma^{-1}$ is not the identity and has *more* fixed points than σ. But this permutation is an element of N! Indeed $\sigma \in N$, so $\sigma^{-1} \in N$ (since N is a subgroup) and $\tau\sigma\tau^{-1} \in N$ (since N is normal).

Therefore, if σ is not a 3-cycle, then σ cannot be a nonidentity element of N with the maximum number of fixed points. Our σ must then be a 3-cycle, and this concludes the proof.   □

Summarizing the situation again: in previous sections we had proved that there are no simple noncommutative groups of order < 60; and now we have found an *infinite* collection of simple noncommutative groups, that is, the alternating groups $\mathcal{A}_n$ for all $n \geq 5$. The smallest of these is $\mathcal{A}_5$, which has order exactly 60.

This is not just a curiosity. In the next and last section of this chapter we will use this fact to show that S_n is not *solvable* for $n \geq 5$. This is a subtle property, with substantial applications. In our brief excursion into Galois theory in the last chapter of this text, it will be key to understanding the issue of 'solvability' of polynomial equations, that is, the existence (or non-existence) of formulas to find the roots of given polynomial equations.

12.7   Solvable Groups

The last class of groups I want to introduce (briefly) is the class of *solvable* groups. There are several equivalent definitions of this concept; the following is the best suited for the application that we will eventually encounter. (See Exercise 12.35 for a popular alternative.)

Going back to our discussion of the classification problem in §12.3, recall that we simplified the question by arguing that if G has a normal subgroup H, then at least to some extent it should be possible to reconstruct G from H and the quotient G/H. This is what led us to consider *simple* groups—if G is not simple, then we can consider a nontrivial normal subgroup H and the quotient G/H; if these are not simple, we can break them down further; and so on. If G is finite to begin with, this process will have to stop, and then all the blocks will be simple.

We can formalize this idea by considering 'normal series', i.e., sequences[3] of subgroups

$$G = G_0 \supsetneq G_1 \supsetneq G_2 \supsetneq \cdots \supsetneq G_r = \{e\}$$

such that G_{i+1} is normal in G_i. Such a series is a 'composition series' if all the quotients G_i/G_{i+1} are simple. In general, the quality of the quotients G_i/G_{i+1} gives some information on G itself.

DEFINITION 12.53   A group G is 'solvable' if it admits a normal series

$$G = G_0 \supsetneq G_1 \supsetneq G_2 \supsetneq \cdots \supsetneq G_r = \{e\}$$

such that the quotients G_i/G_{i+1} are *abelian*.

Example 12.54   Abelian groups are solvable: if A is a nontrivial abelian group, then $A \supsetneq \{0\}$ is a normal series as required.

Example 12.55   Dihedral groups are solvable. Indeed, recall that D_{2n} contains a normal subgroup isomorphic to C_n (Example 11.61); the series

$$D_{2n} \supsetneq C_n \supsetneq \{e\}$$

satisfies the requirement in Definition 12.53: $C_n/\{e\} \cong C_n$ and $D_{2n}/C_n \cong C_2$ are abelian groups.

Example 12.56   In particular, $S_3 \cong D_6$ is solvable. So is S_4: consider the series

$$S_4 \supsetneq \mathcal{A}_4 \supsetneq V \supsetneq \{0\},$$

where V is the *Vierergruppe*. We have seen that V is abelian (it is isomorphic to $C_2 \times C_2$) and normal in $\mathcal{A}_4$ (Example 11.63); since $\mathcal{A}_4/V$ has order 3, this quotient is necessarily cyclic (Proposition 12.5), hence abelian; and $S_4/\mathcal{A}_4 \cong C_2$ (Example 11.62).

Example 12.57   You can show fairly easily that *p*-groups are solvable for every prime p (Exercise 12.33).

Example 12.58   Groups G for which $|G|$ is *odd* are solvable—but this is not 'fairly easy'. In fact, it is the *Feit–Thompson theorem;* the 1963 article proving it runs to about 250 pages, and Thompson went on to win the Fields Medal in 1970. Fortunately, we will not need to prove (or even use) this result.

---

[3] It is occasionally convenient to allow for the possibility that some of the inclusions may not be proper. Repeated terms are then implicitly dropped.

Example 12.59   Simple noncommutative groups are evidently *not* solvable: indeed, if G is simple, then its only proper normal subgroup is $\{e\}$; that is, its only normal series is

$$G \supsetneq \{e\},$$

and the quotient $G/\{e\} \cong G$ is not abelian by hypothesis.

For example, $\mathcal{A}_n$ is not solvable for $n \geq 5$ by Theorem 12.52.

PROPOSITION 12.60   *Subgroups and homomorphic images of solvable groups are solvable.*

Proof   Let G be a solvable group; so G admits a normal series

$$G = G_0 \supsetneq G_1 \supsetneq G_2 \supsetneq \cdots \supsetneq G_r = \{e\}$$

with abelian quotients.

If H is a subgroup of G, then let $H_i = G_i \cap H$ and consider the series

$$H = H_0 \supseteq H_1 \supseteq H_2 \supseteq \cdots \supseteq H_r = \{e\}. \tag{12.3}$$

In general, some of the inclusions will not be proper, and repeated terms can be omitted. In any case, H_{i+1} is normal in H_i because G_{i+1} is normal in G_i; in fact, $H_{i+1} = G_{i+1} \cap H_i$ is the kernel of the composition $H_i \to G_i \to G_i/G_{i+1}$. Thus (by the First Isomorphism Theorem!) we get an *injective* group homomorphism

$$H_i/H_{i+1} \hookrightarrow G_i/G_{i+1}.$$

This implies that H_i/H_{i+1} is abelian, since it is (isomorphic to) a subgroup of an abelian group. Therefore, omitting the repetitions in (12.3) produces a normal series satisfying the requirement in Definition 12.53, and the conclusion is that H is solvable.

Next, assume that G' is a homomorphic image of G, that is, a group such that there exists a surjective homomorphism $\varphi\colon G \to G'$. By the First Isomorphism Theorem, $G' \cong G/K$ for $K = \ker \varphi$. Therefore, in order to prove that homomorphic images of solvable groups are solvable, we have to prove that if K is a normal subgroup of a solvable group G, then G/K is solvable.

Since K is normal, $G_i K$ is a subgroup of G, for every i (Proposition 11.73). Further, K is clearly normal in each $G_i K$. Therefore, we can consider the series

$$G/K \supseteq G_1 K/K \supseteq G_2 K/K \supseteq \cdots \supseteq G_r K/K = K/K = \{e\}. \tag{12.4}$$

For each i, $G_{i+1} K$ is normal in $G_i K$ (verify this!). Therefore, by the Third Isomorphism Theorem, $G_{i+1} K/K$ is normal in $G_i K/K$ and

$$(G_i K/K)/(G_{i+1} K/K) \cong (G_i K)/(G_{i+1} K).$$

I claim that this group is abelian. For this, consider the composition

$$\alpha\colon G_i \to G_i K \to (G_i K)/(G_{i+1} K).$$

Then α is surjective: every $gkG_{i+1}K$ in the quotient $(G_iK)/(G_{i+1}K)$, with $g \in G$ and $k \in K$, equals $\alpha(g) = gG_{i+1}K$, since $g^{-1}gk = k \in K \subseteq G_{i+1}K$. Further, G_{i+1} is contained in $\ker \alpha$ (verify this!). We obtain then an induced surjective homomorphism

$$(G_i/G_{i+1}) \longrightarrow\!\!\!\!\!\rightarrow (G_iK)/(G_{i+1}K)$$

(Exercise 11.39) and since G_i/G_{i+1} is abelian, this implies that $(G_iK)/(G_{i+1}K)$ is abelian. It follows that removing repetitions in (12.4) produces a series satisfying the requirement of Definition 12.53, and this proves that G/K is solvable. □

Remark 12.61 Conversely, if H is a normal subgroup of a group G, and both H and the corresponding homomorphic image G/H of G are solvable, then G is itself solvable. This is easy to prove: you can string together normal series for H and G/H and obtain a normal series for G. In fact, you can string together normal series for all the quotients in a normal series (Exercise 12.28), and this leads to elegant results. ⌋

Solvable groups will be a crucial ingredient in an application of Galois theory, which we will encounter in Chapter 15. In fact, the following consequence of Proposition 12.60 will be important.

THEOREM 12.62 *For $n \geq 5$, S_n is not solvable.*

Proof If it were, then the subgroup $\mathcal{A}_n$ of S_n would be solvable, by Proposition 12.60. But $\mathcal{A}_n$ is noncommutative and simple for $n \geq 5$, by Theorem 12.52; hence it is not solvable (cf. Example 12.59). □

For example, S_5 is a nonsolvable group of order 120. In fact, as it happens, $\mathcal{A}_5$ and S_5 are the two smallest nonsolvable groups.

Preview of coming attractions: Solvable groups are called 'solvable' because they are intimately related to the problem of 'solvability' of polynomial equations. The fact that S_5 is not solvable is the reason why there are no formulas for the solution of a general polynomial equation of degree 5 involving only the ordinary operations and the extraction of roots. The formula for the solution of a degree-2 equation is such a formula (requiring square roots); there are formulas for degree-3 and degree-4 equations. This is a reflection of the fact that S_2, S_3, and S_4 are all solvable groups. We will understand all (or at least some) of this by the end of this text.

Exercises

12.1 ▷ Let G be a group and let $H \subseteq G$ be a subgroup. Prove that $g_1H = g_2H$ if and only if $Hg_1^{-1} = Hg_2^{-1}$. Deduce that the number of right-cosets of H is, if finite, equal to the number of left-cosets.

12.2 ▷ Generalize Fermat's little theorem (which states that $a^{p-1} \equiv 1 \bmod p$ if p is prime and $p \nmid a$; see §2.5). What is the corresponding statement, if the prime p is replaced with an arbitrary integer $n > 0$? (Use Proposition 12.4. The generalization is known as the *Euler–Fermat theorem.*)

12.3 Let G be a finite group, and let H, K be subgroups of G; assume that K is normal, so that HK is a subgroup of G (Proposition 11.73). Prove that $[HK : K] = [H : H \cap K]$ and deduce that

$$|HK| = \frac{|H| \cdot |K|}{|H \cap K|} .$$

(This result actually holds even if both subgroups are not normal; but the proof is a little more involved in that case. Cf. Exercise 11.8, which takes care of the case $H \cap K = \{e\}$.)

12.4 Let G act on a set A, and let a, b be elements in the same orbit: that is, $\exists g \in G$ such that $b = ga$. Prove that $\text{Stab}_G(b) = g\,\text{Stab}_G(a)g^{-1}$.

12.5 Suppose D_{10} acts on a set A with $|A| = 11$, without fixed points. How many orbits must this action have, and what are their sizes?

12.6 Let $p > 2$ be an odd prime, and let G be a p-group acting on a finite set A. For $g \in G$ and $S \subseteq A$, let gS be the set of elements gs for $s \in S$. Say that S is 'fixed' by g if $gS = S$. Prove that the number of fixed subsets is *not* a multiple of p.

12.7 Let G be a finite group acting *transitively* on the set $\{1, \ldots, p\}$, where $p > 0$ is prime. Prove that G contains an element g which cycles through all elements $1, \ldots, p$ in some order. (*Hint:* Cauchy's theorem will be helpful.)

12.8 Prove that the center of a group G is the intersection of all the centralizers of elements of G.

12.9 Let $H \subseteq Z(G)$ be a subgroup of the center of a group G. Prove that H is normal in G.

12.10 A proper subgroup H of a group G is 'maximal' if it is not contained in any other proper subgroup of G. Prove that $Z(G)$ cannot be proper and maximal in G.

12.11 Let G be a noncommutative group. Prove that $G/Z(G)$ cannot be a cyclic group.

12.12 Determine the centers of D_8 and D_{10}.

12.13 Let $p > 0$ be a prime, and let G be a p-group. Prove that $\exists z \in G$ such that (i) $|z| = p$ and (ii) z commutes with every element of G.

12.14 ▷ Let $p > 0$ be a prime, and let G be a p-group, $|G| = p^n$. Prove that G contains subgroups of all orders p^i, $0 \le i \le n$, *without* invoking Sylow's theorems. (Use Exercise 12.13 and an induction.)

12.15 Let G be a finite group. Prove that the number of elements of order 2 in G is odd if $|G|$ is even, even if $|G|$ is odd. (Consider the involution $G \to G$ sending g to g^{-1}.)

12.16 Let G be a commutative group, and let $x \ne y$ be elements of order 2. Prove that $|G|$ is a multiple of 4. Use this fact to deduce that if G is abelian and $|G| = 2n$, with n odd, then G has exactly *one* element of order 2. (This gives a solution to Exercise 10.19 which avoids the classification theorem.)

Is the hypothesis of commutativity necessary?

12.17 ▷ Prove that $D_{12} \ne \mathcal{A}_4$.

12.18 ▷ Let G be a noncommutative group of order 6.

• Prove that $Z(G)$ is trivial. (Use Exercise 12.11.)

- Use the class equation to conclude that the nonidentity elements split into two conjugacy classes, of order 2 and 3.
- Prove that G has some element y of order 3, and that $\langle y \rangle$ is normal.
- Prove that the conjugacy class of y has order 2 and consists of y and y^2.
- Conclude that there is an $x \in G$ such that $yx = xy^2$.
- Prove that $|x| = 2$.
- Prove that $G = \langle x, y \rangle$.
- Prove that $G \cong S_3$.

12.19 Prove that all p-Sylow subgroups of a finite group G are isomorphic to one another.

12.20 Prove once again that a commutative group of order $2n$, with n odd, has exactly one element of order 2. You already did this by using the classification theorem (Exercise 10.19) and Lagrange's theorem (Exercise 12.16); this time, use Sylow III.

12.21 Let $p < q < r$ be positive prime integers. Prove that a group of order pqr cannot be simple.

12.22 Let G be a simple group, and assume that $H \subseteq G$ is a subgroup of index $n > 1$. Prove that $|G|$ divides $n!$.

12.23 ▷ Prove that there are no simple groups of order 36, 40, 45, 48, or 56. (Exercise 12.22 may be helpful.)

12.24 Prove that every group of order 45 is abelian. (Exercise 11.9 may be useful for this.)

12.25 Is there a simple group of order 2019? 2020? 2021?

12.26 Find the class equation for S_5 and S_6. (*Hint:* Use Corollary 12.15, a.k.a. Exercise 11.30. The answer for S_5 is $120 = 1 + 10 + 15 + 20 + 20 + 30 + 24$.)

12.27 Prove that there are no normal subgroups of order 30 in S_5.

12.28 ▷ Let G be a group endowed with a normal series

$$G = G_0 \supsetneq G_1 \supsetneq G_2 \supsetneq \cdots \supsetneq G_r = \{e\}$$

with quotients $Q_i := G_{i-1}/G_i$, $i = 1, \ldots, r$. Assume that each Q_i has a normal series

$$Q_i = Q_{i0} \supsetneq Q_{i1} \supsetneq Q_{i2} \supsetneq \cdots \supsetneq Q_{is_i} = \{e\}$$

with quotients $R_{i1}, R_{i2}, \ldots, R_{is_i}$. Show how to string these series together to get a normal series for G with quotients (isomorphic to) $R_{11}, R_{12}, \ldots, R_{21}, R_{22}, \ldots, R_{rs_r}$.

12.29 Prove that a finite group is solvable if and only if it admits a normal series whose quotients are solvable. (Use Exercise 12.28.)

12.30 Let A be a finite abelian group. Prove that A admits a (normal) series $A = A_0 \supsetneq A_1 \supsetneq \cdots \supsetneq \{e\}$ such that the quotients A_i/A_{i+1} are *cyclic*.

12.31 ▷ Prove that a finite group is solvable if and only if it admits a normal series whose quotients are cyclic. (Use Exercises 12.28 and 12.30.)

12.32 ▷ Let G be a solvable finite group. By Exercise 12.31, G admits a normal series with cyclic quotients; let $m_1, \ldots, m_r$ be the orders of these cyclic quotients.

Let H be a subgroup of G. Prove that H admits a normal series with cyclic quotients, whose orders divide the numbers $m_1, \ldots, m_r$. (*Hint:* Proposition 12.60.)

12.33 ▷ Prove that every p-group is solvable, for every prime p. (*Hint:* Corollary 12.22 and induction.)

12.34 Let G be a finite group. Denote by $G^{(1)}$ the commutator subgroup of G (see Exercise 11.38 for the definition); by $G^{(2)}$ the commutator subgroup of $G^{(1)}$; and inductively let $G^{(k)}$ be the commutator subgroup of $G^{(k-1)}$. The sequence

$$G \supseteq G^{(1)} \supseteq G^{(2)} \supseteq \cdots$$

is called the 'derived series' of G. Prove that if $G^{(n)} = \{e\}$ for some n, then G is solvable.

12.35 ▷ Prove the converse to the result of Exercise 12.34. Therefore, a finite group G is solvable *if and only if* its derived series terminates with the identity. (Use Exercise 11.37 and an induction.)

Thus, this could be taken as the *definition* of 'solvable group'.

12.36 Prove that the derived series (Exercise 12.34) of a simple noncommutative group G is $G \supseteq G \supseteq G \supseteq \cdots$.

12.37 Consider the group S_n, for $n \geq 5$. We have seen that S_n is not solvable (Theorem 12.62); the following is an alternative, and in fact simpler, proof.

Assume by contradiction that

$$S_n = G_0 \supseteq G_1 \supseteq \cdots \supseteq \{e\} \tag{12.5}$$

is a normal series with abelian quotients.

- Prove that if i, j, k, r, s are distinct elements of $\{1, \ldots, n\}$, then

$$(i\, j\, k) = (k\, r\, j)(j\, i\, s)(k\, r\, j)^{-1}(j\, i\, s)^{-1}.$$

- Prove that if $n \geq 5$ and G_0/G_1 is abelian, then G_1 contains all 3-cycles. (*Hint:* Use Exercise 11.37.)
- Prove that if $n \geq 5$ and all the quotients in (12.5) are abelian, then G_i contains all 3-cycles for all i. (Same exercise, and induction.)

Of course this is nonsense, since $\{e\}$ does *not* contain 3-cycles. This contradiction proves the statement.

12.38 Prove that the Feit–Thompson theorem is equivalent to the statement that every noncommutative finite simple group has *even* order.

Part IV

Fields

13 Field Extensions

13.1 Fields and Homomorphisms of Fields

Our next topic is 'fields', or more properly 'field extensions'. You might expect another run through the usual basic setup: maybe we are going to define fields, then field homomorphisms, then explore a canonical decomposition, quotients, isomorphism theorems, and so on. Surprise! This is *not* what we are going to do.

The reason is technical, but simple. We can certainly consider a category of 'fields', where the morphisms would be 'homomorphisms of fields'. In fact, we know already what these gadgets are. Fields are particular cases of rings (Definition 3.29): a field is a nontrivial commutative ring in which every nonzero element has a multiplicative inverse. 'Homomorphisms of fields' are simply *ring* homomorphisms between fields. But then, rewind all the way back to Exercise 5.18: *every ring homomorphism from a field to a nontrivial ring is injective!* Why? Because the only ideals of a field are (0) and (1); the kernel of a homomorphism from a field must then be either (1)—but then the homomorphism sends 1 to 0, so the target is the trivial ring, and in particular not a field—or (0), and then the homomorphism is injective as claimed.

So this new category has a curious feature: *every morphism in it is injective* as a set-function. It follows that the basic problem addressed by canonical decomposition is simply not an issue: we are not interested in decomposing homomorphisms into injective/surjective homomorphisms, since every homomorphism is already injective. Therefore quotients do not arise, and we do not need to bring in any 'isomorphism theorem'. This may look like a drastic simplification, but that does not mean that field theory is 'easy'. We are going to discover that these injective homomorphisms do not all look alike; there is a rich and interesting taxonomy of homomorphisms, which leads to substantial applications.

Since every homomorphism $j: k \to F$ between fields is injective, we can use it to view k as a subfield of F. We will accordingly write $k \subseteq F$. We say that F is an 'extension' of k.

DEFINITION 13.1   Let k be a field. A 'field extension' of k is a field F endowed with a homomorphism $k \hookrightarrow F$ identifying k with a subfield of F.

Our focus will be to single out and examine different types of field extensions. The following examples of extensions of $\mathbb{Q}$ already display some of the richness of the theory, in the sense that they are all different in key qualitative ways.

Example 13.2 Let $f(t) \in \mathbb{Q}[t]$ be a nonzero irreducible polynomial. Then $\mathbb{Q}[t]/(f(t))$ is a field, and $\mathbb{Q} \subseteq \mathbb{Q}[t]/(f(t))$ is a field extension.

Example 13.3 Let $\mathbb{Q}(t)$ be the field of fractions of the polynomial ring $\mathbb{Q}[t]$ (§6.5); that is, $\mathbb{Q}(t)$ consists of quotients $f(t)/g(t)$ where $f(t), g(t) \in \mathbb{Q}[t]$ and $g(t) \neq 0$. Then $\mathbb{Q} \subseteq \mathbb{Q}(t)$ is a field extension.

Example 13.4 $\mathbb{Q} \subseteq \overline{\mathbb{Q}}$, where $\overline{\mathbb{Q}}$ is the field of 'algebraic numbers', i.e., the set of complex numbers which are roots of some nonzero polynomial with rational coefficients.

Example 13.5 $\mathbb{Q} \subseteq \mathbb{R}$.

Indeed, the extension in Example 13.2 is finite-dimensional as a $\mathbb{Q}$-vector space; in fact, its dimension equals the degree of the polynomial $f(t)$ (Theorem 7.11). We will say that such an extension is 'finite', and we will soon understand that *every* element of the larger field in a finite extension of a field k is a root of a polynomial with coefficients in k (Proposition 13.31). This makes these elements, and the extension, 'algebraic'.

By contrast, the extension in Example 13.3 is *not* finite-dimensional; its dimension is countably infinite. The element t is not a root of any polynomial with coefficients in $\mathbb{Q}$; that makes t and the extension 'transcendental'.

What about Example 13.4? By definition, every element of $\overline{\mathbb{Q}}$ is algebraic over $\mathbb{Q}$, so this is again an algebraic extension, as in Example 13.2; but unlike that example, it has infinite (countable) dimension as a $\mathbb{Q}$-vector space. Therefore, it is not a 'finite' extension, although it is 'algebraic'.

Example 13.5 is again different, as the dimension of $\mathbb{R}$ as a $\mathbb{Q}$-vector space is *uncountable*.

We will be most interested in the first type of extension listed above. Even within this type there are subtle (but macroscopic) differences. For example, it will turn out that the extensions $\mathbb{Q} \subseteq \mathbb{Q}[t]/(t^2 - 2)$ and $\mathbb{Q} \subseteq \mathbb{Q}[t]/(t^3 - 2)$ are different in a substantial way: the first is 'Galois' and the second is not. As we will see, one can obtain a complete description of a Galois extension in terms of a *group* associated with the extension.[1] This is the gist of the *Fundamental Theorem of Galois Theory*, which is our main technical goal in this chapter.

There are extensions with (even) more exotic behavior than the ones listed in Examples 13.2–13.5, but to construct such examples we must replace the smaller field $\mathbb{Q}$ with

---

[1] It may be worth mentioning that for the extension $\mathbb{Q} \subseteq \overline{\mathbb{Q}}$ mentioned above, this group is closely related to the *étale fundamental group of the projective line over* $\mathbb{Q}$ *with the three points* 0, 1, ∞ *removed*—which is (according to Milne) 'the most interesting object in mathematics'.

other fields. In fact, the category of fields has a second distinguishing feature, which also makes it look different from the other categories we have run across so far. You may remember from Exercises 3.12 and 6.18 that there is an integer associated with a ring, called its *characteristic*. This number is a particularly important notion for fields, so I will reproduce its definition here.

DEFINITION 13.6   Let k be a field. The 'characteristic' of k, denote char k, is the non-negative generator of the kernel of the unique ring homomorphism $\mathbb{Z} \to k$. That is, it is the smallest positive integer n such that $n \cdot 1_k = 0_k$, or 0 if there is no such positive integer.

If the fact that there is a *unique* ring homomorphism from $\mathbb{Z}$ to any ring sounds rusty, go back to Exercises 4.13 and 4.14 and remind yourself that $\mathbb{Z}$ is 'initial' in the category of rings. Further, in Exercise 3.13 you proved that the characteristic of an integral domain is a *prime* integer. Now that we know our way around a little better, this is no big deal: if a ring R has characteristic n, then by definition (and the First Isomorphism Theorem) there is an injective ring homomorphism $\mathbb{Z}/n\mathbb{Z} \hookrightarrow R$; so we may view $\mathbb{Z}/n\mathbb{Z}$ as a subring of R. If R is an integral domain, then $\mathbb{Z}/n\mathbb{Z}$ is an integral domain, and it follows that n is prime in this case. In particular, the characteristic of a field is a prime integer: it is 0 or a positive irreducible integer.

LEMMA 13.7   *Let* $k \subseteq F$ *be a field extension. Then* char k = char F.

Proof   Since ring homomorphisms from $\mathbb{Z}$ to a ring are uniquely determined, the diagram

is necessarily commutative. As $k \hookrightarrow F$ is injective, this implies that $\ker i = \ker j$ (check this if it sounds less than crystal clear!), and the statement follows since the characteristic is a generator of this ideal.   □

It follows that the category of fields does *not* have an initial object: such a field would have multiple characteristics because of Lemma 13.7, and this is not an option. The category of fields splits into infinitely many categories—the categories of fields of given characteristics—which are oblivious to each other's existence, in the sense that there are no homomorphisms from objects of one of these categories to objects of another. In fact, each of these subcategories does have an initial object, as a consequence of the following elementary observation (cf. Exercise 6.18).

LEMMA 13.8   *Let* F *be a field. Then* F *contains a unique copy of* $\mathbb{Q}$ *or of the field* $\mathbb{Z}/p\mathbb{Z}$, *where* $p > 0$ *is prime.*

Proof   Indeed, if char $F = n$, then either $n = p > 0$ is a positive prime integer or $n = 0$. In the latter case, the unique map $\mathbb{Z} \to F$ is injective; but then F must contain a unique copy of the field of fractions of $\mathbb{Z}$, that is, $\mathbb{Q}$, by Theorem 6.45. If $n = p > 0$ is positive (and prime as observed earlier), then F contains an isomorphic copy of $\mathbb{Z}/p\mathbb{Z}$ by definition of characteristic and the First Isomorphism Theorem for rings. This copy is unique because it is the image of the unique homomorphism $\mathbb{Z} \to F$.   □

The unique copy of $\mathbb{Q}$ or $\mathbb{Z}/p\mathbb{Z}$ contained in a given field is called its 'prime' subfield. Since the fields $\mathbb{Z}/p\mathbb{Z}$ play such a central role in the theory, they get a different notation in the context of field theory.

DEFINITION 13.9   For every positive prime integer p, $\mathbb{F}_p$ denotes the field $\mathbb{Z}/p\mathbb{Z}$.

What Lemma 13.8 shows is that for each nonnegative prime n, the category of fields of characteristic n has an initial object: $\mathbb{Q}$ for $n = 0$, and $\mathbb{F}_p$ for $n = p > 0$ a positive prime. We could separate the study of field theory into the study of each of these categories. In fact we go one step further, and consider the *category of extensions of a given field k:* the objects of this category are the field extensions $k \hookrightarrow F$ (as in Definition 13.1) for a fixed field k, and the morphisms correspond to (ring) homomorphisms $j \colon F' \hookrightarrow F''$ such that the diagram

$$
\begin{array}{ccc}
F' & \xrightarrow{\ \ j\ \ } & F'' \\
\uparrow & & \uparrow \\
k & \underset{\mathrm{id}_k}{=\!=\!=} & k
\end{array}
\tag{13.1}
$$

commutes. We say that j 'extends id_k', or 'restricts to the identity on k'. Otherwise put, we can think of a morphism of extensions of k as a sequence of inclusions $k \subseteq F' \subseteq F''$.

The initial object of this category is the trivial extension $\mathrm{id}_k \colon k \to k$. Letting $k = \mathbb{Q}$ or $k = \mathbb{F}_p$, we get the infinitely many categories making up the whole category of fields as described above in general terms; but *every* field k determines an interesting category in this way.

A suitable choice of k may then give rise to interesting phenomena. For example, it can be shown that 'finite extensions' of $\mathbb{Q}$, such as the type listed in Example 13.2, can only have *finitely many* subfields (this will be a consequence of Theorem 13.29, Proposition 14.23, and Theorem 14.24). By contrast, in positive characteristic one can construct finite extensions with *infinitely many* 'intermediate' fields. For example, take the field $\mathbb{F}_2(u, v)$ of rational functions in two indeterminates over $\mathbb{F}_2$, i.e., the field of fractions of $\mathbb{F}_2[u, v]$. We can consider the extension $\mathbb{F}_2(u^2, v^2) \subseteq \mathbb{F}_2(u, v)$, where $\mathbb{F}_2(u^2, v^2)$ denotes the smallest subfield of $\mathbb{F}_2(u, v)$ containing u^2 and v^2. This is a finite extension, and yet *there are infinitely many subfields of $\mathbb{F}_2(u, v)$ containing $\mathbb{F}_2(u^2, v^2)$.* (This is not so easy to verify.) That is, the field $k = \mathbb{F}_2(u^2, v^2)$ has finite extensions with drastically different behavior than the finite extensions of $k = \mathbb{Q}$.

By and large, we will stay away from such complications.

Also, as in the rest of this text, categorical considerations will not really come to the fore. You will not miss anything of substance if you just think of a field extension in the simplest possible way, that is, as the inclusion of a field into another.

13.2   Finite Extensions and the Degree of an Extension

As we understood a while back (Example 8.6; see also Exercise 8.3), *every* ring homomorphism $R \to S$ makes S into an R-module; in particular, if $k \subseteq F$ is a field extension, then F may be viewed as a vector space over k.

DEFINITION 13.10   The 'degree' of a field extension $k \subseteq F$, denoted $[F : k]$, is the dimension of F as a k-vector space.

For example, the extension $k \subseteq F$ is trivial (i.e., $k = F$) if and only if $[F : k] = 1$.

Spoiler alert: The fact that the notation looks a lot like the notation for the index of a subgroup is *not* a coincidence.

Of course $[F : k]$ could be infinite; we will be primarily interested in the case in which it is *finite*. The corresponding extensions have a rather unimaginative name.

DEFINITION 13.11   A field extension $k \subseteq F$ is 'finite' if F is a finite-dimensional k-vector space.

This does not (necessarily) mean that either k or F is a finite field! For instance, the extension $\mathbb{R} \subseteq \mathbb{C}$ is finite; in fact, $[\mathbb{C} : \mathbb{R}] = 2$ since $\mathbb{C}$ is a two-dimensional vector space over $\mathbb{R}$.

Another non-coincidence is the fact that $[F : k]$ is called the *degree* of an extension. The reason is likely the following very important example of finite extension.

LEMMA 13.12   *Let k be a field and let $f(t) \in k[t]$ be an irreducible polynomial. Then $k[t]/(f(t))$ is a field, and the homomorphism $k \to k[t]/(f(t))$ mapping a to the coset $a + (f(t))$ defines a field extension $k \subseteq k[t]/(f(t))$. This extension is finite, and*

$$[k[t]/(f(t)) : k] = \deg f .$$

Proof   This is just a collection of facts we know already. Irreducible elements in PIDs generate maximal ideals by Proposition 6.33. The resulting field $k[t]/(f(t))$ is a k-vector space of dimension $\deg f$ by Theorem 7.11. Using the terminology recalled in §9.1, the proof of Theorem 7.11 shows that the d cosets $1 + (f(t)), t + (f(t)), \ldots, t^{d-1} + (f(t))$, with $d = \deg f$, form a basis.   □

It will take us a while to understand why, but these are in fact *all* the extensions we really care about in this text. Note that the extension in Example 13.2 is of this type, and that it is the only finite extension among the ones appearing in Examples 13.2–13.5.

One reason why the degree is a useful invariant of an extension is that it is *multiplicative* in the following sense.

PROPOSITION 13.13   *Let $k \subseteq E \subseteq F$ be field extensions. If $k \subseteq E$ and $E \subseteq F$ are both finite extensions, then $k \subseteq F$ is finite, and*

$$[F : k] = [F : E][E : k].$$

Proof   Let $[E : k] = m$, $[F : E] = n$. Since $[E : k] = m$, E has a basis $\epsilon_1, \ldots, \epsilon_m$ over k. Thus, every element of E is (uniquely) a k-linear combination of these elements. Similarly, F has a basis $\varphi_1, \ldots, \varphi_n$ over E, and every element of F is an E-linear combination of the elements φ_j.

I claim that the mn elements $\epsilon_i \varphi_j$, $i = 1, \ldots, m$, $j = 1, \ldots, n$, form a basis of F over k. Indeed, these elements (i) generate F and (ii) are linearly independent over k. To verify (i), let $f \in F$; as observed above, there exist $e_1, \ldots, e_n \in E$ such that

$$f = e_1 \varphi_1 + \cdots + e_n \varphi_n = \sum_{j=1}^{n} e_j \varphi_j.$$

For each j, e_j may be written as a k-linear combination of the elements ϵ_i: there exist elements $k_{ij} \in k$ such that

$$e_j = k_{1j} \epsilon_1 + \cdots + k_{mj} \epsilon_m = \sum_{i=1}^{m} k_{ij} \epsilon_i.$$

Then we have

$$f = \sum_{j=1}^{n} \sum_{i=1}^{m} k_{ij} \epsilon_i \varphi_j,$$

and this confirms that every element $f \in F$ is a k-linear combination of the elements $\epsilon_i \varphi_j$. To verify (ii), that is, that the elements are linearly independent over k, assume

$$\sum_{j=1}^{n} \sum_{i=1}^{m} k_{ij} \epsilon_i \varphi_j = 0.$$

Since the elements φ_j are linearly independent over E, it follows that for all $j = 1, \ldots, n$,

$$\sum_{i=1}^{m} k_{ij} \epsilon_i = 0.$$

Since the elements ϵ_i are linearly independent over k, it further follows that $k_{ij} = 0$ for all i and all j, and this proves linear independence.   □

Proposition 13.13 is a very straightforward observation, yet it is quite powerful. Here is a useful term associated with the situation contemplated in Proposition 13.13.

DEFINITION 13.14   If $k \subseteq E \subseteq F$ are field extensions, we say that E is an 'intermediate field' of the extension $k \subseteq F$.

It is clear that if E is an intermediate field of a finite extension $k \subseteq F$, then both $k \subseteq E$ and $E \subseteq F$ must be finite (Exercise 13.5).

In principle it may seem hard to say anything about the intermediate fields of a given extension. As a vector space over k, F will have *many* subspaces: if k is infinite, there are infinitely many subspaces of any given dimension strictly between 1 and $\dim_k F = [F : k]$. The condition that a given subspace be in fact a field is subtle. Precisely because of this, it is remarkable that such coarse information as the degree of an extension can immediately rule out the existence of many intermediate fields.

Example 13.15   The extension $\mathbb{Q} \subseteq \mathbb{Q}[t]/(t^7 + 2t^2 + 2)$ has no intermediate field other than $\mathbb{Q}$ and $\mathbb{Q}[t]/(t^7 + 2t^2 + 2)$.

(Note that $t^7 + 2t^2 + 2$ is irreducible over $\mathbb{Q}$ by Eisenstein's criterion, Theorem 7.35.)

Do I know this because I am privy to some sophisticated information about the polynomial $t^7 + 2t^2 + 2$? No: I know it because the degree, 7, of this polynomial is *prime*. So if E is an intermediate field,

$$\mathbb{Q} \subseteq E \subseteq \mathbb{Q}[t]/(t^7 + 2t^2 + 2),$$

then by Proposition 13.13 the only possibilities are $[E : \mathbb{Q}] = 1$, and then $E = \mathbb{Q}$, or $[E : \mathbb{Q}] = 7$ and $[\mathbb{Q}[t]/(t^7 + 2t^2 + 2) : E] = 1$, and then $E = \mathbb{Q}[t]/(t^7 + 2t^2 + 2)$.

We will in fact be very interested in the set

$$\{E \text{ field} \mid k \subseteq E \subseteq F\}$$

of intermediate fields of a given extension. For instance, for the extension in Example 13.15 this set contains only two elements: k and F. But I chose a very special extension in order to be able to draw this conclusion. How on earth can one determine *all* intermediate fields of a given extension, if one is a little less lucky? In due time (§15.3) we will set up a correspondence

$$\{E \text{ field} \mid k \subseteq E \subseteq F\} \rightleftarrows \{ \qquad ? \qquad \} \tag{13.2}$$

between this mysterious set and *another* mysterious set associated with an extension. This will be the 'Galois correspondence', and we will determine a class of extensions for which this is a one-to-one correspondence. It will allow us to determine all intermediate fields of an extension, by using group-theoretic methods.

13.3   Simple Extensions

Let $k \subseteq F$ be a field extension, and let $\alpha_1, \ldots, \alpha_r \in F$. I will denote by $k(\alpha_1, \ldots, \alpha_r)$ the smallest subfield of F containing k and the elements α_i. If $F = k(\alpha_1, \ldots, \alpha_r)$, we say that $k \subseteq F$ is 'generated' by $\alpha_1, \ldots, \alpha_r$.

DEFINITION 13.16   An extension $k \subseteq F$ is 'simple' if $F = k(\alpha)$ for some $\alpha \in F$.

The extensions in Examples 13.2 and 13.3 are simple. In fact, *every* simple extension looks like them.

PROPOSITION 13.17   *Let $k \subseteq k(\alpha) = F$ be a simple extension. Then*

- *Either $F \cong k[t]/(p(t))$ for an irreducible monic polynomial $p(t) \in k[t]$ such that $p(\alpha) = 0$.*
- *Or $F \cong k(t)$, the field of fractions of $k[t]$.*

In the first case, $[F : k] = \deg p(t)$. In the second, $[F : k] = \infty$.

Proof   Let $F = k(\alpha)$. Define a homomorphism

$$\varphi \colon k[t] \longrightarrow F$$

extending the inclusion $k \subseteq F$, by mapping t to α; that is, $\varphi(f(t)) = f(\alpha)$. If $\ker \varphi \neq (0)$, then $\ker \varphi = (p(t))$ for some polynomial $p(t) \neq 0$, since $k[t]$ is a PID as k is a field (Corollary 7.7). By construction, $p(\alpha) = 0$; and after multiplying $p(t)$ by the inverse of its leading term, we may in fact assume that it is monic, as required in the statement. By the First Isomorphism Theorem, φ induces an injective homomorphism

$$\tilde{\varphi} \colon k[t]/(p(t)) \hookrightarrow F \,;$$

in particular, $k[t]/(p(t))$ is an integral domain. It follows that $(p(t))$ is prime, and therefore maximal (nonzero prime ideals are maximal in a PID: Proposition 6.30), and hence $k[t]/(p(t))$ is in fact a *field*. The image $\operatorname{im} \tilde{\varphi} \subseteq F$ is a subfield of F containing $\varphi(t) = \alpha$. But then $\operatorname{im} \tilde{\varphi} = F$, since F is the smallest subfield containing k and α. This means that $\tilde{\varphi}$ is also surjective, hence $\tilde{\varphi}$ is an isomorphism. In this case $[F : k] = \dim_k F$ equals $\deg p$, by Theorem 7.11. This takes care of the first case.

If $\ker \varphi = (0)$, then $\varphi \colon k[t] \to F$ is injective. Since $k[t]$ is an integral domain, it has a field of fractions $k(t)$. By the universal property of the field of fractions, i.e., Theorem 6.45, there exists an injective homomorphism

$$k(t) \hookrightarrow F \,.$$

The image of this homomorphism is a subfield of F containing $\varphi(t) = \alpha$, and it follows that $k(t) \cong F$ since F is generated by α.

The dimension of $k(t)$ as a k-vector space is infinite, therefore $[F : k] = \infty$ in this case, concluding the proof.   □

The two cases found in Proposition 13.17 are encoded in the following important definition.

DEFINITION 13.18   Let $k \subseteq F$ be a field extension, and let $\alpha \in F$. Then α is 'algebraic' over k, of degree d, if $[k(\alpha) : k] = d$ is finite. The element α is 'transcendental' if it is not algebraic.

'Algebraic extensions' will be extensions $k \subseteq F$ for which *every* element of F is algebraic over k. We will look more carefully into these extensions in §13.4.

It is worth highlighting the following characterization of algebraic elements, which is a direct consequence of what we have done already.

LEMMA 13.19   *Let $k \subseteq F$ be a field extension, and let $\alpha \in F$. Then α is algebraic over k if and only if α is a root of a nonzero polynomial $f(t) \in k[t]$. In this case, α is a root of a unique monic irreducible polynomial $p(t) \in k[t]$, with $p(t) \mid f(t)$, and $k(\alpha) \cong k[t]/(p(t))$.*

Proof   If α is algebraic, then $k \subseteq k(\alpha)$ is a simple extension of the first kind listed in Proposition 13.17, and in particular $p(\alpha) = 0$ for some nonzero irreducible monic polynomial.

Conversely, assume that $f(\alpha) = 0$ for a nonzero $f(t) \in k[t]$. As in the proof of Proposition 13.17, consider the 'evaluation' homomorphism

$$\varphi \colon k[t] \to F$$

mapping $g(t)$ to $g(\alpha)$. By hypothesis $f(t) \in \ker \varphi$, and in particular $\ker \varphi \neq (0)$. It follows (again as in the proof of Proposition 13.17) that $k(\alpha) \cong k[t]/\ker \varphi$ is a finite extension of k, and therefore α is algebraic; further, $\ker \varphi$ is generated by an irreducible monic polynomial $p(t)$.

Since $f(t) \in \ker \varphi = (p(t))$, $f(t)$ is a multiple of $p(t)$ as stated.

The monic generator of an ideal in $k[t]$ is uniquely determined: if $(p_1(t)) = (p_2(t))$, then $p_1(t)$ and $p_2(t)$ differ by a multiplicative unit (Exercise 6.11), hence they coincide if they are both monic.   □

DEFINITION 13.20   If α is algebraic over k, the 'minimal polynomial' of α over k is the unique monic irreducible polynomial $p(t) \in k[t]$ such that $p(\alpha) = 0$.

With notation as in Definition 13.18, the degree of α is the degree of its minimal polynomial $p(t)$. Again, keep in mind that then $k(\alpha) \cong k[t]/(p(t))$.

Remark 13.21   If α is algebraic over k, then every element of $k(\alpha)$ may be expressed as a *polynomial* in α with coefficients in k. We never have to 'divide by α', since α^{-1} itself may be written as a polynomial in α.

This may be a little counterintuitive, but there is nothing to it. For example, consider $\mathbb{Q}(\sqrt[5]{2})$; $\alpha = \sqrt[5]{2}$ is algebraic over $\mathbb{Q}$ since it is a root of $t^5 - 2$. If $\alpha^5 = 2$, then $\alpha^4 = 2\alpha^{-1}$, and therefore $\alpha^{-1} = \alpha^4/2$ is a polynomial in α with coefficients in $\mathbb{Q}$, as promised.   ⌐

Example 13.22   'Algebraic numbers' are complex numbers that are algebraic over $\mathbb{Q}$ (such as $\sqrt[5]{2}$), and $\overline{\mathbb{Q}}$ denotes the set of algebraic numbers. For example, $i \in \overline{\mathbb{Q}}$; the minimal polynomial of i over $\mathbb{Q}$ is $t^2 + 1$.

It will turn out that $\overline{\mathbb{Q}}$ is itself a field, and further that it is 'algebraically closed' (Exercise 13.19, Example 14.5). Elementary considerations show that $\overline{\mathbb{Q}}$ is *countable* (Exercise 13.9). Since $\mathbb{R}$ and $\mathbb{C}$ are uncountable, it follows that 'almost every' real or complex number is in fact transcendental. Despite this, it is quite hard to prove that a given number is transcendental. We know that π and e are transcendental, but these are serious theorems.[2] I certainly expect $\pi + e$ and πe to be transcendental, but as far as I know nobody has a clue as to how to go about proving this.

Remark 13.23 The notions introduced above all depend on the 'ground' field k. For example, i is algebraic over $\mathbb{R}$, with minimal polynomial $t^2 + 1$. This polynomial is irreducible *over* $\mathbb{R}$; of course it is no longer irreducible in the larger field $\mathbb{C}$. In fact, $t^2 + 1 = (t - i)(t + i)$ over $\mathbb{C}$; the minimal polynomial of i *over* $\mathbb{C}$ is simply $t - i$.

The number π is transcendental over $\mathbb{Q}$, but not over $\mathbb{R}$: π is an element of $\mathbb{R}$, so it is trivially algebraic over $\mathbb{R}$, with minimal polynomial $t - \pi$.   ⌐

> PROPOSITION 13.24 *Let $k \subseteq E \subseteq F$ be field extensions, and let $\alpha \in F$. Assume that α is algebraic over k, with minimal polynomial $p_k(t)$. Then α is also algebraic over E, and its minimal polynomial $p_E(t)$ is a factor of $p_k(t)$.*

Wait—wasn't the minimal polynomial supposed to be *irreducible?* How come it has 'factors'? The minimal polynomial $p_k(t)$ of α over k is irreducible *as a polynomial in $k[t]$*. Over larger fields (such as E) $p_k(t)$ may well have nontrivial factors. For example, $t - \alpha$ is a factor of $p_k(t)$ when this is viewed as a polynomial in $F[t]$.

Proof Since $k \subseteq E$, we can view $p_k(t)$ as a polynomial in $E[t]$. Since $p_k(\alpha) = 0$, we see that α is algebraic over E and its minimal polynomial $p_E(t) \in E[t]$ divides $p_k(t)$, by Lemma 13.19.   □

At this stage we may have gathered the impression that 'simple' extensions must be rare: our previous experience, for example with linear algebra, ideals in rings, and groups, should have primed us to assume that objects generated by *a single* element are very special. On the face of it, something like $\mathbb{Q}(\sqrt{2}, \sqrt{3})$ does not look like it should be a *simple* extension of $\mathbb{Q}$, right? Wrong.

Example 13.25 The extension $\mathbb{Q} \subseteq \mathbb{Q}(\sqrt{2}, \sqrt{3})$ is simple. In fact, we can verify that $\mathbb{Q}(\sqrt{2}, \sqrt{3}) = \mathbb{Q}(\sqrt{2} + \sqrt{3})$.

Indeed, it is clear that $\mathbb{Q}(\sqrt{2} + \sqrt{3}) \subseteq \mathbb{Q}(\sqrt{2}, \sqrt{3})$, since $\sqrt{2} + \sqrt{3} \in \mathbb{Q}(\sqrt{2}, \sqrt{3})$; we have to show that both $\sqrt{2}$ and $\sqrt{3}$ are in $\mathbb{Q}(\sqrt{2} + \sqrt{3})$, that is, they may be expressed as polynomials in $\sqrt{2} + \sqrt{3}$ with coefficients in $\mathbb{Q}$.

This is no big deal. Let $\alpha = \sqrt{2} + \sqrt{3}$. Then

$$\alpha^3 = (\sqrt{2})^3 + 3(\sqrt{2})^2\sqrt{3} + 3\sqrt{2}(\sqrt{3})^2 + (\sqrt{3})^3 = 2\sqrt{2} + 6\sqrt{3} + 9\sqrt{2} + 3\sqrt{3} = 11\sqrt{2} + 9\sqrt{3}.$$

We have the relations

$$\begin{cases} \alpha = \sqrt{2} + \sqrt{3}, \\ \alpha^3 = 11\sqrt{2} + 9\sqrt{3}, \end{cases}$$

---

[2] Both follow from an 1882 result due to Lindemann, to the effect that e^α is transcendental if $\alpha \neq 0$ is algebraic. For $\alpha = 1$, this proves directly that e is transcendental. For $\alpha = i\pi$, the fact that $e^{i\pi} = 1$ is *not* transcendental implies that $i\pi$ cannot be algebraic, and hence π must be transcendental.

and we may deal with this as if it were a system of linear equations in the 'unknowns' $\sqrt{2}$, $\sqrt{3}$. Solving it, we discover that

$$\sqrt{2} = \frac{\alpha^3 - 9\alpha}{2}, \quad \sqrt{3} = \frac{11\alpha - \alpha^3}{2},$$

confirming that $\mathbb{Q}(\sqrt{2}, \sqrt{3}) \subseteq \mathbb{Q}(\alpha)$.

In fact, whole sweeping classes of extensions turn out to be simple, sometimes for reasons that do not seem to immediately have to do with field theory.

Example 13.26   Finite extensions of finite fields are simple.

Indeed, let $k \subseteq F$ be a field extension, with k finite. If the extension is finite, then the field F is also a finite field. But then the multiplicative group F^* is cyclic *as a group:* this was one highlight in our applications of the structure theorem for finite abelian groups (Theorem 10.24). If $F^* = \langle \alpha \rangle$ as a cyclic group, then every nonzero element of F is a power of α, and in particular it is in $k(\alpha)$. This shows that $F = k(\alpha)$ is simple.

There is a rather nontrivial characterization of simple algebraic extensions, which is, at least contextually, compelling. Half of the characterization is the following consequence of Proposition 13.24, which again gives a surprisingly strong constraint on possible intermediate fields of an extension.

> PROPOSITION 13.27   *Let $k \subseteq F = k(\alpha)$ be a simple extension, with α algebraic over k. Then $k \subseteq F$ admits only finitely many intermediate fields.*

Proof   Let $p(t) \in k[t]$ be the minimal polynomial of α over k, and let E be an intermediate field of the extension $k \subseteq k(\alpha)$. We may view $p(t)$ as a polynomial in $E[t]$; and as proved in Proposition 13.24, the minimal polynomial of α over E is a factor $p_E(t)$ of $p(t)$ in $E[t]$. Let $p_E(t) = e_0 + e_1 t + \cdots + e_{d-1} t^{d-1} + t^d \in E[t]$, so $[k(\alpha) : E] = d$.

> CLAIM 13.28 $E = k(e_0, \ldots, e_{d-1})$.

To prove the claim, let $E' = k(e_0, \ldots, e_{d-1})$ and observe that we have the sequence of extensions

$$E' \subseteq E \subseteq k(\alpha).$$

Since E' contains all the coefficients of $p_E(t)$, we can view $p_E(t)$ as a polynomial in $E'[t]$. Since $p_E(t)$ is irreducible over E, it must be irreducible over E'. But then $p_E(t)$ must be the minimal polynomial of α *over E';* it follows that $[k(\alpha) : E'] = d = [k(\alpha) : E]$. By Proposition 13.13

$$d = [k(\alpha) : E'] = [k(\alpha) : E][E : E'] = d \cdot [E : E'],$$

and it follows that $[E : E'] = 1$. This shows $E = E'$ and proves the claim.

The claim shows that E is determined by the factor $p_E(t)$ of $p(t)$. Since $p(t)$ has only finitely many monic factors in $k(\alpha)[t]$, there can only be finitely many intermediate fields.   □

Amazingly, Proposition 13.27 has a converse! Let's package both implications into one statement.

THEOREM 13.29   *Let $k \subseteq F$ be a field extension. Then $F = k(\alpha)$, with α algebraic over k, if and only if the extension has only finitely many intermediate fields.*

Proof   The 'only if' implication was proved in Proposition 13.27; we have to prove that if $k \subseteq F$ admits only finitely many intermediate fields, then it is simple. The extension $k \subseteq F$ is finitely generated: if not, we could construct an infinitely ascending sequence

$$k \subsetneq k(u_1) \subsetneq k(u_1, u_2) \subsetneq k(u_1, u_2, u_3) \subsetneq \cdots \subseteq F,$$

giving infinitely many intermediate fields. Therefore we may assume that F is generated by n elements $u_1, \ldots, u_n$: $F = k(u_1, \ldots, u_n)$. We may also assume that each u_i is algebraic over $k(u_1, \ldots, u_{i-1})$: indeed, if u_i is transcendental, then there are infinitely many intermediate fields between $k(u_1, \ldots, u_{i-1})$ and $k(u_1, \ldots, u_i)$ (Exercise 13.15), and hence infinitely many between k and F. In particular, $k \subseteq F$ is a *finite* extension, as you should verify right away (Exercise 13.16; we will prove a stronger result in the next section).

Induction reduces the arbitrary case to the case $n = 2$. Indeed, for $n \geq 3$, and assuming that the result is known for $n - 1$ generators,

$$F = k(u_1, \ldots, u_n) = k(u_1, \ldots, u_{n-1})(u_n) = k(u)(u_n) = k(u, v)$$

for some u and with $v = u_n$.

Therefore, we may assume $F = k(u, v)$. If k is a finite field, then so is F (since $k \subseteq F$ is finite), so the extension is simple (Example 13.26). Therefore, we may assume that k is infinite.

Consider then the infinitely many elements $u + cv$ with $c \in k$, and the corresponding intermediate fields $k(u + cv)$. By hypothesis there are only finitely many intermediate fields, so necessarily $\exists c' \neq c'' \in k$ such that

$$k(c'u + v) = k(c''u + v).$$

I claim that this field equals $k(u, v) = F$: so F is indeed simple. To verify my claim, we have to verify that both u and v are in $k(u + c'v) = k(u + c''v)$. But this field contains both $u + c'v$ and $u + c''v$, therefore it contains

$$v = \frac{(u + c'v) - (u + c''v)}{c' - c''} \quad \text{and} \quad u = (u + c'v) - c'v, \tag{13.3}$$

so indeed it equals F.

This shows that the extension is simple: $F = k(\alpha)$, with $\alpha = u + c'v$. Since $k \subseteq F = k(\alpha)$ is a finite extension, α is algebraic as stated, and this completes the proof.   □

We are chipping away at the question we loosely raised at the end of §13.2, i.e., understanding the set of intermediate fields of an extension. At this point we see that the left-hand side of the mysterious correspondence (13.2), i.e., the set

$$\{E \text{ field} \mid k \subseteq E \subseteq F\}$$

is *finite* for a large class of interesting examples—all simple extensions $k \subseteq F = k(\alpha)$ with α algebraic over k. We will see later on that this class of examples is even larger than we may suspect at this stage. For example, a strong result we will prove in the next chapter (Theorem 14.24) will imply that *all finite extensions of $\mathbb{Q}$ are simple.*

13.4   Algebraic Extensions

In the previous section I mentioned the following important definition.

DEFINITION 13.30   A field extension $k \subseteq F$ is 'algebraic' if every element of F is algebraic over k.

As we have seen, an element $\alpha \in F$ is algebraic over k if $k \subseteq k(\alpha)$ is finite (Definition 13.18), or equivalently if α is a root of a nonzero polynomial with coefficients in k (Lemma 13.19). Is it clear that if α is a root of a polynomial, then so is every element which can be expressed as a polynomial in α? Maybe this is not immediately 'clear' to me, but as it turns out, it is very easy to prove.

PROPOSITION 13.31   *Finite extensions are algebraic. In fact, if $k \subseteq F$ is finite and $[F : k] = d$, then every $\alpha \in F$ is algebraic over k, and its degree divides d.*

Proof   Let $k \subseteq F$ be a finite extension, and let $\alpha \in F$. Then $k(\alpha)$ is an intermediate field: $k \subseteq k(\alpha) \subseteq F$. Since $k \subseteq F$ is finite, so is $k \subseteq k(\alpha)$. But then α is algebraic according to Definition 13.18.

Concerning the degree of α, apply Proposition 13.13 to $k \subseteq k(\alpha) \subseteq F$ to get

$$d = [F : k] = [F : k(\alpha)][k(\alpha) : k]$$

and conclude that $[k(\alpha) : k]$ divides d.   □

Maybe even more transparently: if $\alpha \in F$, and $[F : k] = d$ is finite, then the monomials $1, \alpha, \ldots, \alpha^d$ are necessarily linearly dependent, so there exist $\lambda_0, \ldots, \lambda_d$ such that

$$\lambda_0 + \lambda_1 \alpha + \cdots + \lambda_d \alpha^d = 0.$$

Therefore α is algebraic, and its minimal polynomial divides $\lambda_0 + \lambda_1 t + \cdots + \lambda_d t^d$. (But Proposition 13.13 is still needed to conclude that the degree of the minimal polynomial divides d.)

A full converse to Proposition 13.31 does not hold, but a more restrictive statement does, and is useful.

PROPOSITION 13.32 *If $k \subseteq F$ is algebraic and finitely generated, then it is finite.*
    In fact, let $F = k(\alpha_1, \ldots, \alpha_r)$ be a finitely generated extension, with α_i algebraic of degree d_i over k. Then $[F : k] \le d_1 \cdots d_r$, so $k \subseteq F$ is finite (and hence algebraic).
    In particular, $k \subseteq k(\alpha_1, \ldots, \alpha_r)$ is algebraic if and only if each α_i is algebraic over k.

Proof   If $F = k(\alpha_1, \ldots, \alpha_r)$ as in the statement, then we have the sequence of extensions

$$k \subseteq k(\alpha_1) \subseteq k(\alpha_1, \alpha_2) \subseteq \cdots \subseteq k(\alpha_1, \ldots, \alpha_r) = F.$$

We have

$$[k(\alpha_1, \ldots, \alpha_i) : k(\alpha_1, \ldots, \alpha_{i-1})] = [k(\alpha_1, \ldots, \alpha_{i-1})(\alpha_i) : k(\alpha_1, \ldots, \alpha_{i-1})]$$
$$\le [k(\alpha_i) : k] = d_i$$

by Proposition 13.24. Proposition 13.13 then gives

$$[F : k] = [k(\alpha_1, \ldots, \alpha_r) : k(\alpha_1, \ldots, \alpha_{r-1})] \cdots [k(\alpha_1) : k] \le d_r \cdots d_1$$

as stated.   □

The last assertion in Proposition 13.32 is also not 'obvious' to me (notwithstanding the fact that its proof is so very easy!). For example, it says that if α is a root of a nonzero polynomial $p_\alpha(t) \in k[t]$ and β is a root of a nonzero polynomial $p_\beta(t) \in k[t]$, then $\alpha + \beta$ and $\alpha \cdot \beta$ also are roots of polynomials in $k[t]$. Even after proving Proposition 13.32, I would not really know a reasonable way to construct polynomials for $\alpha + \beta$ and $\alpha \cdot \beta$ from $p_\alpha(t)$ and $p_\beta(t)$.

Proposition 13.32 has useful consequences, such as the following.

PROPOSITION 13.33   *Let $k \subseteq F$ be a field extension, and let E be the set of elements of F which are algebraic over k. Then E is a field.*

Proof   If α and β are elements of E, i.e., elements of F that are algebraic over k, then $k \subseteq k(\alpha, \beta)$ is an algebraic extension by Proposition 13.32, and in particular $\alpha + \beta$, $\alpha \cdot \beta$, and α^{-1} (if $\alpha \ne 0$) are also elements of E. This proves the statement.   □

Example 13.34   By Proposition 13.33, the set $\overline{\mathbb{Q}} \subseteq \mathbb{C}$ of 'algebraic numbers', i.e., complex numbers that are algebraic over $\mathbb{Q}$, is a field (as promised in Examples 13.4 and 13.22). The extension $\mathbb{Q} \subseteq \overline{\mathbb{Q}}$ is the focus of large swaths of number theory, particularly algebraic number theory.[3] The field $\overline{\mathbb{Q}}$ contains all 'number fields', i.e., all finite extensions of $\mathbb{Q}$.

Of course $\mathbb{Q} \subseteq \overline{\mathbb{Q}}$ is *not* itself finite: indeed, as we know (Corollary 7.38) there are irreducible polynomials $f(t)$ of arbitrarily high degree in $\mathbb{Q}[t]$, whose roots (in $\mathbb{C}$) are then algebraic numbers of arbitrarily high degree.

---

[3] Items 11R and 11S in the Mathematics Subject Classification.

Another consequence of Proposition 13.32 is that compositions of algebraic extensions are algebraic extensions. This is clear for finite extensions, but it is true without any finiteness hypothesis.

> PROPOSITION 13.35 *Let $k \subseteq E$ and $E \subseteq F$ be algebraic extensions. Then the extension $k \subseteq F$ is algebraic.*

Proof Let $\alpha \in F$; we have to prove that α is algebraic over k. Since α is algebraic over E by hypothesis, it is a root of a polynomial $f(t) = e_0 + e_1 t + \cdots + e_n t^n \in E[t]$. It follows that α is 'already' algebraic over the field $k(e_0, \ldots, e_n)$ generated by the coefficients of f. The extension $k \subseteq k(e_0, \ldots, e_n)$ is algebraic and finitely generated, so it is finite by Proposition 13.32. Therefore the composition

$$k \subseteq k(e_0, \ldots, e_n) \subseteq k(e_0, \ldots, e_n, \alpha) = k(e_0, \ldots, e_n)(\alpha)$$

is finite, hence algebraic (Proposition 13.31), and in particular α is algebraic over k as stated. □

13.5 Application: 'Geometric Impossibilities'

We know *very* little field theory at this point, but enough to dispose handily of questions that puzzled people for hundreds if not thousands of years.

Lots of interesting geometric shapes may be drawn by only using straightedge and compass. Start with the ordinary plane, with two points O and P marked for you. For example, here is how you would 'construct' an equilateral triangle if armed with straightedge and compass:

— Center the compass at O, and draw a circle through P.
— Center the compass at P, and draw a circle through O.
— These circles will meet at two points; let Q be either.
— Use the straightedge to draw lines through O and P, O and Q, and P and Q.

Here is a picture summarizing the construction.

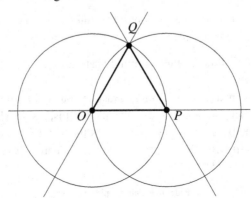

And here is how you would construct a line perpendicular to the line OP and through the point O:

— Draw the line through O and P.
— Center the compass at O, and draw a circle through P.
— This line meets the line OP at another point; call this point Q.
— Center the compass at P, draw the circle through Q.
— Center the compass at Q, draw the circle through P.
— The new circles meet at two points; let one of these points be R.
— The line through O and R is perpendicular to OP and goes through O.

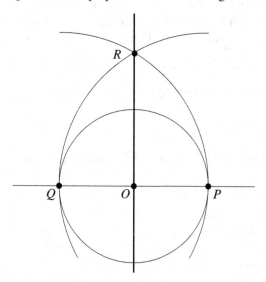

It is pleasant to try and construct different figures by the same method. You start from a plane and two marked points O and P. You are allowed to mark ('construct') new points, lines, circles by the following actions.

• Draw a line through two points you have already constructed.
• Draw a circle with center at a point you have constructed, through another point you have constructed.
• Mark a point of intersection of two lines, a line and a circle, or two circles you have already constructed.

You are not allowed other seemingly natural actions, such as measuring distances. Your straightedge is a *straightedge,* not a *ruler:* there are no marks on it, and you are not allowed to mark any point on it. Practice your skills a bit by just using these operations to construct a square with a given side (that is, a side whose vertices you have already constructed), or the line which bisects the angle between two lines you have already constructed (Exercise 13.20). If you want a harder task, construct a regular *pentagon* with a given side.[4]

---

[4] If this seems hard, consider that Gauss explained how to construct a regular 17-gon at age 19.

Believe it or not, people puzzled for centuries over particular constructions that no one had been able to figure out. Specifically, can you construct

- The side of a square with the same area as a circle you have constructed?
- The side of a cube whose volume is 2?
- A line *trisecting* the angle formed by two lines you have constructed?

To this date, some people spend their free time trying to 'square the circle'. This is too bad, because the few notions in field theory that we have covered so far amply suffice[5] to prove that no matter how clever you are, you cannot perform the tasks I have just listed.

Why?

All constructions start from the given points O and P. Set a coordinate system so that $O = (0, 0)$ and $P = (1, 0)$.

DEFINITION 13.36   A real number is 'constructible' if it is a coordinate of a point you can construct by straightedge and compass.

Example 13.37   The numbers $\frac{1}{2}$ and $\frac{\sqrt{3}}{2}$ are constructible: indeed, the top vertex Q in the equilateral triangle we constructed above has coordinates $(\frac{1}{2}, \frac{\sqrt{3}}{2})$.

It is easy to see that all rational numbers are constructible (Exercise 13.21).

From this point of view, the three challenges listed above translate to the questions of whether certain numbers are constructible:

- Is $\sqrt{\pi}$ constructible?
- Is $\sqrt[3]{2}$ constructible?
- If you have constructed an angle θ, is $\cos(\frac{\theta}{3})$ constructible?

Just to make the third task (even) more concrete, since we constructed a 60° angle in constructing the equilateral triangle, can you construct an angle of 20°—or, equivalently, is $\cos(20°)$ constructible? A little patience with the addition formulas for cosine and sine[6] shows that $\cos(3\theta) = 4\cos^3(\theta) - 3\cos(\theta)$, and hence (plugging in $\theta = 20°$) that $\cos(20°)$ is a root of the polynomial $8t^3 - 6t - 1$. This polynomial is irreducible over $\mathbb{Q}$, for example by the 'rational root test'.

- Is a root of the polynomial $8t^3 - 6t - 1$ constructible?

The punch line is now that these numbers are simply *not* constructible.

---

[5] There is a small white lie here: dealing with 'squaring the circle' hinges on the fact that π is transcendental, which we have not proved.

[6] There actually are efficient alternatives to remembering trigonometry; see Exercise 13.24.

THEOREM 13.38   *Let α be a constructible number. Then $\alpha \in F$, where F is an extension of $\mathbb{Q}$ such that $[F : \mathbb{Q}]$ is a power of 2. In particular, α is an algebraic number and the degree of α over $\mathbb{Q}$ is a power of 2.*

Proof   At any stage of a construction, we consider the set of numbers appearing as coordinates of points that have been constructed, or as coefficients in the equations of the lines and circles that have been constructed. For circles

$$x^2 + y^2 + ax + by + c = 0$$

we take the numbers a, b, c; for lines

$$x = r \quad \text{or} \quad y = mx + r$$

we take r, or m and r, as the case may be.

CLAIM 13.39   *It suffices to show that if at any stage in a construction all these numbers lie in a field E, then after applying one of the basic operations the same data lies in a field $E(u)$, where u is algebraic of degree at most 2 over E.*

Indeed, at the beginning of the construction we just have the points $O = (0,0)$, $P = (1,0)$, so the numbers lie in $\mathbb{Q}$. If the claim is verified, this will show that every constructible number lies in a field at the top of a tower of fields

$$\mathbb{Q} \subseteq \mathbb{Q}(u_1) \subseteq \cdots \subseteq \mathbb{Q}(u_1, \ldots, u_n) = F,$$

with u_i of degree ≤ 2 over $\mathbb{Q}(u_1, \ldots, u_{i-1})$; and $[F : \mathbb{Q}]$ will be a power of 2 by the multiplicativity of degrees (Proposition 13.13). This will prove the statement.

Therefore, we just have to prove the claim. This is really just analytic geometry! In fact, most of the needed constructions do not require adding elements to E (in which case we may take $u = 1$). For example, if (x_0, y_0) and (x_1, y_1) are two points (and x_0, y_0, x_1, y_1 are all in a field E), then the circle with center at (x_0, y_0) and through (x_1, y_1) has equation

$$(x - x_0)^2 + (y - y_0)^2 = (x_1 - x_0)^2 + (y_1 - y_0)^2$$

and all the coefficients in this equation lie in E because they are just (very simple) polynomials in x_0, x_1, y_0, y_1.

The operations which do require enlarging the field are those involving intersecting with circles. For example, the intersection of a line and a circle with equations as above amounts to solving the system

$$\begin{cases} x^2 + y^2 + ax + by + c = 0 \\ y = mx + r, \end{cases}$$

which leads to the equation

$$x^2 + (mx + r)^2 + ax + b(mx + r) + c = 0,$$

that is,

$$(m^2 + 1)x^2 + (2mr + bm + a)x + (r^2 + br + c) = 0.$$

The x-coordinates of the intersection points satisfy this equation. By the quadratic formula, these numbers are given by the uninspiring expression

$$\frac{-(2mr + bm + a) \pm \sqrt{(2mr + bm + a)^2 - 4(m^2 + 1)(r^2 + br + c)}}{2(m^2 + 1)},$$

barring copying mistakes on my part. Even such mistakes would not affect the only content we need, which is that these numbers belong to $E(u)$, where u is the square root appearing in the quadratic formula. If

$$(2mr + bm + a)^2 - 4(m^2 + 1)(r^2 + br + c)$$

is a perfect square in E, then u has degree 1 over E; if not, then u has degree 2 and the statement is verified in this case. Indeed, we have verified that the x-coordinates of the new points are in $E(u)$, and the y-coordinates satisfy $y = mx + r$ with $m, r \in E$, so they are in $E(u)$ as well.

It might seem that the same method would require the intersection of two *circles* to introduce higher-degree numbers, but in fact the system

$$\begin{cases} x^2 + y^2 + a_1x + b_1y + c_1 = 0 \\ x^2 + y^2 + a_2x + b_2y + c_2 = 0 \end{cases}$$

is equivalent to the system

$$\begin{cases} x^2 + y^2 + a_1x + b_1y + c_1 = 0 \\ (a_2 - a_1)x + (b_2 - b_1)y + (c_2 - c_1) = 0 \end{cases}$$

so this case may be treated in the same way as the previous one. □

COROLLARY 13.40 *No construction by straightedge and compass can 'square the circle', construct the side of a cube of volume 2, or trisect an arbitrary angle.*

Proof Assume we have constructed the radius r of a circle. The area of the circle is πr^2, so the side of a square with the same area is $r\sqrt{\pi}$. Therefore, squaring the circle requires constructing $r\sqrt{\pi}$, and in particular both r and $r\sqrt{\pi}$ would be algebraic over $\mathbb{Q}$; and it would follow that $\sqrt{\pi}$ is an algebraic number. But if $\sqrt{\pi}$ were algebraic, then so would be π; and π is *not* algebraic, as was mentioned in Example 13.22. So $r\sqrt{\pi}$ is not constructible, and this proves that our circle cannot be 'squared'. Constructing the side of a cube of volume 2 means constructing $\sqrt[3]{2}$; the degree of this number over $\mathbb{Q}$ is 3, not a power of 2, since $t^3 - 2$ is irreducible over $\mathbb{Q}$. (You have already thought along these lines if you worked out Exercise 13.10.) If one could trisect arbitrary angles, one could construct the angle 20°, and as we have seen, this would mean constructing a root of $8t^3 - 6t - 1$. Since $8t^3 - 6t - 1$ is irreducible over $\mathbb{Q}$, this would again be a number of degree 3 over $\mathbb{Q}$, contradicting Theorem 13.38. □

A more thorough analysis can refine Theorem 13.38 considerably. It is not hard to show that the set of constructible numbers is actually a *subfield* of $\mathbb{R}$, and it can be characterized as the set of real numbers α that are elements of an extension $\mathbb{Q}(u_1, \ldots, u_n)$, where each u_i has degree ≤ 2 over $\mathbb{Q}(u_1, \ldots, u_{i-1})$. A list of the needed numbers u_i can be turned into an explicit construction by straightedge and compass giving α. So this whole business has to do with certain intermediate fields of certain extensions of $\mathbb{Q}$. The 'Galois correspondence' mentioned at the end of §13.2 may be used in studying specific constructibility questions, and we will come back to this when we learn more about this correspondence (see, e.g., Proposition 15.55). For example, it can be shown that a regular n-gon is constructible *if and only if* n is a power of 2 times a product of distinct 'Fermat primes': in fact, this will be the last exercise of the last chapter of this last part of this text! (Exercise 15.30). Fermat primes are primes of the form $2^{(2^r)} + 1$: the first five are $3, 5, 17, 257, 65537$, and ... *no other number of the form* $2^{(2^r)} + 1$ *is known to be prime.* Prove that there are no other 'Fermat primes' (or find a new one), and you will become famous.

Exercises

13.1   Does the category of fields have a final object? (Cf. Exercise 4.14.)

13.2   Prove that every homomorphism $F' \to F''$ between two fields of characteristic 0 is a homomorphism $F' \to F''$ of extensions of $\mathbb{Q}$. That is, show that the corresponding diagram

necessarily commutes.

13.3   Show that in general *not* every homomorphism $F' \to F''$ between two fields is a homomorphism $F' \to F''$ of extensions of some subfield k. For example, consider $F' = F'' = k = \mathbb{Q}[t]/(t^2 - 2)$. Construct a homomorphism $f : \mathbb{Q}[t]/(t^2 - 2) \to \mathbb{Q}[t]/(t^2 - 2)$ such that the diagram

$$\mathbb{Q}[t]/(t^2 - 2) \xrightarrow{\ f\ } \mathbb{Q}[t]/(t^2 - 2)$$

$$\mathrm{id} \uparrow \qquad\qquad \nearrow\ \mathrm{id}$$

$$\mathbb{Q}[t]/(t^2 - 2)$$

does *not* commute.

13.4   Let F be a finite field. Prove that the order of F is a prime power: $|F| = p^r$ for a positive prime and an integer $r \geq 0$. (*Hint:* The prime subfield of F is some finite field $\mathbb{F}_p$, for p prime; and the extension $\mathbb{F}_p \subseteq F$ is finite.)

After working this out, look again at Exercise 7.17.

13.5 ▷ Let $k \subseteq F$ be a finite extension, and let E be one of its intermediate fields. Prove that both $k \subseteq E$ and $E \subseteq F$ are finite extensions.

13.6 Let $k \subseteq F$ be a finite extension, and let E be one of its intermediate fields. Prove that $E = F$ if and only if $[E : k] = [F : k]$.

13.7 ▷ Let $k \subseteq F$ be a finite field extension such that $[F : k]$ is a *prime* integer. Prove that the only intermediate fields of the extension are k and F.

13.8 ▷ Let $k \subseteq F$ be a finite extension, and let $\varphi \colon F \to F$ be a ring homomorphism extending the identity on k, i.e., such that $\forall c \in k, \varphi(c) = c$. Prove that φ is an isomorphism. (Use Exercise 13.6.)

Does the conclusion necessarily hold if $k \subseteq F$ is *not* assumed to be finite?

13.9 ▷ Prove that $\overline{\mathbb{Q}}$ is countable.

13.10 ▷ Let $\mathbb{Q} \subseteq F$ be an extension of degree 2^r for some $r \geq 0$, with F a subfield of $\mathbb{C}$. Prove that F contains *no* cube root of 2.

13.11 Prove that if a rational number is not a perfect square in $\mathbb{Q}$, it is also not a perfect square in $\mathbb{Q}[t]/(t^3 - 2)$.

13.12 Let $k \subseteq F = k(\alpha)$ be a simple algebraic extension. Suppose $\beta \in k(\alpha)$ is a root of the minimal polynomial of α over k. Prove that $k(\beta) = k(\alpha)$.

13.13 Let $k \subseteq k(\alpha)$ be a simple extension of degree p, where $p > 2$ is prime. Prove that α can be written as a polynomial in α^2 with coefficients in k.

13.14 Show that the extension $\mathbb{Q} \subseteq \mathbb{Q}(\sqrt{2}, \sqrt[3]{2})$ is simple, by proving that $\mathbb{Q}(\sqrt{2}, \sqrt[3]{2}) = \mathbb{Q}(\sqrt{2} + \sqrt[3]{2})$. (*Hint:* Let $\alpha = \sqrt{2} + \sqrt[3]{2}$, and note that it suffices to show that $\sqrt{2}$ and $\sqrt[3]{2}$ may be written as rational functions in α. What is $(\alpha - \sqrt{2})^3$?)

13.15 ▷ Prove that the simple extension $k \subseteq k(t)$, with t transcendental over k, has infinitely many intermediate fields.

13.16 ▷ Let $k \subseteq F$ be a finitely generated extension: $F = k(\alpha_1, \ldots, \alpha_n)$. Assume that for $i = 1, \ldots, n$, α_i is algebraic over $k(\alpha_1, \ldots, \alpha_{i-1})$. Prove that $k \subseteq F$ is a finite extension.

13.17 Let $k \subseteq F$ be an algebraic extension, and let R be an 'intermediate ring': $k \subseteq R \subseteq F$. Prove that R is a field.

13.18 Let $k \subseteq F$ be a field extension, and let E be the set of elements of F that are algebraic over k; so E is a field (by Proposition 13.33). Let $\alpha \in F$ be algebraic over E. Prove that $\alpha \in E$.

13.19 ▷ Let $k \subseteq F$ be a field extension, with F *algebraically closed*. (That is: every polynomial in $F[t]$ has a root in F. For example, $\mathbb{C}$ is algebraically closed, see Theorem 7.17.) Let E be the field consisting of all elements of F that are algebraic over k. Prove that E is algebraically closed. (Use Exercise 13.18.)

In particular, $\overline{\mathbb{Q}}$ is algebraically closed.

13.20 ▷ Show how to construct squares and bisect angles by straightedge and compass.

13.21 ▷ Let p and q be positive integers. Explain how to construct the rational number $\frac{p}{q}$.

13.22 Formalize the analytic geometry considerations in the proof of Theorem 13.38.

13.23 Say that a *complex* number is 'constructible' if and only if its real and imaginary components are constructible. Prove that if $\gamma \in \mathbb{C}$ is constructible, then it is algebraic over $\mathbb{Q}$, and its degree over $\mathbb{Q}$ is a power of 2.

13.24 ▷ Prove that $\cos(20°) + i\sin(20°)$ is *not* constructible (in the sense of Exercise 13.23), without using any trigonometry. (*Hint:* The factorization of $t^9 + 1$ in $\mathbb{Q}[t]$ is $(t + 1)(t^2 - t + 1)(t^6 - t^3 + 1)$). This again proves that arbitrary constructible angles cannot be 'trisected'.

13.25 Prove that $\cos(72°) + i\sin(72°)$ is algebraic of degree $4 = 2^2$ over $\mathbb{Q}$. (And indeed, regular pentagons are constructible.)

13.26 Prove that regular 7-gons cannot be constructed by straightedge and compass, again without using trigonometry.

14 Normal and Separable Extensions, and Splitting Fields

14.1 Simple Extensions, Again

Let's get back to simple algebraic extensions, identified in Proposition 13.17: if $k \subseteq F$ is an extension, $\alpha \in F$ is algebraic over k, and $p(t)$ is the minimal polynomial of α, then the extension $k \subseteq k(\alpha)$ is isomorphic to $k \subseteq k[t]/(p(t))$.

We can view this situation a little differently. Suppose k is a field, and $p(x)$ is an *irreducible* polynomial in $k[x]$, which we may assume to be monic. Then *the field $E = k[t]/(p(t))$ is an extension of k which contains a root of $p(x)$.* Indeed, k maps (injectively) to E, by $c \mapsto c + (p(t))$; so the polynomial $p(x) \in k[x]$ may be viewed as a polynomial in $E[x]$. If α denotes the coset $t + (p(t))$ of t, then

$$p(\alpha) = p(t + (p(t))) = p(t) + (p(t)) = 0_E,$$

so that α is a root of $p(x)$ in E. Mod-ing out $k[t]$ by $(p(t))$ is a way to 'solve' the equation $p(x) = 0$, or more accurately to 'add a root' of this polynomial to our field k; this root turns out to be the coset of t. So for example $\mathbb{R}[t]/(t^2 + 1) \cong \mathbb{C}$ contains i, a root of $x^2 + 1$; i corresponds to the coset $t + (t^2 + 1)$. We have been aware of this since our first encounter with the quotient $\mathbb{R}[x]/(x^2 + 1)$ (Example 5.34).

In fact, $k[t]/(p(t))$ is the most efficient way to construct such a root, in the sense that this field will map to *any* field containing roots of the polynomial $p(x)$. More precisely, we have the following result (which sounds in part like yet another 'universal property').

> **PROPOSITION 14.1** *Let $k \subseteq F$ be a field extension, and let $p(x)$ be an irreducible polynomial in $k[x]$. Then for each root λ of $p(x)$ in F there is exactly one embedding (i.e., injective homomorphism)*
>
> $$k[t]/(p(t)) \subseteq F$$
>
> *extending the inclusion $k \subseteq F$ and mapping the coset of t to λ.*
>
> *Conversely, if $k[t]/(p(t)) \subseteq F$ is an embedding extending the inclusion $k \subseteq F$, then the image of the coset of t is a root of $p(x)$ in F.*
>
> *Thus, there is a bijection between the set of roots of $p(x)$ in F and the set of embeddings $k[t]/(p(t)) \subseteq F$ extending the inclusion $k \subseteq F$.*

Proof Let

$$p(x) = a_0 + a_1 x + \cdots + a_d x^d,$$

with coefficients in k; as $k \subseteq F$, we also view $p(x)$ as a polynomial with coefficients in F. Denote by α the coset of t in $k[t]/(p(t))$.

Let λ be a root of $p(x)$ in F. Consider the unique homomorphism

$$\hat{\jmath} \colon k[t] \to F$$

extending the inclusion $k \subseteq F$ and mapping t to λ. That is, $\hat{\jmath}(f(t)) = f(\lambda)$:

$$\hat{\jmath}(b_0 + b_1 t + \cdots + b_r t^r) = b_0 + b_1 \lambda + \cdots + b_r \lambda^r.$$

Since λ is a root of $p(x)$, $\hat{\jmath}(p(t)) = p(\lambda) = 0$: therefore $p(t) \subseteq \ker \hat{\jmath}$. We get (by Proposition 5.39) an induced homomorphism

$$j \colon k[t]/(p(t)) \to F$$

sending $\alpha = t + (p(t))$ to λ. Since $k[t]/(p(t))$ is a field, j must be injective, so we do get an embedding $k[t]/(p(t)) \subseteq F$ as stated.

Conversely, let $j \colon k[t]/(p(t)) \hookrightarrow F$ be a homomorphism extending the inclusion $k \subseteq F$; that is, such that $j(a) = a$ for $a \in k$, where we identify $a \in k$ with its image $a + (p(t))$ in $k[t]/(p(t))$. Then

$$
\begin{aligned}
p(j(\alpha)) &= a_0 + a_1 j(\alpha) + \cdots + a_d j(\alpha)^d \\
&= j(a_0 + a_1 \alpha + \cdots + a_d \alpha^d) = j(p(\alpha)) = j(0) \\
&= 0
\end{aligned}
$$

since α is a root of $p(x)$. Therefore $j(\alpha)$ is a root of $p(x)$ in F, and we are done. ☐

This should all be clear—the arguments leading to these considerations are all very straightforward—and yet something a little counterintuitive is going on. Suppose the polynomial $p(x)$ is irreducible over k, but has *several* roots $\lambda_1, \lambda_2, \ldots$ in some larger field F. By Proposition 14.1, we get *several different* embeddings of the same field $k[t]/(p(t))$ in F: these are the different subfields $k(\lambda_1), k(\lambda_2), \ldots$ of F. These fields are all isomorphic, but may look completely different within F.

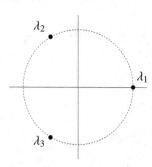

For example, let $p(x) = x^3 - 2 \in \mathbb{Q}[t]$. The field $\mathbb{C}$ contains a full set of roots of $x^3 - 2$: they are the real root $\lambda_1 = \sqrt[3]{2}$ and two complex roots $\lambda_2 = \frac{-1+i\sqrt{3}}{2}\lambda_1$, $\lambda_3 = \frac{-1-i\sqrt{3}}{2}\lambda_1$. By what we have just observed, $\mathbb{Q}(\lambda_1)$, $\mathbb{Q}(\lambda_2)$, and $\mathbb{Q}(\lambda_3)$ are all *isomorphic as fields:* they are just realizations of the field $\mathbb{Q}[t]/(t^3 - 2)$. But these fields look very different within $\mathbb{C}$! The field $\mathbb{Q}(\lambda_1)$ is contained in $\mathbb{R}$, while the fields $\mathbb{Q}(\lambda_2)$ and $\mathbb{Q}(\lambda_3)$ are topologically 'dense' in $\mathbb{C}$: if you are any complex number, some element of the field $\mathbb{Q}(\lambda_2)$ is very close to you.

One lesson to learn here is that we should not let our topological intuition guide us when studying field theory (at least at this level). We also learn that something subtle is going on. In this example, the embeddings $\mathbb{Q}(\lambda_1)$, $\mathbb{Q}(\lambda_2)$ are obviously different as subfields of $\mathbb{C}$. But consider instead the polynomial $x^2 - 2$. This is also irreducible over $\mathbb{Q}$, and it has two roots in $\mathbb{C}$, namely $\sqrt{2}$ and $-\sqrt{2}$. Proposition 14.1 tells us that there are *two* distinct embeddings of $\mathbb{Q}[t]/(t^2 - 2)$ in $\mathbb{C}$: one which sends the coset of t to $\sqrt{2}$, and one which sends the coset of t to $-\sqrt{2}$. This time, however,

$$-\sqrt{2} = -1 \cdot \sqrt{2} \in \mathbb{Q}(\sqrt{2}):$$

that is, $\mathbb{Q}(-\sqrt{2}) = \mathbb{Q}(\sqrt{2})$. The two different embeddings determine *the same* subfield of $\mathbb{C}$.

Therefore, there is a fundamental difference between the extensions $\mathbb{Q} \subseteq \mathbb{Q}(\sqrt{2})$ and $\mathbb{Q} \subseteq \mathbb{Q}(\sqrt[3]{2})$: there is *only one* subfield of $\mathbb{C}$ isomorphic to $\mathbb{Q}(\sqrt{2})$, while there are several subfields of $\mathbb{C}$ isomorphic to $\mathbb{Q}(\sqrt[3]{2})$. *Preview of coming attractions:* The first type of extension will be 'Galois', the second will not.

There is more. Choose one of the embeddings of $\mathbb{Q}[t]/(t^3 - 2)$ in $\mathbb{C}$, say $\mathbb{Q}(\sqrt[3]{2})$. Since $\mathbb{Q}(\sqrt[3]{2}) \subseteq \mathbb{R}$, this embedding does *not* contain the other roots of $x^3 - 2$: the polynomial $x^3 - 2$ does not factor as a product of linear terms in $\mathbb{Q}(\sqrt[3]{2})$. One way to view this is as follows. Let α be any root of $x^3 - 2$. We have

$$x^3 - 2 = (x - \alpha)(x^2 + \alpha x + \alpha^2). \tag{14.1}$$

The polynomial $x^2 + \alpha x + \alpha^2$ is *irreducible* in $\mathbb{Q}(\alpha)$ (can you prove this directly? Exercise 14.2), so we can't factor $x^3 - 2$ any further over $\mathbb{Q}(\alpha)$.

By contrast, if you add $\sqrt{2}$ to $\mathbb{Q}$, that is, one root of $x^2 - 2$, then as pointed out above the *other* root also belongs to the extension: $-\sqrt{2} \in \mathbb{Q}(\sqrt{2})$. This may seem to be an artifact of the circumstance that $x^2 - 2$ has very low degree, but that is not all that is going on.

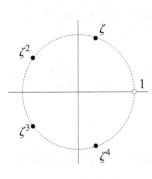

For example, consider the 'cyclotomic' polynomial $p(x) = 1 + x + x^2 + x^3 + x^4$. You may feel that you are not particularly familiar with this polynomial, but in fact you know a lot about it. First of all, you have verified that it is *irreducible* over $\mathbb{Q}$, as a nice application of Eisenstein's criterion. Therefore, the roots of $p(x)$ are the four fifth roots of 1 other than 1 itself: $e^{2\pi i/5}$, $e^{4\pi i/5}$, $e^{6\pi i/5}$, and $e^{8\pi i/5}$. (In the complex plane, they are vertices of a regular pentagon centered at 0.) As a consequence, there are four embeddings of $\mathbb{Q}[t]/(p(t))$ into $\mathbb{C}$. But again, note that once you have $\zeta = e^{2\pi i/5}$, then you get all the other roots

for free:

$$e^{4\pi i/5} = \zeta^2, \quad e^{6\pi i/5} = \zeta^3, \quad e^{8\pi i/5} = \zeta^4.$$

Similarly, if a field contains ζ^2, then it must contain $(\zeta^2)^3 = \zeta^6 = \zeta$, and hence it must contain ζ^3 and ζ^4 as well. A moment's thought reveals that $\mathbb{Q}(\zeta) = \mathbb{Q}(\zeta^2) = \mathbb{Q}(\zeta^3) = \mathbb{Q}(\zeta^4)$: the embeddings differ, but their images *coincide*. The field $\mathbb{Q}(\zeta)$ contains *all* roots of $1 + x + x^2 + x^3 + x^4$: therefore, this polynomial factors as a product of linear polynomials in $\mathbb{Q}(\zeta)[x]$. (Unlike $x^3 - 2$ in $\mathbb{Q}(\sqrt[3]{2})[x]$.) Further, $\mathbb{Q}(\zeta)$ is the smallest field over which this can happen: if a subfield of $\mathbb{C}$ contains the roots of $1 + x + x^2 + x^3 + x^4$, then in particular it must contain ζ, so it must contain $\mathbb{Q}(\zeta)$.

We will be especially interested in extensions of this type: algebraic extensions $k \subseteq F$

such that some polynomial in $k[x]$ 'splits completely' in F, and further such that F is as small as possible subject to this requirement. The extensions $\mathbb{Q} \subseteq \mathbb{Q}(\sqrt{2})$ and $\mathbb{Q} \subseteq \mathbb{Q}(\zeta)$ are such extensions, while $\mathbb{Q} \subseteq \mathbb{Q}(\sqrt[3]{2})$ is not.

A 'Galois extension' will be an extension of this type for some polynomial $f(x)$ in $k[x]$, and satisfying one further technical condition which we will encounter in a little while (§14.4).

14.2   Splitting Fields

First, let's focus on the requirement that 'polynomials split completely'. We will say that $f(x)$ 'splits' over a field F if it is a product of linear terms in $F[x]$: $f(x) = c \prod_{i=1}^{d}(x-\alpha_i)$, for $c \neq 0$ in k and $\alpha_1, \ldots, \alpha_d$ in F.

DEFINITION 14.2   A field extension $k \subseteq F$ is a 'splitting field' for a polynomial $f(x) \in k[x]$ if $f(x)$ factors as a product of linear terms in $F[x]$: $f(x) = c \prod_{i=1}^{d}(x - \alpha_i)$, and further $F = k(\alpha_1, \ldots, \alpha_d)$.

It is immediately clear that splitting fields exist for polynomials over, e.g., $\mathbb{Q}$. Indeed, if $f(x) \in \mathbb{Q}[x]$, then as a consequence of the Fundamental Theorem of Algebra $f(x)$ splits *over* $\mathbb{C}$ (Corollary 7.19): $f(x) = c \prod_{i=1}^{d}(x - \lambda_i)$ with $\lambda_i \in \mathbb{C}$. The subfield $\mathbb{Q}(\lambda_1, \ldots, \lambda_d)$ of $\mathbb{C}$ is then a splitting field for $f(x)$.

In fact, the same reasoning works over *every* field: one can prove that every field may be embedded in some algebraically closed field. In fact, we have the following more precise result.

THEOREM 14.3   *Let k be a field. Then there exists an algebraic field extension $k \subseteq \overline{k}$ such that $\overline{k}$ is algebraically closed. This extension is unique up to isomorphism.*

The field $\overline{k}$ has a name.

DEFINITION 14.4   The unique (up to isomorphism) algebraically closed field $\overline{k}$ that is algebraic over k is the 'algebraic closure' of k.

This field will play an important role in the development of the theory. It follows from the definition that every polynomial $f(x) \in k[x]$ has roots in $\overline{k}$, and that is useful in itself, as already pointed out above.

The proof of Theorem 14.3 is not difficult, but it is somewhat involved, and I will omit it. The 'hard' part is the construction of an algebraically closed field K containing k; again, for the case that we may be most interested in, i.e., $k = \mathbb{Q}$, we do not need to do any work as we can take $K = \mathbb{C}$. Once this field is constructed, the algebraic closure $\overline{k}$ may be defined as the set of elements of K which are algebraic over k; see Exercise 13.19. A little more work is needed to prove that the algebraic closure is unique up to isomorphism.

The requirement that $k \subseteq \bar{k}$ be algebraic makes $\bar{k}$ the 'smallest' algebraically closed field containing k. This is not hard to see, and a good exercise for you (Exercise 14.5).

If we accept Theorem 14.3, then it is clear that splitting fields always exist: if k is any field, and $f(x) \in k[x]$ is any polynomial, then $f(x)$ splits over the algebraic closure $\bar{k}$: $f(x) = c \prod_{i=1}^{d}(x - \lambda_i)$ with $\lambda_i \in \bar{k}$. Then $k(\lambda_1, \ldots, \lambda_d)$ is a splitting field for $f(x)$.

Example 14.5   The algebraic closure of $\mathbb{Q}$ is $\overline{\mathbb{Q}}$ (Example 13.34). Indeed, in this case we do know of an algebraically closed field containing $\mathbb{Q}$, namely $\mathbb{C}$; since $\overline{\mathbb{Q}}$ consists of all complex numbers that are algebraic over $\mathbb{Q}$, $\overline{\mathbb{Q}}$ is algebraically closed and the extension $\mathbb{Q} \subseteq \overline{\mathbb{Q}}$ is algebraic (cf. Exercise 13.19). Note that the complex roots λ_i of a polynomial $f(x) \in \mathbb{Q}[x]$ lie in $\overline{\mathbb{Q}}$, so $\overline{\mathbb{Q}}$ contains the splitting field $\mathbb{Q}(\lambda_1, \ldots, \lambda_d)$.

We can actually give a construction of a splitting field which does not rely on the existence of an algebraic closure, and in the process we obtain an estimate of 'how big' a splitting field may be.

THEOREM 14.6   *Let k be a field, and let $f(x) \in k[t]$ be a polynomial of degree d. Then there exists a splitting field $k \subseteq F$ for $f(x)$, and further $[F : k] \leq d!$.*

Proof   Note that we may assume $f(x)$ to be monic; indeed, if $c \in k$ is the leading coefficient of $f(x)$, then an extension $k \subseteq F$ is a splitting field for $f(x)$ if and only if it is a splitting field for $c^{-1}f(x)$, which is monic.

To prove the statement, we argue by induction on d. The statement is true for $d = 1$: if $f(x) = x - c$ with $c \in k$, then $k = k(c)$ is a splitting field for $f(x)$, of degree $1 = 1!$ over k.

For $d > 1$, assume that the statement holds for all fields and all polynomials of degree $< d$. Given $f(x) \in k[x]$ monic of degree d, let $p(x)$ be a monic irreducible factor of $f(x)$, and let $E = k[t]/(p(t))$; so $E = k(\alpha_1)$, where α_1 is a root of $p(x)$ in E. It follows that $p(x)$, and hence $f(x)$, has a factor $x - \alpha_1$ in $E[x]$: $f(x) = (x - \alpha_1)g(x)$ in $E[x]$. The degree of $g(x)$ is $d - 1$, so by the induction hypothesis there exists a splitting field extension $E \subseteq F$: $g(x)$ factors as

$$g(x) = \prod_{i=2}^{d}(x - \alpha_i)$$

in $F[x]$, $F = E(\alpha_2, \ldots, \alpha_d)$, and $[F : E] \leq (d - 1)!$. We are done, since

$$f(x) = (x - \alpha_1)g(x) = (x - \alpha_1)\prod_{i=2}^{d}(x - \alpha_i) = \prod_{i=1}^{d}(x - \alpha_i)$$

splits completely in $F[x]$, $F = F(\alpha_1)(\alpha_2, \ldots, \alpha_d) = F(\alpha_1, \ldots, \alpha_d)$, and

$$[F : k] = [F : E][E : k] = [F : E] \cdot d \leq (d - 1)! \cdot d = d!$$

as needed.   □

Example 14.7 As always, the best way to absorb a moderately complex argument is to run it on a concrete example. Take the polynomial $x^3 - 2 \in \mathbb{Q}[x]$. In order to construct a splitting field by the method shown in the proof of Theorem 14.6, we first look for an irreducible factor of $x^3 - 2$; and since $x^3 - 2$ is irreducible over $\mathbb{Q}$, we can choose $x^3 - 2$ itself. Then we let $E = k[t]/(t^3 - 2)$, and consider the polynomial $x^3 - 2$ *as a polynomial in* $E[x]$. By construction, E contains a root α_1 of $x^3 - 2$, so

$$x^3 - 2 = (x - \alpha_1) g(x);$$

in fact, we already observed in §14.1 that if α_1 is a root of $x^3 - 2$, then

$$x^3 - 2 = (x - \alpha_1)(x^2 + \alpha_1 x + \alpha_1^2)$$

(see (14.1)), and that the polynomial $g(x) = x^2 + \alpha_1 x + \alpha_1^2$ is *irreducible* in $E[x]$. We then let $F = E[u]/(u^2 + \alpha_1 u + \alpha_1^2)$. This field may be viewed as $E(\alpha_2)$, where α_2 is a root of $u^2 + \alpha_1 u + \alpha_1^2$ (and hence another root of $x^3 - 2$, different from α_1). In fact, since F contains one root of this degree-2 polynomial, the remaining factor has degree 1, i.e., the polynomial must factor completely in $F[x]$:

$$u^2 + \alpha_1 u + \alpha_1^2 = (u - \alpha_2)(u - \alpha_3)$$

with α_2, α_3 *both* in F. Then we have

$$x^3 - 2 = (x - \alpha_1)(x - \alpha_2)(x - \alpha_3)$$

in $F[x]$, and $F = E(\alpha_2, \alpha_3) = k(\alpha_1, \alpha_2, \alpha_3)$. Therefore F is a splitting field for $x^3 - 2$. Further,

$$[F : k] = [F : E][E : k] = 2 \cdot 3 = 6 = 3!\,,$$

so this degree is as large as it can be according to Theorem 14.6.

Of course, the degree of a splitting field for $f(x) \in k[x]$ may be much less than $(\deg f)!$. For one thing, if $k \subseteq F$ is a splitting field for $f(x) \in k[x]$ and E is *any* intermediate field, then F is also a splitting field for $f(x)$ over E, and its degree over E is strictly less than $[F : k] \leq d!$ if $k \subsetneq E$. As an extreme example, if F is a splitting field for $f(x)$ over k, it is also a splitting field for $f(x)$ *over* F, and its degree over F is 1.

If $f(x)$ is irreducible over k, then the degree of a splitting field for $f(x)$ over k will be *at least* $\deg f$: indeed, the splitting field must contain a simple extension $k(\alpha)$, where α is a root of $f(x)$, and this extension has degree $\deg f$ over k by Proposition 13.17.

Example 14.8 This is the case for the polynomial $f(x) = x^4 + x^3 + x^2 + x^2 + 1 \in \mathbb{Q}[x]$, as we have seen in §14.1. We found that the 'first step' in the construction of a splitting field, that is, $E = \mathbb{Q}[t]/(t^4 + t^3 + t^2 + t + 1)$, is already a splitting field for $f(x)$: $E = \mathbb{Q}(\zeta)$ with notation as in §14.1, and all the roots of $f(x)$ are polynomials in ζ (and in fact they are simply $\zeta, \zeta^2, \zeta^3, \zeta^4$).

In this case, the degree of the splitting field is $[E : k] = 4 = \deg f$.

The inductive construction given in Theorem 14.6 appears to depend on some arbitrary choice—for example, on the irreducible factor $p(x)$ extracted from $f(x)$ at the beginning of the construction. Remarkably, the end result is actually independent of these choices, in the sense that the splitting field of a polynomial is *unique up to isomorphism*. This is particularly clear if we accept the realization of a splitting field as a subfield of the algebraic closure of k.

PROPOSITION 14.9   *Let k be a field, and let $\bar{k}$ be its algebraic closure. Let $f(x) \in k[x]$ be a polynomial, and let $f(x) = c \prod_{i=1}^{d}(x - \lambda_i)$ be the splitting of $f(x)$ in $\bar{k}[x]$. Further, let F be a splitting field for $f(x)$ according to Definition 14.2. Then there exists a homomorphism $F \to \bar{k}$ extending the inclusion $k \subseteq \bar{k}$ and whose image is $k(\lambda_1, \ldots, \lambda_d)$.*

Proof   We prove that, with notation as in the statement and in Definition 14.2, we can define a homomorphism $F \to \bar{k}$ extending the inclusion $k \subseteq \bar{k}$ and mapping α_i to λ_i for $i = 1, \ldots, d$, possibly after a reordering of the elements λ_i.

For this, we again argue by induction on the degree d of $f(x)$. If $d = 1$, then $f(x) = x - c$ for some $c \in k$; the splitting field is k itself, and there is nothing to prove.

For $d > 1$, let $F = k(\alpha_1, \ldots, \alpha_d)$ be a splitting field for $f(x)$ according to Definition 14.2. Then α_1 is algebraic over k; the minimal polynomial of α_1 is an irreducible factor $p(t)$ of $f(t)$. The subfield $k(\alpha_1)$ of F is isomorphic to $k[t]/(p(t))$, and α_1 corresponds to the coset of t under this isomorphism. As $p(t)$ is a factor of $f(t)$, one of the roots of $f(t)$ in $\bar{k}$ is a root of $p(t)$; up to a reordering, we may assume that this root is λ_1. By Proposition 14.1, we have a homomorphism from $k(\alpha_1) \cong k[t]/(p(t))$ to $\bar{k}$, mapping α_1 to λ_1. We may use this homomorphism to identify $k(\alpha_1)$ with $k(\lambda_1)$, and hence view $\bar{k}$ as an extension of $k(\alpha_1)$; note that $\bar{k}$ is also an algebraic closure of $k(\alpha_1) = k(\lambda_1)$ (Exercise 14.6).

Now, $f(x) = (x - \alpha_1)g(x)$ in $k(\alpha_1)[x]$. By definition, the extension $k(\alpha_1) \subseteq F$ is a splitting field for $g(x)$; $g(x)$ factors as $\prod_{i=2}^{d}(x - \alpha_i)$ in $F[x]$, and as $\prod_{i=2}^{d}(x - \lambda_i)$ in $\bar{k}[x]$. Since $\deg g = d - 1 < d$, by the induction hypothesis we have an embedding homomorphism $F \to \bar{k}$ mapping α_i to λ_i for $i = 2, \ldots, d$ (possibly after a reordering of the roots λ_i) and extending the inclusion $k(\alpha_1) \subseteq \bar{k}$. This homomorphism satisfies the requirement in the statement, and we are done.   □

COROLLARY 14.10   *The splitting field of a polynomial $f(x) \in k[x]$ is unique up to isomorphism. In fact, let $k \subseteq F_1$, $k \subseteq F_2$ be two splitting fields for $f(x)$. Then there is an isomorphism $\iota : F_1 \to F_2$ extending the identity on k.*

Proof   By the proposition, we have isomorphisms $j_1 : F_1 \to k(\lambda_1, \ldots, \lambda_d)$, $j_2 : F_2 \to k(\lambda_1, \ldots, \lambda_d)$ extending the inclusion of k. Then $\iota = j_2^{-1} \circ j_1 : F_1 \to F_2$ is an isomorphism

extending the identity on k,

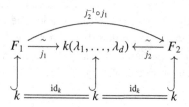

as needed.   □

We will in fact need the following minor upgrade of Corollary 14.10.

COROLLARY 14.11   *Let $i\colon k_1 \to k_2$ be an isomorphism of fields; let $f_1(x) \in k_1[x]$ be a polynomial, $f(x) = a_0 + \cdots + a_d x^d$, and let $f_2(x) = i(a_0) + \cdots + i(a_d)x^d \in k_2[x]$ be its image in $k_2[x]$.*

Let F_1 be a splitting field for $f_1(x)$, and let F_2 be a splitting field for $f_2(x)$. Then there exists an isomorphism $\iota\colon F_1 \to F_2$ extending $i\colon k_1 \to k_2$.

Proof   The composition $k_1 \to k_2 \subseteq F_2$ lets us view F_2 as an extension of k_1, and as such F_2 is also a splitting field for $f_1(x)$. By Corollary 14.10 there is an isomorphism $\iota\colon F_1 \to F_2$ such that the diagram

$$
\begin{array}{ccc}
k_1 & \xrightarrow{\ i\ } k_2 \hookrightarrow & F_2 \\
\big\| & & \uparrow{\scriptstyle \iota} \\
k_1 & \hookrightarrow & F_1
\end{array}
$$

commutes, and this is the statement.   □

Remark 14.12   We can complement Proposition 14.9 with the observation that in fact the image of *every* homomorphism of the splitting field F of $f(x)$ to $\bar{k}$ extending the inclusion of k into $\bar{k}$ must be the subfield $k(\lambda_1, \ldots, \lambda_r)$ generated by the roots of $f(x)$. Indeed, the image of such a homomorphism must contain all roots of $f(x)$, since $f(x)$ splits in $F[x]$, and must be generated by them since F is.

This explains one difference we noticed in §14.1 between $\mathbb{Q}(\sqrt[3]{2})$ (which is not a splitting field) and $\mathbb{Q}(\sqrt{2})$ or $\mathbb{Q}[t]/(p(t))$ with $p(t) = 1 + t + t^2 + t^3 + t^4$: we found several distinct images of $\mathbb{Q}(\sqrt[3]{2})$ in $\mathbb{C}$, and hence in the algebraic closure $\overline{\mathbb{Q}}$ of $\mathbb{Q}$, while different embeddings of, e.g., $\mathbb{Q}[t]/(p(t))$ all had the same image. The reason is that both $\mathbb{Q}(\sqrt{2})$ and $\mathbb{Q}[t]/(p(t))$ are splitting fields.   ⌟

Let's give one more look at the examples we studied in §14.1, from the point of view of the embeddings in an algebraic closure. Can we compute the number of such embeddings?

Example 14.13   Let $p(x) = 1 + x + x^2 + x^3 + x^4$. We can summarize what we know already from previous encounters:

- The splitting field E for $p(t)$ is $\mathbb{Q}[t]/(p(t))$.
- $[E : \mathbb{Q}] = 4$.
- There are four distinct embeddings of E into $\overline{\mathbb{Q}}$, and their image is the subfield $\mathbb{Q}(\zeta)$ generated by a complex fifth root $\zeta \neq 1$ of 1.

Indeed, in §14.1 we found that there are four embeddings of E into $\mathbb{C}$; they are contained in $\overline{\mathbb{Q}}$ since ζ is algebraic over $\mathbb{Q}$.

Example 14.14   As for $x^3 - 2$, we have constructed a splitting field in Example 14.7. The first step in the construction is $E = \mathbb{Q}[t]/(t^3 - 2)$; this can be mapped to $\overline{\mathbb{Q}}$ in three distinct ways—one for each of the three complex roots $\lambda_1, \lambda_2, \lambda_3$ of $x^3 - 2$. Each of these embeddings $E \subseteq \overline{\mathbb{Q}}$ contains *one* of the three roots, the image of the coset α of t. The second stage in the construction is $F = E[u]/(u^2 + \alpha u + \alpha^2)$. For each of the three embeddings $\iota\colon E \hookrightarrow \overline{\mathbb{Q}}$, there are two ways to map F to $\overline{\mathbb{Q}}$ extending ι, according to the choice of a *second* root. The image of F in $\overline{\mathbb{Q}}$ consists of $\mathbb{Q}(\alpha_1, \alpha_2, \alpha_3)$. In total, there are $3 \cdot 2 = 6$ embeddings $F \hookrightarrow \overline{\mathbb{Q}}$.

Summarizing:

- The splitting field F for $x^3 - 2$ is $E[u]/(u^2 + \alpha u + \alpha^2)$, where $E = \mathbb{Q}[t]/(t^3 - 2)$ and $\alpha \in E$ is the coset $t + (t^3 - 2)$.
- $[F : \mathbb{Q}] = 6$.
- There are six distinct embeddings of F into $\overline{\mathbb{Q}}$. Their image is the subfield $\mathbb{Q}(\lambda_1, \lambda_2, \lambda_3)$ generated by the three complex roots of $t^3 - 2$.

Incidentally, we can realize this splitting field (even) more explicitly. In $\mathbb{C}$, we have $\lambda_1 = \sqrt[3]{2}$ and $\lambda_2 = \frac{-1+i\sqrt{3}}{2}\lambda_1$, $\lambda_3 = \frac{-1-i\sqrt{3}}{2}\lambda_1$. It follows that $\mathbb{Q}(\lambda_1, \lambda_2, \lambda_3) \subseteq \mathbb{Q}(\sqrt[3]{2}, i\sqrt{3})$, and in fact this must be an equality since both fields have degree 6 over $\mathbb{Q}$.

These examples capture important features of the theory. First, note that in both cases the degree of the extension turns out to be the same as the number of distinct embeddings of the splitting field in an algebraic closure. This mysterious coincidence will turn out to be a general feature of 'separable' extensions, as we will prove when the time comes (Corollary 14.25).

Second, in Example 14.14 the number of embeddings is the maximum allowed by Theorem 14.6, i.e., $\deg(f(t))!$. We saw that in this example the roots are in some sense independent of each other: we get an embedding by choosing arbitrarily one of the three roots of $x^3 - 2$, and then by choosing arbitrarily one of the remaining two. In Example 14.6, the story is different: once a root of $1 + x + x^2 + x^3 + x^4$ is chosen (for example, ζ), then the images of the others are determined (they must be $\zeta^2, \zeta^3, \zeta^4$). The roots are not independent of one another, and ultimately this is what limits the number of different embeddings of the splitting field in $\overline{\mathbb{Q}}$.

A way to interpret this situation is that in the example of $x^3 - 2$, we are allowed to permute the three roots at will: if $\alpha_1, \alpha_2, \alpha_3$ are the three roots of $x^3 - 2$ in the splitting field F, we can obtain an embedding $F \subseteq \overline{\mathbb{Q}}$ by mapping these roots in any of the 3! possible ways to the roots $\lambda_1, \lambda_2, \lambda_3$ of $x^3 - 2$ in $\overline{\mathbb{Q}}$. In the example of $1 + x + x^2 + x^3 + x^4$, this is not the case. The roots $\alpha_1, \ldots, \alpha_4$ of this polynomial in its splitting field are all powers of one root: for example, we can let $\alpha_2 = \alpha_1^2$, etc. These relations imply that if an embedding maps α_1 to $\zeta = e^{2\pi i/5}$, then necessarily α_2 must map to ζ^2, etc. We cannot send the roots α_i to the roots λ_i according to an arbitrary permutation. The allowed permutations must preserve certain relations among the roots.

Summarizing, the different embeddings of the splitting field of a polynomial $f(x)$ are constrained by different 'symmetries' among the roots of $f(x)$.

Galois theory will fully explain these experimental observations. The very core of Galois theory is the powerful idea that the possible symmetries among the roots of a polynomial capture essential features of that polynomial. This will be our guiding theme in Chapter 15.

14.3   Normal Extensions

The following definition appears to be a substantial strengthening of the notion of splitting field.

DEFINITION 14.15   An algebraic field extension $k \subseteq F$ is 'normal' if whenever F contains a root of an irreducible polynomial $p(x) \in k[x]$, then $p(x)$ splits in F.

This looks like a very strong requirement! In §14.2 we have learned how to split *one* polynomial; in normal extensions, *every* polynomial that has a chance of doing so *must* split. If $k \subseteq F$ is a normal algebraic extension and $\alpha \in F$, then the minimal polynomial of α must split over F, so *all* other roots of this polynomial must belong to F.

Amazingly, splitting fields do satisfy this strong requirement!

THEOREM 14.16   *A field extension* $k \subseteq F$ *is a splitting field for some polynomial* $f(x) \in k[x]$ *if and only if it is a* finite normal extension.

Proof ($\Leftarrow$) This implication is not surprising. If $k \subseteq F$ is finite, then $F = k(\alpha_1, \ldots, \alpha_r)$ for some elements $\alpha_i \in F$. Let $p_i(x) \in k[x]$ be the minimal polynomial of α_i. As $k \subseteq F$ is normal, each $p_i(x)$ splits over F, and therefore so does $f(x) = \prod_{i=1}^r p_i(x)$. It follows that F is a splitting field for $f(x)$.

($\Rightarrow$) *This* implication is surprising, at least to me. Let $k \subseteq F$ be a splitting field for a polynomial $f(x) \in k[x]$. In particular, it is a finite extension; and by Proposition 14.9 we can in fact identify F with the subfield $k(\lambda_1, \ldots, \lambda_d)$ of $\bar{k}$ generated by all roots of $f(x)$. Let $p(x) \in k[x]$ be an irreducible polynomial, and assume that F contains a root α of $p(x)$. If β is *another* root of $p(x)$ in $\bar{k}$, we have to show that $\beta \in F$. We make several observations, which lead to a proof of this fact.

- The two simple extensions $k \subseteq k(\alpha)$, $k \subseteq k(\beta)$ (within $\bar{k}$) are both isomorphic to $k[t]/(p(t))$, and in particular there is an isomorphism $k(\alpha) \xrightarrow{\sim} k(\beta)$ extending the identity on k and sending α to β. Therefore, $[k(\beta):k] = [k(\alpha):k]$
- Since F is a splitting field of $f(x)$ over k, and $k(\alpha) \subseteq F$, F is a splitting field of $f(x)$ *over* $k(\alpha)$.

- On the other hand, $F(\beta) = k(\beta, \lambda_1, \ldots, \lambda_d) \subseteq \overline{k}$ is a splitting field for $f(x)$ *over* $k(\beta)$: the polynomial $f(x)$ splits in $F(\beta)$ since it splits in F, and its roots generate $F(\beta)$ over $k(\beta)$ since they generate F over k.
- We are then in the situation presented in Corollary 14.11: $F = F(\alpha)$ and $F(\beta)$ are splitting fields for $f(x)$ over $k(\alpha)$, $k(\beta)$, respectively. By Corollary 14.11, there is an isomorphism $\iota \colon F \to F(\beta)$ extending the isomorphism $k(\alpha) \overset{\sim}{\to} k(\beta)$:

$$
\begin{array}{ccccc}
k & \lhook\joinrel\longrightarrow & k(\beta) & \lhook\joinrel\longrightarrow & F(\beta) \\
\| & & \big\uparrow{\scriptstyle\sim} & & \big\uparrow{\scriptstyle\iota} \\
k & \lhook\joinrel\longrightarrow & k(\alpha) & \lhook\joinrel\longrightarrow & F = F(\alpha).
\end{array}
$$

- It follows that $[F(\beta) : k] = [F : k]$.
- We also have the inclusion

$$F = k(\lambda_1, \ldots, \lambda_d) \subseteq k(\beta, \lambda_1, \ldots, \lambda_d) = F(\beta)$$

within $\overline{k}$: $k \subseteq F \subseteq F(\beta)$. By multiplicativity of degrees (Proposition 13.13),

$$[F(\beta) : k] = [F(\beta) : F][F : k].$$

- Therefore $[F(\beta) : F] = 1$, that is, $F(\beta) = F$. In particular $\beta \in F$, and we are done. □

Remark 14.17 This is possibly one of the trickiest argument in this whole text. We need to obtain some information about the inclusion $F \subseteq F(\beta)$ within $\overline{k}$. The isomorphism $\iota \colon F \to F(\beta)$ constructed in the proof is *not* this inclusion; but the fact that there is an isomorphism is enough to deduce that the dimensions of F and $F(\beta)$ as k-vector spaces are equal. And then the inclusion $F \subseteq F(\beta)$ must necessarily be an equality, by dimension considerations. ⌟

Example 14.18 Let $\zeta = e^{2\pi i/5}$, as in §14.1, and let $g(x) \in \mathbb{Q}[x]$ be any polynomial. Then I claim that (i) the complex number $\rho = g(\zeta)$ satisfies an irreducible polynomial $q(x) \in \mathbb{Q}[x]$ of degree at most 4; and (ii) *every* other root of $q(x)$ may be written as a polynomial in ζ with rational coefficients.

Indeed, we have seen that $\mathbb{Q}(\zeta)$ is a splitting field over $\mathbb{Q}$, and $[\mathbb{Q}(\zeta) : \mathbb{Q}] = 4$ (Example 14.13). By assumption $\rho \in \mathbb{Q}(\zeta)$, so $[\mathbb{Q}(\rho) : \mathbb{Q}] \le 4$, giving (i). By Theorem 14.16 the extension $\mathbb{Q} \subseteq \mathbb{Q}(\zeta)$ is normal, and this gives (ii).

It is rather satisfying that we can draw such precise conclusions *without* doing any computation. Even if you were in the mood to compute, how would you go about *finding* a polynomial satisfied by something like $\rho = 1 + 3\zeta^2 + 3\zeta^3$? And how on earth could you say anything *a priori* about its *other* roots?

These will actually be perfectly reasonable questions once we learn a little more (cf. Example 15.26).

14.4   Separable Extensions; and Simple Extensions Once Again

We need one more ingredient before we can tackle Galois theory.

DEFINITION 14.19   Let k be a field. A polynomial $f(x) \in k[x]$ is 'separable' if its factorization over its splitting field (or, equivalently, over $\bar{k}$) has no multiple factors.

An algebraic extension $k \subseteq F$ is 'separable' if the minimal polynomial of every $\alpha \in F$ is separable.

In other words, $f(x)$ is separable if it has $\deg f$ distinct roots in its splitting field. If $f(x)$ is *not* separable, it is said to be 'inseparable'.

An obvious way to produce inseparable polynomials, over any field, is simply to take a power of any nonconstant polynomial: x^2 is inseparable. It may come as a surprise, however, that there exist *irreducible* inseparable polynomials: a polynomial $f(x)$ may be irreducible over a field k, and then be a perfect power over the splitting field.

Example 14.20   Consider the field $\mathbb{F}_2(u)$ of *rational functions* over $\mathbb{F}_2$: this is the field of fractions of the polynomial ring $\mathbb{F}_2[u]$, a transcendental simple extension of $\mathbb{F}_2$ (Proposition 13.17). Let $f(x) = x^2 - u$. This polynomial is *irreducible* over $\mathbb{F}_2(u)$, since there is no rational function whose square is u (Exercise 14.13).

The splitting field for this polynomial is $F = \mathbb{F}_2(u)[t]/(t^2 - u)$. The coset $\underline{t}$ of t is a root of $x^2 - u$, that is, $\underline{t}^2 = u$. But now something remarkable happens:

$$(x - \underline{t})^2 = x^2 - 2x\underline{t} + \underline{t}^2 = x^2 - u \tag{14.2}$$

since $2 = 0$ in F. That is, $f(x)$ has a multiple factor in its splitting field, and a single root! It is irreducible *and* inseparable.

The same construction can be performed over $\mathbb{F}_p$ for every prime $p > 0$, and shows that $x^p - u$ is irreducible and inseparable in $\mathbb{F}_p(u)[t]$.

Luckily, verifying whether a polynomial is separable or inseparable does *not* require constructing its splitting field as in Example 14.20. Define the following formal 'derivative': for

$$f(x) = a_0 + a_1 x + a_2 x^2 + \cdots + a_n x^n \in k[x],$$

let

$$f'(x) = a_1 + 2a_2 x + \cdots + na_n x^{n-1}.$$

Of course we are not appealing to any limit (as we do in calculus) in order to define f'. Still, this operation satisfies the basic formal properties of a derivative: $(f + g)' = f' + g'$ and the Leibniz rule $(fg)' = f'g + fg'$ (Exercise 14.12). One difference with respect to ordinary derivatives is that the derivative of a polynomial may be identically 0 even if the polynomial is not constant: for example, the derivative of $x^3 \in \mathbb{F}_3[x]$ is 0, because 3 equals 0 in $\mathbb{F}_3$.

PROPOSITION 14.21   *A polynomial $f(x)$ is separable if and only if the gcd of $f(x)$ and $f'(x)$ is 1.*

Remark 14.22   For this, it is good to keep in mind that if the gcd of two polynomials $f(x), g(x) \in k[x]$ is 1, then it is also 1 if $f(x), g(x)$ are viewed as polynomials in $F[x]$ for a larger field F. Indeed, if $\gcd(f(x), g(x)) = 1$ in $k[x]$, then there exist polynomials $\alpha(x), \beta(x) \in k[x]$ such that $\alpha(x)f(x) + \beta(x)g(x) = 1$. Since $\alpha(x), \beta(x)$ are also polynomials in $F[x]$, $\gcd(f(x), g(x)) = 1$ in $F[x]$ as well. (You proved a natural generalization of this fact in Exercise 6.26.)   ⌐

Proof   Let F be the splitting field of $f(x)$.

Assume that $f(x)$ is *inseparable*, so that $f(x)$ has a multiple factor in $F[x]$: there exists $\alpha \in F$, $g(x) \in F[x]$, such that $f(x) = (x - \alpha)^2 g(x)$. Therefore

$$f'(x) = 2(x - \alpha)g(x) + (x - \alpha)^2 g'(x),$$

and as a consequence $f(x)$, $f'(x)$ have a nontrivial common factor in $F[x]$. Therefore $\gcd(f(x), f'(x)) \neq 1$ in $F[x]$, and it follows that $\gcd(f(x), f'(x)) \neq 1$ in $k[x]$ as pointed out in Remark 14.22.

Conversely, assume that $\gcd(f(x), f'(x)) \neq 1$. If $p(x)$ is a common factor, let α be a root of $p(x)$ in F; then $(x - \alpha)$ divides both $f(x)$ and $f'(x)$ in $F[x]$. We have

$$f(x) = (x - \alpha)\, g(x)$$

in $F[x]$, and hence

$$f'(x) = g(x) + (x - \alpha)\, g'(x).$$

Since $(x - \alpha) \mid f'(x)$, necessarily $(x - \alpha) \mid g(x)$. Therefore

$$f(x) = (x - \alpha)^2\, h(x)$$

for some $h(x) \in F[x]$. This shows that $f(x)$ has a multiple factor over the splitting field, hence it is inseparable.   □

We can use Proposition 14.20 to verify again that $x^2 - u$ is inseparable over the field $k = \mathbb{F}_2(u)$: the derivative of $x^2 - u$ is 0 since $2x = 0$ in characteristic 2 and u is a constant, i.e., an element of k. By the same token, $x^p - u$ is inseparable over $\mathbb{F}_p(u)$.

As a consequence of Proposition 14.21, we can verify that the fact that the characteristic of the field is positive in these examples is not a coincidence. In fact, the next result is a major simplification from our point of view: it says that if we are only dealing with, say, $\mathbb{Q}$, we do not need to worry about (in)separability issues.

PROPOSITION 14.23   *Let k be a field of characteristic 0, and let $f(x) \in k[x]$ be an irreducible polynomial. Then $f(x)$ is separable.*

As a consequence, algebraic extensions of fields of characteristic 0 are separable.

Proof   Let $f(x)$ be an irreducible polynomial. In particular, $f(x)$ is not constant, $d = \deg f(x) \geq 1$:

$$f(x) = a_0 + a_1 x + \cdots + a_d x^d, \quad d \geq 1, a_d \neq 0.$$

It follows that

$$f'(x) = a_1 + \cdots + d a_d x^{d-1} \neq 0$$

since $d a_d \neq 0$ by the assumption on the characteristic. Since $f(x)$ is irreducible, and $\deg f'(x) < \deg f(x)$, $f(x)$ and $f'(x)$ have no common irreducible factor. Therefore $\gcd(f(x), f'(x)) = 1$, and $f(x)$ is separable by Proposition 14.21.

For the last assertion, just note that if $k \subseteq F$ is an algebraic extension, and $\alpha \in F$, then the minimal polynomial of α is an irreducible polynomial, hence separable if char $k = 0$ by what we just proved.   □

By and large, we will keep inseparable polynomials at arm's length. What's wrong with them? We can look at $x^2 - u$ over $\mathbb{F}_2(u)$ from the same viewpoint as in Examples 14.13 and 14.14:

- The splitting field F for $x^2 - u$ is $\mathbb{F}_2(u)[t]/(x^2 - u)$.
- $[F : \mathbb{F}_2(u)] = 2$.
- There is a single embedding of F into $\overline{\mathbb{F}_2(u)}$.

Indeed, in a field where $x^2 - u$ has a root (be it the splitting field or $\overline{\mathbb{F}_2(u)}$), it has a *single* root (as (14.2) shows). By Proposition 14.1, there is only one embedding of F into the algebraic closure.

Therefore, for this inseparable polynomial, unlike in the examples we have analyzed in §14.2, the degree of the splitting field over the ground field does *not* equal the number of embeddings. This may not strike you as problematic, but it is a point of departure with respect to those (separable) examples, and it may be viewed as a complication best left to more advanced texts (and more knowledgeable writers). As we will prove in short order, for separable extensions the degree does equal the number of embeddings into the algebraic closure (Corollary 14.25).

This will follow from the following striking fact, which in itself is probably the most convincing reason why the separability condition is desirable.

> THEOREM 14.24 (Primitive element theorem)   *Finite separable extensions are simple. That is, every finite separable extension $k \subseteq F$ has a 'primitive element' $\alpha \in F$ such that $F = k(\alpha)$.*

This fact explains phenomena such as the one analyzed in Example 13.25: *every* extension $\mathbb{Q}(\alpha_1, \ldots, \alpha_n)$ of $\mathbb{Q}$, with each α_i algebraic, turns out to equal $\mathbb{Q}(\alpha)$ for some cleverly found α. This is because such extensions are finite, and since the characteristic of $\mathbb{Q}$ is 0, every algebraic extension of $\mathbb{Q}$ is automatically separable, by virtue of Proposition 14.23. By Theorem 14.24, such extensions are necessarily simple.

The argument proving Theorem 14.24 is possibly one of the most involved in this text. Proceed with caution.

Proof of Theorem 14.24   Let $k \subseteq F$ be a finite extension. If k is finite, then so is F, and then F^* is cyclic (as a group) by Theorem 10.24. If α is a generator of this group, then $F = k(\alpha)$ is a simple extension (cf. Example 13.26).

Therefore, we may assume that k is infinite. Since $k \subseteq F$ is a finite extension, it is in particular finitely generated: $F = k(u_1, \ldots, u_n)$. Arguing inductively (as in the proof of Theorem 13.29) we only need to prove the statement for $n = 2$.

Therefore, we are reduced to showing that if $F = k(u, v)$ is a finite *separable* extension of an infinite field k, then F is simple. We will show that if $k \subseteq k(u, v)$ is *not* simple, then the extension is *not* separable. In fact, we will show that if $k \subseteq k(u, v)$ is not simple, then the minimal polynomial of v is *inseparable*.

Let $f(x)$, resp. $g(x)$, be the minimal polynomials for u, resp. v, over k. Consider the intermediate fields $k(u + cv)$, as $c \in k$. If there exists any c for which $v \in k(u + cv)$, then this field also contains $u = (u+cv)-cv$, and it follows that $k(u, v) = k(\alpha)$ with $\alpha = u+cv$. Therefore, it suffices to show that *if $v \notin k(u + cv)$ for all c, then $g(x)$ is inseparable.*

Assume then that $\forall c \in k$, $v \notin k(u + cv)$, and let $p_c(x)$ be the minimal polynomial of v over $k(u + cv)$; thus, the degree of each $p_c(x)$ is at least 2. Also, let K be the splitting field for $g(x)$ over $k(u, v)$; in particular, K contains $k(u, v)$. It suffices to show that $g(x)$ has multiple factors in $K[x]$ (Exercise 14.15). In fact, we will show that v is a multiple root of $g(x)$ in $K[x]$.

Since $g(v) = 0$ and $p_c(x)$ is the minimal polynomial for v over $k(u + cv)$, $p_c(x)$ divides $g(x)$ in $k(u + cv)[x]$, and hence in $K[x]$. Since c ranges over k, which is infinite, while $g(x)$ has only finitely many factors in $K[x]$, it follows that for infinitely many c the polynomial $p_c(x)$ equals some fixed polynomial $p(x) \in K[x]$ dividing $g(x)$. The polynomial $p(x)$ has degree at least 2 (since all $p_c(x)$ have degree at least 2), so it has at least two linear factors in $K[x]$ (which may coincide).

Let v' be any root of $p(x)$ in K, i.e., let $x - v'$ be any linear factor of $p(x)$ in $K[x]$. We are done if we can show that $v' = v$: this will show that $(x - v)$ is a multiple factor of $p(x)$, and hence $g(x)$, confirming that $g(x)$ is inseparable.

For this, consider the polynomials $f(u + c(v - x)) \in k(u + cv)[x]$. Since $f(u) = 0$ by assumption, this polynomial vanishes at $x = v$, hence it is a multiple of $p_c(x)$. Therefore $p(x)$ divides $f(u + c(v - x))$ for infinitely many c, and it follows that $f(u + c(v - v')) = 0$ for infinitely many c.

We are almost done! The conclusion in the previous paragraph shows that *the polynomial $f(u + y(v - v')) \in K[y]$ has infinitely many roots.* Nonzero polynomials over a field can only have finitely many roots (cf., e.g., Exercise 7.14), so this shows that the polynomial $f(u + y(v - v'))$ is identically equal to zero:

$$f(u + y(v - v')) \equiv 0.$$

It follows that $v - v' = 0$, that is, $v' = v$, as needed.   □

The primitive element theorem is as powerful as a classification theorem: now we can establish facts about arbitrary finite separable extensions by reducing them to statements about simple extensions. For example, we can now explain the agreement between the

degree of an extension and the number of its embeddings in the algebraic closure, which
we observed in Examples 14.13 and 14.14.

COROLLARY 14.25   *Let $k \subseteq F$ be a finite separable extension. Then there are exactly
$[F : k]$ embeddings of F into the algebraic closure $\overline{k}$ extending the inclusion $k \subseteq \overline{k}$.*

Proof   By Theorem 14.24, $F = k(\alpha)$ for some element α. The minimal polynomial of α
has degree $d := [F : k]$ and is separable, so it has d distinct roots in $\overline{k}$. There is exactly
one embedding of $F = k(\alpha)$ into $\overline{k}$ for each root, by Proposition 14.1, and this gives the
statement.   □

The extensions in Examples 14.13 and 14.14 were both separable, since char $\mathbb{Q} = 0$
(cf. Proposition 14.23), so Corollary 14.25 explains why the numbers should coincide
for these extensions.

One lesson learned from Corollary 14.25 is that the number of embeddings into an
algebraic closure is closely related to separability. Chasing this idea leads to a useful
characterization of separability, which will allow us to verify, for example, that a simple
extension $k \subseteq F = k(\alpha)$ is separable as soon as the minimal polynomial *of the generator*
α over k is separable.

For this, it is convenient to introduce one more piece of notation.

DEFINITION 14.26   The 'separability degree' of an algebraic extension $k \subseteq F$, denoted
$[F : k]_s$, is the number of distinct embeddings of F in the algebraic closure $\overline{k}$ extending
the inclusion $k \subseteq \overline{k}$.

Example 14.27   If $k \subseteq k(\alpha)$ is a simple algebraic extension, then $[k(\alpha) : k]_s$ equals the
number of distinct roots in $\overline{k}$ of the minimal polynomial of α. (This follows from Proposition 14.1.)

In particular, $[k(\alpha) : k]_s \leq [k(\alpha) : k]$ in any case, and equality holds if and only if the
minimal polynomial of α over k is separable.

PROPOSITION 14.28   *The separability degree is multiplicative: if $k \subseteq E \subseteq F$ are
algebraic extensions, then*

$$[F : k]_s = [F : E]_s[E : k]_s .$$

Proof   We have $[E : k]_s$ embeddings j of E in $\overline{k}$ extending the inclusion $k \subseteq \overline{k}$: $k \subseteq$
$E \overset{j}{\hookrightarrow} \overline{k}$. For each of these, $\overline{k}$ is then realized as an algebraic closure *of* E (Exercise 14.6);
so there are $[F : E]_s$ embeddings of F into $\overline{E} = \overline{k}$ extending each embedding j. In total
we have $[F : E]_s[E : k]_s$ ways to embed F into $\overline{k}$ extending the inclusion $k \subseteq \overline{k}$, and this
is the statement.   □

COROLLARY 14.29 *A simple algebraic extension $k \subseteq k(\alpha)$ is separable if and only if the minimal polynomial of α over k is separable.*

Proof If $k \subseteq k(\alpha)$ is a separable extension, then the minimal polynomial of *every* element of $k(\alpha)$ is separable; in particular, the minimal polynomial of α is separable.

The interesting implication is the converse. Assume that the minimal polynomial of α is separable, and let $\beta \in k(\alpha)$; we have to show that the minimal polynomial of β is separable. We have $k \subseteq k(\beta) \subseteq k(\alpha)$, therefore by Proposition 14.28

$$[k(\alpha) : k]_s = [k(\alpha) : k(\beta)]_s [k(\beta) : k]_s .$$

Now the minimal polynomial for α over k is separable, therefore (see Example 14.27)

$$[k(\alpha) : k] = [k(\alpha) : k]_s = [k(\alpha) : k(\beta)]_s [k(\beta) : k]_s \le [k(\alpha) : k(\beta)][k(\beta) : k] = [k(\alpha) : k] .$$

Necessarily, the inequality in the middle must be an equality. In particular, this implies $[k(\beta) : k]_s = [k(\beta) : k]$, proving that the minimal polynomial of β is separable, as needed. $\qquad \square$

Remark 14.30 Corollary 14.29 has a straightforward generalization to finitely generated extensions: an algebraic extension $k \subseteq k(\alpha_1, \ldots, \alpha_r)$ is separable if and only if the minimal polynomial of each α_i over k is separable (Exercise 14.24).

For example, if $k \subseteq F$ is the splitting field of a separable polynomial $f(x) \in k[x]$, then it is a separable extension. ⌐

Remark 14.31 It also follows from the preceding results that the equality $[F : k]_s = [F : k]$ characterizes separability for *all* finite extensions. See if you can work out the details needed to prove this assertion (Exercise 14.23). ⌐

14.5 Application: Finite Fields

At this point we know enough to classify all *finite fields*. Of course we are very familiar with the fields $\mathbb{F}_p = \mathbb{Z}/p\mathbb{Z}$, where $p > 0$ is a prime integer. We also know how to construct more: if $f(x) \in \mathbb{F}_p[x]$ is an irreducible polynomial, then $\mathbb{F}_p[x]/(f(x))$ is a field because irreducible elements generate maximal ideals in a PID (Proposition 6.33). If $f(x)$ has degree d, then this field has $q = p^d$ elements. In fact, if F is a finite field, then it contains a copy of $\mathbb{F}_p$ for some prime integer p (Lemma 13.8); if $[F : \mathbb{F}_p] = d$, then F is a d-dimensional vector space over $\mathbb{F}_p$, and it follows that its order $|F|$ equals p^d. Therefore, the order of a finite field is necessarily a prime power.

This is good solid information, but it does not immediately answer more refined questions such as:

- If $q = p^d$ is a prime power, is there a finite field F such that $|F| = q$?
- How many isomorphism classes of finite fields of a given order are there?

'Classifying' finite fields requires at the minimum an answer to these questions. We will find that for every $q = p^d$ there is *exactly one* field of order q up to isomorphism. In fact, we will prove the following more precise statement.

THEOREM 14.32   *Let $p > 0$ be a prime integer, and let $d \geq 1$ be an integer. Let $q = p^d$. There is exactly one isomorphism class of fields $\mathbb{F}_q$ of order q. In fact, $\mathbb{F}_q$ is a splitting field for the polynomial $x^q - x \in \mathbb{F}_p[x]$.*

The extension $\mathbb{F} \subseteq \mathbb{F}_q$ is finite, normal, and separable.

The only ingredient we are missing for the proof of this theorem is the *Frobenius homomorphism.*

DEFINITION 14.33   Let F be a field of characteristic $p > 0$. The 'Frobenius homomorphism' on F is the function Fr: $F \to F$ given by $\mathrm{Fr}(x) = x^p$.

Wait—this function does not look like a homomorphism. Yet, it is! Indeed,

$$\mathrm{Fr}(xy) = (xy)^p = x^p y^p = \mathrm{Fr}(x)\,\mathrm{Fr}(y)$$

since fields are commutative, and, more interestingly,[1]

$$\mathrm{Fr}(x + y) = (x + y)^p = x^p + \binom{p}{1}x^{p-1}y + \binom{p}{2}x^{p-2}y^2 + \cdots + \binom{p}{p-1}xy^{p-1} + y^p$$

$$= x^p + y^p = \mathrm{Fr}(x) + \mathrm{Fr}(y)$$

since $\binom{p}{i}$ is a multiple of p for $i = 1, \ldots, p - 1$. (We verified this in Example 7.39.)

Proof of Theorem 14.32   We have to prove two things:

(i) The splitting field F of the polynomial $x^{p^d} - x \in \mathbb{F}_p[x]$ has order $|F| = p^d = q$.

(ii) Conversely, if F is a field and $|F| = p^d$, then F is a splitting field for the polynomial $x^{p^d} - x \in \mathbb{F}_p[x]$.

The statement follows, since splitting fields are unique up to isomorphism (Corollary 14.10).

(i) Let F be the splitting field for $f(x) = x^{p^d} - x \in \mathbb{F}_p[x]$. The polynomial $f(x)$ is separable: indeed, its derivative is -1, so the criterion of Proposition 14.21 is verified. Therefore, $f(x)$ has $q = p^d$ distinct roots in F; let $E \subseteq F$ be the set of roots, that is, the set of elements $u \in F$ such that $u^q = u$. I claim that E is a field! And then $f(x)$ splits completely in E, and E is generated by the roots of $f(x)$, so E is the splitting field: $E = F$. In particular, $|F| = |E| = p^d = q$ as stated.

To verify that E is a field, we verify that it is closed with respect to multiplication and subtraction (which makes it a subring of F by Exercise 4.5), and further that every nonzero $u \in E$ has an inverse in E (making it a field). Closure of E with respect to multiplication is evident:

$$u, v \in E \implies u^q = u, v^q = v \implies (uv)^q = u^q v^q = uv$$

---

[1] This fact is known as *Freshman's dream*, do you see why?

since E is commutative. This shows that if u and v are roots of $f(x)$, then so is uv. It is also clear that nonzero elements in E have inverses: if $u^q = u$ and $u \neq 0$, then u^{-1} exists in F, and

$$(u^{-1})^q = u^{-q} = (u^q)^{-1} = u^{-1},$$

so $u^{-1} \in E$. More amusing is the observation that E is closed under subtraction. For this, note that

$$(x + y)^q = (\cdots((x + y)^p)^p \cdots)^p = \underbrace{\text{Fr} \circ \cdots \circ \text{Fr}}_{d \text{ times}} (x + y) = x^q + y^q.$$

Therefore, if $u, v \in E$, so that $u^q = u$ and $v^q = v$, then

$$(u - v)^q = u^q + (-1)^q v^q = u - v,$$

proving that $u - v \in E$. (Note that $(-1)^q = -1$ if q is odd, and $(-1)^q = 1 = -1$ if q is even, since in the latter case the characteristic is 2.)

Therefore E is indeed a field, concluding the proof of (i).

(ii) Conversely, we have to verify that *every* field F with $|F| = q = p^d$ is a splitting field for $f(x) = x^q - x$. I claim that if $|F| = q$, then $f(u) = 0$ for *every* element $u \in F$. Therefore, F contains q roots of $f(x)$ (that is, $f(x)$ splits completely in $F[x]$), and no smaller field does, so that F is indeed a splitting field.

To verify my claim, note that F^* is a group of order $q - 1$ under multiplication: so $u^{q-1} = 1$ for every $u \neq 0$ (Proposition 12.4), therefore $u^q = u$, i.e., $f(u) = 0$, for all $u \neq 0$. Since $f(0) = 0$, we have that $f(u) = 0$ for all $u \in F$ and we are done.

Concerning the last assertion: splitting fields are normal (Theorem 14.16), and the polynomial $x^{p^d} - x$ is separable over $\mathbb{F}_p$, as noted in the proof of (i).   □

DEFINITION 14.34   For $q > 1$ a prime power, we denote by $\mathbb{F}_q$ the unique field (up to isomorphism) with q elements.

COROLLARY 14.35   *For every prime integer $p > 0$ and every $d > 0$ there exist irreducible polynomials of degree d in $\mathbb{F}_p[x]$.*

Proof   The extension $\mathbb{F}_p \subseteq \mathbb{F}_q$ is simple: this follows from the primitive element theorem, Theorem 14.24, since the extension is finite and separable; or, more directly, recall once again that $\mathbb{F}_q^*$ is cyclic (Theorem 10.24), so $\mathbb{F}_q = \mathbb{F}_p(\alpha)$ for a generator α of the group $\mathbb{F}_q^*$.

Since $[\mathbb{F}_q : \mathbb{F}_p] = d$, the minimal polynomial for α over $\mathbb{F}_p$ has degree d, and is irreducible.   □

The right place for Corollary 14.35 might have been §7.5, but back then we did not have the tools to prove this simple statement. In fact, we can refine Corollary 14.35 considerably, as follows. Since $x^{p^d} - x \in \mathbb{F}_p[x]$ is separable, its factorization in $\mathbb{F}_p[x]$ has no multiple factors. What are the irreducible factors of this polynomial?

PROPOSITION 14.36   *The monic irreducible factors of $x^{p^d} - x$ in $\mathbb{F}_p[x]$ are all and only the monic irreducible polynomials of degree $e \mid d$ in $\mathbb{F}_p[x]$.*

Proof   First we verify that if $f(x)$ is a monic irreducible factor of $x^{p^d} - x$, then its degree e divides d. By Theorem 14.32, $x^{p^d} - x$ splits completely in $\mathbb{F}_{p^d}$. It follows that $f(x)$ splits completely in $\mathbb{F}_{p^d}$; let β be a root of $f(x)$ in $\mathbb{F}_{p^d}$, so that $f(x)$ is the minimal polynomial for β over $\mathbb{F}_p$. The subfield $\mathbb{F}_p(\beta)$ generated by β within $\mathbb{F}_{p^d}$ is an intermediate field: $\mathbb{F}_p \subseteq \mathbb{F}_p(\beta) \subseteq \mathbb{F}_{p^d}$, and $|\mathbb{F}_p(\beta)| = p^e$. By multiplicativity of degrees (Proposition 13.13) we have

$$d = [\mathbb{F}_{p^d} : \mathbb{F}_p] = [\mathbb{F}_{p^d} : \mathbb{F}_p(\beta)][\mathbb{F}_p(\beta) : \mathbb{F}_p] = [\mathbb{F}_{p^d} : \mathbb{F}_p(\beta)] \cdot e$$

and this verifies $e \mid d$.

Conversely, we have to show that if $f(x)$ is a monic irreducible polynomial of degree $e \mid d$, then $f(x)$ is a factor of $x^{p^d} - x$. If $f(x)$ is irreducible of degree e, then $\mathbb{F}_p \subseteq \mathbb{F}_p[t]/(f(t)) = \mathbb{F}_p(\beta)$ is a simple field extension of degree e, therefore $\mathbb{F}_p(\beta) \cong \mathbb{F}_{p^e}$ is the splitting field for $x^{p^e} - x$ by Theorem 14.32. In particular, $\beta^{p^e} - \beta = 0$ so the minimal polynomial $f(x)$ of β divides $x^{p^e} - x$. It is easy to see that $(x^{p^e} - x) \mid (x^{p^d} - x)$ (Exercise 14.27), so this verifies that $f(x)$ divides $x^{p^d} - x$ and concludes the proof.   □

Remark 14.37   As a subproduct of the argument proving Proposition 14.36, we see that there is an extension $\mathbb{F}_{p^e} \subseteq \mathbb{F}_{p^d}$ if and only if $e \mid d$.

Armed with Proposition 14.36, we can inductively determine all irreducible polynomials of a given degree over $\mathbb{F}_p[x]$, or at least the number of such polynomials. This is actually rather fun.

Example 14.38   It is already entertaining to look at some examples for small primes p. For example, take $p = 2$. (So all the irreducible polynomials are monic, since the only possible nonzero coefficient is 1. Of course this is not the case for $p > 2$.)

- $d = 1$: There are two irreducible polynomials of degree 1, namely x and $x - 1$.
- $d = 2$: By Proposition 14.36, the factorization of $x^{2^2} - x = x^4 - x$ in $\mathbb{F}_2[x]$ consists of all irreducible polynomials of degree 1 and 2. Since we know the polynomials of degree 1, we can conclude that there is exactly one irreducible polynomial $f_2(x)$ of degree 2:

$$x^4 - x = x(x - 1)f_2(x)$$

and performing the division shows that $f_2(x) = x^2 + x + 1$.

- $d = 3$: Again applying Proposition 14.36, we see that the factorization of $x^8 - x$ must consist of all irreducible polynomials of degree 1 and 3. Therefore

$$x^8 - x = x(x - 1)f_3(x)g_3(x) :$$

there is room for two distinct irreducible polynomials of degree 3. A bit of trial and error shows that $x^3 + x + 1$ and $x^3 + x^2 + 1$ are irreducible, and we need look no further since we know there are exactly two.

- $d = 4$: We proceed in the same way. The factorization of $x^{16} - x$ must consist of all irreducible polynomials of degree $1, 2$, and 4. Therefore

$$x^{16} - x = x(x - 1)(x^2 + x + 1)f_4(x)g_4(x)h_4(x) :$$

there must be three distinct irreducible polynomials of degree 4 in $\mathbb{F}_2[x]$. (Can you find them?)

- $d = 5$: The polynomial $x^{32} - x$ must factor as the product of all irreducible polynomials of degree 1 and 5. Taking away the factors x and $x - 1$ shows that the product of the degree-5 irreducible polynomial has degree 30. Therefore there are six irreducible polynomials of degree 5 in $\mathbb{F}_2[x]$.

- $d = 6$: The polynomial $x^{64} - x$ factors as the product of all irreducible polynomials of degree $1, 2, 3$, and 6. The product of those of degree 1, 2, and 3 is

$$x(x - 1)f_2(x)f_3(x)g_3(x) = (x^8 - x)(x^2 + x + 1) ;$$

the product of those of degree 6 must be a polynomial of degree $64 - 1 - 1 - 2 - 3 - 3 = 54$, so now we know that there are $54/6 = 9$ irreducible polynomials of degree 6 in $\mathbb{F}_2[x]$.

- etc., etc.

I hope you are duly amazed. Counting the number of irreducible polynomials of a given degree over $\mathbb{Z}/p\mathbb{Z}$ would have been a completely unreasonable project back in Chapter 7, even though we had all the right definitions in front of us. (Imagine solving Exercise 14.28 *without* knowing what we just learned....) Studying splitting fields and simple extensions has reduced this type of question to very straightforward combinatorics.

Exercises

14.1 Prove that there is no embedding of $\mathbb{Q}[t]/(t^3 - 2)$ in $\mathbb{Q}(i)$.

14.2 ▷ Let α be any root of $t^3 - 2 \in \mathbb{Q}[t]$. Prove that the polynomial $t^2 + \alpha t + \alpha^2$ is irreducible over $\mathbb{Q}(\alpha)$. Conclude that the images of the three embeddings of $\mathbb{Q}[t]/(t^3 - 2)$ into $\mathbb{C}$ are all different.

14.3 Prove that a splitting field of $x^3 - 2 \in \mathbb{Q}[x]$ is $\mathbb{Q}[t, u]/(t^3 - 2, t^2 + ut + u^2)$, and find the three roots of $x^3 - 2$ in this field as cosets of explicit polynomials in t and u.

14.4 Let F be a field. Prove that F is algebraically closed if and only if it does not admit proper algebraic extensions, that is, if the following property holds: if $F \subseteq K$ is an algebraic extension, then $F = K$.

14.5 ▷ Let $k \subseteq F$ be an algebraic extension, and assume that F is algebraically closed. Let E be an intermediate field, smaller than F: $k \subseteq E \subsetneq F$. Prove that E is *not* algebraically closed. (Use Exercise 14.4.)

14.6 ▷ Let $k \subseteq \bar{k}$ be the algebraic closure of a field k, and let E be an intermediate field: $k \subseteq E \subseteq \bar{k}$. Prove that $\bar{k}$ is an algebraic closure of E.

14.7   Describe the splitting field of $x^3 - 1 \in \mathbb{Q}[x]$ and its embeddings in $\overline{\mathbb{Q}}$.

14.8   Describe the splitting field of $x^4 + 4 \in \mathbb{Q}[x]$ and its embeddings in $\overline{\mathbb{Q}}$.

14.9   Let $\omega = e^{2\pi i/3}$, and let $\rho = \omega^2 - \omega + 1$. Find a nonzero polynomial with rational coefficients satisfied by ρ, and the other root(s) of this polynomial. (*Hint:* Do Exercise 14.7 first.)

14.10   Prove that the field $\mathbb{Q}(\sqrt[4]{2})$ is not the splitting field of any polynomial in $\mathbb{Q}[t]$.

14.11   Let $k \subseteq F \subseteq K$ be algebraic extensions, and assume $k \subseteq F$ is normal. Let $\sigma: K \to K$ be any isomorphism such that $\sigma|_k = \mathrm{id}_k$. Prove that $\sigma(F) = F$.

14.12   ▷ Let k be a field, and $f(x), g(x) \in k[x]$. Prove that $(f + g)'(x) = f'(x) + g'(x)$ and $(fg)'(x) = f'(x)g(x) + f(x)g'(x)$.

14.13   ▷ Let k be a field, and let $k(u)$ be the field of rational functions with coefficients in k. Prove that $x^2 - u$ is irreducible over $k(u)$. (*Hint:* If not, there would be polynomials $f(u), g(u) \neq 0$, such that $(f(u)/g(u))^2 = u$. We would then have $f(u)^2 = ug(u)^2$ in $k[u]$. What's wrong with this?)

14.14   Verify 'directly' (i.e., by considering the splitting field) that $x^p - u$ is inseparable over the field $\mathbb{F}_p(u)$.

14.15   ▷ Let k be a field, and $f(x) \in k[x]$. Assume that there is an extension $k \subseteq F$ such that $f(x)$ has a multiple root in F. Prove that $f(x)$ is inseparable.

14.16   Let $f(x) = a_0 + a_1 x + \cdots + a_d x^d \in k[x]$ be an irreducible polynomial over a field k of characteristic $p > 0$. Assume that p does not divide the degree of some nonzero term in $f(x)$, i.e., $\exists i$ s.t. $a_i \neq 0$ and $p \nmid i$. Prove that $f(x)$ is separable.

14.17   Let $k \subseteq F$ be a finite extension in characteristic $p > 0$. Assume that $p \nmid [F : k]$. Prove that the extension is separable.

14.18   Let $p > 0$ be a prime integer. Prove that the Frobenius homomorphism on $\mathbb{F}_p$ is the identity. (What 'famous' theorem is equivalent to this statement?)

14.19   Let $p > 0$ be a prime integer, and let $c \in \mathbb{F}_{p^d}$ be a root of a polynomial $f(x) \in \mathbb{F}_p[x]$. Prove that c^p is also a root of $f(x)$.

14.20   ▷ A field k of characteristic $p > 0$ is 'perfect' if its Frobenius homomorphism is an *isomorphism* (i.e., if it is surjective). Prove that finite fields are perfect.

14.21   Prove that irreducible polynomials over perfect fields are separable.

14.22   (With notation as in Definition 14.26.) Prove that $[F : k]_s \leq [F : k]$ for all finite extensions $k \subseteq F$. (*Hint:* Use multiplicativity of both ordinary and separable degrees to reduce to the case of a simple extension.)

14.23   ▷ (With notation as in Definition 14.26.) Prove that a finite extension $k \subseteq F$ is separable if and only if $[F : k]_s = [F : k]$. (*Hint:* Half of this is Corollary 14.25. For the other half, work as in Corollary 14.29, making good use of Exercise 14.22.)

14.24   ▷ Let $k \subseteq k(\alpha_1, \ldots, \alpha_r)$ be an algebraic extension. Prove that it is separable if and only if the minimal polynomial of each α_i over k is separable. (Use Exercise 14.23.)

14.25   Let $p > 0$ be a prime integer, and let $d > 0$ be an integer. What is $(\mathbb{F}_{p^d}, +)$, as an abelian group?

14.26   Prove that there are *no* irreducible polynomials of the form $x^2 - c$ in $\mathbb{F}_{2^d}[x]$, for $c \in \mathbb{F}_{2^d}$.

14.27   ▷ Let d, e, p be positive integers, and assume $e \mid d$.

- Prove that $x^e - 1$ divides $x^d - 1$ in $\mathbb{Z}[x]$.

- Prove that $p^e - 1$ divides $p^d - 1$.
- Prove that $x^{p^e} - x$ divides $x^{p^d} - x$ in $\mathbb{Z}[x]$.

14.28 ▷ Find the number of monic irreducible polynomials of degree 12 over $\mathbb{F}_7$.

14.29 Let $I_{p,d}$ be the number of monic irreducible polynomials of degree d in $\mathbb{F}_p[x]$. Prove that

$$\sum_{1 \le k \mid d} k I_{p,k} = p^d \,.$$

Use this fact to find an explicit formula for the number of monic irreducible polynomials of degree 6 in $\mathbb{F}_p[x]$.

15 Galois Theory

15.1 Galois Groups and Galois Extensions

With a few inevitable exceptions, groups have not played a major role in the story for fields so far. That is going to change now. There is a group naturally associated with an extension, just as there is a group naturally associated with every object of every category: the group of *automorphisms* of that object. I made this point in §11.2: in a loose sense, this is the reason why we study groups in the first place.

Recall from §13.1 that a morphism from an extension $k \subseteq F'$ to an extension $k \subseteq F''$ corresponds to a homomorphism $\varphi \colon F' \to F''$ such that the diagram (13.1)

$$
\begin{array}{ccc}
F' & \xrightarrow{\ \varphi\ } & F'' \\
\uparrow & & \uparrow \\
k & \underset{\mathrm{id}_k}{=\!=\!=\!=} & k
\end{array}
$$

commutes, that is, such that φ restricts to the identity on k: $\varphi|_k = \mathrm{id}_k$. (In other words, $\varphi(c) = c$ for all $c \in k$.) An automorphism of an extension $k \subseteq F$ corresponds then to a ring *isomorphism* $\varphi \colon F \to F$ restricting to the identity on k. It is clear that the composition of two automorphisms is an automorphism, and the inverse of an automorphism is an automorphism, so indeed we get a group. This group has a fancy name associated with it.

DEFINITION 15.1   The 'Galois group' of a field extension $k \subseteq F$, denoted $\mathrm{Gal}_k(F)$, is the group of automorphisms of the extension. That is, it is the group of isomorphisms $\varphi \colon F \to F$ which restrict to the identity on k.

Why should we study this group? Galois theory is an amazing interplay of group theory and field theory, with important applications to number theory (which we will not be able to explore, however). Its impact actually goes beyond 'algebra': linear ordinary differential equations have a Galois group; the theory of covering spaces in topology is a close analogue of Galois theory; and manifestations of the same underlying principle occur in many other fields. Our motivation in the rather elementary context of this text may be gleaned by comparing Examples 14.13 and 14.14. These are both examples of splitting fields of polynomials in $\mathbb{Q}[x]$, but they differ in important qualitative ways. As we have seen, as we map the splitting field of $x^3 - 2$ to $\overline{\mathbb{Q}}$, we can choose the image of any root independently of the image of the others. By contrast, the roots $z_1 = \zeta, \cdots, z_4 = \zeta^4$

of the polynomial $1 + x + x^2 + x^3 + x^4$ are entangled, in the sense that if we map, e.g., $z_1 = \zeta$ to some element u of another field, then the images of the other roots will be determined (for example, $z_2 = \zeta^2$ will necessarily have to map to u^2). In general, the roots of a polynomial in a splitting field may display specific interdependencies, from which subtle properties of the polynomial can be deduced. We will try to understand this connection.

The interdependencies are encoded in the different ways the splitting field $k \subseteq F$ for a polynomial $f(x)$ maps to the algebraic closure $\overline{k}$, isomorphically onto the subfield $k(\lambda_1, \ldots, \lambda_d)$ spanned by the roots of $f(x)$ in $\overline{k}$ (cf. Proposition 14.9). The Galois group captures this information. This will be (even) clearer down the road, but for now consider that there is a one-to-one correspondence between the Galois group of the splitting field $k \subseteq F$ and the set of embeddings of F in $\overline{k}$ extending the inclusion $k \subseteq \overline{k}$. Indeed, we may take F to be $k(\lambda_1, \ldots, \lambda_d)$ itself; and then every isomorphism $\gamma \colon F \to F$ extending id_k determines an embedding of F into $\overline{k}$:

$$ F \xrightarrow{\;\gamma\;} F \subseteq \overline{k} $$

extending the inclusion $k \subseteq \overline{k}$, and conversely. Incidentally, if you have absorbed Definition 14.26, then you will see that this means that $|\mathrm{Gal}_k(F)| = [F : k]_s$ for a splitting field. The relation between the (ordinary) degree of an extension and the size of its Galois group will be a *leitmotif* in what follows.

Also note that each embedding $F \hookrightarrow \overline{k}$ as above determines and is determined by a permutation of the roots $\lambda_1, \ldots, \lambda_d$; the Galois group may be identified with this group of permutations, a specific subgroup of S_d. We will soon formalize this observation (Corollary 15.8).

Thus, our exploration of the number of embeddings of a splitting field into $\overline{k}$ in the (very few) examples we have seen really was an exploration of the size of the corresponding Galois groups. In the examples we have studied, which happened to be splitting fields F of *separable* polynomials in $k[x]$, this size turned out to equal $[F : k]$; and eventually we were able to establish that this is a general feature of finite separable extensions (Corollary 14.25). Roughly, one of the goals of this chapter is to understand this fact more deeply.

In the end, the Galois group of the splitting field of a polynomial $f(x)$, or the 'Galois group of the polynomial $f(x)$' for short, encodes sophisticated information about the roots of $f(x)$, and studying this Galois group can reveal very subtle features of the polynomial. This will be the theme in the second half of this chapter.

One moral we can gather from the above discussion is that extensions obtained as *splitting fields of separable polynomials* are 'special' from the point of view of Galois groups. These extensions also have a fancy name.

DEFINITION 15.2　A field extension $k \subseteq F$ is a 'Galois extension' if F is the splitting field of a separable polynomial $f(x) \in k[x]$. Equivalently, a Galois extension is a finite normal separable extension.

(The equivalence of the two definitions is the content of Theorem 14.16, plus Remark 14.30 for separability.) In fact, the choice of normal, separable extensions as primary objects of study will be forced upon us by the interaction of field theory and group theory, as we will see in, for example, Proposition 15.16.

Example 15.3   The extensions $\mathbb{F}_p \subseteq \mathbb{F}_{p^d}$ studied in §14.5 are Galois; this was proved in Theorem 14.32. In fact, the finite fields $\mathbb{F}_q$ are also known as 'Galois fields'.

First we will establish general facts about Galois groups, and look more carefully at the case of simple extensions.

LEMMA 15.4   *If $f(x) \in k[x]$ and $\gamma \in \mathrm{Gal}_k(F)$, then for all $\alpha \in F$ we have $f(\gamma(\alpha)) = \gamma(f(\alpha))$.*

Proof   Say $f(x) = c_0 + c_1 x + \cdots + c_n x^n$, with all coefficients c_i in k; then

$$
\begin{aligned}
\gamma(f(\alpha)) &= \gamma(c_0 + c_1\alpha + \cdots + c_n\alpha^n) \\
&= \gamma(c_0) + \gamma(c_1)\gamma(\alpha) + \cdots + \gamma(c_n)\gamma(\alpha)^n \\
&= c_0 + c_1\gamma(\alpha) + \cdots + c_n\gamma(\alpha)^n \\
&= f(\gamma(\alpha))
\end{aligned}
$$

since γ is a ring homomorphism and $\gamma|_k = \mathrm{id}_k$.   □

PROPOSITION 15.5   *Let $k \subseteq F = k(\alpha_1, \ldots, \alpha_r)$ be a finitely generated extension. Then every element $\gamma \in \mathrm{Gal}_k(F)$ is determined by $\gamma(\alpha_1), \ldots, \gamma(\alpha_r)$. In particular, if $F = k(\alpha)$ is a simple extension, then γ is determined by $\gamma(\alpha)$.*

Proof   Every element of $F = k(\alpha_1, \ldots, \alpha_r)$ may be written as a rational function $Q(\alpha_1, \ldots, \alpha_r)$ with coefficients in k, and by the same computation as in Lemma 15.4 we see that $\gamma(Q(\alpha_1, \ldots, \alpha_r)) = Q(\gamma(\alpha_1), \ldots, \gamma(\alpha_r))$.

(In fact, we will only be interested in finite algebraic extensions, for which the generators α_i are algebraic and Q may be taken to be a polynomial.)   □

Proposition 15.5 is helpful in determining Galois groups of algebraic extensions, because of the following observation.

PROPOSITION 15.6   *Let $k \subseteq F$ be a field extension, and let $\alpha \in F$ be a root of a polynomial $f(x) \in k[x]$. Then for every $\gamma \in \mathrm{Gal}_k(F)$, $\gamma(\alpha)$ is also a root of $f(x)$.*

Proof   By Lemma 15.4 we have that $f(\varphi(\alpha)) = \varphi(f(\alpha)) = \varphi(0) = 0$.   □

COROLLARY 15.7   *Let $k \subseteq F = k(\alpha_1, \ldots, \alpha_r)$ be a finite algebraic extension. For $i = 1, \ldots, r$ let d_i be the degree of α_i over k. Then $|\mathrm{Gal}_k(F)| \le d_1 \cdots d_r$.*
In particular, if $k \subseteq F$ is a simple extension, then $|\mathrm{Gal}_k(F)| \le [F : k]$.

Proof By Proposition 15.5, each $\gamma \in \mathrm{Gal}_k(F)$ is determined by the images of the generators, $\gamma(\alpha_1), \ldots, \gamma(\alpha_r)$. By assumption α_i is a root of a polynomial of degree d_i, and by Proposition 15.6 $\gamma(\alpha_i)$ must also be a root of the same polynomial. Therefore each $\gamma(\alpha_i)$ can only be one of at most d_i elements of F, and the bound follows. $\square$

In particular, the Galois group of a finite extension is finite, and 'not too large'. The same observation realizes $\mathrm{Gal}_F(k)$ as a subgroup of a symmetric group, provided F is generated by the roots of a polynomial (for example, if F is a splitting field).

> COROLLARY 15.8 *Let $k \subseteq F$ be an algebraic extension. Assume $F = k(\lambda_1, \ldots, \lambda_d)$, where the elements λ_i are the roots in F of a polynomial $f(x) \in k[x]$. Then $\mathrm{Gal}_k(F)$ is isomorphic to a subgroup of S_d.*

Proof By Proposition 15.6, every $\gamma \in \mathrm{Gal}_k(k(\alpha))$ sends roots of $f(x)$ to roots of $f(x)$; therefore γ acts on the set $\{\lambda_1, \ldots, \lambda_d\}$. This defines a group homomorphism $\iota: \mathrm{Gal}_k(k(\alpha)) \to S_d$. Since γ is determined by the images of the generators λ_i (Proposition 15.5), it is determined by $\iota(\gamma)$; therefore ι is injective, and identifies $\mathrm{Gal}_k(k(\alpha))$ with a subgroup of S_d as stated. $\square$

These considerations already suffice to determine Galois groups for many extensions.

Example 15.9 The Galois group of the extension $\mathbb{Q} \subseteq \mathbb{Q}(\sqrt[3]{2})$ is trivial.

Indeed, $\sqrt[3]{2}$ is a root of $x^3 - 2$; by Proposition 15.6, if $\gamma \in \mathrm{Gal}_{\mathbb{Q}}(\mathbb{Q}(\sqrt[3]{2}))$, then $\gamma(\sqrt[3]{2})$ must also be a root of $x^3 - 2$. However, $\sqrt[3]{2}$ is the *only*[1] root of $x^3 - 2$ in $\mathbb{Q}(\sqrt[3]{2})$, so necessarily $\gamma(\sqrt[3]{2}) = \sqrt[3]{2}$. It follows that γ is the identity.

Example 15.10 Let $k \subseteq F$ be a *quadratic* extension, i.e., assume $[F : k] = 2$, and assume that the characteristic of k is not 2. Then $k \subseteq F$ is Galois, and $\mathrm{Gal}_k(F) \cong C_2$.

Indeed, let $\alpha \in F, \alpha \notin k$; so $F = k(\alpha)$ (because $[k(\alpha) : k] > 1$ implies $[k(\alpha) : k] = 2$, forcing $k(\alpha) = F$) and α is a root of a degree-2 irreducible polynomial $f(x) = x^2 + ax + b \in k[x]$. Since the characteristic is not 2, $f'(x) \neq 0$, and it follows that $f(x)$ is separable (Proposition 14.21). Therefore $f(x) = (x - \alpha)(x - \beta)$ in $F[x]$, with $\beta \neq \alpha$ in F; in particular, F is the splitting field of a separable polynomial, so the extension is a Galois extension. There is an element $\gamma \in \mathrm{Gal}_k(F)$ sending α to β (by Proposition 14.1), so $\mathrm{Gal}_k(F)$ is not trivial. On the other hand, $\mathrm{Gal}_k(F) \subseteq S_2 \cong C_2$ by Corollary 15.8, therefore $\mathrm{Gal}_k(F) \cong C_2$ as stated.

Example 15.11 The Galois group of $x^3 - 2 \in \mathbb{Q}[x]$ is S_3.

Indeed, the splitting field F for this polynomial is generated by the three complex roots $\alpha_1, \alpha_2, \alpha_3$ of $x^3 - 2$ (Example 14.7), therefore $\mathrm{Gal}_{\mathbb{Q}}(F)$ may be identified with a subgroup G of S_3 by Corollary 15.8. Arguing as in the discussion following Definition 15.1, we know that $|G| = |\mathrm{Gal}_{\mathbb{Q}}(F)|$ equals the number of embeddings of F in $\overline{\mathbb{Q}}$, i.e., 6 (Example 14.14). It follows that $G = S_3$.

---

[1] As a reminder, we can embed $\mathbb{Q}(\sqrt[3]{2})$ in $\mathbb{R}$, and there is only one cube root of 2 in $\mathbb{R}$.

For a different way to draw this conclusion: I claim that G contains the three copies of C_2 in S_3 generated by transpositions. Indeed, for $i = 1, 2, 3$ we have the intermediate field $\mathbb{Q}(\alpha_i)$: $\mathbb{Q} \subseteq \mathbb{Q}(\alpha_i) \subseteq F$. So $\mathrm{Gal}_\mathbb{Q}(F)$ contains a subgroup $\mathrm{Gal}_{\mathbb{Q}(\alpha_i)}(F)$. Since $\mathbb{Q}(\alpha_i) \subseteq F$ is a quadratic extension, this subgroup is isomorphic to C_2; it is generated by an element $\gamma_i \in \mathrm{Gal}_\mathbb{Q}(F)$ swapping the two roots other than α_i. In S_3, the elements $\gamma_1, \gamma_2, \gamma_3$ are the three transpositions.

The only subgroup of S_3 containing the three transpositions is S_3 itself, therefore $\mathrm{Gal}_\mathbb{Q}(F) \cong S_3$.

Example 15.12   The Galois group of $f(x) = x^4 + x^3 + x^2 + x^2 + 1 \in \mathbb{Q}[x]$ is cyclic of order 4.

Indeed, we know that a splitting field F of $f(x)$ is $\mathbb{Q}(\zeta)$, where $\zeta = e^{2\pi i/5}$ (Example 14.8). By Proposition 15.5, elements $\gamma \in \mathrm{Gal}_\mathbb{Q}(F)$ are determined by $\gamma(\zeta)$, and by Proposition 15.6, $\gamma(\zeta)$ must be another root of $f(x)$: that is, it must be ζ^2, ζ^3, or ζ^4. By Proposition 14.1, every choice of a root will determine an element of the Galois group. Therefore we can label the elements of $\mathrm{Gal}_\mathbb{Q}(F)$ as follows:

$$\gamma_1 : \zeta \to \zeta, \quad \gamma_2 : \zeta \to \zeta^2, \quad \gamma_3 : \zeta \to \zeta^3, \quad \gamma_4 : \zeta \to \zeta^4.$$

The identity is γ_1; and the group is cyclic, generated by γ_2, since

$$\gamma_2 \circ \gamma_2(\zeta) = \gamma_2(\zeta^2) = \zeta^4,$$

so $\gamma_2^2 = \gamma_4$; and

$$\gamma_2 \circ \gamma_2 \circ \gamma_2(\zeta) = \gamma_2 \circ \gamma_2(\zeta^2) = \gamma_2(\zeta^4) = \zeta^8 = \zeta^3,$$

so $\gamma_2^3 = \gamma_3$.

What this argument really shows is that the function $(\mathbb{Z}/5\mathbb{Z})^* \to \mathrm{Gal}_\mathbb{Q}(F)$ sending $[i]_5$ to γ_i for $i = 1, \dots, 4$ is an isomorphism. Thus, $\mathrm{Gal}_\mathbb{Q}(F)$ turns out to be cyclic because $(\mathbb{Z}/5\mathbb{Z})^*$ is cyclic. This fact extends to all primes $p > 0$ (Exercise 15.3).

Focusing on simple algebraic extensions, our previous work in Chapters 13 and 14 allows us to establish very easily a strong characterization of Galois extensions.

> PROPOSITION 15.13   *Let $k \subseteq F = k(\alpha)$ be a simple algebraic extension, and let $p(x) \in k[x]$ be the minimal polynomial of α. Then $|\mathrm{Gal}_k(F)|$ equals the number of roots of $p(x)$ in F.*
>
> *In particular, if $k \subseteq F$ is a simple algebraic extension, then $|\mathrm{Gal}_k(F)| = [F : k]$ if and only if $k \subseteq F$ is normal and separable, i.e., if and only if $k \subseteq F$ is Galois.*

Proof   By Proposition 13.17, $F = k(\alpha) \cong k[t]/(p(t))$; and then by Proposition 14.1 there is exactly one homomorphism (necessarily an isomorphism, cf. Exercise 13.8) $F \to F$ extending id_k for each root of $p(x)$ in F. This establishes the first assertion.

As for the second assertion, let $d = [F : k] = \deg p(x)$. By the first part, $|\mathrm{Gal}_k(F)| = d$

if and only if F contains d distinct roots of $p(x)$. This is the case if and only if $p(x)$ splits as a product of d distinct linear factors in F. As F is generated by a root of $p(x)$, this means that F is a splitting field for $p(x)$ and that $p(x)$ is a separable polynomial. This makes the extension separable (Corollary 14.29). The fact that $k \subseteq F$ is normal follows from Theorem 14.16. $\qquad\square$

The last part of Proposition 15.13 is quite insightful: it tells us that we can detect sophisticated information, such as whether an extension is normal, by 'numerical' means. As it happens, the hypothesis that the extension be *simple* is an overshoot: it suffices to assume that the extention is *finite*. Verifying this fact completely will require a certain amount of effort, which we will carry out in the next section, leading to an efficient characterization of Galois extensions (Theorem 15.21).

Example 15.14 (Cf. §14.5.) The Galois group of the extension $\mathbb{F}_p \subseteq \mathbb{F}_{p^d}$ is cyclic of order d, generated by the Frobenius isomorphism.

Recall that the Frobenius homomorphism is defined in characteristic p by

$$\mathrm{Fr}(x) = x^p$$

(Definition 14.33). For finite fields, it is an isomorphism (Exercise 14.20); so Fr defines an element of $\mathrm{Gal}_{\mathbb{F}_p}(\mathbb{F}_{p^d})$.

The extension $\mathbb{F}_p \subseteq \mathbb{F}_{p^d}$ is finite, normal, and separable (Theorem 14.32). In particular it is simple (Theorem 14.24), and $|\mathrm{Gal}_{\mathbb{F}_p}(\mathbb{F}_{p^d})| = [\mathbb{F}_{p^d} : \mathbb{F}_p] = d$ by Proposition 15.13. In order to prove the statement it suffices to show that $|\mathrm{Fr}| = d$; and since $|\mathrm{Fr}| \leq |\mathrm{Gal}_{\mathbb{F}_p}(\mathbb{F}_{p^d})| = d$ trivially, it suffices to show that $|\mathrm{Fr}| \geq d$.

Let $e = |\mathrm{Fr}|$: then $\mathrm{Fr}^e = \mathrm{id}$, that is, $(x^p)^e \equiv x$. In other words, every element in $\mathbb{F}_{p^d}$ is a root of the polynomial $x^{p^e} - x$. Since the number of roots of a nonzero polynomial over a field is bounded by its degree, this gives $p^d \leq p^e$, i.e., $|\mathrm{Fr}| = e \geq d$, and we are done.

15.2 Characterization of Galois Extensions

We will now sharpen Proposition 15.13, by generalizing its characterization of Galois extensions in terms of their degrees to arbitrary finite extensions: we will show that a finite extension $k \subseteq F$ is Galois if and only if $|\mathrm{Gal}_k(F)| = [F : k]$. For *separable* extensions, this should be clear at this point: finite separable extensions are simple by Theorem 14.24, so it suffices to invoke Proposition 15.13 in this case. But this characterization holds even without the *a priori* assumption of separability.

In the process of understanding this, we will establish other important characterizations of Galois extensions. In fact, we will see that 'finite, normal, separable' extensions arise in an extremely natural way from the interplay of field theory and group theory; Proposition 15.16 will formalize this observation.

First, we introduce another piece of terminology. Every subgroup G of $\mathrm{Gal}_k(F)$ determines a specific intermediate field F^G of the extension.

DEFINITION 15.15   Let $k \subseteq F$ be a field extension. The 'fixed field' of a subgroup G of $\mathrm{Gal}_k(F)$ is the field

$$F^G := \{u \in F \mid \forall \gamma \in G, \gamma(u) = u\}.$$

Clearly, $k \subseteq F^G \subseteq F$. Take a moment to verify that F^G is indeed a field (Exercise 15.2). The surprise is that the extensions $F^G \subseteq F$ are necessarily *Galois*.

PROPOSITION 15.16   *Let $k \subseteq F$ be a finite extension, and let G be a subgroup of $\mathrm{Gal}_k(F)$. Then the extension $F^G \subseteq F$ is a finite, normal, separable (i.e., Galois) extension. Further, $\mathrm{Gal}_{F^G}(F) = G$ and $|G| = [F : F^G]$.*

Proof   Since $k \subseteq F$ is finite and $k \subseteq F^G \subseteq F$, the extension $F^G \subseteq F$ is finite; in particular, it is algebraic. Let $u \in F$, and let $p(x)$ be the minimal polynomial of u over k. For every $\gamma \in G$, $\gamma(u)$ is also a root of $p(x)$ in F, by Proposition 15.6; let $u = u_1, u_2, \ldots, u_r$ be the roots of $p(x)$ obtained in this way. Keep in mind that for all $\gamma \in G$ the set

$$\{\gamma(u_1), \ldots, \gamma(u_r)\}$$

equals the set $\{u_1, \ldots, u_r\}$: indeed, if $u_i = \rho(u)$ for some $\rho \in G$, then $\gamma(u_i) = \gamma(\rho(u)) = (\gamma \circ \rho)(u)$ is also obtained by applying an element of G to u. The effect of applying γ is just to permute the list $(u_1, \ldots, u_r)$. Consider then the polynomial

$$q(x) := (x - u_1) \cdots (x - u_r) = a_0 + a_1 x + \cdots + a_{r-1} x^{r-1} + x^r \in F[x].$$

I claim that the coefficients of $q(x)$ are *fixed* by every $\gamma \in G$. Indeed, for $\gamma \in G$ we have

$$\gamma(a_0) + \gamma(a_1)x + \cdots + \gamma(a_{r-1})x^{r-1} + x^r = (x - \gamma(u_1)) \cdots (x - \gamma(u_r))$$
$$= (x - u_1) \cdots (x - u_r)$$
$$= a_0 + a_1 x + \cdots + a_{r-1} x^{r-1} + x^r$$

since γ permutes the elements u_i. Therefore $\gamma(a_j) = a_j$ for all j, as claimed. In other words, $q(x)$ has coefficients in F^G.

Since $q(x)$ is separable by construction, and u is a root of $q(x) \in F^G[x]$, this proves that every element of F is the root of a separable polynomial in $F^G[x]$: the extension $F^G \subseteq F$ is separable.

Since $F^G \subseteq F$ is finite and separable, it is simple by Theorem 14.24: $F = F^G(\alpha)$ for some $\alpha \in F$. Then α is a root of the polynomial $f(x) = (x - \alpha_1) \cdots (x - \alpha_d)$, where the elements α_i are obtained by applying all $\gamma \in G$ to α. Incidentally, make a note of the fact that the construction implies $d \leq |G|$. By the argument we applied in the first part of the proof, $f(x)$ has coefficients in F^G. Since $f(x)$ splits completely in F, and F is generated by the roots α_i (in fact, α already generates F), it follows that F is a splitting field for $f(x) \in F^G[x]$, therefore $F^G \subseteq F$ is normal, by Theorem 14.16.

Thus, we now know that $F^G \subseteq F$ is indeed a Galois extension, as stated.

We still have to prove that $\mathrm{Gal}_{F^G}(F) = G$ and $|G| = [F : F^G]$. Note that G is a subgroup of $\mathrm{Gal}_{F^G}(F)$: if $\gamma \in G$, then γ is an isomorphism $F \to F$ and $\gamma|_{F^G} = \mathrm{id}\,|_{F^G}$ by the very definition of F^G. On the other hand, the degree of $F = F^G(\alpha)$ is the degree of the minimal polynomial of α; and this is $\leq d$ since $f(\alpha) = 0$ (with notation as in the previous paragraph). Since $F^G \subseteq F$ is a (simple) Galois extension, $|\text{Gal}_{F^G}(F)| = [F : F^G]$ by Proposition 15.13. Putting everything together,

$$|G| \leq |\text{Gal}_{F^G}(F)| = [F : F^G] \leq d \leq |G|.$$

The conclusion is that $|G| = |\text{Gal}_{F^G}(F)|$, and therefore $G = \text{Gal}_{F^G}(F)$ since one is a subgroup of the other, and $|G| = [F : F^G]$ as stated.   □

Remark 15.17 There is an interesting side-product of the proof of this proposition: in the course of the proof we have verified that if G is a subgroup of $\text{Gal}_k(F)$, and u is any element of F, then u is a root of the polynomial $q(x)$ obtained as the product of all distinct linear factors $(x - \gamma(u))$ as γ ranges in G; and this polynomial $q(x)$ has coefficients in F^G.   ⌐

The correspondence between subgroups of $\text{Gal}_k(F)$ and intermediate fields $k \subseteq F$ is what Galois theory is really about. We will study this 'Galois correspondence' in the next section, and use it to provide yet another characterization of Galois extensions. Proposition 15.16 will be a key ingredient in that result.

For the moment, Proposition 15.16 will allow us to carry out the program announced at the beginning of the section, that is, prove that a finite extension $k \subseteq F$ is Galois if and only if $|\text{Gal}_k(F)| = [F : k]$. One of the implications we need follows immediately from Proposition 15.13.

LEMMA 15.18   *Let $k \subseteq F$ be a Galois extension. Then $|\text{Gal}_k(F)| = [F : k]$.*

Proof Galois extensions are finite and separable, hence simple by the primitive element theorem (Theorem 14.24). The statement follows then directly from Proposition 15.13.   □

Proposition 15.16 allows us to make progress on the other implication.

LEMMA 15.19   *Let $k \subseteq F$ be a finite field extension, and assume that $|\text{Gal}_k(F)| = [F : k]$. Then the fixed field of $\text{Gal}_k(F)$ is k itself.*

Proof By the last part of Proposition 15.16, $|\text{Gal}_k(F)| = [F : F^{\text{Gal}_k(F)}]$. Since the elements of $\text{Gal}_k(F)$ restrict to the identity on k, we have $k \subseteq F^{\text{Gal}_k(F)}$; by multiplicativity of degrees,

$$[F : k] = [F : F^{\text{Gal}_k(F)}][F^{\text{Gal}_k(F)} : k] = |\text{Gal}_k(F)|[F^{\text{Gal}_k(F)} : k].$$

Under the assumption that $[F : k]$ equals $|\text{Gal}_k(F)|$, it follows that $[F^{\text{Gal}_k(F)} : k] = 1$, and this proves that $k = F^{\text{Gal}_k(F)}$ as stated.   □

LEMMA 15.20   *Let $k \subseteq F$ be a finite field extension, and assume that $F^{\text{Gal}_k(F)} = k$. Then $k \subseteq F$ is Galois.*

Proof   This is an immediate consequence of Proposition 15.16: $F^{\mathrm{Gal}_k(F)} \subseteq F$ is a Galois extension, therefore $k \subseteq F$ is Galois if $k = F^{\mathrm{Gal}_k(F)}$.   □

We are ready to upgrade Proposition 15.13 as promised at the beginning of the section, and with room to spare.

> THEOREM 15.21   *Let $k \subseteq F$ be a finite field extension. Then the following are equivalent.*
>
> - *$k \subseteq F$ is Galois.*
> - *$|\mathrm{Gal}_k(F)| = [F : k]$.*
> - *The fixed field of $\mathrm{Gal}_k(F)$ is k.*

Proof   Just stitch together Lemmas 15.18–15.20.   □

In view of the third condition listed in Theorem 15.21, the content of Remark 15.17 for a Galois extension $k \subseteq F$ is that every $u \in F$ is the root of the polynomial $q(x)$ *with coefficients in k* obtained as the product of all distinct linear terms $(x - \gamma(u))$, as γ ranges over $\mathrm{Gal}_k(F)$. These elements $\gamma(u)$ have a name.

DEFINITION 15.22   Let $k \subseteq F$ be a Galois extension, and let $u \in F$. The distinct elements $\gamma(u)$ obtained as $\gamma \in \mathrm{Gal}_k(F)$ are the 'Galois conjugates' of u.

Example 15.23   The extension $\mathbb{R} \subseteq \mathbb{C}$ is Galois, with Galois group C_2. The nontrivial element in $\mathrm{Gal}_{\mathbb{R}}(\mathbb{C})$ is the ordinary *complex conjugation*. Thus, the complex conjugate $\bar{z}$ of a complex number z is an example of a Galois conjugate in the sense introduced in Definition 15.22.

How many conjugates does a given element u have? It is easy to see (Exercise 15.6) that this number equals the index of the subgroup $\mathrm{Gal}_{k(u)}(F)$ in $\mathrm{Gal}_k(F)$, and this fact has a nice consequence.

> PROPOSITION 15.24   *Let $k \subseteq F$ be a Galois extension. For $u \in F$, let $u = u_1, \ldots, u_r$ be the Galois conjugates of u, and let $q(x) = \prod_{i=1}^r (x - u_i)$. Then $q(x)$ is the minimal polynomial of u over k.*

Proof   We already know that $q(x) \in k[x]$ (Remark 15.17), u is a root of $q(x)$, and $q(x)$ is monic; it suffices to show that $\deg q(x) = [k(u) : k]$. Now, the degree of $q(x)$ equals the number r of conjugates of u, and $r = [\mathrm{Gal}_k(F) : \mathrm{Gal}_{k(u)}(F)]$ as pointed out a moment ago. Since both $k \subseteq F$ and $k(u) \subseteq F$ are Galois (they are both splitting fields of a separable polynomial!), we have

$$[\mathrm{Gal}_k(F) : \mathrm{Gal}_{k(u)}(F)] = \frac{|\mathrm{Gal}_k(F)|}{|\mathrm{Gal}_{k(u)}(F)|} = \frac{[F : k]}{[F : k(u)]} = [k(u) : k].$$

Thus $\deg q(x) = [k(u) : k]$, as needed.   □

In particular, $q(x)$ is automatically *irreducible* in $k[x]$.

We now know enough to look back at a few examples encountered earlier and see what was 'really' going on.

Example 15.25   Let's go back to Example 13.25, where we checked by hand that the extension $\mathbb{Q} \subseteq \mathbb{Q}(\sqrt{2}, \sqrt{3})$ was simple, and in fact generated by $\sqrt{2} + \sqrt{3}$. The extension $\mathbb{Q} \subseteq \mathbb{Q}(\sqrt{2}, \sqrt{3})$ is a splitting field for the separable polynomial $f(t) = (t^2 - 2)(t^2 - 3)$, so in particular it is Galois. The extension has degree 4 (since $\sqrt{3} \notin \mathbb{Q}(\sqrt{2})$), so its Galois group G has order 4 (Theorem 15.21). In fact, the roots of $f(t)$ come in pairs: $\pm\sqrt{2}$ and $\pm\sqrt{3}$, and the elements of G can only permute the elements in each pair. There are exactly four such permutations, so these must be the elements of G. (Incidentally, it follows that $G \cong C_2 \times C_2$.) The conjugates of $\sqrt{2} + \sqrt{3}$ are $\pm\sqrt{2} \pm \sqrt{3}$; these are all distinct. By Proposition 15.24, the minimal polynomial of $\sqrt{2} + \sqrt{3}$ over $\mathbb{Q}$ must be

$$\left(x - \sqrt{2} - \sqrt{3}\right)\left(x + \sqrt{2} - \sqrt{3}\right)\left(x - \sqrt{2} + \sqrt{3}\right)\left(x + \sqrt{2} + \sqrt{3}\right). \tag{15.1}$$

In particular $[\mathbb{Q}(\sqrt{2} + \sqrt{3}) : \mathbb{Q}] = 4$, and this suffices to conclude that $\mathbb{Q}(\sqrt{2} + \sqrt{3}) = \mathbb{Q}(\sqrt{2}, \sqrt{3})$, without having to perform the computations worked out in Example 13.25.

(If you wish, you can check that the product in (15.1) equals $x^4 - 10x^2 + 1$, so it has coefficients in $\mathbb{Q}$ as promised.)

Example 15.26   What about Example 14.18? There we had $\zeta = e^{2\pi i/5}$; $\mathbb{Q}(\zeta)$ is a splitting field over $\mathbb{Q}$, and $[\mathbb{Q}(\zeta) : \mathbb{Q}] = 4$. We were able to conclude that every element $\rho = g(\zeta)$ must satisfy an irreducible polynomial $q(x)$ of degree at most 4, and the other roots of $q(x)$ must also be expressible as polynomials in ζ with rational coefficients.

However, back then we had no tools to find this information efficiently. Now we do. Take $\rho = 1 + 3\zeta^2 + 3\zeta^3$ (as suggested in the discussion following Example 14.18). In Example 15.12 we found that the Galois group of the extension is C_4, and its elements may be described as follows:

$$\gamma_1 : \zeta \to \zeta, \quad \gamma_2 : \zeta \to \zeta^2, \quad \gamma_3 : \zeta \to \zeta^3, \quad \gamma_4 : \zeta \to \zeta^4.$$

Applying these elements to ρ gives

$$\begin{cases} 1 + 3\zeta^2 + 3\zeta^3, \\ 1 + 3\zeta^4 + 3\zeta^6 = 1 + 3\zeta + 3\zeta^4, \\ 1 + 3\zeta^6 + 3\zeta^9 = 1 + 3\zeta + 3\zeta^4, \\ 1 + 3\zeta^8 + 3\zeta^{12} = 1 + 3\zeta^2 + 3\zeta^3, \end{cases}$$

where I have used the relation $\zeta^5 = 1$. Therefore, ρ has *two* distinct conjugates: itself and $1 + 3\zeta + 3\zeta^4$. According to Proposition 15.24, its minimal polynomial over $\mathbb{Q}$ is

$$(x - (1 + 3\zeta^2 + 3\zeta^3)) (x - (1 + 3\zeta + 3\zeta^4)),$$

and expanding this product gives the polynomial $q(x) = x^2 + x - 11$. So we found a polynomial $q(x) \in \mathbb{Q}[x]$ of which ρ is a root, and in the process we obtained an explicit expression for the other root of $q(x)$.

15.3   The Fundamental Theorem of Galois Theory

Towards the beginning of our discussion of fields, in §13.2, I advertised that we would be interested in a certain 'correspondence':

$$\{E \text{ field} \mid k \subseteq E \subseteq F\} \rightleftarrows \{ \qquad ? \qquad \}.$$

This was (13.2). We are finally ready to fill in this diagram. Given an intermediate field E of an extension, we can consider the set of automorphisms $\gamma \in \mathrm{Gal}_k(F)$ which restrict to the identity *on E:* this is a subgroup of $\mathrm{Gal}_k(F)$, naturally identified with $\mathrm{Gal}_E(F)$. Conversely, given a subgroup G of $\mathrm{Gal}_k(F)$, we can consider the subfield F^G consisting of elements fixed by G; this 'fixed field' (Definition 15.15) played an important role in §15.2.

DEFINITION 15.27   Let $k \subseteq F$ be a field extension. The 'Galois correspondence' consists of the pair of functions

$$\{E \text{ field} \mid k \subseteq E \subseteq F\} \underset{\tau}{\overset{\sigma}{\rightleftarrows}} \{\text{subgroups of } \mathrm{Gal}_k(F)\}$$

associating with each intermediate field E the subgroup $\sigma(E) = \mathrm{Gal}_E(F)$ of $\mathrm{Gal}_k(F)$, and with each subgroup G of $\mathrm{Gal}_k(F)$ the fixed field $\tau(G) = F^G$.

> PROPOSITION 15.28   *The Galois correspondence is inclusion-reversing. Further, for every intermediate field $k \subseteq E \subseteq F$ and every subgroup G of $\mathrm{Gal}_k(F)$:*
>
> - $E \subseteq F^{\mathrm{Gal}_E(F)}$,
> - $G \subseteq \mathrm{Gal}_{F^G}(F)$.

Proof   It is clear that the correspondence reverses inclusions (Exercise 15.8), and the two stated relations are straightforward. Indeed:

   • If $e \in E$ and $\gamma \in \mathrm{Gal}_E(F)$, then by definition of Galois group we have $\gamma(e) = e$. Therefore e is fixed by all elements of $\mathrm{Gal}_E(F)$. This is the first assertion.

   • Similarly, if $\gamma \in G$, then γ restricts to the identity on the fixed field F^G, by definition of the latter. This says that $\gamma \in \mathrm{Gal}_{F^G}(F)$, and establishes the second assertion.   □

If the inclusions $\subseteq$ in Proposition 15.28 could be replaced by =, then we would deduce that the two functions defining the Galois correspondence are inverses of each other, and it would follow that they are bijections. However, this is not necessarily the case.

Example 15.29   For the extension $\mathbb{Q} \subseteq \mathbb{Q}(\sqrt[3]{2})$:

$$\{E \text{ field} \mid \mathbb{Q} \subseteq E \subseteq \mathbb{Q}(\sqrt[3]{2})\}$$

consists of two elements: $\mathbb{Q}$ and $\mathbb{Q}(\sqrt[3]{2})$. Indeed, the degree of the extension is prime, so there is no room for other intermediate fields (Exercise 13.7).

On the other hand, $\mathrm{Gal}_{\mathbb{Q}}(\mathbb{Q}(\sqrt[3]{2}))$ is *trivial:* see Example 15.9. Therefore,

$$\{\text{subgroups of } \mathrm{Gal}_{\mathbb{Q}}(\mathbb{Q}(\sqrt[3]{2}))\}$$

is a singleton.

The two sides of the Galois correspondence have different cardinalities, so the Galois correspondence is not a bijection for this extension.

Not all is lost, though. We have already proved that for *finite* extensions one of the two $\subseteq$ in Proposition 15.28 *is* an equality.

> **PROPOSITION 15.30** *Let $k \subseteq F$ be a finite extension, and let G be a subgroup of $\mathrm{Gal}_k(F)$. Then $G = \mathrm{Gal}_{F^G}(F)$ and $|G| = [F : F^G]$.*

Proof   This is part of Proposition 15.16!   □

That is, $\sigma \circ \tau = \mathrm{id}$ in the Galois correspondence. It follows that the 'right-to-left' function τ in the Galois correspondence is *injective* and the 'left-to-right' function σ is *surjective*. (Right?) Every subgroup G is the Galois group of the extension over an intermediate field; indeed, it is the Galois group of the extension $F^G \subseteq F$.

Still, Example 15.29 shows that this function σ is not always *injective:* in the example, $\sigma(\mathbb{Q}(\sqrt[3]{2})) = \sigma(\mathbb{Q})$, i.e., $\mathrm{Gal}_{\mathbb{Q}(\sqrt[3]{2})}(\mathbb{Q}(\sqrt[3]{2})) = \mathrm{Gal}_{\mathbb{Q}}(\mathbb{Q}(\sqrt[3]{2}))$, for the rather mundane reason that both groups are trivial.

Well, there surely ought to be some condition on an extension guaranteeing that the Galois correspondence *is* a bijection. . .

> **THEOREM 15.31 (Fundamental Theorem of Galois Theory, I)** *Let $k \subseteq F$ be a finite field extension. The extension $k \subseteq F$ is Galois if and only if its Galois correspondence is a bijection.*

Proof   Assume that $k \subseteq F$ is a Galois extension. In particular it is finite, so by Proposition 15.30, $\sigma \circ \tau = \mathrm{id}$ in the Galois correspondence. In order to show that the Galois correspondence is bijective, it suffices to verify that $\tau \circ \sigma = \mathrm{id}$, that is, that $E = F^{\mathrm{Gal}_E(F)}$ for all intermediate fields E. Now, since $k \subseteq F$ is Galois, F is a splitting field for a separable polynomial $f(x)$ with coefficients in k. Viewing the coefficients in the larger field E, we see that $E \subseteq F$ is also a Galois extension. By Theorem 15.21, E is the fixed field of $\mathrm{Gal}_E(F)$, and this is precisely what we had to show.

For the converse implication, assume that the Galois correspondence is a bijection. In particular, the 'left-to-right' function σ in the correspondence is *injective*. According to Proposition 15.16, $\mathrm{Gal}_{F^G}(F) = G$ for all subgroups G of $\mathrm{Gal}_k(F)$. Applying this to $G = \mathrm{Gal}_k(F)$ gives

$$\mathrm{Gal}_{F^G}(F) = \mathrm{Gal}_k(F).$$

That is, $\sigma(F^{\mathrm{Gal}_k(F)}) = \sigma(k)$. Since σ is injective by hypothesis, this implies $F^{\mathrm{Gal}_k(F)} = k$: the fixed field of $\mathrm{Gal}_k(F)$ is k. By Theorem 15.21, $k \subseteq F$ is Galois, and we are done. $\square$

Example 15.32 Let's revisit once more the splitting field $\mathbb{Q} \subseteq F$ for the polynomial $x^3 - 2 \in \mathbb{Q}[x]$ (Example 15.11). We have found that its Galois group is S_3, simply acting by permutations on the roots $\alpha_1 = \sqrt[3]{2}$, $\alpha_2 = \frac{-1+i\sqrt{3}}{2}\alpha_1$, $\alpha_3 = \frac{-1-i\sqrt{3}}{2}\alpha_1$. We know all subgroups of S_3: using cycle notation, these form the lattice

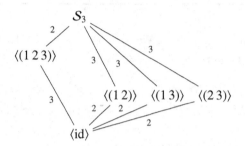

where the number on each segment indicates the index of the corresponding subgroup—for example, $[S_3 : \langle (1\,2\,3) \rangle] = 2$.

By the Galois correspondence the intermediate fields of the extension $\mathbb{Q} \subseteq F$ must form an isomorphic lattice, where inclusions are reversed:

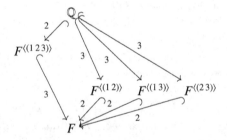

and now the number marking an inclusion arrow indicates the degree of the corresponding extension.[2] These are *all* the intermediate fields! They can be identified more concretely, as follows.

The field $F^{\langle (2\,3) \rangle}$ is the fixed field for the subgroup $\langle (2\,3) \rangle$ swapping α_2 and α_3. Since $\sqrt[3]{2} = \alpha_1$ is fixed by this subgroup, $\mathbb{Q}(\sqrt[3]{2}) \subseteq F^{\langle (2\,3) \rangle}$. On the other hand, computing degrees: $[F : F^{\langle (2\,3) \rangle}] = |\langle (2\,3) \rangle| = 2 = [F : \mathbb{Q}(\sqrt[3]{2})]$, so we can conclude that $F^{\langle (2\,3) \rangle} = \mathbb{Q}(\sqrt[3]{2})$.

Similarly, $F^{\langle (1\,3) \rangle} = \mathbb{Q}(\alpha_2) = \mathbb{Q}\left(\frac{-1+i\sqrt{3}}{2}\sqrt[3]{2} \right)$, $F^{\langle (1\,2) \rangle} = \mathbb{Q}(\alpha_3) = \mathbb{Q}\left(\frac{-1-i\sqrt{3}}{2}\sqrt[3]{2} \right)$.

Concerning $F^{\langle (1\,2\,3) \rangle}$, this must be generated by an element μ of degree 2 over $\mathbb{Q}$, since $[F : \mathbb{Q}(\mu)] = |\langle (1\,2,3) \rangle| = 3 = \frac{6}{2}$; and μ must be fixed by cyclic permutations

---

[2] As promised, the fact that we use the same notation for the index of a subgroup and for the degree of an extension is not a coincidence!

$\alpha_1 \mapsto \alpha_2 \mapsto \alpha_3 \mapsto \alpha_1$. If you look carefully, you will see that such an element is

$$\frac{\alpha_2}{\alpha_1} = \frac{\alpha_3}{\alpha_2} = \frac{\alpha_1}{\alpha_3} = \frac{-1 + i\sqrt{3}}{2}.$$

Equivalently, $F^{\langle(1\,2\,3)\rangle} = \mathbb{Q}(i\sqrt{3})$. We can redraw the lattice of *all* intermediate fields of the splitting field of $x^3 - 2$ as follows.

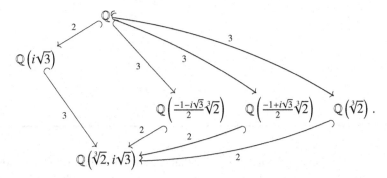

Example 15.33   We can carry out the same program for the splitting field of $x^5 - 1$, a.k.a. the splitting field of $x^4 + x^3 + x^2 + x + 1$. We know (cf. Example 15.12) that the splitting field is $\mathbb{Q}(\zeta)$, where $\zeta = e^{2\pi i/5}$, and that the Galois group is C_4, generated by the automorphism γ defined by sending ζ to ζ^2. We also know the structure of subgroups of a cyclic group C_n: there is one subgroup for every divisor of n (Proposition 10.10). For $n = 4$, the lattice of subgroups is particularly simple:

$$C_4 = \{e, \gamma, \gamma^2, \gamma^3\}$$

$$\Big|\, 2$$

$$\{e, \gamma^2\}$$

$$\Big|\, 2$$

$$\{e\}.$$

By the Galois correspondence, the extension $\mathbb{Q} \subseteq \mathbb{Q}(\zeta)$ must have precisely one proper intermediate field E, the fixed field of the subgroup $\{e, \gamma^2\}$; and $[\mathbb{Q}(\zeta) : E] = 2$. What is this field E?

Every element of $\mathbb{Q}(\zeta)$ may be written uniquely as a $\mathbb{Q}$-linear combination of powers $\zeta^i, i = 1, \ldots, 4$:

$$z = q_1\zeta + q_2\zeta^2 + q_3\zeta^3 + q_4\zeta^4.$$

The element γ^2 of the Galois group is the automorphism defined by mapping ζ to ζ^4. Therefore

$$\gamma^2(z) = q_1\zeta^4 + q_2\zeta^8 + q_3\zeta^9 + q_4\zeta^{16} = q_4\zeta + q_3\zeta^2 + q_2\zeta^3 + q_1\zeta^4,$$

where I have implemented the fact that $\zeta^5 = 1$. Therefore, z is in the fixed field if and only if $q_1 = q_4$ and $q_2 = q_3$. For example, $\zeta + \zeta^4 \in E$. This element has degree 2 over $\mathbb{Q}$:

it has two distinct conjugates, $\zeta + \zeta^4$ and $\zeta^2 + \zeta^3$, so we can argue as in Example 15.26. In fact, we can use this to compute that its minimal polynomial over $\mathbb{Q}$ is $x^2 + x - 1$, and this reveals that $\zeta + \zeta^4 = \frac{-1+\sqrt{5}}{2} = 0.618\ldots$, the (inverse of the) *golden ratio*. The conclusion is that $E = \mathbb{Q}(\zeta + \zeta^4) = \mathbb{Q}(\sqrt{5})$. The lattice of intermediate fields is

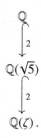

You have probably figured out that there will be a part II to the Fundamental Theorem of Galois Theory. What is it going to be about? If $k \subseteq F$ is Galois, and E is an intermediate field, then it is clear that $E \subseteq F$ is Galois: indeed, Galois extensions are splitting fields of separable polynomials, and a polynomial with coefficients in k is in particular a polynomial with coefficients in E if $k \subseteq E$. On the other hand, the extension $k \subseteq E$ is *not* necessarily Galois: our standard example $\mathbb{Q} \subseteq \mathbb{Q}(\sqrt[3]{2})$ of a non-Galois extension is a subfield of the splitting field of $x^3 - 2$ over $\mathbb{Q}$. As the next result shows, this example is, in fact, typical.

> PROPOSITION 15.34 *Let $k \subseteq E$ be a finite separable extension. Then E may be identified with an intermediate field of a Galois extension.*

Proof Finite separable extensions are simple (Theorem 14.24), so $E = k(\alpha)$ for some algebraic element α. Let $p(x) \in k[x]$ be the minimal polynomial for α over k; since $k \subseteq E$ is separable, $p(x)$ is a separable polynomial. Let then F be the splitting field of $p(x)$. The extension $k \subseteq F$ is Galois, since it is the splitting field of a separable polynomial. If α' is any root of $p(x)$ in F, there is a homomorphism $E \to F$ sending α to α' (Proposition 14.1); this homomorphism identifies E with an intermediate field of $k \subseteq F$, as needed. $\square$

Therefore, there is a natural question on the table: under what circumstances is a sub-extension $k \subseteq E$ of a Galois extension also Galois? A hint to the answer is provided by simple numerology. By multiplicativity of degrees (Proposition 13.13) and Proposition 15.30, if $k \subseteq E \subseteq F$, we have

$$[E : k] = \frac{[F : k]}{[F : E]} = \frac{|\mathrm{Gal}_k(F)|}{|\mathrm{Gal}_E(F)|} = [\mathrm{Gal}_k(F) : \mathrm{Gal}_E(F)] = |\mathrm{Gal}_k(F)/\mathrm{Gal}_E(F)|,$$

where $\mathrm{Gal}_k(F)/\mathrm{Gal}_E(F)$ denotes the set of left-cosets of the subgroup $\mathrm{Gal}_E(F)$ of $\mathrm{Gal}_k(F)$ (Definition 11.50). Therefore, by Theorem 15.21 *the extension $k \subseteq E$ is Galois if and* *only if*

$$|\operatorname{Gal}_k(E)| = |\operatorname{Gal}_k(F)/\operatorname{Gal}_E(F)|.$$

The only reasonable explanation for this observation is that if $k \subseteq E$ is Galois, then its Galois group should turn out to be isomorphic to the quotient $\operatorname{Gal}_k(F)/\operatorname{Gal}_E(F)$, and in particular the latter should be a *group*. That is, $\operatorname{Gal}_E(F)$ should be a *normal* subgroup in this case. The second part of the Fundamental Theorem of Galois Theory states that this is indeed the case.

> THEOREM 15.35 (Fundamental Theorem of Galois Theory, II) *Let $k \subseteq F$ be a Galois extension, and let E be an intermediate field: $k \subseteq E \subseteq F$. Then the extension $k \subseteq E$ is Galois if and only if $\operatorname{Gal}_E(F)$ is a* normal *subgroup of $\operatorname{Gal}_k(F)$, and in this case*
>
> $$\operatorname{Gal}_k(E) \cong \operatorname{Gal}_k(F)/\operatorname{Gal}_E(F).$$

Proof We have to prove that if $k \subseteq E$ is a subextension of a Galois extension, and in particular it is finite and separable (since so are Galois extensions), then the extension $k \subseteq E$ is normal if and only if the subgroup $\operatorname{Gal}_E(F)$ of $\operatorname{Gal}_k(F)$ is normal.[3]

First, let's assume that $\operatorname{Gal}_E(F)$ is normal in $\operatorname{Gal}_k(F)$. Therefore, $\varphi^{-1}\operatorname{Gal}_E(F)\varphi \subseteq \operatorname{Gal}_E(F)$ for all $\varphi \in \operatorname{Gal}_k(F)$. That is, $\forall \varphi \in \operatorname{Gal}_k(F)$, $\forall \psi \in \operatorname{Gal}_E(F)$, $\exists \psi' \in \operatorname{Gal}_E(F)$ such that $\psi\varphi = \varphi\psi'$.

We have to prove that the extension $k \subseteq E$ is normal, i.e., that if an irreducible polynomial $q(x) \in k[x]$ has a root u in E, then *all* roots of $q(x)$ are in E. Note that since F is Galois over k, and in particular normal, $q(x)$ splits completely in $F[x]$: therefore, F contains all roots of $q(x)$. In fact, the roots are the Galois conjugates of u (Proposition 15.24).

Let then $u \in E$ be a root of $q(x) \in k[x]$, and let $v \in F$ be another root of $q(x)$. Since v is a Galois conjugate of u, $\exists \varphi \in \operatorname{Gal}_k(F)$ such that $v = \varphi(u)$. We have to prove that $v \in E$.

Since $E \subseteq F$ is a Galois extension, E is the fixed field of $\operatorname{Gal}_E(F)$ (Theorem 15.21). Therefore, in order to prove that v is an element of E, it suffices to prove that it is fixed by every element of $\operatorname{Gal}_E(F)$. So we let $\psi \in \operatorname{Gal}_E(F)$, and we try to gather some information about $\psi(v)$.

By normality of $\operatorname{Gal}_E(F)$ there exists a $\psi' \in \operatorname{Gal}_E(F)$ such that $\psi\varphi = \varphi\psi'$. Therefore

$$\psi(v) = \psi(\varphi(u)) = \varphi(\psi'(u)) = \varphi(u) = v :$$

indeed u is an element of E, therefore u is fixed by every element of $\operatorname{Gal}_E(F)$. Since $\psi(v) = v$ for all $\psi \in \operatorname{Gal}_E(F)$, this verifies that $v \in E$ as needed.

This concludes the proof that $\operatorname{Gal}_E(F)$ normal $\implies k \subseteq E$ normal.

For the converse inclusion, consider $k \subseteq E \subseteq F$ with $k \subseteq F$ Galois *and* $k \subseteq E$ normal, i.e., Galois. The isomorphism proposed in the statement is a strong hint that we should try to set up an application of the First Isomorphism Theorem. Therefore, we seek a *surjective* homomorphism $\operatorname{Gal}_k(F) \to \operatorname{Gal}_k(E)$.

---

[3] If you have wondered why normal extensions are called normal, now you know.

A moment's thought reveals that the only reasonable homomorphism between these two groups acts by *restriction:* given $\varphi \in \mathrm{Gal}_k(F)$, i.e., an isomorphism $F \to F$ restricting to the identity on k, we can consider the restriction $\varphi|_E$ of φ to E. I claim that

- The image of $\varphi|_E$ is E: therefore, $\varphi|_E \in \mathrm{Gal}_k(E)$.
- The resulting homomorphism $\mathrm{Gal}_k(F) \to \mathrm{Gal}_k(E)$ is surjective.

The first claim follows directly from the normality of the extension $k \subseteq E$. Indeed, let $u \in E$ and let $q(x) \in k[x]$ be the minimal polynomial of u. By Proposition 15.6, $\varphi(u)$ must also be a root of $q(x)$. Since E contains one root of $q(x)$ (that is, u), and the extension $k \subseteq E$ is normal, it follows that E contains $\varphi(u)$ as well. This verifies that $\varphi|_E(E) \subseteq E$. As you verified in Exercise 13.8, it follows that $\varphi|_E$ is an *isomorphism* $E \to E$, and it restricts to the identity on k because so does φ. This verifies the first claim.

For the second claim, we have to go back to the minor upgrade of the statement proving the uniqueness of splitting fields: in Corollary 14.11 we observed that field isomorphisms lift to isomorphisms of splitting fields. Let $\psi \colon E \to E$ be an element of $\mathrm{Gal}_k(E)$. Viewing F as a splitting field over E (which we may do, since $E \subseteq F$ is Galois), Corollary 14.11 implies that there exists an isomorphism $\varphi \colon F \to F$ such that $\varphi|_E = \psi$. Since ψ restricts to the identity on k, so does φ; so $\varphi \in \mathrm{Gal}_k(F)$, and this verifies the second claim.

Therefore, restriction to E gives a surjective function $\mathrm{Gal}_k(F) \to \mathrm{Gal}_k(E)$. It is clear that this function preserves composition, so we get a *surjective group homomorphism* $\rho \colon \mathrm{Gal}_k(F) \to \mathrm{Gal}_k(E)$. By the First Isomorphism Theorem for groups,

$$\mathrm{Gal}_k(E) \cong \mathrm{Gal}_k(F)/\ker\rho.$$

What is $\ker\rho$?

$$\ker\rho = \{\varphi \in \mathrm{Gal}_k(F) \,|\, \rho(\varphi) = \mathrm{id}_E\} = \{\varphi \in \mathrm{Gal}_k(F) \,|\, \varphi|_E = \mathrm{id}_E\}$$
$$= \mathrm{Gal}_E(F).$$

We are done. This proves that $\mathrm{Gal}_E(F)$ is a normal subgroup of $\mathrm{Gal}_k(F)$, and establishes the stated isomorphism. □

Example 15.36   The splitting field of $x^3 - 2 \in \mathbb{Q}[x]$ provides us again with a good illustration of the material we have covered. The Galois group of $x^3 - 2$ is S_3. The only proper normal subgroup of S_3 other than the identity and S_3 itself is $\mathcal{A}_3 \cong C_3 = \langle(123)\rangle$. Therefore, the only intermediate field which is Galois over $\mathbb{Q}$, other than $\mathbb{Q}$ and the splitting field itself, is the fixed field $\mathbb{Q}(i\sqrt{3})$ of $\langle(123)\rangle$. The other three subgroups $\langle(12)\rangle, \langle(13)\rangle, \langle(23)\rangle$ are not normal in S_3, therefore the corresponding fixed fields are not Galois over $\mathbb{Q}$. The fixed field of $\langle(23)\rangle$ is $\mathbb{Q}(\sqrt[3]{2})$, so we are already quite familiar with this fact.

15.4   Galois Groups of Polynomials

At this point we have a beautiful machine associating a finite group with every polynomial with coefficients in a field: if k is a field, $f(x) \in k[x]$ is a polynomial, and
F is the splitting field of $f(x)$, then we have defined and studied to some extent the
group $\mathrm{Gal}_k(F)$. In the next section we will see the most famous application of this technology, according to which one can decide whether there exist formulas of a certain
type for the roots of a given polynomial in terms of group-theoretic qualities of the
corresponding group. This will be the *Galois criterion,* Theorem 15.47.

How to compute the Galois group of a polynomial is a different business, and even
naïve questions such as whether a given group is or is not the Galois group of some
polynomial become very subtle very quickly. It is conjectured that *every* finite group
may be realized as the Galois group of a polynomial with coefficients in $\mathbb{Q}$; this is the
inverse Galois problem. Solve it, and you will become famous.

The fact that every finite *abelian* group is the Galois group of a polynomial with rational coefficients is not too challenging. In fact, it is not difficult to prove that every finite
abelian group may be realized as $\mathrm{Gal}_\mathbb{Q}(E)$ for an intermediate field E of a 'cyclotomic'
extension $\mathbb{Q} \subseteq \mathbb{Q}(\zeta_n)$ by an nth root of 1. You can prove this yourself (Exercise 15.19),
if you accept a theorem of Dirichlet on 'arithmetic progressions'.

In fact, much more is true. Abelian groups are particular cases of *solvable* groups,
and Igor Shafarevich proved that every finite solvable group is the Galois group of a
polynomial with rational coefficients. We dealt briefly with solvable groups in §12.7,
and I mentioned that every group of *odd* order is solvable (this is the Feit–Thompson
theorem). So every group of odd order may also be realized as the Galois group of a
polynomial with rational coefficients—but this is a very hard, very deep result.

The requirement that the ground field be $\mathbb{Q}$ is very substantial in these questions: it
is relatively easy to prove that every finite group may be realized as the Galois group
of some extension $k \subseteq F$. Indeed, by Cayley's theorem (Theorem 11.29), every finite
group is a subgroup of a symmetric group S_n, so it suffices to produce an extension
whose Galois group is S_n.

For this, let $F = \mathbb{Q}(t_1, \ldots, t_n)$ be the field of rational functions in n indeterminates
$t_1, \ldots, t_n$. We define polynomials with rational (in fact, integer) coefficients,

$$s_1 = s_1(t_1, \ldots, t_n), \quad \ldots, \quad s_n = s_n(t_1, \ldots, t_n),$$

by requiring that the following identity holds:

$$x^n + s_1 x^{n-1} + \cdots + s_{n-1}x + s_n = (x + t_1) \cdots (x + t_n). \tag{15.2}$$

This equality implies

$$\begin{cases} s_1 = t_1 + \cdots + t_n, \\ s_2 = t_1 t_2 + t_1 t_3 + \cdots + t_{n-1}t_n, \\ \cdots \\ s_n = t_1 \cdots t_n. \end{cases}$$

The polynomials $s_i(t_1, \ldots, t_n)$ are the 'elementary symmetric functions' in the indeterminates $t_1, \ldots, t_n$. We may view them as elements of the field $F = \mathbb{Q}(t_1, \ldots, t_n)$, so they span a subfield $k = \mathbb{Q}(s_1, \ldots, s_n)$ of F.

PROPOSITION 15.37   *With notation as above, $k \subseteq F$ is the splitting field of the polynomial*

$$S(x) := x^n + s_1 x^{n-1} + \cdots + s_{n-1} x + s_n \in k[x].$$

The extension $k \subseteq F$ is Galois, and $\mathrm{Gal}_k(F) \cong S_n$.

Proof   By the very definition of the elements s_i, the polynomial $S(x)$ splits into n distinct linear factors; its roots $-t_i$ generate $F = \mathbb{Q}(t_1, \ldots, t_n)$. Therefore F is indeed the splitting field of $S(x)$, and $S(x)$ is separable. It follows that the extension is Galois, as stated, and we only have to prove that $\mathrm{Gal}_k(F) \cong S_n$.

For all permutations $\sigma \in S_n$, we may define an isomorphism $\iota_\sigma : F \to F$ by mapping any element $f(t_1, \ldots, t_n) \in F = \mathbb{Q}(t_1, \ldots, t_n)$ (i.e., a rational function in the indeterminates t_i) to $f(t_{\sigma(1)}, \ldots, t_{\sigma(n)})$. Observe that for $i = 1, \ldots, n$, $\iota_\sigma(s_i) = s_i$: indeed, permuting the factors of the polynomial in the right-hand side of (15.2) does not change this polynomial. (This is why the functions s_i are called 'symmetric'.) This means that each ι_σ restricts to the identity on $k = \mathbb{Q}(s_1, \ldots, s_n)$, i.e., it is an element of the Galois group $\mathrm{Gal}_k(F)$.

Thus, we have defined a function $\iota : S_n \to \mathrm{Gal}_k(F)$, mapping σ to ι_σ. You should verify that ι is a group homomorphism. It is clearly injective, since its kernel is the identity (the only permutation σ for which $\sigma(t_i) = t_i$ for all i is the identity). Therefore, we may identify S_n with a subgroup of $\mathrm{Gal}_k(F)$: $S_n \subseteq \mathrm{Gal}_k(F)$.

On the other hand, since $S(x)$ has n roots, $\mathrm{Gal}_k(F)$ may be identified with a subgroup of S_n by Corollary 15.8. Therefore $S_n \subseteq \mathrm{Gal}_k(F) \subseteq S_n$, and the conclusion is that $\mathrm{Gal}_k(F)$ is isomorphic to S_n as stated. □

Remark 15.38   One way to view Proposition 15.37 is the following. In general, a function of n variables $f(t_1, \ldots, t_n)$ is 'symmetric' if $f(t_{\sigma(1)}, \ldots, t_{\sigma(n)}) = f(t_1, \ldots, t_n)$ for all permutations $\sigma \in S_n$. Thus, the field of symmetric rational functions is the fixed field of S_n in $\mathbb{Q}(t_1, \ldots, t_n)$. What we have proved is that this fixed field is precisely the field $\mathbb{Q}(s_1, \ldots, s_n)$ of rational functions in the elementary symmetric functions.

In other words, we have proved that every symmetric rational function may be expressed as a rational function in the elementary symmetric functions. This is known as the *fundamental theorem on symmetric functions*. We worked over $\mathbb{Q}$, but the argument transfers without modification to any base field. Refining it proves that in fact every symmetric *polynomial* is a *polynomial* in the elementary symmetric functions.   ⌟

COROLLARY 15.39   *For every finite group G there exists a field E and a separable polynomial $f(x) \in E[x]$ such that the Galois group of $f(x)$ is isomorphic to G.*

Proof   By Cayley's theorem, we may identify G with a subgroup of S_n, for some n.

(For example, we could take $n = |G|$.) By Proposition 15.37 there exists a Galois extension $k \subseteq F$ whose Galois group is isomorphic to S_n. For the fixed field E of the subgroup G of S_n, $E \subseteq F$ is Galois and $\operatorname{Gal}_E(F) \cong G$ according to the Galois correspondence. Since F is Galois over E, it is a splitting field for a separable polynomial[4] $f(x)$, and we are done.   □

Remark 15.40   It almost feels as if we have cheated a little. Following the argument leading to the proof of this corollary, it turns out that we can take $F = \mathbb{Q}(t_1, \ldots, t_n)$ and $f(x) = S(x) = x^n + s_1 x^{n-1} + \cdots + s_n$. Neither F nor this polynomial are very interesting; the burden of realizing G as the Galois group of $E \subseteq F$ rests all on the shoulders of the mysterious subfield E. By contrast, the *inverse Galois problem* fixes the ground field to be $\mathbb{Q}$, and demands the construction of an interesting polynomial $f(x) \in \mathbb{Q}[x]$ (or of the corresponding splitting field F) realizing the group G. This is many orders of magnitude deeper; as I mentioned, the 'inverse Galois problem' is still open.   ⌐

Is there a 'direct' Galois problem? That would be the problem of finding the Galois group of a given polynomial. We have seen a few examples—Proposition 15.37 a moment ago, but also more specific examples such as $x^3 - 2 \in \mathbb{Q}[x]$, which we have analyzed in depth (Example 15.11 and 15.32), or $x^4 + x^3 + x^2 + x + 1$ (Example 15.12); and the reader will see more in the exercises. The example of $x^3 - 2 \in \mathbb{Q}[x]$ has been a constant refrain, and we are now in a position to generalize it very substantially. This generalization is a good illustration of the type of tools needed for this type of result.

Example 15.41   Let k be a field, and let

$$f(x) = x^3 + ax^2 + bx + c \in k[x]$$

be a separable irreducible polynomial of degree 3. What is its Galois group?

Let $\lambda_1, \lambda_2, \lambda_3$ be the roots of $f(x)$ in $\bar{k}$, so that $k \subseteq F = k(\lambda_1, \lambda_2, \lambda_3)$ is a splitting field for $f(x)$. By Theorem 14.6, $[F : k] \leq 3! \leq 6$; by Corollary 15.8, $\operatorname{Gal}_k(F)$ is a subgroup of S_3. Since

$$x^3 + ax^2 + bx + c = (x - \lambda_1)(x - \lambda_2)(x - \lambda_3)$$

in $F[x]$, we see that $\lambda_1 + \lambda_2 + \lambda_3 = a \in k$: this tells us that $\lambda_3 \in k(\lambda_1, \lambda_2)$, so in fact $F = k(\lambda_1, \lambda_2)$. We have the sequence of extensions

$$k \subseteq k(\lambda_1) \subseteq k(\lambda_1, \lambda_2) = F,$$

and $[k(\lambda_1) : k] = 3$ (since $f(x)$ is irreducible). Therefore, there are two possibilities:

- Either $k(\lambda_1) = k(\lambda_1, \lambda_2)$: in this case $F = k(\lambda_1)$ is the splitting field of $f(x)$, and we have $[F : k] = 3$, hence $|\operatorname{Gal}_k(F)| = 3$ and $\operatorname{Gal}_k(F) \cong C_3$ is cyclic of order 3.
- Or $k(\lambda_1) \subsetneq k(\lambda_1, \lambda_2)$: in this case $[F : k] = 6$ (since $3 < [F : k] \leq 6$, and $[F : k]$ is a multiple of 3 by multiplicativity of degrees), hence $|\operatorname{Gal}_k(F)| = 6$ and $\operatorname{Gal}_k(F) \cong S_3$ is the full symmetric group.

---

[4] In fact, with notation as in Proposition 15.37, we could take $f(x) = S(x)$.

This is already progress: the Galois group of $f(x)$ is either $\mathcal{A}_3$ (the only subgroup of S_3 isomorphic to C_3) or S_3. The first case occurs when there are no proper intermediate fields $k \subsetneq E \subsetneq F$. By the Fundamental Theorem of Galois Theory, in the second case the extension must have an intermediate field $k \subsetneq E \subsetneq F$ corresponding to the alternating group $\mathcal{A}_3 \subsetneq S_3$. In this case $E = k(\delta)$, with δ fixed by $\mathcal{A}_3$ but not by S_3 (since $\delta \notin k$). So the Galois group is S_3 precisely when there is such an element δ. Consider then

$$\delta = (\lambda_1 - \lambda_2)(\lambda_1 - \lambda_3)(\lambda_2 - \lambda_3).$$

Take a moment to verify that every transposition changes the sign of this element; it follows that δ is fixed by $\mathcal{A}_3$ but not by S_3. Therefore:

the Galois group of $f(x)$ is $\mathcal{A}_3$ or S_3 according to whether $\delta \in k$ or $\delta \notin k$.

We also see that

$$D := \delta^2 = (\lambda_1 - \lambda_2)^2(\lambda_1 - \lambda_3)^2(\lambda_2 - \lambda_3)^2$$

is fixed by S_3; it follows that $D \in k$ in both cases. Therefore:

the Galois group of $f(x)$ is $\mathcal{A}_3$ or S_3 according to whether D is, or is not,

a square in k.

The quantity D is called the 'discriminant' of $f(x)$: it is 0 precisely when two of the roots coincide (which does not happen in our case, since we are assuming the polynomial to be separable).

There is a formula expressing D explicitly in terms of the coefficients a, b, c of $f(x)$, but it is a little messy; a clever trick simplifies the situation considerably, at least if the characteristic of k is not 3. Note that D does not change if we shift $\lambda_1, \lambda_2, \lambda_3$ by a fixed quantity. We shift by $-a/3$, or equivalently we perform the change of variables $x \mapsto x - a/3$ in $f(x)$. We obtain

$$f(x - a/3) = (x - a/3)^3 + a(x - a/3)^2 + b(x - a/3) + c = x^3 - ax^2 + \cdots + ax^2 + \cdots$$
$$= x^3 + px + q$$

for suitable p and q: that is, we have eliminated the term in x^2. The shift does not modify D, since D only depends on the *differences* of the roots, and does not change the Galois group (since it does not change the splitting field); we have no loss of generality. One can then verify that

$$D = -4p^3 - 27q^2.$$

We are done! We have proved that:

the Galois group of an irreducible polynomial $x^3 + px + q$ is $\mathcal{A}_3$ or S_3

according to whether $D = -4p^3 - 27q^2$ is, or is not, a square in k,

and up to a shift this is as good as the general case.

For instance, $D = -27 \cdot 4 = -108$ for $x^3 - 2$; since -108 is not a square in $\mathbb{Q}$, this confirms that the Galois group of $x^3 - 2$ over $\mathbb{Q}$ is S_3. On the other hand, the discriminant of $x^3 - 3x + 1$ is

$$-4 \cdot (-3)^3 - 27 \cdot 1^2 = 108 - 27 = 81,$$

a perfect square in $\mathbb{Q}$, therefore the Galois group of $x^3 - 3x + 1$ over $\mathbb{Q}$ is $\mathcal{A}_3$.

Similar (but of course more demanding) analyses may be carried out for polynomials of higher degree. A helpful observation is that if $f(x) \in k[x]$ is an *irreducible* polynomial, then its Galois group G must act *transitively* (Definition 11.24) on the set of roots of $f(x)$ in the splitting field (why? Exercise 15.22). On the other hand, by Corollary 15.8, G must be isomorphic to a subgroup of S_d, where $d = \deg f$. This restricts the possibilities for G. For example, one may verify that the only subgroups of S_4 acting transitively on $\{1, 2, 3, 4\}$ are (isomorphic to) C_4, $C_2 \times C_2$, D_8, $\mathcal{A}_4$, and S_4. Therefore, the Galois group of an irreducible quartic must be isomorphic to a group on this list.

15.5 Solving Polynomial Equations by Radicals

Polynomial equations of degree 2 over some field k,

$$x^2 + bx + c = 0,$$

may be solved (in characteristic $\neq 2$) using the *quadratic formula*

$$x = \frac{-b \pm \sqrt{b^2 - 4c}}{2}.$$

From the point of view of field theory, what this shows is that the splitting field of $f(x) = x^2 + bx + c$ over k is $k(\sqrt{D})$, where $D = b^2 - 4c$ is the discriminant of $f(x)$: by the quadratic formula, $k(\sqrt{D})$ contains the roots of $f(x)$ and it is generated by them over k. Thus, the splitting field of $f(x)$ is contained in (and actually equal to) an extension

$$k \subseteq k(u),$$

where u is a square root of $D \in k$. More generally, we say that a polynomial equation

$$f(x) = x^n + a_{n-1}x^{n-1} + \cdots + a_2 x^2 + a_1 x + a_0 = 0 \tag{15.3}$$

with coefficients in a field k is 'solvable by radicals' if the splitting field F of $f(x)$ is contained in the top field of a 'tower'

$$k \subseteq k(u_1) \subseteq k(u_1, u_2) \subseteq \cdots \subseteq k(u_1, \ldots, u_r)$$

where each u_i is an m_ith root of an element of $k(u_1, \ldots, u_{i-1})$. In this case every element of F, and in particular the roots of $f(x)$, may be written as some formula starting from elements of k and using the usual field operations and the extraction of roots, that is, 'radicals'. The quadratic formula is the simplest example. The solutions of the cubic equation

$$x^3 + ax^2 + bx + c = 0$$

may also be expressed in this fashion: for example, one solution for the simpler case $x^3 + px + q = 0$ is

$$\sqrt[3]{-\frac{q}{2} + \sqrt{\frac{p^3}{27} + \frac{q^2}{4}}} + \sqrt[3]{-\frac{q}{2} - \sqrt{\frac{p^3}{27} + \frac{q^2}{4}}},$$

where p and q are assumed to be real and the cube roots are the real cube roots. The other solutions have a similar form.

The fact that formulas in radicals exist for the solutions of polynomial equations of degree 2, 3, and (as it happens) 4 says that the corresponding splitting fields are contained in *radical* extensions.

DEFINITION 15.42   A field extension $k \subseteq K$ is 'radical' if there exist $u_1, \ldots, u_r \in K$ and positive integers $m_1, \ldots, m_r$ such that

$$k \subseteq k(u_1) \subseteq k(u_1, u_2) \subseteq \cdots \subseteq k(u_1, \ldots, u_r) = K$$

and $u_i^{m_i} \in k(u_1, \ldots, u_{i-1})$ for $i = 1, \ldots, r$.

A polynomial $f(x) \in k[x]$ is 'solvable by radicals' if its splitting field is contained in a radical extension.

Example 15.43   For every $n \geq 1$, the splitting field E of $x^n - 1 \in k[x]$ is a radical extension of k. Indeed, the set of nth roots of 1 is clearly closed under products and taking inverses, so it is a subgroup of the multiplicative group of E. As such, it must be *cyclic*, by Theorem 10.24. Assuming for simplicity that $x^n - 1$ is separable (this is the case if k has characteristic 0, for example), then there are exactly n roots, so this group is isomorphic to $\mathbb{Z}/n\mathbb{Z}$. If ζ_n is a generator of this cyclic group, then $E = k(\zeta_n)$. (If $k = \mathbb{Q}$, we can take $\zeta_n = e^{2\pi i/n}$ in $\mathbb{C}$.) The extension $k \subseteq k(\zeta_n) = E$ is radical, since $\zeta_n^n = 1 \in k$.

The polynomial $x^n - 1$ is solvable by radicals: and indeed, its roots may be found by a 'formula' involving the radical $\sqrt[n]{1}$.

What can we say about $\mathrm{Gal}_k(E)$? Every element σ of $\mathrm{Gal}_k(E)$ is determined by the image of ζ_n; this image must be another root of $x^n - 1$, so $\sigma(\zeta_n) = \zeta_n^i$ for some i. If $\sigma, \tau \in \mathrm{Gal}_k(E)$, say $\tau(\zeta_n) = \zeta_n^j$, then $\sigma \circ \tau$ and $\tau \circ \sigma$ must agree, since they both map the generator ζ_n to ζ_n^{ij}. Therefore, $\mathrm{Gal}_k(E)$ is abelian.

(In fact, $\mathrm{Gal}_k(E) \cong (\mathbb{Z}/n\mathbb{Z})^*$ if $x^n - 1$ is separable: Exercise 15.4.)

We will use the result of this example below. The generator ζ_n featured in this example has a name.

DEFINITION 15.44   Let F be a field. An element $\zeta_n \in F$ is a 'primitive' nth root of 1 if the order $|\zeta_n|$ of ζ_n in F^* is n.

That is, ζ_n is a primitive root of 1 if $\zeta_n^n = 1$ and $\zeta_n^m \neq 1$ for all $0 < m < n$. It follows that the n elements ζ_n^i are all the roots of the polynomial $x^n - 1$ in F. (In the case of interest to us, this polynomial will be separable, so its n roots are distinct.)

With solvability of polynomials as our goal, we can pose a specific question: *given a Galois extension $k \subseteq F$, does there exist a radical extension $k \subseteq K$ containing it?*

One useful technical observation is that we may in fact assume that the radical extension is itself a Galois extension.

> **LEMMA 15.45**   *Every separable radical extension is contained in a Galois radical extension.*

Proof   Let

$$k \subseteq k(u_1) \subseteq k(u_1, u_2) \subseteq \cdots \subseteq k(u_1, \ldots, u_r) = K$$

be a radical extension, as in Definition 15.42, and assume it is separable. As a finite separable extension, $k \subseteq K$ is *simple* (Theorem 14.24): $K = k(\alpha)$, where α is a root of an irreducible separable polynomial $p(x) \in k[x]$. I claim that the splitting field $k \subseteq L$ of $p(x)$ is also a radical extension. As the splitting field of a separable polynomial, L is Galois; so this will establish the lemma.

To verify my claim, let $\alpha = \alpha_1, \alpha_2, \ldots, \alpha_r$ be the roots of $p(t)$ in L, so that $L = k(\alpha_1, \ldots, \alpha_r)$. Each extension $k \subseteq k(\alpha_j)$ is isomorphic to $k \subseteq k(\alpha)$, so it is a radical extension: we have

$$k \subseteq k(u_1^{(j)}) \subseteq k(u_1^{(j)}, u_2^{(j)}) \subseteq \cdots \subseteq k(u_1^{(j)}, \ldots, u_r^{(j)}) = k(\alpha_j)$$

with $u_i^{(j)^{m_i}} \in k(u_1^{(j)}, \ldots, u_{i-1}^{(j)})$. We can then present L as a radical extension: we have

$$k \subseteq k(\alpha_1) \subseteq k(\alpha_1, \alpha_2) \subseteq \cdots \subseteq k(\alpha_1, \ldots, \alpha_r) = L$$

and the jth extension in this sequence is obtained by adjoining $u_1^{(j)}, \ldots, u_r^{(j)}$; each step is radical, so $k \subseteq L$ is radical.   □

With this understood, and working in characteristic 0 to avoid separability issues, we have a very neat answer to the question raised a moment ago. We will prove the following theorem later on in this section.

> **THEOREM 15.46**   *Let k be a field of characteristic 0, and let $k \subseteq F$ be a Galois extension. Then $k \subseteq F$ is contained in a Galois radical extension if and only if $\mathrm{Gal}_k(F)$ is solvable.*

Solvable groups were the object of §12.7. Now you know why they are called 'solvable'! The following celebrated theorem is a direct consequence.

> **THEOREM 15.47 (Galois criterion)**   *Let k be a field of characteristic 0, and let $f(x) \in k[x]$ be an irreducible polynomial. Then $f(x)$ is solvable by radicals if and only if its Galois group is solvable.*

Proof This follows from Theorem 15.46. Indeed, by definition a polynomial is solvable by radicals if and only if its splitting field F is contained in a radical extension. By Lemma 15.45 this is the case if and only if F is contained in a *Galois* radical extension; and by Theorem 15.46, F is contained in a Galois radical extension if and only if $\mathrm{Gal}_k(F)$ is solvable.   □

COROLLARY 15.48   *There are no formulas in radicals for the solutions of a general polynomial equation of degree ≥ 5.*

Here, by a 'general polynomial' equation I mean an equation of the form (15.3), with $a_0, \ldots, a_{n-1}$ indeterminates. The quadratic formula is a formula in radicals for the solution of $x^2 + bx + c = 0$ with indeterminates b and c. Solutions of a given equation would be obtained by plugging in the coefficients for the indeterminates in such a formula.

Proof The existence of a formula in radicals for the general polynomial equation

$$f(x) = x^n + a_{n-1}x^{n-1} + \cdots + a_2x^2 + a_1x + a_0 = 0$$

would imply that the splitting field of this polynomial, as an extension of the field $\mathbb{Q}(a_0, \ldots, a_{n-1})$, is contained in a radical extension. The Galois group of $f(x)$ is S_n, by Proposition 15.37. By Theorem 12.62, S_n is *not* solvable for $n \geq 5$, so the statement follows from Theorem 15.47.   □

Corollary 15.48 is historically significant. The quadratic formula was known since antiquity. Formulas for the solution of equations of degree 3 and 4 were found by Cardano and Tartaglia in the sixteenth century. It took another 250+ years before it was proved that there are no formulas in radicals for the solutions of equations of degree 5 or higher (that is, Corollary 15.48): this is known as the *Abel–Ruffini theorem*. The original proof did not use Galois theory, and did not provide the precise criterion of Theorem 15.47. This was due to Galois, who obtained it at age 18, developing group theory and field theory from scratch for this purpose.[5]

It is not difficult to produce explicit polynomials of degree $n \geq 5$ with integer coefficients and Galois group S_n. For example, you will verify that $x^5 - 8x + 2$ is such a polynomial in Exercise 15.27: therefore, this concrete quintic cannot be solved by radicals. This means that its roots are complex numbers that cannot be written in terms of rational numbers by applying the usual operations and extractions of roots.

The proof of Theorem 15.46 does not involve any tool that we have not covered, but it is somewhat technical. The statement itself is not implausible. By the Galois correspondence, sequences of intermediate fields of an extension $k \subseteq K$ correspond to sequences of subgroups of the Galois group $\mathrm{Gal}_k(K)$. Under suitable hypotheses, these sequences of subgroups are in fact *normal series* in the sense of 12.7. One then proves a structure theorem showing that, again under suitable hypotheses, extensions with *cyclic* Galois groups are precisely the extensions $E \subseteq E(u)$ obtained by adjoining a root of the polynomial $x^m - D$, for some $D \in E$. Therefore, Galois radical extensions correspond

---

[5] Yes, this is just as jaw-dropping as it sounds.

to Galois groups admitting normal series with cyclic quotients, i.e., solvable groups (Exercise 12.31).

Fleshing out full details of this quick sketch requires some care, and will take the rest of this section. Let us first prove one implication in the theorem, that is, the statement that if a Galois extension $k \subseteq F$ is contained in a Galois radical extension, then $\mathrm{Gal}_k(F)$ is solvable. This implication suffices in order to prove Corollary 15.48, and it is more straightforward than the converse implication.

Proof Note that since k is assumed to be of characteristic 0, polynomials $x^m - D$ with $D \neq 0$ are separable, and a finite extension of k is Galois over k if and only if it is the splitting field of a polynomial (Exercise 15.1). This will come in handy in the course of the proof.

Let $k \subseteq F$ be a Galois extension. Assume it is contained in a Galois radical extension $k \subseteq K$. By definition, we have a chain of extensions

$$k \subseteq k(u_1) \subseteq k(u_1, u_2) \subseteq \cdots \subseteq k(u_1, \ldots, u_r) = K$$

with $u_i^{m_i} \in k(u_1, \ldots, u_{i-1})$ for $i = 1, \ldots, r$.

As K is Galois, it is the splitting field of a polynomial. We can then view K as a subfield of the algebraic closure $\overline{k}$ of k, and the elements $u_1, \ldots, u_r$ as elements of $\overline{k}$. We enlarge $k \subseteq K$ further, as follows. Let $n = m_1 \cdots m_r$, and let E be the subfield of $\overline{k}$ generated by k and all nth roots of 1. Then $k \subseteq E$ is a radical Galois extension (cf. Example 15.43), and E contains all m_ith roots of 1, for all i. We replace $k \subseteq K$ with $k \subseteq L$, where $L = E(u_1, \ldots, u_r)$. We have

$$k \subseteq E \subseteq E(u_1) \subseteq E(u_1, u_2) \subseteq \cdots \subseteq E(u_1, \ldots, u_r) = L, \tag{15.4}$$

so that $k \subseteq L$ is a radical extension. Note that $k \subseteq L$ is still Galois: since $k \subseteq K$ is Galois, it is the splitting field of a polynomial $f(x)$; L is then the splitting field of $(x^n - 1)f(x)$.

By the Galois correspondence, the chain of extensions (15.4) corresponds to a chain of subgroups of $\mathrm{Gal}_k(L)$:

$$\mathrm{Gal}_k(L) \supseteq G_0 \supseteq G_1 \supseteq G_2 \supseteq \cdots \supseteq G_r = \{e\}, \tag{15.5}$$

with $G_i = \mathrm{Gal}_{E(u_1, \ldots, u_i)}(L)$. I claim that the nontrivial inclusions in this chain form a normal series, with abelian quotients: therefore $\mathrm{Gal}_k(L)$ is solvable.

To verify my claim, use the second part of the Fundamental Theorem of Galois Theory, Theorem 15.35. As we have verified in Example 15.43, $k \subseteq E$ is Galois, and $\mathrm{Gal}_k(E)$ is abelian. By the Fundamental Theorem, $G_0 = \mathrm{Gal}_E(L)$ is normal in $\mathrm{Gal}_k(L)$, and $\mathrm{Gal}_k(L)/\mathrm{Gal}_E(L) \cong \mathrm{Gal}_k(E)$ is abelian. By the same token, in order to prove that G_{i+1} is normal in G_i, with abelian quotient, we have to prove that

$$E(u_1, \ldots, u_i) \subseteq E(u_1, \ldots, u_{i+1})$$

is Galois, and $\mathrm{Gal}_{E(u_1, \ldots, u_i)}(E(u_1, \ldots, u_{i+1}))$ is abelian.

Therefore, in order to prove the claim we are reduced to proving that extensions

$$E' \subseteq E'(u),$$

with $u \notin E'$, $u^m = D \in E'$, and such that E' contains all m-roots of 1, are Galois and have abelian Galois group. (You will prove something a little more precise in Exercise 15.28.)

For this, let $\zeta_m \in E'$ be a primitive mth root of 1, so that

$$1, \quad \zeta_m, \quad \zeta_m^2, \quad \ldots, \quad \zeta_m^{m-1}$$

are the m distinct roots of $x^m - 1$ in E'. Consider the m distinct elements of $E'(u)$:

$$u, \quad \zeta_m u, \quad \zeta_m^2 u, \quad \ldots, \quad \zeta_m^{m-1} u.$$

Since u is a root of $x^m - D$, so is each $\zeta_m^i u$:

$$(\zeta_m^i u)^m = \zeta_m^{mi} u^m = 1^i u^m = D.$$

Since a polynomial of degree m over a field can have at most m roots (Exercise 7.14), it follows that $\zeta_m^i u$, $i = 0, \ldots, m-1$, are *all* the roots of $x^m - D$, and that

$$E'(u) = E'(u, \zeta_m u, \zeta_m^2 u, \cdots, \zeta_m^{m-1} u)$$

is the splitting field of $x^m - D$ over E'. This proves that $E' \subseteq E'(u)$ is Galois, as stated.

Concerning $\mathrm{Gal}_{E'}(E'(u))$, argue as in Example 15.43. Elements $\sigma, \tau \in \mathrm{Gal}_{E'}(E'(u))$ are determined by the images $\sigma(u)$, $\tau(u)$, and these must be roots of $x^m - D$: therefore $\sigma(u) = \zeta_m^i u$, $\tau(u) = \zeta_m^j u$ for some integers i, j. But then $\sigma \circ \tau$ and $\tau \circ \sigma$ must agree, since both of them map u to $\zeta_m^{i+j} u$, as you should verify.

Therefore $\mathrm{Gal}_{E'}(E'(u))$ is abelian as stated, and this completes the verification of my claim: at this stage we have proved that $\mathrm{Gal}_k(L)$ admits a normal series with abelian quotients. Therefore $\mathrm{Gal}_k(L)$ is solvable.

By assumption, $k \subseteq F$ is contained in the extension $k \subseteq L$: we have $k \subseteq F \subseteq L$, with $k \subseteq F$ Galois. By the second part of the Fundamental Theorem of Galois Theory again,

$$\mathrm{Gal}_k(F) \cong \mathrm{Gal}_k(L)/\mathrm{Gal}_F(L).$$

In particular, $\mathrm{Gal}_k(F)$ is a homomorphic image of a solvable group. By Proposition 12.60 this implies that $\mathrm{Gal}_k(F)$ is solvable, and we are done. □

This proves 'half' of the key Theorem 15.46—which is enough to prove the unsolvability of general polynomial equations of degree 5 or higher, i.e., Corollary 15.48. (So we already know more about this than humankind did before Ruffini, Abel, and Galois came along!)

The general blueprint for the proof of the converse implication is as you may expect. We need to show that if $k \subseteq F$ is a Galois extension such that $\mathrm{Gal}_k(F)$ is solvable, then it is contained in a Galois radical extension. Since $\mathrm{Gal}_k(F)$ is solvable, it admits a normal series where each quotient is abelian; in fact, we may assume that the quotients are *cyclic* (Exercise 12.31). If the cyclic quotients associated with the normal series have orders $m_1, \ldots, m_r$, we can enlarge $k \subseteq F$ so that it contains all m_ith roots of 1, for all i. Then the task amounts to showing that if the roots of 1 are present, an extension with cyclic Galois group is radical. This will show that $k \subseteq F$ is contained in a (separable) radical extension, and we already know (Lemma 15.45) that this implies that $k \subseteq F$ is contained in a *Galois* radical extension.

Making mathematics out of this sketch requires a couple of preliminary observations, which for whatever reason appear to be more technical than what we have seen so far. As they say at the dentist's, this may cause some discomfort. Brace yourself.

The first observation concerns the behavior of the Galois group as we 'enlarge' an extension. Work in an algebraic closure $\bar{k}$ of the field k. Let E be a subfield of $\bar{k}$ containing k, and let $\alpha \in \bar{k}$. Assume that $k \subseteq k(\alpha)$ is a Galois extension. We are interested in the extension $E \subseteq E(\alpha)$.

> **LEMMA 15.49** *With notation as above, assume $k \subseteq k(\alpha)$ is a Galois extension. Then $E \subseteq E(\alpha)$ is a Galois extension, and $\mathrm{Gal}_E(E(\alpha))$ may be identified with a subgroup of $\mathrm{Gal}_k(k(\alpha))$.*

Proof Since $k \subseteq k(\alpha)$ is a Galois extension, it is a splitting field of a separable polynomial $f(x) \in k[x]$. It follows that $E \subseteq E(\alpha)$ is a splitting field for the same polynomial, viewed in $E[x]$. This shows that $E \subseteq E(\alpha)$ is Galois.

We are going to define an injective homomorphism $\rho \colon \mathrm{Gal}_E(E(\alpha)) \to \mathrm{Gal}_k(k(\alpha))$. For this, let $\varphi \in \mathrm{Gal}_E(E(\alpha))$ be an isomorphism $E(\alpha) \to E(\alpha)$ restricting to the identity on E. Since $k \subseteq k(\alpha)$ is a Galois extension, the restriction $\varphi|_{k(\alpha)}$ maps $k(\alpha)$ isomorphically to itself. (Cf. Remark 14.12; in fact, this is a characterization of Galois extensions, see Exercise 15.7.) Define $\rho(\varphi) := \varphi|_{k(\alpha)}$.

It is clear that ρ is a group homomorphism. I claim it is injective. Indeed, suppose $\varphi|_{k(\alpha)} = \mathrm{id}_{k(\alpha)}$. Then $\varphi(\alpha) = \alpha$, while $\varphi|_E = \mathrm{id}_E$ since $\varphi \in \mathrm{Gal}_E(E(\alpha))$. It follows that φ is the identity on $E(\alpha)$, and this proves that $\ker \rho$ is trivial.

As we have an injective homomorphism $\rho \colon \mathrm{Gal}_E(E(\alpha)) \to \mathrm{Gal}_k(k(\alpha))$, $\mathrm{Gal}_E(E(\alpha))$ may be identified with a subgroup of $\mathrm{Gal}_k(k(\alpha))$ as promised. $\qquad\square$

Note that, by Proposition 12.60, $\mathrm{Gal}_E(E(\alpha))$ is solvable if $\mathrm{Gal}_k(k(\alpha))$ is solvable.

A second preliminary observation deals with characters of a group. A 'character' of a group G is a group homomorphism from G to the multiplicative group of a field. The following result is attributed to Dedekind.

> **LEMMA 15.50** *Distinct characters of a group G in a field F are linearly independent over F.*

This statement should be interpreted as follows. Let $\varphi_1, \dots, \varphi_r$ be distinct characters $G \to F^*$. For coefficients $\lambda_i \in F$ we can consider the expression

$$\lambda_1 \varphi_1 + \cdots + \lambda_r \varphi_r$$

as a function from G to F (of course this function need not be a character). The statement is that if the φ_i are distinct, then the only such expression giving the function identically equal to 0 is the one obtained for $\lambda_1 = \cdots = \lambda_r = 0$.

Proof Arguing by contradiction, assume that there are distinct characters $\varphi_i \colon G \to F^*$,

$i = 1, \ldots, r$, and nonzero elements $\lambda_i \in F$ such that

$$\forall g \in G \qquad \lambda_1 \varphi_1(g) + \lambda_2 \varphi_2(g) + \cdots + \lambda_r \varphi_r(g) = 0. \tag{15.6}$$

We may assume that r is as small as possible with this requirement. We are seeking a contradiction.

We manipulate (15.6) in two ways. First, we apply it to arbitrary products $hg \in G$:

$$\forall g, h \in G \qquad \lambda_1 \varphi_1(hg) + \lambda_2 \varphi_2(hg) + \cdots + \lambda_r \varphi_r(hg) = 0.$$

Note that for all $g, h \in G$, $\varphi_i(hg) = \varphi_i(h)\varphi_i(g)$: by definition, characters are homomorphisms. Therefore, we get

$$\forall g, h \in G \qquad \lambda_1 \varphi_1(h)\varphi_1(g) + \lambda_2 \varphi_2(h)\varphi_2(g) + \cdots + \lambda_r \varphi_r(h)\varphi_r(g) = 0. \tag{15.7}$$

On the other hand, we can also multiply (15.6) through by $\varphi_1(h)$, for any $h \in H$. This gives

$$\forall g, h \in G \qquad \lambda_1 \varphi_1(h)\varphi_1(g) + \lambda_2 \varphi_1(h)\varphi_1(g) + \cdots + \lambda_r \varphi_1(h)\varphi_r(g) = 0. \tag{15.8}$$

Subtracting (15.8) from (15.7) gives

$$\forall g, h \in G \qquad \lambda_2(\varphi_2(h) - \varphi_1(h))\varphi_2(g) + \cdots + \lambda_r(\varphi_r(h) - \varphi_1(h))\varphi_r(g) = 0.$$

Now since $\varphi_1 \neq \varphi_2$, there must be an $h \in G$ such that $\varphi_2(h) - \varphi_1(h) \neq 0$. Letting $\lambda'_i = \lambda_i(\varphi_i(h) - \varphi_1(h))$, we obtain

$$\forall g \in G \qquad \lambda'_2 \varphi_2(g) + \cdots + \lambda'_r \varphi_r(g) = 0,$$

and not all coefficients λ'_i are 0. (In fact, $\lambda'_2 \neq 0$.) This contradicts our choice of r as the *smallest* number of linearly dependent distinct characters, and this contradiction proves the statement.   □

We need Dedekind's lemma in order to obtain the following key particular case of the statement we are aiming to prove.

COROLLARY 15.51   *Let $E \subseteq F$ be a Galois extension. Assume that $\mathrm{Gal}_E(F)$ is cyclic, of order m, and that E contains all mth roots of 1. Then $E \subseteq F$ is radical. In fact, $F = E(u)$, where $D = u^m \in E$.*

Proof   Let φ be a generator of $\mathrm{Gal}_E(F)$; so $\mathrm{id}_F, \varphi, \ldots, \varphi^{m-1}$ are distinct isomorphisms $F \to F$, and in particular they are group homomorphisms of F^* to itself: we may view them as characters of F^* in F. By Lemma 15.50, they are linearly independent over F, and in particular

$$\mathrm{id}_F + \zeta^{-1}\varphi + \zeta^{-2}\varphi^2 + \cdots + \zeta^{-(m-1)}\varphi^{m-1}$$

is not identically 0, where $\zeta \in E$ is a primitive mth root of 1. Therefore, there exists some $v \in F$ such that

$$u := v + \zeta^{-1}\varphi(v) + \zeta^{-2}\varphi^2(v) + \cdots + \zeta^{-(m-1)}\varphi^{m-1}(v) \neq 0.$$

Since $\zeta \in E$, $\varphi(\zeta) = \zeta$. Therefore

$$\varphi(u) = \varphi(v + \zeta^{-1}\varphi(v) + \zeta^{-2}\varphi^2(v) + \cdots + \zeta^{-(m-1)}\varphi^{m-1}(v))$$
$$= \varphi(v) + \zeta^{-1}\varphi^2(v) + \zeta^{-2}\varphi^3(v) + \cdots + \zeta^{-(m-1)}\varphi^m(v)$$
$$= \zeta u$$

since $\varphi^m = \mathrm{id}_F$ and $\zeta^{-(m-1)} = \zeta^{-m}\zeta = \zeta$.

This implies that $\varphi^i(u) = \zeta^i u$, and in particular we see that u is not fixed by any non-identity element of $\mathrm{Gal}_E(F)$. This means that $\mathrm{Gal}_{E(u)}(F) = \{\mathrm{id}_F\}$, and it follows that $F = E(u)$ since the Galois correspondence is a bijection for Galois extensions.

At the same time,

$$\varphi(u^m) = \zeta^m u^m = u^m :$$

this implies that $D = u^m$ is in the fixed field of $\mathrm{Gal}_E(F)$. This fixed field is E itself, again since $E \subseteq F$ is a Galois extension. Therefore $D \in E$, and we are done.   □

Together with Exercise 15.28, Corollary 15.51 gives a characterization of cyclic extensions over fields with 'enough' roots of 1.

With these preliminaries out of the way, we can finally carry out the proof of the converse implication in Theorem 15.46.

Proof We are assuming that $k \subseteq F$ is a Galois extension such that $\mathrm{Gal}_k(F)$ is solvable, and our task is to show that the extension is contained in a Galois radical extension. In fact, by Lemma 15.45 it suffices to show that it is contained in *any* radical extension, since radical extensions are contained in Galois radical extensions.

Since $\mathrm{Gal}_k(F)$ is solvable, it admits a normal series where each quotient is abelian; in fact, we may assume that the quotients are *cyclic* (Exercise 12.31). If the cyclic quotients associated with the normal series have orders $m_1, \ldots, m_r$, we can enlarge $k \subseteq F$ so that it contains all m_ith roots of 1 for all i, as we did in the proof of the other implication in Theorem 15.46: since F is a finite separable extension, $F = k(\alpha)$ for some α; we can embed F in $\bar{k}$, and view $\alpha \in \bar{k}$; we can let $k \subseteq E := k(\zeta)$ be the extension obtained by adding to k a primitive nth root of 1 in $\bar{k}$, where $n = m_1 \cdots m_r$; and then $k \subseteq F$ is contained in

$$k \subseteq E = k(\zeta) \subseteq F(\zeta) = k(\alpha, \zeta) = E(\alpha).$$

Since $k \subseteq E$ is a radical extension (Example 15.43), we are reduced to showing that $E \subseteq E(\alpha)$ is contained in a radical extension.

By Lemma 15.49, $\mathrm{Gal}_E(E(\alpha))$ is solvable. In fact, we have (by Exercise 12.32) a normal series

$$\mathrm{Gal}_E(E(\alpha)) = G_0 \supsetneq G_1 \supsetneq \cdots \supsetneq G_s = \{e\}$$

where G_{i+1} is normal in G_i, G_i/G_{i+1} is cyclic, and the orders n_i of each of these quotients divide some of $m_1, \ldots, m_r$. So E contains all n_ith roots of 1, for all i. By the Galois correspondence, this normal series corresponds to a sequence of field extensions

$$E = E_0 \subsetneq E_1 \subsetneq \cdots \subsetneq E_s = E(\alpha)$$

such that $E_i \subseteq E_{i+1}$ is Galois, with cyclic Galois group of order n_i, and E_i contains all n_ith roots of 1. Such extensions are radical, by Corollary 15.51, so we are done.   □

15.6   Other Applications

The application of Galois theory discussed in §15.5 is striking, and it was in fact Galois's own motivation in developing this material. In this last section I would like to advertise two other applications, chosen because they are sharp and beautiful. They will both depend on a particular case of the following observation.

> PROPOSITION 15.52   *Let $k \subseteq F$ be a Galois extension, and assume $[F : k] = p^r$ for some positive prime integer p. Then there exist intermediate fields $E_1, \ldots, E_{r-1}$, giving a sequence of extensions*
>
> $$k \subsetneq E_1 \subsetneq E_2 \subsetneq \cdots \subsetneq E_{r-1} \subsetneq F$$
>
> *where each successive extension has degree p.*

Proof   Since the extension is Galois, $|\operatorname{Gal}_k(F)| = p^r$. By the first Sylow theorem, Theorem 12.29 (or even more direct arguments, cf. Exercise 12.14), $\operatorname{Gal}_k(F)$ has subgroups of all orders p^i for $i = 0, \ldots, r$, and in particular it has a subgroup G_1 of order p^{r-1}; G_1 has a subgroup G_2 of order p^{r-2}; and so on. There exists a chain of subgroups

$$\{e\} \subsetneq G_{r-1} \subsetneq G_{r-2} \subsetneq \cdots \subsetneq G_2 \subsetneq G_1 \subsetneq \operatorname{Gal}_k(F),$$

where each subgroup has index p in the subgroup which follows in the chain. By the Galois correspondence, the fixed fields E_i of G_i satisfy the requirement in the statement.   □

We will use this fact for the prime $p = 2$.

—*The Fundamental Theorem of Algebra, again.*   We have run across the Fundamental Theorem of Algebra, that is, the fact that $\mathbb{C}$ is algebraically closed, as early as §7.3 (Theorem 7.17). Back then I summarized very briefly the standard proof of this theorem, appealing to complex analysis. Using Galois theory, we can derive a proof that only relies on (perhaps) simpler ingredients: mainly, the fact that every polynomial in $\mathbb{R}[x]$ of odd degree has a root in $\mathbb{R}$. As I recalled in §7.3, this is a consequence of the Intermediate Value Theorem. Let's rephrase this observation in field-theoretic terms.

> LEMMA 15.53   *Let $\mathbb{R} \subseteq F$ be a finite extension of odd degree. Then $F = \mathbb{R}$. That is, there are no extensions $\mathbb{R} \subseteq F$ of odd degree > 1.*

Proof   Every extension is separable in characteristic 0, and every finite separable extension is simple by Theorem 14.24; therefore $F = \mathbb{R}(\alpha)$, where α is the root of an irreducible polynomial $p(x) \in \mathbb{R}[x]$ of degree $[F : \mathbb{R}]$. Since every polynomial of odd degree ≥ 3 in $\mathbb{R}[x]$ has a root in $\mathbb{R}$, no such polynomial is irreducible (cf. the paragraph preceding Proposition 7.20). It follows that $\deg p(x) = 1$, i.e., $\alpha \in \mathbb{R}$. Therefore the inclusion $\mathbb{R} \subseteq F = \mathbb{R}(\alpha)$ is an equality. $\square$

Note that the proof of this lemma does *not* use the Fundamental Theorem of Algebra. Neither does the following statement.

LEMMA 15.54   *There are no extensions $\mathbb{C} \subseteq F$ of degree 2.*

Proof   There are no irreducible polynomials of degree 2 in $\mathbb{C}[x]$, since every polynomial of degree 2 has a root in $\mathbb{C}$ by the quadratic formula.[6] $\square$

Now I will show that every nonconstant polynomial $f(x) \in \mathbb{C}[x]$ has a root in $\mathbb{C}$, thereby proving the Fundamental Theorem of Algebra.

• Let $f(x) \in \mathbb{C}[x]$ be a nonconstant polynomial. If $f(x)$ has no roots in $\mathbb{C}$, then $f(x)\overline{f(x)} \in \mathbb{R}[x]$ also has no roots in $\mathbb{C}$; therefore we may in fact assume that $f(x)$ has real coefficients.

• Let F be a splitting field of $f(x)$ over $\mathbb{R}$, and let $F(i)$ be the splitting field of $f(x)(x^2 + 1)$, that is, the field obtained by adding to F a root of $x^2 + 1$. We have

$$\mathbb{R} \subseteq \mathbb{C} = \mathbb{R}(i) \subseteq F(i).$$

• The extension $\mathbb{R} \subseteq F(i)$ is Galois: in characteristic 0, every splitting field is the splitting field of a separable polynomial. Let G be its Galois group. Note that $\mathbb{C} \subseteq F(i)$ is also a Galois extension.

• Let H be a 2-Sylow subgroup of G, so that the index $[G : H]$ is *odd*. By the Galois correspondence, the fixed field of H is an intermediate field E, and we have $\mathbb{R} \subseteq E \subseteq F(i)$, with $[E : \mathbb{R}] = [G : H]$ odd. By Lemma 15.53, $E = \mathbb{R}$, hence $H = G$.

• Therefore, G is a 2-group, and consequently $[F(i) : \mathbb{R}]$ is a power of 2. By multiplicativity of degrees (Proposition 13.13), $[F(i) : \mathbb{C}] = 2^r$ for some r.

• By Proposition 15.52, if $r > 0$, then there is a sequence of degree-2 extensions

$$\mathbb{C} \subsetneq E_1 \subsetneq E_2 \subsetneq \cdots \subsetneq E_{r-1} \subsetneq F(i).$$

However, by Lemma 15.54 there are *no* degree-2 extensions of $\mathbb{C}$. This contradiction shows that $r = 0$.

• It follows that $F(i) = \mathbb{C}$, and this proves that $\mathbb{C}$ contains the roots of $f(x)$, since F does. $\square$

Constructibility of regular polygons, again.   In §13.5 we studied the problem of constructing geometric figures in the plane 'by straightedge and compass', according to a strict set of rules. We proved (Theorem 13.38) that if α is a 'constructible' real number, then α is an algebraic number and $[\mathbb{Q}(\alpha) : \mathbb{Q}]$ is a power of 2.

We have *not* proved in §13.5 that if $\alpha \in \mathbb{R}$ and $[\mathbb{Q}(\alpha) : \mathbb{Q}]$ is a power of 2, then α is constructible. We can do that now, under the hypothesis that $\mathbb{Q}(\alpha)$ is Galois over $\mathbb{Q}$.

---

[6] Of course, applying the quadratic formula requires knowing that every complex number has a complex square root; and this fact is easily reduced to the fact that positive real numbers have real square roots. The Fundamental Theorem of Algebra is not needed for these considerations.

In fact, it is convenient to extend this observation to *complex* numbers. Say that $\alpha \in \mathbb{C}$ is 'constructible' if both the real and imaginary parts of α are constructible. Equivalently, α is constructible if it can be constructed as a point of the (complex) plane by straightedge and compass.

> PROPOSITION 15.55   *Let $\alpha \in \mathbb{C}$ be an algebraic number. Assume that the extension $\mathbb{Q} \subseteq \mathbb{Q}(\alpha)$ is Galois, and that $[\mathbb{Q}(\alpha) : \mathbb{Q}]$ is a power of 2. Then α is constructible.*

Proof Let $[\mathbb{Q}(\alpha) : \mathbb{Q}] = 2^r$. By Proposition 15.52, there exist intermediate fields $E_1, \ldots, E_{r-1}$,

$$\mathbb{Q} = E_0 \subsetneq E_1 \subsetneq E_2 \subsetneq \cdots \subsetneq E_{r-1} \subsetneq E_r = \mathbb{Q}(\alpha),$$

such that $E_{i-1} \subseteq E_i$ is quadratic for $i = 1, \ldots, r$. This means that $E_i = E_{i-1}(u_i)$, where u_i is a root of an irreducible polynomial $p_i(x) \in E_{i-1}$ of degree 2. If $p_i(x) = x^2 + bx + c$, then (by the quadratic formula) $E(u_i) = E(\delta)$, where δ is the square root of the discriminant $D = b^2 - 4c$. Therefore we may assume that $u_i^2 \in E_{i-1}$. Arguing by induction, we are reduced to showing that we can construct square roots by straightedge and compass.

Constructing the square root of a complex number z amounts to halving the argument of z and extracting the square root of $r = |z|$.

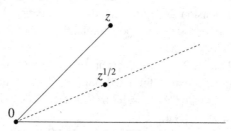

It is easy to halve an angle by straightedge and compass (Exercise 13.20), so we are reduced to constructing the square root of a real number r. The construction is encapsulated in the following picture (given here for $r > 1$ for illustration).

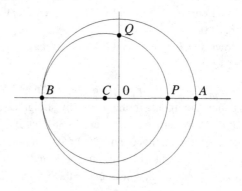

Here $P = (1, 0)$ and $A = (r, 0)$, $B = (-r, 0)$; C is the midpoint of the segment BP (midpoints are constructible), with coordinates $(-\frac{1}{2}(r - 1), 0)$. The circle with center C

and containing P has radius $\frac{1}{2}(r + 1)$ and intersects the positive y-axis at a point $Q = (0, u)$. By the Pythagoras theorem,

$$u^2 + \frac{1}{4}(r - 1)^2 = \frac{1}{4}(r + 1)^2,$$

and it follows that $u^2 = r$. This proves that $u = \sqrt{r}$ is constructible, as needed. □

Using Proposition 15.55, we can refine the results obtained in §13.5. It follows from Theorem 13.38 that one cannot construct, e.g., a regular 18-gon: indeed, this would require constructing a 20° angle, and we have shown that this cannot be done (see the proof of Corollary 13.40). From our new vantage point, we can prove the following precise criterion.

THEOREM 15.56 *The regular n-gon is constructible if and only if the totient function $\phi(n)$ is a power of 2.*

Proof Constructing the regular n-gon amounts to constructing $\zeta_n := e^{2\pi i/n}$. This complex number generates a cyclotomic extension, which is Galois of degree $|(\mathbb{Z}/n\mathbb{Z})^*| = \phi(n)$ (Exercise 15.4). If $\phi(n)$ is not a power of 2, then ζ_n cannot be constructed, by Theorem 13.38. If $\phi(n)$ is a power of 2, then ζ_n can be constructed, by Proposition 15.55. □

For example, this again shows that the regular 18-gon can*not* be constructed by straightedge and compass: indeed, $\phi(18) = 6$ is not a power of 2. On the other hand, we have now proved that the regular 17-gon *can* be constructed: $\phi(17) = 16 = 2^4$. In fact, studying the Galois correspondence for the cyclotomic extension $\mathbb{Q} \subseteq \mathbb{Q}(e^{2\pi i/17})$ produces explicit numbers $\delta_1, \delta_2, \delta_3$ such that we have a sequence of extensions

$$\mathbb{Q} \subseteq \mathbb{Q}(\delta_1) \subseteq \mathbb{Q}(\delta_2) \subseteq \mathbb{Q}(\delta_3) \subseteq \mathbb{Q}(e^{2\pi i/17})$$

and such that each step in the sequence is quadratic. This knowledge can be turned into an explicit sequence of steps constructing a regular 17-gon with straightedge and compass; as Gauss did at age 19, in 1796, more than 30 years before Galois did his work.

Are you brave enough to try this yourself?

Exercises

15.1 ▷ Let $k \subseteq F$ be a field extension and assume that the characteristic of k is 0. Prove that the extension is Galois if and only if it is the splitting field of a polynomial $f(x) \in k[x]$. (That is, prove that in characteristic 0 we can omit the requirement that $f(x)$ be separable.)

15.2 ▷ Let $k \subseteq F$ be a field extension and let G be a subgroup of $\mathrm{Gal}_k(F)$. Verify that F^G is an intermediate field of the extension.

15.3 ▷ Let $p > 0$ be a prime integer and consider the cyclotomic polynomial $f(x) = x^{p-1} + \cdots + x + 1 \in \mathbb{Q}[x]$. Prove that the Galois group of $f(x)$ is isomorphic to $(\mathbb{Z}/p\mathbb{Z})^*$ (hence it is cyclic, of order $p - 1$). (*Hint:* The case $p = 5$ is worked out in Example 15.12.)

15.4 ▷ Upgrade the result of Exercise 15.3 to all positive integer n: let F be the splitting field of the polynomial $x^n - 1 \in \mathbb{Q}[x]$ and prove that $\mathrm{Gal}_\mathbb{Q}(F)$ is isomorphic to $(\mathbb{Z}/n\mathbb{Z})^*$.

These extensions are called 'cyclotomic extensions'. For a fixed n, the extension is generated by $\zeta_n = e^{2\pi i/n}$. The minimal polynomial of ζ_n is the nth 'cyclotomic' polynomial' $\phi_n(x)$, generalizing the case for $n = p$ prime considered in Example 7.39.

15.5 Prove that the degree of the nth cyclotomic polynomial $\phi_n(x)$ (see Exercise 15.4) is Euler's totient function $\phi(n)$. Compute the fourth cyclotomic polynomial.

Challenge: Prove that $\prod_{1 \le d | n} \phi_d(x) = x^n - 1$ (and then reprove the result of Exercise 10.8 as a corollary).

15.6 ▷ Let $k \subseteq F$ be a Galois extension and let $\alpha \in F$. Prove that the number of distinct Galois conjugates of α equals the index $[\mathrm{Gal}_k(F) : \mathrm{Gal}_{k(\alpha)}(F)]$. (*Hint:* Theorem 12.10.)

Deduce that $F = k(\alpha)$ if and only if the number of Galois conjugates of α equals the degree $[F : k]$ of the extension.

15.7 ▷ Let $k \subseteq F$ be a finite separable extension. Consider all homomorphisms $\varphi \colon F \to \bar{k}$ extending the inclusion $k \subseteq \bar{k}$. Prove that $k \subseteq F$ is Galois if and only if the image of φ is independent of the homomorphism φ (cf. Remark 14.12).

15.8 ▷ Prove that both functions in the Galois correspondence reverse inclusions.

15.9 In §15.3 we have listed all intermediate fields for $\mathbb{Q} \subseteq \mathbb{Q}(\sqrt[3]{2}, i\sqrt{3})$. Which of them is $\mathbb{Q}(\sqrt[3]{2} + i\sqrt{3})$?

15.10 Let $a_1, a_2, \ldots, a_r$ be integers. Prove that the extension $\mathbb{Q} \subseteq \mathbb{Q}(\sqrt{a_1}, \ldots, \sqrt{a_r})$ is Galois. (*Hint:* Realize it as a splitting field.) Prove that every element of its Galois group has order 2. What more can you say about its Galois group?

15.11 Describe the Galois correspondence for the extension $\mathbb{Q} \subseteq \mathbb{Q}(\sqrt{2}, \sqrt{3})$ studied in Examples 13.25 and 15.25.

15.12 What is the minimal polynomial of $\zeta = e^{2\pi i/5}$ over $\mathbb{Q}(\sqrt{5})$? (Cf. Example 15.33. This will be very fast if you think in terms of conjugates.)

15.13 Describe the Galois correspondence for the extension $\mathbb{Q} \subseteq \mathbb{Q}(\zeta)$, $\zeta = e^{2\pi i/7}$. (How explicitly can you describe it for $\mathbb{Q} \subseteq \mathbb{Q}(\eta)$, $\eta = e^{2\pi i/17}$?)

15.14 Let $k \subseteq F$ be a Galois extension and assume $\mathrm{Gal}_k(F)$ is *abelian*. Prove that every intermediate field of $k \subseteq F$ is Galois over k.

15.15 Let $k \subseteq F$ be a Galois extension and let $p > 0$ be a prime dividing $[F : k]$. Prove that there is a nontrivial isomorphism $\sigma \colon F \to F$ restricting to the identity on k, such that $\sigma^p = \mathrm{id}_F$.

15.16 Let $k \subseteq F$ be a nontrivial Galois extension, $d = [F : k]$ and assume that d is not prime and $1 < d < 60$. Prove that there exists an intermediate field E, $E \ne k$, $E \ne F$, such that E is Galois over k. (*Hint:* Proposition 12.42.)

15.17 Formulate a theorem about field extensions which you can prove given your knowledge of Sylow's theorems.

15.18 Let $n > 0$ be an integer and let α be any complex number obtained as a linear combination of nth roots of 1, with rational coefficients. Prove that $\mathbb{Q} \subseteq \mathbb{Q}(\alpha)$ is Galois, with abelian Galois group. (*Hint:* Use Exercise 15.4. Remarkably, the converse also holds; it follows from the Kronecker–Weber theorem, a highly nontrivial result.)

15.19 ▷ Solve the 'inverse Galois problem' for abelian groups: prove that every finite abelian group may be realized as the Galois group of an intermediate field of a cyclotomic extension. (Use Exercise 10.25. Cyclotomic extensions are defined in Exercise 15.4.)

15.20 Verify that the discriminant of $x^3 + px + q$ is $-4p^3 - 27q^2$.

15.21 Determine the Galois group of $x^3 - 2$ over $\mathbb{F}_5$.

15.22 ▷ Let $f(x) \in k[x]$ be a separable irreducible polynomial. Prove that the Galois group of $f(x)$ acts *transitively* on the set of roots of $f(x)$ in its splitting field. (Use Proposition 14.1 and Corollary 14.11.)

15.23 Let $p > 0$ be a prime integer, and let F be a field of characteristic p. Prove that F has *no* primitive nth roots of 1 if n is a multiple of p.

15.24 Let $k \subseteq E \subseteq F$ be field extensions and assume that $k \subseteq F$ is radical. Prove that $E \subseteq K$ is radical.

15.25 Let $f(x) \in k[x]$ be a separable irreducible polynomial. Prove that the order of the Galois group of $f(x)$ is a multiple of $\deg f$.

15.26 Let $f(x) \in k[x]$ be a separable irreducible polynomial of degree p, where $p > 0$ is prime. Prove that its Galois group contains a p-cycle—i.e., there is an isomorphism of the splitting field of $f(x)$ which cyclically permutes the p roots of $f(x)$. (*Hint:* Use Exercise 15.25 and Cauchy's theorem.)

15.27 ▷ Let $f(x) \in \mathbb{Q}[x]$ be a separable irreducible polynomial of degree p, where $p > 0$ is prime, and assume that $f(x)$ has exactly $p - 2$ real roots. Prove that the Galois group of $f(x)$ is isomorphic to S_p. (*Hint:* Identify the Galois group of $f(x)$ with a subgroup of S_p. Prove that this subgroup contains a p-cycle and a transposition. Therefore....)

Use this to prove that the Galois group of $x^5 - 8x + 2 \in \mathbb{Q}[x]$ is S_5.

15.28 ▷ Let $E \subseteq E(u)$ be a separable extension of degree m, and assume that $u^m = D \in E$ and that E contains a primitive mth root ζ of 1.

- Prove that the roots of $x^m - D$ in $E(u)$ are $u, \zeta u, \ldots, \zeta^{m-1} u$.
- Prove that $E \subseteq E(u)$ is Galois.
- Prove that the polynomial $x^m - D \in E[x]$ is irreducible.
- For all $i = 0, \ldots, m - 1$, prove that there is an element $\sigma \in \mathrm{Gal}_E(E(u))$ mapping u to $\zeta^i u$.
- Use the previous point to set up a function $\mathbb{Z}/m\mathbb{Z} \to \mathrm{Gal}_E(E(u))$ and prove that it is an isomorphism.

Thus, such extensions have a *cyclic* Galois group.

15.29 Let k be a field, $\bar{k}$ its algebraic closure, E a subfield of $\bar{k}$ containing k, and let $\alpha \in \bar{k}$ and $F = k(\alpha)$. Assume that $k \subseteq F$ is a Galois extension. In Lemma 15.49 we have proved that $E \subseteq E(\alpha)$ is Galois, and that $\mathrm{Gal}_E(E(\alpha))$ may be identified with a subgroup G of $\mathrm{Gal}_k(F)$ via the restriction homomorphism $\rho(\varphi) = \varphi|_F$.

- Prove that $F^G \supseteq E \cap F$.
- Let $u \in F^G$. Prove that $\varphi(u) = u$ for all $\varphi \in \mathrm{Gal}_E(E(\alpha))$.

- Prove that $F^G \subseteq E$, and conclude that $F^G = E \cap F$.

Conclude that $\text{Gal}_E(E(\alpha)) \cong \text{Gal}_{E \cap F}(F)$.

15.30 ▷ Prove that the regular n-gon can be constructed by straightedge and compass if and only if n is a product of distinct Fermat primes times a power of 2. (Use Theorem 15.56. Fermat primes are defined at the end of §13.5.)

Appendix A Background

This appendix collects several notions that are assumed throughout the text. It is not a replacement for a solid text or course on the language of abstract mathematics, but it will hopefully serve as a reminder of some basic facts and as a way to establish the notation used in this book. The proofs in the first several sections of this text are given with an abundance of details, and a reader who is familiar with the notions summarized in this appendix should be able to absorb these proofs without difficulty. It is hoped that, in doing so, the reader will acquire the facility with the language that is assumed in more advanced texts and in later chapters of this book.

The reader should be warned that logic and set theory are fields of mathematics in themselves,[1] and we are certainly not dealing with these fields in this appendix. To really learn about set theory or logic, the reader will have to look elsewhere. At the level intended in this appendix, the subject is conventionally known as *naive* set theory.

A.1   Sets—Basic Notation

From the naive perspective, a 'set' is simply a collection of elements. What is an 'element'? This term is left undefined in naive set theory.[2] It is assumed that everything can be an element: numbers, geometric shapes, molecules, people, galaxies, and colorless green ideas sleeping furiously can be elements of sets. A set is specified by its elements, and two sets A and B are 'equal' (meaning that they are the same set) precisely when they consist of the same elements.

We use the notation $a \in A$ to denote that a is an element of the set A, and the notation $a \notin A$ to denote that a is *not* an element of the set A.

The standard notation for a set is as follows:

$$A = \{\langle \text{a precise description of the elements of } A \rangle\}.$$

Pay attention to the type of parentheses we use: $\{\cdots\}$. Other types, such as $(\cdots)$, are used with a different meaning.

Example A.1   To indicate that A is the set consisting of the numbers 1 through 5, we can write $A = \{1, 2, 3, 4, 5\}$.

---

[1] Item 03 in the AMS Mathematics Subject Classification.
[2] This is one reason why it is 'naïve'.

We could also write $A = \{5, 2, 3, 1, 4\}$, or $A = \{5, 5, 3, 1, 1, 4, 2\}$: the *order* in which the elements are listed, or possible repetitions of one or more elements, are not part of the information of a 'set'. Thus, as sets, $\{1, 2, 3, 4, 5\} = \{5, 2, 3, 1, 4\}$.

(By contrast, $(1, 2, 3, 4, 5) \neq (5, 2, 3, 1, 4)$: the notation with round parentheses stands for an *ordered* list, so in using this notation one implies that the order *does* matter.)

Example A.2 The 'empty set' $\emptyset$ is the set consisting of no elements: $\emptyset = \{\}$.

Is there more than one 'empty set'? No, because of our convention that sets are determined by their elements. If $\emptyset'$ and $\emptyset''$ are two sets consisting of no elements, then $\emptyset' = \emptyset''$ according to this principle.

Listing explicitly the elements in a set, as I have done in Example A.1, is only viable if the set is 'finite', that is, it consists of finitely many elements. The reader is already familiar with many infinite sets, for example different types of sets of numbers.

Example A.3 Several sets of numbers have a standard notation:

- $\mathbb{N}$ is the set of nonnegative integers, that is, ordinary whole numbers, including[3] 0.
- $\mathbb{Z}$ is the set of integers, including negative integers.
- $\mathbb{Q}$ is the set of 'rational numbers', that is, numbers that can be expressed as fractions $\frac{p}{q}$, where p and q are integers and $q \neq 0$.
- $\mathbb{R}$ is the set of 'real numbers'. Real numbers admit a decimal expansion, such as $\pi = 3.1415926539\ldots$.
- $\mathbb{C}$ is the set of 'complex numbers'. These are numbers of the form $a + bi$, where a and b are real numbers and i is a complex number with the property that $i^2 = -1$.

Defining precisely some of these sets is actually somewhat challenging, and it is not our concern here. For example, 'real numbers' may be defined in terms of 'Dedekind cuts', or 'Cauchy sequences'; these definitions may be given (for example) in courses on analysis. I will simply assume that the reader has a working familiarity with these sets of numbers, from exposure in elementary algebra and calculus courses. In any case, in this book we will acquire knowledge that will clarify considerably the relations between some of these sets.

Once we have a good catalogue of sets, we can define more sets using the following variation on the notation introduced above:

$$A = \{\langle\text{elements in a known set}\rangle \mid \langle\text{a property defining which of these belong to } A\rangle\}.$$

The | in the middle of this notation stands for 'such that', which may be written out in its place or summarized in some other way (s.t. and : are popular alternatives).

---

[3] Some prefer to *not* include 0 in the set $\mathbb{N}$. In this text, 0 is considered a natural number.

Example A.4 The same set we defined in Example A.1 could be denoted as follows:

$$A = \{x \in \mathbb{Z} \mid 0 < x < 6\}.$$

In words, this is the set of

integers x such that x is larger than 0 and smaller than 6,

or simply 'all integers larger than 0 and smaller than 6': that is, $1, 2, 3, 4, 5$. Note that we could also denote the same set by

$$\{x \in \mathbb{N} \mid 0 < x < 6\},$$

while the set

$$\{x \in \mathbb{R} \mid 0 < x < 6\}$$

is different: this set contains numbers such as 0.5 or π, which are not elements of our set A.

Example A.5 Several sets can be defined in this way from those listed in Example A.3 by imposing a simple requirement on the *sign* of the corresponding numbers, and in that case it is almost standard to denote the corresponding set by placing the requirement as a superscript. For example, one could write $\mathbb{Q}^{\geq 0}$ for the set of *nonnegative* rational numbers, which now we have learned can be defined precisely as follows:

$$\mathbb{Q}^{\geq 0} = \{q \in \mathbb{Q} \mid q \geq 0\}.$$

Another example:

$$\mathbb{R}^{\neq 0} = \{r \in \mathbb{R} \mid r \neq 0\}$$

is the set of nonzero real number. With the convention adopted in this text, $\mathbb{N} = \mathbb{Z}^{\geq 0}$ while, as I mentioned in the footnote to Example A.3, other texts may denote by $\mathbb{N}$ the set $\mathbb{Z}^{>0}$.

Also useful is the notion of 'indexed set', where the elements s_α may depend on an 'index' α ranging on another set A. For example, one may view a sequence of elements $s_0, s_1, s_2, \ldots$ as comprising an indexed set, which may be denoted $\{s_n\}_{n \in \mathbb{N}}$.

A.2 Logic

I 'translated' the symbols $0 < x < 6$ with the sentence 'x is larger than 0 and smaller than 6'. You may notice that the single condition $0 < x < 6$ is expressed as a combination of *two* different conditions: $0 < x$ *and* $x < 6$. The '*and*' in between encodes an important logic operation, which may be denoted by the symbol $\wedge$. If p and q denote 'statements' (that is, assertions which are necessarily either T(rue) or F(alse), such as 'this integer x is less than 6, or 'triangles have four sides'), then

$$p \wedge q$$

is another statement, which is taken to be T if both p and q are T, and F otherwise; that is, if either p or q or both are F. Thus, we may view the statement $0 < x < 6$ as shorthand for $(0 < x) \wedge (x < 6)$.

It is convenient to have notation taking care of other such 'logical connectives'. The symbol $\vee$ is used for '*or*'. Thus,

$$p \vee q$$

is a statement if p and q are statements, and it is taken to be T as soon as one of p or q or both are T. It is F precisely when both p and q are F. So, $(0 < x) \vee (x < 6)$ is T for all $x \in \mathbb{Z}$ that are either larger than 0, or less than 6, or both. If you think about it a moment, you will realize that *every* integer satisfies this requirement. So for all $x \in \mathbb{Z}$, the statement $(0 < x) \vee (x < 6)$ is T.

The symbol $\neg$ denotes *negation:* that is, if p is a statement, then $\neg p$ is the statement that is T when p is F and F when p is T. If $x \in \mathbb{Z}$ and p is the statement $0 < x$, then the statement $\neg p$ is $0 \geq x$.

There is a convenient visual way to encode the information presented above: we can write 'truth tables' spelling out the result of applying one of these logical connectives.

p	q	$p \wedge q$
T	T	T
T	F	F
F	T	F
F	F	F

p	q	$p \vee q$
T	T	T
T	F	T
F	T	T
F	F	F

p	$\neg p$
T	F
F	T

It is not difficult to see that *every* possible truth table can be realized by a combination of these connectives, and often in many different ways.

Example A.6   Say we are interested in expressing a condition which should have the following behavior on *three* statements p, q, r:

p	q	r	?
T	T	T	T
T	T	F	T
T	F	T	T
T	F	F	F
F	T	T	F
F	T	F	F
F	F	T	F
F	F	F	F

(A.1)

Then I claim that $(p \wedge q) \vee (p \wedge r)$ realizes precisely this operation. You can check that this is the case by working out all possibilities, and in fact a truth table would allow us to do this quite efficiently:

p	q	r	$p \wedge q$	$p \wedge r$	$(p \wedge q) \vee (p \wedge r)$
T	T	T	T	T	T
T	T	F	T	F	T
T	F	T	F	T	T
T	F	F	F	F	F
F	T	T	F	F	F
F	T	F	F	F	F
F	F	T	F	F	F
F	F	F	F	F	F

As it happens, there are many other ways to achieve precisely the same effect: for instance,

$$p \wedge (q \vee r)$$

has again the same truth table (A.1). Take a moment to check this!

As this example shows, different expressions involving statements and logical expressions may end up being 'logically equivalent', in the sense that they have the same truth value for all possible choices of the truth values of the individual statements appearing in the expression.

It is convenient to introduce more connectives to indicate that two statements may be logically equivalent, or that the truth of one statement implies the truth of another statement. The connectives 'if and only if', $\iff$, and 'implies', $\implies$, are defined by the following truth tables:

p	q	$p \iff q$
T	T	T
T	F	F
F	T	F
F	F	T

p	q	$p \implies q$
T	T	T
T	F	F
F	T	T
F	F	T

The first connective is probably self-explanatory: we want to record whether it is true that 'p is true *if and only if* q is true, and this should be the case precisely when p and q have the same truth value. That is what the table implements. If $p \iff q$ is T, we say that p and q are (logically) 'equivalent' statements.

The second connective may look a little more puzzling. We want a statement that is T precisely when 'p implies q', that is, of the type '*if p, then q*'. What this means in ordinary language may be a little open to interpretation, but it certainly makes sense that if p is T and yet q is F, then it is not true that p implies q; that is the reason why the second line of the last column in the table is F. The reason why the third and fourth lines are T is that if p implies q, and p is not T to begin with (so that it is F), then no requirement is placed on q: it can be T or F. This is what the truth table is encapsulating.

Note that if p implies q, *and* q implies p, then it should be the case that p and q are equivalent. And indeed, here is the truth table for $(p \implies q) \land (q \implies p)$:

p	q	$p \implies q$	$q \implies p$	$(p \implies q) \land (q \implies p)$
T	T	T	T	T
T	F	F	T	F
F	T	T	F	F
F	F	T	T	T

We see that the result agrees with $p \iff q$, as the 'natural language' interpretation for these symbols would suggest.

Using these symbols, we can rephrase the observation we made in Example A.6 by stating that

$$((p \land q) \lor (p \land r)) \iff (p \land (q \lor r)) \tag{A.2}$$

is T for all truth values of p, q, and r. This makes it a 'tautology'.

If we simply state an expression such as (A.2) without comments, we imply that this expression is a tautology, i.e., it is T in all possible cases.

The opposite of a tautology is a 'contradiction', that is, a statement that is F for all truth values of its constituents.

Example A.7   The simplest tautology is $p \lor (\neg p)$, which is T no matter whether p is T or F.

The simplest contradiction is $p \land (\neg p)$, which is F no matter whether p is T or F.

It is good to get used to a few tautologies, which amount to an 'algebraic' way to manipulate logical expression. (This type of algebra is called 'Boolean algebra'.)

Example A.8   You can verify that the expressions

- $(p \land (q \land r)) \iff ((p \land q) \land r)$
- $(p \lor (q \lor r)) \iff ((p \lor q) \lor r)$
- $((p \land q) \lor (p \land r)) \iff (p \land (q \lor r))$
- $((p \lor q) \land (p \lor r)) \iff (p \lor (q \land r))$

are all tautologies.

The following useful tautologies are called 'De Morgan's laws'.

- $\neg(p \lor q) \iff (\neg p) \land (\neg q)$
- $\neg(p \land q) \iff (\neg p) \lor (\neg q)$

Thus, negating an expression negates its constituents and swaps $\land$ and $\lor$.

Also very useful is the tautology

$$(p \implies q) \iff ((\neg p) \lor q) \tag{A.3}$$

translating the connective $\implies$ into an expression that only uses the basic connectives $\lor$ and $\neg$. Finally, note that

$$(p \implies q) \iff ((\neg q) \implies (\neg p)). \tag{A.4}$$

You can verify this quickly enough with a truth table, or by appealing to De Morgan's laws and (A.3):

$$(p \implies q) \iff ((\neg p) \vee q) \iff (\neg(\neg q) \vee (\neg p)) \iff ((\neg q) \implies (\neg p))$$

(by which we mean that each $\iff$ appearing here is a tautology). This observation will be at the root of 'proofs by contrapositive', see §A.4 below.

A.3   Quantifiers

We now have an efficient language to string together properties into more complex expressions, and to check the 'truth value' of such expressions. We could already use this language to define relatively complicated sets rather concisely. But we can add one more type of notation to our dictionary, expanding further the expressive power of the language. The purpose of this additional piece of notation is to 'limit the scope' of the variables in an expression.

To see what I mean, consider something like

$$n > 0.$$

This is not a statement, even if we specify that n is (for example) an integer: this expression is not True or False until we know something about n. For example, it is T if $n = 1$, and it is F if $n = -1$. We may want to express the fact that this expression is T for *some* integer n: the notation for that is

$$\exists n \in \mathbb{Z} \quad n > 0. \tag{A.5}$$

In words, 'There exists an integer n such that n is greater than 0.' This statement is T, and to convince you I just need to produce an n for which $n > 0$ is true: $n = 1$ is one such integer. The 'rotated' symbol $\exists$ stands for 'there exists'; this is the 'existential quantifier'. Incidentally, if we say 'there exists an integer', or even 'there exists one integer', we usually assume that there may be many (as in this example). If we really mean that there is *only one* item, we write $\exists!$, with an exclamation mark following the $\exists$ symbol.

We may instead want to make a statement about *all* n in a certain set. For instance, I could consider the following statement:

$$\forall n \in \mathbb{Z} \quad n > 0. \tag{A.6}$$

In words, 'For all integers n, n is greater than 0.' This statement is F, since there are integers n for which $n > 0$ is false: $n = -1$ is one such integer. The rotated symbol $\forall$ stands for 'for all'; this is the 'universal quantifier'.

Example A.9 I could denote the set of even integers as follows:

$$\{n \in \mathbb{Z} \,|\, \exists k \in \mathbb{Z}, \, n = 2k\}.$$

Indeed, this is the set of integers which are obtained by doubling *some* other integers. Note that something like

$$\{n \in \mathbb{Z} \mid \forall k \in \mathbb{Z},\, n = 2k\}$$

defines the empty set: there is no integer n which is simultaneously obtained by doubling *all* integers. And I am simply not able to parse

$$\{n \in \mathbb{Z} \mid \quad k \in \mathbb{Z},\, n = 2k\}$$

because for a given n, '$k \in \mathbb{Z},\, n = 2k$' is not a statement as it is missing a quantifier to determine the scope of k.

One moral of Example A.9 is that quantifiers are quite essential: if they are missing, then a given expression may fail to be a statement; and the choice of a quantifier is an important part of the 'meaning' of a statement.

This should not be surprising. The *only* difference between (A.5) and (A.6) is the very first character: $\exists$ in one case, $\forall$ in the other. And yet (A.5) is T and (A.6) is F. These symbols are not mere decorations.

One easily missed subtlety about quantifiers is that the *order* in which they are listed makes a difference. For example, the statement

$$(\forall n \in \mathbb{Z})\,(\exists m \in \mathbb{Z}) \qquad m > n \tag{A.7}$$

is T: it is indeed the case that for every given integer n there exists some integer m larger than n. (For example, $m = n + 1$ will do.) On the other hand, the statement

$$(\exists m \in \mathbb{Z})\,(\forall n \in \mathbb{Z}) \qquad m > n \tag{A.8}$$

is F: it is not the case that there is a special integer m that is larger than every integer n. (If you think '$+\infty$' would be such a gadget, think again: ∞ is *not* an integer.) The only difference between (A.7) and (A.8) is the order in which the quantifiers are listed, and yet one of the statements is T and the other one is F. This tells us that the order matters.

The 'logical connectives' we considered in §A.2 interact with quantifiers in an interesting way. Suppose that we have a statement $p(x)$ that depends on a variable x. For instance, '$x > 0$' is such a statement: once x is bound in some way, by being assigned some value or by being quantified by means of $\exists$ or $\forall$, then $x > 0$ becomes a statement. The following tautologies are very useful:

$$\begin{aligned} \neg(\exists x \quad p(x)) &\iff (\forall x \quad \neg p(x)), \\ \neg(\forall x \quad p(x)) &\iff (\exists x \quad \neg p(x)). \end{aligned} \tag{A.9}$$

(*Challenge:* Understand in what sense these are an extension of the De Morgan laws we encountered in Example A.8.) That is, negating a statement swaps the existential and universal quantifiers.

Why is, e.g., $\neg(\forall x,\, p(x))$ equivalent to $\exists x,\, \neg p(x)$? Think about it. To say that it is not the case that some property holds *for all* x, is the same as saying that the property does not hold *for some* x. This is how at the beginning of this section I tried to convince you that the statement '$\forall n,\ n > 0$' is F: by producing *one* example, $n = -1$, for which $n > 0$ is F. That is, I showed that the statement '$\exists n,\ n \leq 0$' is T.

If you find that you have to 'memorize' something like (A.9), then I suggest that you pause and form your own mental image of what is going on. All we are doing is formalizing common sense; an appeal to common sense should suffice to remember how the expressions (A.9) work. Once you absorb (A.9), you will be able to speedily negate complicated expressions without having to parse what they 'mean'.

Example A.10   You have likely run into the following calculus definition of 'limit' (even though you may have never used it concretely): $\lim_{x \to c} f(x) = L$ is shorthand for the logical expression

$$(\forall \epsilon > 0)\,(\exists \delta > 0)\,(\forall x) \qquad 0 < |x - c| < \delta \implies |f(x) - L| < \epsilon.$$

Here all the variables are assumed to be real numbers. It does not matter now whether you 'understand' this definition at a deep level. I propose that we formally negate it, that is, produce a precise mathematical translation of the assertion $\lim_{x \to c} f(x) \neq L$:

$$\neg(\,(\forall \epsilon > 0)\,(\exists \delta > 0)\,(\forall x) \qquad 0 < |x - c| < \delta \implies |f(x) - L| < \epsilon\,),$$

but without any $\neg$ left in the expression. What we have learned so far makes this a very reasonable exercise. First, let the $\neg$ go through your quantifiers, as we have seen in (A.9). This flips all quantifiers, giving

$$(\exists \epsilon > 0)\,(\forall \delta > 0)\,(\exists x) \qquad \neg(\,0 < |x - c| < \delta \implies |f(x) - L| < \epsilon\,).$$

Second, we are dealing now with a statement of the form $\neg(p \implies q)$. As we have seen in Example A.8, $p \implies q$ is logically equivalent to $(\neg p) \vee q$. By applying one of De Morgan's laws,

$$\neg(p \implies q) \iff \neg((\neg p) \vee q) \iff (p \wedge (\neg q)). \tag{A.10}$$

Therefore

$$\neg(\,0 < |x - c| < \delta \implies |f(x) - L| < \epsilon\,)$$

is equivalent to

$$(0 < |x - c| < \delta) \wedge \neg(|f(x) - L| < \epsilon)$$

or simply (disposing of the last $\neg$)

$$(0 < |x - c| < \delta) \wedge (|f(x) - L| \geq \epsilon).$$

In conclusion, $\lim_{x \to c} f(x) \neq L$ must mean

$$(\exists \epsilon > 0)\,(\forall \delta > 0)\,(\exists x) \qquad (0 < |x - c| < \delta) \wedge (|f(x) - L| \geq \epsilon).$$

If you really *know* about limits, this will make perfect sense to you. But it is impressive that we can perform these seemingly complex logical manipulations even without any special inside knowledge of the 'meaning' of these expressions.

A.4   Types of Proof: Direct, Contradiction, Contrapositive

The little logic we have learned may clarify the way some standard proofs work.

A 'theorem' is a true mathematical statement. Labels such as 'lemma', 'proposition', 'corollary' are also used for true mathematical statements—the distinction between them is psychological, not mathematical. If you label a statement 'lemma', you are telling the reader that the statement's main purpose will be to provide some intermediate step in a later proof. A 'corollary' is a true statement that follows directly from another theorem. A 'proposition' may simply be a true statement that is useful in the context of a particular discussion, but is not the main or the most important conclusion in that discussion.

A 'proof' of a theorem (or lemma, corollary, proposition, ...) consists of a sequence of logical steps which starts from facts whose truth has already been established and concludes by establishing the truth of the fact stated in the theorem. A good proof should convey clearly the reason why one can draw a certain conclusion. This is usually done by a combination of engaging prose in natural language and the use of precise mathematical notation. Writing nice, 'elegant' proofs is a difficult task, best learned by reading proofs written by professional mathematicians and practicing writing (and rewriting, and rewriting again) one's own proofs.

The *logic* underlying a proof should be transparent; it should clarify the argument, not distract the reader from it. In practice, most arguments will use certain template formats, which go under different names. Giving a complete taxonomy of all possible 'methods of proof' would be a futile (and unbearably tedious) task. However, a few general considerations are possibly useful.

- *On the use of examples.* Most theorems are of the form $\forall x, P(x)$ or $\exists x, P(x)$, where $P(x)$ is a statement depending on some variable x. (Of course, more variables may be used.) For example, the following is a theorem of the first type:

$$\forall n \in \mathbb{Z} \quad n^2 \geq 0. \tag{A.11}$$

It states that a certain property ($n^2 \geq 0$) holds *for all* integers. And here is an example of the second type:

$$\exists n \in \mathbb{Z} \quad n^2 - 2 < 0. \tag{A.12}$$

This is not a statement about all integers; it claims that some property ($n^2 - 2 < 0$) is satisfied by *some* integer.

Would an example, e.g., $n = 1$, 'prove' (A.11)?

No. If you point out that the property $n^2 \geq 0$ is satisfied for $n = 1$, you are proving

$$\exists n \in \mathbb{Z} \quad n^2 \geq 0,$$

and this is a different statement than (A.11), because it involves a different quantifier.

Would an example, e.g., $n = 1$, 'prove' (A.12)? Yes. The existence of an example satisfying the given property is precisely what (A.12) claims.

What if I claim that

$$\forall n \in \mathbb{Z} \quad n^2 \geq 2 \quad ?$$

Would the example $n = 1$, for which $n^2 \geq 2$ is not true, *dis*prove this statement? (Therefore proving that it is *not* a theorem?)

Yes. If you want to convince someone that

$$\forall n \in \mathbb{Z} \quad n^2 \geq 2$$

is F, then you have to show that

$$\neg (\forall n \in \mathbb{Z} \quad n^2 \geq 2)$$

is T, and as we have learned in §A.3, this statement is equivalent to

$$\exists n \in \mathbb{Z} \quad \neg (n^2 \geq 2),$$

so indeed producing one n for which the statement $n^2 \geq 2$ is not true does the job.

Summarizing, if you are wondering whether an example would or would not do what you want (prove a theorem, disprove a statement,...) you just have to interpret carefully what your task is—which will likely involve using what you know about the way quantifiers behave with respect to logic operations. That should clarify the issue without any room for uncertainties.

Very often, the main statement $P(x)$ appearing in a theorem is of the form 'If x satisfies these conditions, then the following occurs'. That is: it is an 'if... then...' statement, $p \implies q$. Proofs for statements of this type can roughly be categorized as follows.

- *Direct proofs.* A 'direct proof' simply starts from the premise p, applies a sequence of deductions, and concludes q.

Example A.11   Let's go back to the simple example (A.11):

$$\forall n \quad n \in \mathbb{Z} \implies n^2 \geq 0,$$

which I rewrote to make it fit the general template we are considering.

Proof   If n is an integer, then either $n \geq 0$, in which case $n^2 \geq 0$, or $n < 0$, that is, $n = -m$ for some integer $m > 0$. In this case

$$n^2 = (-m)^2 = (-1)^2 m^2 = m^2 \geq 0$$

since $(-1)^2 = 1$. We conclude that $n^2 \geq 0$ for all integers n, as needed.   □

This is not the greatest example, because it is too simple; but you will agree with me that the logic is transparent and 'direct': we just started with our assumption ($n \in \mathbb{Z}$), performed some known mathematical operations, and deduced our conclusion ($n^2 \geq 0$).

- *Proofs by contrapositive.* 'Proofs by contrapositive' are based on tautology (A.4), to the effect that the statement $p \implies q$ is equivalent to the statement $(\neg q) \implies (\neg p)$. Thus, the task of proving

$$p \implies q$$

may be carried out by proving

$$(\neg q) \implies (\neg p).$$

This is called the 'contrapositive' of the given statement.

Example A.12   Let's give a 'proof by contrapositive' of the following theorem:

$$\forall a, b, n \in \mathbb{N} \qquad a + b = n \implies \left(a \le \frac{n}{2}\right) \vee \left(b \le \frac{n}{2}\right). \qquad (A.13)$$

(In words, if the sum of two integers is an integer n, then at least one of the two integers is at most half as large as n.)

Proof   The contrapositive statement is

$$\forall a, b, n \in \mathbb{N} \qquad \neg\left(\left(a \le \frac{n}{2}\right) \vee \left(b \le \frac{n}{2}\right)\right) \implies a + b \ne n,$$

that is (using De Morgan's laws)

$$\forall a, b, n \in \mathbb{N} \qquad \left(\left(a > \frac{n}{2}\right) \wedge \left(b > \frac{n}{2}\right)\right) \implies a + b \ne n. \qquad (A.14)$$

Thus, proving (A.13) is equivalent to proving (A.14). To prove (A.14), it suffices to observe that

$$\left(a > \frac{n}{2}\right) \wedge \left(b > \frac{n}{2}\right) \implies a + b > \frac{n}{2} + \frac{n}{2} = n.$$

If $a + b > n$, then in particular $a + b \ne n$, so this indeed proves (A.14). $\qquad\square$

To recap: (A.14) is equivalent to (A.13). Since we are able to show that (A.14) is true, it follows that the proposed theorem (A.13) must also be true.

- *Proofs by contradiction.* Proofs by contradiction rely on different logical considerations. In order to prove

$$p \implies q$$

'by contradiction', we *assume that p is true and q is false,* and derive from these assumptions a statement that is known to be *false* (a 'contradiction').

Why does this work? Recall that the only F appearing in the truth table for $p \implies q$ occurs when p is T and q is F. By proving that this possibility implies a false statement, that is, a *contradiction,* we prove that it does *not* occur, and we can conclude that the statement $p \implies q$ is necessarily T.

In other words, we rely on the equivalence between $p \implies q$ and $(\neg p) \vee q$ (Example A.8). By one of the De Morgan laws, the negation of this statement is $p \wedge \neg q$. Therefore, we can prove that $p \implies q$ is *true* by proving that the statement $p \wedge \neg q$ is *false,* and this is what we do when we carry out a proof by contradiction.

Example A.13 We can prove the following statement by contradiction:

$$\forall n \quad n \in \mathbb{Z} \implies n(n-1) \geq 0 \,.$$

Proof   Arguing by contradiction, assume that $n \in \mathbb{Z}$ *and* that $n(n-1) < 0$. Since the product of n and $n-1$ is negative, n and $n-1$ must have opposite signs: that is, one of the following two possibilities must occur:

(i) $n < 0$ and $n - 1 > 0$, or

(ii) $n > 0$ and $n - 1 < 0$.

In case (i) we would have $n < 0$ and $n > 1$, which is a contradiction. In case (ii), we would have $0 < n < 1$. Since there are no integers strictly between 0 and 1, this is also a contradiction.

Summarizing, the assumptions that $n \in \mathbb{Z}$ and $n(n-1) < 0$ lead to a contradiction, and we can conclude that $n \in \mathbb{Z}$ implies $n(n-1) \geq 0$, as stated.   □

The reader should be able to produce a 'direct' proof of the same statement, for example by considering separately the case $n \leq 0$ and $n \geq 1$.

It is often the case that a theorem can be proved in several different ways, and the choice of a method over another is mostly dictated by expository considerations. There is perhaps a tendency to favor direct arguments when they are available, since their logic tends to be more transparent.

The examples given above are very simple, due to the need to only deal with elementary mathematics in this appendix. In the text we will frequently use the methods reviewed here, in more interesting situations. For example, in Exercise 1.26 the reader will use a contradiction argument to prove the famous theorem, due to Euclid, stating that there are *infinitely many* prime numbers.

A.5   Set Operations

The 'logical connectives' we have defined in §A.2 have counterparts in naive set theory. The very definition of set relies on the principle that two sets are equal precisely when they consist of the same elements (as mentioned in §A.1). This principle asserts that two sets A and B are equal, i.e., $A = B$, if and only if

$$\forall x \quad (x \in A) \iff (x \in B) \,.$$

Thus, we may view equality of sets as a counterpart of the logical connective $\iff$. Similarly, the connective $\implies$ corresponds to the notion of 'subset'. We say that a set A is a subset of a set B, and use the notation $A \subseteq B$, if and only if

$$\forall x \quad (x \in A) \implies (x \in B) \,. \tag{A.15}$$

Intuitively, A is a subset of B if it is 'contained' in B (but this is just a natural-language interpretation of (A.15); (A.15) is the actual definition).

The connectives ∧ and ∨ correspond to two basic operations involving sets. The 'union' of two sets A and B, denoted $A \cup B$, is the set defined by

$$A \cup B = \{x \mid (x \in A) \vee (x \in B)\}.$$

The 'intersection' of two sets A and B, denoted $A \cap B$, is the set defined by

$$A \cap B = \{x \mid (x \in A) \wedge (x \in B)\}.$$

There is also an operation corresponding to the negation: the 'difference' $A \setminus B$ is defined by

$$A \setminus B = \{x \in A \mid \neg(x \in B)\}.$$

If A is a subset of a set S, the difference set $S \setminus A$ is called the 'complement' of A (in S). One may visualize these basic set operations by drawing so-called 'Venn diagrams':

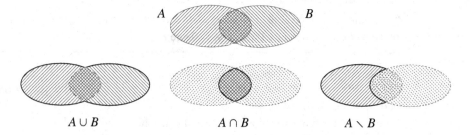

$$A \cup B \qquad\qquad A \cap B \qquad\qquad A \setminus B$$

Such pictures do not prove theorems, but they may provide a good intuitive sense for the result of performing set-theoretic operations.

Since the set operations mentioned above are precise counterparts of the logical connectives we have reviewed earlier, we see that every tautology has a corresponding set-theoretic manifestation. For example, let A and B be subsets of a set S, and denote by A^c, B^c, etc. the complements of A, B, etc. in S. Then

$$A^c \cap B^c = (A \cup B)^c,$$
$$A^c \cup B^c = (A \cap B)^c.$$

Indeed, these set-theoretic identities may be proven by applying De Morgan's laws. The reader should be able to carry out this verification.

It is natural to ask whether the quantifiers ∃, ∀ also have counterparts in some operations among sets. This is indeed the case. These quantifiers allow us to extend to *infinite* 'families' of sets (that is, sets of sets) the operations of union and intersection. For this, suppose we are given a set S_α for every α in some set A; in other words, let $\{S_\alpha\}_{\alpha \in A}$ be an indexed set of sets. Then we can give the following definitions:

$$\bigcup_{\alpha \in A} S_\alpha = \{x \mid \exists \alpha \in A, x \in S_\alpha\},$$
$$\bigcap_{\alpha \in A} S_\alpha = \{x \mid \forall \alpha \in A, x \in S_\alpha\}.$$

This notation may seem quite abstract, but it is a natural generalization of the ordinary union and intersection. Indeed, when α ranges on a set $A = \{1, 2\}$ with just two elements, these definitions simply reduce to the usual $S_1 \cup S_2$, $S_1 \cap S_2$, as you should verify. You could also enjoy verifying that if all sets S_α are subsets of a given set S, and we use ^{c} as above to denote complement in S, then

$$\left(\bigcup_{\alpha \in A} S_\alpha \right)^c = \bigcap_{\alpha \in A} S_\alpha^c \quad \text{and} \quad \left(\bigcap_{\alpha \in A} S_\alpha \right)^c = \bigcup_{\alpha \in A} S_\alpha^c \,.$$

These are the set-theoretic version of the useful tautologies (A.9).

Example A.14 For q a nonnegative rational number, let $(-q, q)$ denote the open interval in the real line with endpoints $-q$ and q (as in calculus!). You can view this as a set S_q determined by the chosen $q \in \mathbb{Q}^{\geq 0}$. (Here $\mathbb{Q}^{\geq 0}$ denotes the set of nonnegative rational numbers.) Then you may verify that

$$\bigcup_{q \in \mathbb{Q}^{\geq 0}} (-q, q) = \mathbb{R} \quad \text{and} \quad \bigcap_{q \in \mathbb{Q}^{\geq 0}} (-q, q) = \{0\} \,.$$

Some information concerning sets is not directly reduced to elementary logic. For instance, if a set consists of finitely many elements, we denote by $|A|$ the number of these elements. Thus, if $A = \{1, 2, 3, 4, 5\}$, then $|A| = 5$. This is called the 'cardinality' of the set. It can be given a precise meaning even for infinite sets (cf. §A.7).

Also, there are other ways to produce new sets from given sets, which do not correspond directly to logical connectives.

The 'Cartesian product' of two sets A and B, denoted $A \times B$, is the set of ordered pairs (a, b) with $a \in A$ and $b \in B$:

$$A \times B = \{(a, b) \,|\, (a \in A) \wedge (b \in B)\} \,.$$

The 'ordered pair' (a, b) carries the information of a and b and of the order in which they appear. Thus, $(a_1, b_1) = (a_2, b_2)$ precisely when $a_1 = a_2$ and $b_1 = b_2$. (Unfortunately we use the same notation in calculus for 'open intervals' in the real line; I have done so myself a moment ago in Example A.14. Despite appearances, the meanings of this notation in these different contexts are completely different.)

Example A.15 If $B = \{x, y, z\}$ and $A = \{\circ, \bullet\}$, then

$$A \times B = \{(x, \circ), (y, \circ), (z, \circ), (x, \bullet), (y, \bullet), (z, \bullet)\} \,.$$

In general, if A and B are both finite sets, then $|A \times B| = |A| \cdot |B|$.

The 'plane $\mathbb{R}^2$' you used in calculus is nothing but the Cartesian product $\mathbb{R} \times \mathbb{R}$ of ordered pairs of real numbers. Every point of the plane is determined by its coordinates, i.e., by the terms of the corresponding ordered pair (x, y).

One more standard construction is the 'power set' (or 'set of parts') of a given set A, denoted $\mathcal{P}(A)$ or 2^A. This is the set consisting of all *subsets* of the given set.

Example A.16   If $A = \{x, y, z\}$, then

$$2^A = \{\emptyset, \{x\}, \{y\}, \{z\}, \{x, y\}, \{x, z\}, \{y, z\}, \{x, y, z\}\}.$$

It is a good exercise to verify that if A is a finite set, then $|2^A| = 2^{|A|}$.

A.6   Functions

From the point of view of naive set theory, *functions* are certain subsets of Cartesian products. If A and B are sets, then a 'function' f from A to B is determined by a subset $\Gamma_f \subseteq A \times B$ such that

$$(\forall a \in A)(\exists! b \in B) \qquad (a, b) \in \Gamma_f. \tag{A.16}$$

(Recall that $\exists!$ stands for 'there exists a unique...'.) The set Γ_f is called the 'graph' of the function. If $(a, b) \in \Gamma_f$, then by (A.16) the element b is uniquely determined by a, so we may use a notation for it which records this fact. That notation is $f(a)$. Using this notation, we could say that a function f is defined by prescribing *one and one only* element $f(a) \in B$ for every $a \in A$, and this is probably the point of view you took in Calculus or other previous encounters with functions. The element $f(a)$ is called the 'image' of a.

To denote that f is a function from the set A to the set B, we write $f : A \to B$, or draw the diagram

$$A \xrightarrow{\;f\;} B.$$

If $f : A \to B$ is a function, then A is said to be the 'domain' of f, and B is the 'codomain' or 'target' of f.

The 'image' of a function $f : A \to B$ is the subset $f(A)$ of B consisting of all images of elements of A by f:

$$f(A) = \{b \in B \mid (\exists a \in A), b = f(a)\}.$$

Example A.17   Writing '$f(x) = x^2$' determines a function $f : \mathbb{R} \to \mathbb{R}$: for every $r \in \mathbb{R}$, there is one and only one $r^2 \in \mathbb{R}$ obtained by plugging in r for x in this formula.

The source and the target of this function are both $\mathbb{R}$. The image $f(\mathbb{R})$ of the function is the set $\mathbb{R}^{\geq 0}$ of nonnegative real numbers: indeed, every nonnegative real number has a square root, while negative real numbers do not.

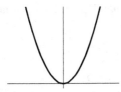

By definition, the graph corresponding to the function in Example A.17 is the set of pairs (x, y) in the Cartesian product $\mathbb{R} \times \mathbb{R}$ for which $y = x^2$. You have drawn this graph many times, so you are most certainly familiar with the way it looks. It is a pleasantly symmetric curve in the plane $\mathbb{R}^2$.

For every set A there is a function $\mathrm{id}_A : A \to A$, called the 'identity function', determined by setting $\mathrm{id}_A(a) = a$ for all $a \in A$.

If $f : A \to B$ and $g : B \to C$ are functions, then we have a 'composition' function $g \circ f$ defined by letting for all $a \in A$

$$(g \circ f)(a) = g(f(a)).$$

To convey that the function h is the composition $g \circ f$, we may draw diagrams such as

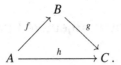

or

$$
\begin{array}{ccc}
 & B & \\
 {\scriptstyle f}\nearrow & & \searrow{\scriptstyle g} \\
A & \xrightarrow{\ \ h\ \ } & C.
\end{array}
$$

Such diagrams are said to 'commute': this means that if we can travel from one point of the diagram to another by following the arrows in different ways, the corresponding compositions of functions are equal. In the above examples we can travel from A to C directly by h or along the arrows f and g. The diagram commutes when for all a in A, $h(a)$ equals $g(f(a))$, and this means that $h = g \circ f$.

Sometimes commutative diagrams are an efficient way to encode complicated equalities involving compositions.

The following remarks are easy to verify, but they are important.

(i) Let $f : A \to B$ be a function. Then

$$f \circ \mathrm{id}_A = f = \mathrm{id}_B \circ f.$$

(ii) Let $f : A \to B$, $g : B \to C$, and $h : C \to D$ be functions. Then

$$(h \circ g) \circ f = h \circ (g \circ f).$$

For example, to verify the second formula, let $a \in A$ be any element; then

$$((h \circ g) \circ f)(a) = (h \circ g)(f(a)) = h(g(f(a))),$$
$$(h \circ (g \circ f))(a) = h((g \circ f)(a)) = h(g(f(a))).$$

The results of applying the two functions $(h \circ g) \circ f$ and $h \circ (g \circ f)$ to an arbitrary element

$a \in A$ are equal, so the functions themselves are equal. Here is a commutative diagram expressing this fact:

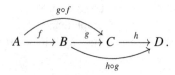

$$A \xrightarrow{\ f\ } B \xrightarrow{\ g\ } C \xrightarrow{\ h\ } D \ .$$

Property (i) says that the identity function is an *identity* with respect to composition (this is what gives it its name). Property (ii) says that the composition operation is *associative*. These two properties recur in all the algebraic constructions examined in this text.

Also, properties (i) and (ii) essentially say that the collections of sets and of functions between sets form a 'category'. This concept underlies implicitly much of the material treated in this text, but the reader will not need to have a technical understanding of it.

The set of functions $f: A \to B$ is denoted B^A. It is not hard to verify that if A and B are both finite, then $|B^A| = |B|^{|A|}$. As a good exercise, you could figure out why the power set of a set A (cf. §A.5) is denoted 2^A. In what sense can the power set 2^A be understood as the set of functions from A to a set with two elements?

A.7 Injective/Surjective/Bijective Functions

A function $f: A \to B$ is

- 'injective' (or 'one-to-one') if

$$\forall a_1, a_2 \in A \qquad a_1 \neq a_2 \implies f(a_1) \neq f(a_2) ;$$

- 'surjective' (or 'onto') if

$$(\forall b \in B)(\exists a \in A) \qquad f(a) = b ;$$

- 'bijective' (or a 'bijection'; or a 'one-to-one correspondence') if it is both injective and surjective.

A bijective function $f: A \to B$ 'identifies' the two sets, in the sense that we could use f to label every element $a \in A$ with the element $f(a) \in B$: different elements of A would get different labels, since f is injective, and every label $b \in B$ would be used, since f is surjective. In particular, if A and B are finite, then there is a bijection from A to B if and only if A and B have the same number of elements.

Example A.18 Let $A = \{\circ, \bullet, *\}$ and let $B = \{1, 2, 3\}$. Then the function $f: A \to B$ defined by setting

$$f(\circ) = 1, \quad f(\bullet) = 2, \quad f(*) = 3$$

is a bijection. The function $g: A \to B$ defined by setting

$$g(\circ) = 3, \quad g(\bullet) = 1, \quad g(*) = 2$$

is another bijection. In fact, there are exactly six bijections between these two sets A and B. (Why?) As stated in §A.6, there are $3^3 = 27$ functions from A to B.

A bijection $f: A \rightarrow B$ may also be called an 'isomorphism of sets'. If there is a bijection $f: A \rightarrow B$, we say that the sets A and B are 'isomorphic as sets', and write $A \cong B$. We may also say that A and B 'have the same cardinality'. Thus, two *finite* sets A and B have the same cardinality if and only if they have the same number of elements: $A \cong B$ if and only if $|A| = |B|$. The notion of 'cardinality' extends to possibly infinite sets the naïve notion of 'number of elements' for finite sets.

There is a compelling alternative description of bijections. A function $g: B \rightarrow A$ is the 'inverse' of f if $g \circ f = \text{id}_A$ and $f \circ g = \text{id}_B$. It is easy to check that a function f is a bijection *if and only if* it has an inverse, and that the inverse of f is a bijection. This also implies that $A \cong B$ if and only if $B \cong A$.

Thus, the isomorphisms (of sets) are precisely the functions that have an inverse. This is the 'categorical' notion of isomorphism; there is a counterpart of this notion for all the structures (rings, groups, etc.) that we consider in this text.

It is also easy to verify that if $f: A \rightarrow B$ and $g: B \rightarrow C$ are bijections, then so is $g \circ f$. Therefore, if $A \cong B$ and $B \cong C$, then $A \cong C$.

A.8   Relations; Equivalence Relations and Quotient Sets

As stated in §A.6, 'functions' from a set A to a set B are certain subsets of the Cartesian product $A \times B$. More generally, a 'relation' between elements of A and elements of B is *any* subset $R \subseteq A \times B$. We say that $a \in A$ and $b \in B$ are 'related by R' if $(a, b) \in R$. Thus, functions are certain types of relations, satisfying the requirement expressed in (A.16).

There are other important types of relations, and particularly relations defined between elements of the same set A—i.e., subsets of the Cartesian product $A \times A$. An 'order' relation R satisfies the following three properties. Writing $a \, R \, b$ for $(a, b) \in R$, we have that R is

- *reflexive*, i.e., $a \, R \, a$ for every $a \in A$;
- *transitive*, i.e., $a \, R \, b$ and $b \, R \, c$ imply $a \, R \, c$ for all $a, b, c \in A$; and
- *antisymmetric*, i.e., $a \, R \, b$ and $b \, R \, a$ imply $a = b$, for all $a, b \in A$.

Example A.19   The ordinary relation $\leq$ on $\mathbb{Z}$ is an order relation. (To view $\leq$ as a relation on $\mathbb{Z}$, i.e., as a subset of $\mathbb{Z} \times \mathbb{Z}$, simply let (a, b) be in the subset if and only if $a \leq b$.) Indeed

- For all $a \in \mathbb{Z}$, $a \leq a$; so $\leq$ is reflexive.
- For all $a, b, c \in \mathbb{Z}$, if $a \leq b$ and $b \leq c$, then $a \leq c$; so $\leq$ is transitive.
- For all $a, b \in \mathbb{Z}$, if $a \leq b$ and $b \leq a$, then $a = b$; so $\leq$ is antisymmetric.

By contrast, $<$ is not an order relation, since it is not reflexive: in fact, $a < a$ is not true for any $a \in \mathbb{Z}$.

Of paramount importance in many constructions is the notion of equivalence relation. A relation $\sim$ on a set A is an 'equivalence relation' when it is *reflexive, symmetric,* and *transitive,* that is:

- $\forall a \in A, a \sim a$;
- $\forall a, b \in A, a \sim b \implies b \sim a$; and
- $\forall a, b, c \in A, a \sim b \land b \sim c \implies a \sim c$.

Example A.20   The 'equality' relation $=$ is an equivalence relation, on any set A. Indeed

- $\forall a \in A, a = a$;
- $\forall a, b \in A, a = b \implies b = a$; and
- $\forall a, b, c \in A, a = b \land b = c \implies a = c$.

In fact, one may view equivalence relations as a generalization of the equality relation, obtained by abstracting these three natural properties of equality.

Example A.21   Another example may be given by letting $A = \mathbb{Z}$ and defining $a \sim b$ to mean that $b - a$ is even.

- $\forall a \in \mathbb{Z}, a - a = 0$ is even, so $a \sim a$;
- $\forall a, b \in \mathbb{Z}$, if $b - a$ is even, then so is $a - b$, thus $a \sim b$ implies $b \sim a$; and
- $\forall a, b, c \in \mathbb{Z}$, if $a \sim b$, that is, $b - a$ is even, and $b \sim c$, that is, $c - b$ is even, then $c - a = (c - b) + (b - a)$ is also even, implying $a \sim c$.

Equivalence relations generalizing this simple example give rise to many interesting constructions in this text.

If a set A is endowed with an equivalence relation $\sim$, we can define an associated set $A/\sim$, which we call the 'quotient of A modulo $\sim$'. This is a prototype for all the 'quotient' constructions for rings, modules, groups, that are among the most important notions developed in this text. The elements of this set are the 'equivalence classes' of the elements of A 'modulo $\sim$', defined by

$$[a]_\sim = \{b \in A \mid b \sim a\}.$$

It is very easy to verify that $[a]_\sim = [b]_\sim \iff a \sim b$. The element a is called a 'representative' of the equivalence class $[a]_\sim$. If a is a representative of an equivalence class, then any other element b such that $a \sim b$ is *also* a representative for the same equivalence class.

This point deserves to be stressed. Notice that while if $a = b$, then $[a]_\sim = [b]_\sim$, the converse does not hold in general. One equivalence class will in general have many different representatives.

With this notation, we define

$$A/\sim = \{[a]_\sim \mid a \in A\} :$$

the quotient of A modulo $\sim$ is the set of all equivalence classes modulo $\sim$.

Example A.22 Consider the equivalence relation defined on $\mathbb{Z}$ in Example A.21, by setting $a \sim b$ if and only if $b - a$ is even. The equivalence class of 0 is

$$[0]_\sim = \{n \in \mathbb{Z} \mid n - 0 \text{ is even}\} = \text{the set of even integers}$$

and similarly

$$[1]_\sim = \text{the set of odd integers.}$$

We can write $[0]_\sim = [2]_\sim = [-98]_\sim$: these are just different names for the set of even integers. The numbers 0, 2, −98, and every other even number, are *representatives* for the single equivalence class consisting of all even integers. Similarly, 1, −35, 1773, and every odd number are representatives for the other equivalence class, consisting of all odd integers.

Using the terminology introduced above, the quotient set $\mathbb{Z}/\sim$ consists of *just two elements*. Notice that the corresponding two subsets of $\mathbb{Z}$ (that is, the set of even integers and the set of odd integers) are not empty, they are disjoint, and their union is $\mathbb{Z}$.

The last observation in this example has a straightforward generalization. It is easy to verify that the equivalence classes modulo an equivalence relation $\sim$ on a set A always form a 'partition' of A: this means that the equivalence classes are

- nonempty
- disjoint
- and their union is the whole set A.

Conversely, if $\mathcal{P}$ is a partition of a set A, that is, a collection of subsets of A satisfying the conditions listed in the previous paragraph, then we can define an equivalence relation $\sim_\mathcal{P}$ on A by setting

$$a \sim_\mathcal{P} b \iff a \text{ and } b \text{ belong to the same element of } \mathcal{P}.$$

It is a good (and easy) exercise to verify that this relation $\sim_\mathcal{P}$ is reflexive, symmetric, and transitive, that is, it indeed is an equivalence relation. Further, if the partition $\mathcal{P}$ arises from an equivalence relation $\sim$, i.e., it equals the quotient $A/\sim$, then

$$a \sim_\mathcal{P} b \iff [a]_\sim = [b]_\sim \iff a \sim b :$$

that is, the equivalence relation $\sim_\mathcal{P}$ corresponding to this partition is nothing but the equivalence relation $\sim$ we started from.

Summarizing, 'equivalence relations' and 'partitions' are different ways to encode the same type of information. The information carried by the choice of an equivalence relation ~ on a set A is 'the same as' the information carried by the corresponding quotient set $A/\sim$.

This is an important observation. In the context of rings, modules, groups, we introduce in the text corresponding notions of 'quotient' (quotient ring, quotient module, quotient group) by choosing specific types of equivalence relations. Understanding how this works in the simpler context of sets is very good preparatory work to approach these more demanding constructions.

One can visualize a partition as cutting up a set into 'chunks' ('parts'). The partition is a set in its own right, with one element for each part.

A.9   Universal Property of Quotients and Canonical Decomposition

Let ~ be an equivalence relation on a set A, and consider the corresponding quotient set $A/\sim$ defined in §A.8. In this situation, we have a function

$$\pi : A \to A/\sim$$

defined by setting $\pi(a)$ to equal the equivalence class $[a]_\sim$ (which is an element of the quotient $A/\sim$, by definition of the latter). This function is clearly surjective: indeed, every element of $A/\sim$ is the equivalence class of some element of A.

The quotient $A/\sim$, along with this function π, satisfies the following property. Suppose $f : A \to B$ is a function such that for all $a', a'' \in A$,

$$a' \sim a'' \implies f(a') = f(a'').$$

Then I claim that there exists a function $\overline{f} : (A/\sim) \to B$ that makes the following diagram commute:

that is, such that $\overline{f} \circ \pi = f$.

Example A.23   Again let us consider the equivalence relation ~ defined on $\mathbb{Z}$ by declaring that $a \sim b$ if and only if $b - a$ is even (cf. Examples A.21 and A.22). We can define a function $f : \mathbb{Z} \to \{0, 1\}$ by setting $\forall a \in \mathbb{Z}$

$$f(a) = \text{the remainder of the division of } a \text{ by 2}.$$

If $a \sim b$, then $b - a$ is even, and it follows easily that $f(a) = f(b)$. So this function satisfies the condition specified above. According to the property we just stated, there exists a function $\overline{f} \colon (\mathbb{Z}/\sim) \to \{0, 1\}$ such that the following diagram commutes:

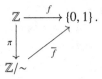

A moment's thought should confirm that this is the case. Indeed, the set $\mathbb{Z}/\sim$ consists of two elements: one of these elements is the set of even numbers, and the other element is the set of odd numbers (cf. Example A.22). Let $\overline{f} \colon (\mathbb{Z}/\sim) \to \{0, 1\}$ be the function defined by sending the first element to 0 and the second element to 1. Then we will indeed have $f = \overline{f} \circ \pi$: if a is even, then $\pi(a) = [a]_\sim$ = the set of even numbers, so $\overline{f} \circ \pi(a) = 0 = f(a)$; and similarly if a is odd.

A reader who thoroughly understands this simple example will have no difficulty understanding the general case. Given A, $\sim$, and $f \colon A \to B$ as above, define $\overline{f} \colon (A/\sim) \to B$ by setting $\forall a \in A$

$$\overline{f}([a]_\sim) = f(a) . \tag{A.17}$$

To verify that this is indeed a function $(A/\sim) \to B$, we have to verify that the proposed value for $\overline{f}([a]_\sim)$ *only depends on the equivalence class* $[a]_\sim$, not on the specific chosen representative a for this class. As we observed in §A.8, in general an equivalence class has many different representatives, and we have to verify that the result of the definition given in (A.17) is not affected by this choice.

This verification is straightforward. If a and b are representatives for the same equivalence class, i.e., $[a]_\sim = [b]_\sim$, then $a \sim b$. By the hypothesis on f, this implies that $f(a) = f(b)$. So indeed the result of (A.17) is the same whether we choose a or b to represent the equivalence class.

We say that $\overline{f}$ is 'well-defined': the definition given in (A.17) appears to depend on a specific choice (the representative for the equivalence class) but, as we just verified, the result of the definition is in fact independent of this choice.

The definition of the function $\overline{f} \colon (A/\sim) \to B$ proves that it satisfies its main requirement: if $a \in A$, then

$$\overline{f} \circ \pi(a) = \overline{f}([a]_\sim) = f(a) ,$$

proving that $\overline{f} \circ \pi = f$ as required.

In fact, we could observe that the requirement $f = \overline{f} \circ \pi$ actually *forces* the definition of $\overline{f}$ given in (A.17). This implies that the function $\overline{f}$ is actually the *unique* function satisfying this requirement. We say that $\overline{f}$ is 'induced' by f.

The property we described in the previous several paragraphs is an example of a 'universal property'. Many universal properties are encountered in this text.

A compelling application of the universal property of quotients is a natural way to decompose *every* function of sets $f: A \to B$ as a composition of functions with pre-scribed properties. I claim that if $f: A \to B$ is a function of sets, then there exist two sets A' and B' and

- a *surjective* function $\pi: A \to A'$,
- a *bijective* function $\tilde{f}: A' \to B'$, and
- an *injective* function $i: B' \to B$

such that

$$f = i \circ \tilde{f} \circ \pi.$$

That is, such that the diagram

$$A \xrightarrow{\pi} A' \xrightarrow{\tilde{f}} B' \xrightarrow{i} B$$

with f the arc over the top.

commutes. This is a template of results leading to the so-called 'isomorphism' theorems in the contexts of rings, modules, groups.

With our preparatory work, verifying this statement is not difficult.

- We let $B' = f(A)$ be the *image of f* (cf. §A.6). Since B' is a subset of B, we may define $i: B' \to B$ by simply setting $i(b) = b$ for every $b \in B'$. It is clear that this function is injective.
- In order to construct A', we consider the equivalence relation $\sim$ defined on A by setting $\forall a', a'' \in A$

$$a' \sim a'' \iff f(a') = f(a'').$$

The reader should verify that this is indeed an equivalence relation. We can then let A' be the quotient set $A/\sim$, and let $\pi: A \to A'$ be the function defined by setting $f(a) = [a]_\sim$ for all $a \in A$. As observed at the beginning of this section, this function is surjective.

- Finally, we now have an equivalence relation $\sim$ on A and a function $f: A \to B$ such that $a' \sim a'' \implies f(a') = f(a'')$. By the universal property, f induces a function $\overline{f}: A \to B$ such that $f = \overline{f} \circ \pi$. The image of $\overline{f}$ equals the image B' of f; so we may view $\overline{f}$ as a function $\tilde{f}: A' \to B'$. Explicitly (recalling (A.17))

$$\tilde{f}([a]_\sim) = f(a) \in f(A) = B'. \tag{A.18}$$

With this definition we have $\overline{f} = i \circ \tilde{f}$, therefore

$$f = \overline{f} \circ \pi = i \circ \tilde{f} \circ \pi$$

as needed. All that is left to do is to verify that $\tilde{f}$ is a bijection, that is, that it is injective and surjective.

To verify that $\tilde{f}$ is surjective, let $b \in B' = f(A)$. Then there exists $a \in A$ such that $b = f(a)$. By (A.18), we have

$$\tilde{f}([a]_\sim) = f(a) = b,$$

and this proves that $\tilde{f}$ is surjective.

To verify that $\tilde{f}$ is injective, assume that

$$\tilde{f}([a']_\sim) = \tilde{f}([a'']_\sim).$$

By (A.18), this implies

$$f(a') = f(a''),$$

and by definition of $\sim$ this means

$$a' \sim a''$$

and therefore

$$[a']_\sim = [a'']_\sim.$$

This shows that $\tilde{f}$ satisfies the requirement to be injective, and concludes the verification that $\tilde{f}$ is a bijective function as stated.

If f is surjective, then $B' = B$, so according to the result we just verified it follows that B is isomorphic to $A/\sim$ in this case.

The 'canonical decomposition' we just obtained for every function between sets upgrades to the useful *First Isomorphism Theorem* in the various algebraic contexts considered in this text. This will be one of our main tools in the study of rings, groups, etc. and of functions between these objects. From the point of view taken here, the result is a consequence of the universal property of quotients, which will also have a counterpart in these different contexts.

Appendix B Solutions to Selected Exercises

B.1   The Integers

1.2 (i) $a \mid a$ since $a = a \cdot 1$.

(ii) If $a \mid b$ and $b \mid c$, then there are integers ℓ, m such that $b = a\ell$ and $c = bm$. Then $c = bm = (a\ell)m = a(\ell m)$, and this proves that $a \mid c$.

(iii) If $a \mid b$ and $b \mid a$, then there exist integers ℓ, m such that $b = a\ell$ and $a = bm$. Then $a = bm = (a\ell)m = a(\ell m)$. Since $a \neq 0$ (a is assumed to be positive), this implies $\ell m = 1$. As both ℓ and m are integers, this forces $\ell = m = \pm 1$; and since a and b are positive, we conclude $\ell = m = 1$. But then $a = bm = b \cdot 1 = b$, as we had to verify.

This relation is not a *total* ordering. For example, 2 does not divide 3 and 3 does not divide 2.

It *does* have a 'maximum', and that maximum is 0. Indeed, for every integer a, $a \mid 0$.

1.8 First assume that a and b have the same remainder when divided by n: that is, $a = nq_1 + r$ and $b = nq_2 + r$ for some integers q_1, q_2, r, such that $0 \leq r < n$. But then

$$a - b = (nq_1 + r) - (nq_2 + r) = n(q_1 - q_2),$$

so indeed $a - b = nk$ for some integer k: $k = q_1 - q_2$ works here.

Conversely, assume $a - b = nk$ for some integer k, and write $b = nq + r$ with $0 \leq r < n$. Then we have

$$a = nk + b = nk + nq + r = n(k + q) + r.$$

By the uniqueness part of division with remainder (Theorem 1.3), r is the remainder of the division of a by n (and the quotient equals $k + q$). Therefore a and b have the same remainder after division by n, as stated.

1.10 Since $\gcd(a, b) = 1$, by Corollary 1.9 there exist integers m, n such that $ma + nb = 1$. Then $mac + nbc = c$. Since b divides c, ab divides mac; since a divides c, ab divides nbc. It follows that ab divides c, as needed.   ⌟

There are of course other solutions, particularly if we can use irreducible factorizations (but the reader was instructed to first solve this problem by only using the material in §1.3). For example: let $ab = 2^{\epsilon_2}3^{\epsilon_3}5^{\epsilon_5} \cdots$ and $c = 2^{\gamma_2}3^{\gamma_3}5^{\gamma_5} \cdots$. Since $\gcd(a, b) = 1$, a and b have no common irreducible factor (Corollary 1.27). Therefore, if $\epsilon_q \neq 0$, then q^{ϵ_q} is either a factor of a or a factor of b. In either case it must divide c, since $a \mid c$ and $b \mid c$. It follows that $\epsilon_i \leq \gamma_i$ for all i, hence $ab \mid c$ as needed.

1.17 We will use the well-ordering principle. Let p be prime, and let S be the set of positive integers s for which there exists a product $a_1 \cdots a_s$ such that $p \mid a_1 \cdots a_s$ and p does *not* divide any of the integers a_i. We need to prove that $S = \emptyset$. If S is not empty, then by the well-ordering principle it has a minimum m: that is, there exists a product $a_1 \cdots a_m$ *with a minimum number of factors*, which is a multiple of p and such that p does not divide any of the integers a_i. Let then $b = a_1 \cdots a_{m-1}$, $c = a_m$. Then p divides bc, and $p \nmid c$ by assumption. Since p is prime, necessarily $p \mid b$: p divides the product $a_1 \cdots a_{m-1}$. Since m is the least element of S, $m - 1 \notin S$: it follows that p must divide one of the factors $a_1, \ldots, a_{m-1}$, and this contradicts our assumption.

Therefore the hypothesis that S be nonempty leads to a contradiction. This shows that S is necessarily empty, as needed.   ⌟

The same argument, using induction: We have to prove that if p is prime, then for all $s \geq 0$, if p divides a product $a_1 \cdots a_s$, then p divides one of the factors. This is trivially true for $s = 0$ and $s = 1$. Arguing by induction, we have to prove that for all $s \geq 2$, the truth of this statement for $s - 1$ implies the truth for s. Assume then that $s \geq 2$ and that p divides $a_1 \cdots a_s$; we have to prove that p divides one of the factors a_i. Let $b = a_1 \cdots a_{s-1}$ and $c = a_s$, so that p divides bc. If p divides $c = a_s$, we are done; so we may assume that p does not divide c. Since p is prime, $p \mid bc$, and $p \nmid c$, necessarily p divides $b = a_1 \cdots a_{s-1}$. This product has $s - 1$ factors, so by the induction hypothesis b must divide one of $a_1, \ldots, a_{s-1}$, and again we are done.

1.21 If $c \mid n$, then there exists an integer m such that $n = cm$. Writing $m = 2^{\mu_2} 3^{\mu_3} 5^{\mu_5} \cdots$, we obtain

$$n = cm = 2^{(\gamma_2 + \mu_2)} 3^{(\gamma_3 + \mu_3)} 5^{(\gamma_5 + \mu_5)} 7^{(\gamma_7 + \mu_7)} \cdots.$$

By uniqueness of factorizations, we can conclude that $v_2 = \gamma_2 + \mu_2$, $v_3 = \gamma_3 + \mu_3$, etc. This implies that $\gamma_i \leq v_i$ for all i. (Note that if we did not know that factorizations are unique, we could not draw this conclusion!)

Conversely, assume that $\gamma_i \leq v_i$ for all i, and let $\mu_i = v_i - \gamma_i$. Since $\mu_i = 0$ for all but finitely many i, we can consider the integer $m = 2^{\mu_2} 3^{\mu_3} 5^{\mu_5} 7^{\mu_7} \cdots$. Then

$$cm = 2^{(\gamma_2 + \mu_2)} 3^{(\gamma_3 + \mu_3)} 5^{(\gamma_5 + \mu_5)} 7^{(\gamma_7 + \mu_7)} \cdots = 2^{v_2} 3^{v_3} 5^{v_5} 7^{v_7} \cdots = n :$$

it follows that c divides n, as needed.

1.23 Let $a = 2^{\alpha_2} 3^{\alpha_3} 5^{\alpha_5} 7^{\alpha_7} 11^{\alpha_{11}} \cdots$ and $b = 2^{\beta_2} 3^{\beta_3} 5^{\beta_5} 7^{\beta_7} 11^{\beta_{11}} \cdots$. Then

$$\gcd(a, b) = 2^{\min(\alpha_2, \beta_2)} 3^{\min(\alpha_3, \beta_3)} 5^{\min(\alpha_5, \beta_5)} 7^{\min(\alpha_7, \beta_7)} 11^{\min(\alpha_{11}, \beta_{11})} \cdots,$$

$$\operatorname{lcm}(a, b) = 2^{\max(\alpha_2, \beta_2)} 3^{\max(\alpha_3, \beta_3)} 5^{\max(\alpha_5, \beta_5)} 7^{\max(\alpha_7, \beta_7)} 11^{\max(\alpha_{11}, \beta_{11})} \cdots,$$

$$a b = 2^{\alpha_2 + \beta_2} 3^{\alpha_3 + \beta_3} 5^{\alpha_5 + \beta_5} 7^{\alpha_7 + \beta_7} 11^{\alpha_{11} + \beta_{11}} \cdots$$

(Proposition 1.25 and the comments following it). So we just need to observe that $\min(\alpha, \beta) + \max(\alpha, \beta) = \alpha + \beta$, which is immediately checked by running through all possibilities $\alpha < \beta$, $\alpha = \beta$, $\alpha > \beta$.

1.25 If $\sqrt{p} = \frac{a}{b}$ for some integers a and b, then we must have $a^2 = pb^2$. The uniqueness part of the Fundamental Theorem tells us that the left-hand side and the right-hand side must have *the same* factorization into irreducibles. But all the exponents of the factorization of a^2 are even, while the exponent of p (which is assumed to be irreducible) in pb^2 is odd. This is a contradiction.

Therefore we can*not* have $\sqrt{p} = \frac{a}{b}$ with a, b integers. This says precisely that $\sqrt{p}$ is irrational.

1.26 Assume that there are only finitely many primes $p_1, \ldots, p_k$, and consider the number $n = p_1 \cdots p_k + 1$. Note that $n > 1$. By the Fundamental Theorem of Arithmetic, n has a factorization $n = p_1^{v_1} \cdots p_k^{v_k}$, and some exponent v_i is positive. But then $p_i \mid n$. We have that $p_i \mid p_1 \cdots p_k$ and $p_i \mid n = p_1 \cdots p_k + 1$, so we can conclude $p_i \mid 1$, and we have reached a contradiction.

Thus the assumption that there are only finitely many primes leads to a contradiction, and the conclusion is that there must be *infinitely many* primes.

B.2   Modular Arithmetic

2.4 The stated fact does not hold in general. For a counterexample, take $n = 4$, $a = 2$, $b = 1$, $c = 3$. Then indeed $[2]_4 \neq [0]_4$ and $[2]_4[1]_4 = [2]_4 = [6]_4 = [2]_4[3]_4$, but $[1]_4 \neq [3]_4$.

2.8 The point here is that the last digit of a number n is the number c, $0 \leq c \leq 9$, such that $[n]_{10} = [c]_{10}$. (If this is not super-clear, you have not picked up the main content of Lemma 2.6.) Therefore, we have to compute $[7]_{10}^{1000000}$.

For this, note that $[7]_{10}^2 = [49]_{10} = [9]_{10}$, therefore

$$[7]_{10}^4 = ([7]_{10}^2)^2 = [9]_{10}^2 = [81]_{10} = [1]_{10},$$

and

$$[7]_{10}^{1000000} = ([7]_{10}^4)^{250000} = [1]_{10}^{250000} = [1]_{10}.$$

The conclusion is that the last digit of $7^{1000000}$ is 1.

2.9 This can be done quickly by trial and error. Of course $[1]_7^k = [1]_7$ for all k, so $a = 1$ won't do. We have $[2]_7^2 = [4]_7$, $[2]_7^3 = [1]_7$, and then the powers cycle through $[2]_7$, $[4]_7$, $[1]_7$ over and over again, so $a = 2$ also won't do. But $[3]_7$ works:

$$[3]_7^0 = [1]_7,$$
$$[3]_7^1 = [3]_7,$$
$$[3]_7^2 = [9]_7 = [2]_7,$$
$$[3]_7^3 = [3]_7^2[3]_7 = [2]_7[3]_7 = [6]_7,$$
$$[3]_7^4 = [3]_7^3[3]_7 = [6]_7[3]_7 = [18]_7 = [4]_7,$$
$$[3]_7^5 = [3]_7^4[3]_7 = [4]_7[3]_7 = [12]_7 = [5]_7,$$

and we see that all classes mod 7 except for $[0]_7$ may be obtained as powers of $[3]_7$.

2.10 Assume $n > 0$ is odd and $n = a^2 + b^2$ for two integers a and b. Then $[n]_4 =$

$[a]_4^2 + [b]_4^2$. The class $[n]_4$ equals $[1]_4$ or $[3]_4$ since n is odd; we have to show that it *cannot* equal $[3]_4$.

For this, note that $[a]_4^2$ is one of $[0]_4^2 = [0]_4$, $[1]_4^2 = [1]^2$, $[2]_4^2 = [4]_4 = [0]_4$, $[3]_4^2 = [9]_4 = [1]_4$: the class mod 4 of a perfect square can only be $[0]_4$ or $[1]_4$. It follows that the sum of two such classes can only be

$$[0]_4 + [0]_4 = [0]_4\,, \quad [0]_4 + [1]_4 = [1]_4\,, \quad [1]_4 + [0]_4 = [1]_4\,, \quad [1]_4 + [1]_4 = [2]_4 :$$

the class $[3]_4$ is simply not an option.

2.11 Let k be an integer such that $a = [k]_n$. Then

$$a \cdot 0 = [k]_n [0]_n = [k \cdot 0]_n = [0]_n = 0\,,$$

as needed.

2.18 The formulations are as follows. Let $p > 0$ be a prime integer.

(i) If a is any integer, then $[a]_p^p = [a]_p$.

(ii) If a is an integer such that $p \nmid a$, then $a^{p-1} \equiv 1 \bmod p$.

To see that (ii) $\Longrightarrow$ (i): If $p \mid a$, then $[a]_p = [0]_p$, so $[a]_p^p = [0]_p^p = [0]_p = [a]_p$, as it should according to (i). If $p \nmid a$, then $a^{p-1} \equiv 1 \bmod p$ by (ii), so (multiplying by a) $a^p \equiv a \bmod p$; this says precisely that $[a]_p^p = [a]_p$, verifying (i) in this case as well.

To see that (i) $\Longrightarrow$ (ii), let a be such that $p \nmid a$. By (i), $[a]_p[a]_p^{p-1} = [a]_p[1]_p$. Since $p \nmid a$, we have $[a]_p \neq [0]_p$, therefore by cancellation (Remark 2.17) we can conclude that $[a]_p^{p-1} = [1]_p$. This says that $a^{p-1} \equiv 1 \bmod p$, verifying (i) as needed.

2.19 (i) The classes $[1 \cdot 3]_{11}, [2 \cdot 3]_{11}, \dots, [10 \cdot 3]_{11}$ are

$$[1 \cdot 3]_{11} = [3]_{11}\,, \quad [2 \cdot 3]_{11} = [6]_{11}\,, \quad [3 \cdot 3]_{11} = [9]_{11}\,, \quad [4 \cdot 3]_{11} = [1]_{11}\,,$$
$$[5 \cdot 3]_{11} = [4]_{11}\,, \quad [6 \cdot 3]_{11} = [7]_{11}\,, \quad [7 \cdot 3]_{11} = [10]_{11}\,, \quad [8 \cdot 3]_{11} = [2]_{11}\,,$$
$$[9 \cdot 3]_{11} = [5]_{11}\,, \quad [10 \cdot 3]_{11} = [8]_{11}\,.$$

As promised, these are again the classes $[1]_{11}, \dots, [10]_{11}$, just in a different order.

(ii) The classes $[1 \cdot 3]_{12}, [2 \cdot 3]_{12}, \dots, [11 \cdot 3]_{12}$ are

$$[1 \cdot 3]_{12} = [3]_{12}\,, \quad [2 \cdot 3]_{12} = [6]_{12}\,, \quad [3 \cdot 3]_{12} = [9]_{12}\,, \quad [4 \cdot 3]_{12} = [0]_{12}\,,$$
$$[5 \cdot 3]_{12} = [3]_{12}\,, \quad [6 \cdot 3]_{12} = [6]_{12}\,, \quad [7 \cdot 3]_{12} = [9]_{12}\,, \quad [8 \cdot 3]_{12} = [0]_{12}\,,$$
$$[9 \cdot 3]_{12} = [3]_{12}\,, \quad [10 \cdot 3]_{12} = [6]_{12}\,, \quad [11 \cdot 3]_{12} = [9]_{12}\,.$$

These are the classes $[0]_{12}, [3]_{12}, [6]_{12}, [9]_{12}$, with some repetitions. The situation is quite different from (i).

2.20 For the first point: we are assuming that $a^{n-1} \not\equiv 1 \bmod n$ and $b^{n-1} \equiv 1 \bmod n$. Then $(ab)^{n-1} = a^{n-1}b^{n-1} \not\equiv 1 \bmod n$, so ab is a witness.

For the second, assume $[b_i]_n \neq [b_j]_n$; we have to prove that $[ab_i]_n \neq [ab_j]_n$. Equivalently, we have to prove that if $[ab_i]_n = [ab_j]_n$, then $[b_i]_n = [b_j]_n$. We are also assuming that $\gcd(a, n) = 1$, so by Proposition 2.16 the class $[a]_n$ has a multiplicative inverse: there exists a class $[a']_n$ such that $[a']_n[a]_n = [1]_n$. It follows that

$$[ab_i]_n = [ab_j]_n \implies [a']_n[a]_n[b_i]_n = [a']_n[a]_n[b_j]_n \implies [b_i]_n = [b_j]_n\,,$$

as needed.

Now assume that there are m distinct non-witnesses $[b_1]_n, \ldots, [b_m]_n$ with $1 < b_i < n$. Then $[ab_1]_n, \ldots, [ab_m]_n$ are distinct and all distinct from the classes $[b_i]_n$ (because they are witnesses, while each $[b_i]_n$ is a non-witness). It follows that $n - 1 \geq 2m$ (note that $[0]_n$ is not one of the integers b_i by assumption, and not one of the $[ab_i]_n$ since these are witnesses). We can conclude that $m < \frac{n}{2}$, that is, n has fewer than $n/2$ non-witnesses.

(Keep in mind that to draw this conclusion we need to have a witness a such that $\gcd(a, n) = 1$. This is an actual requirement: for example, no such a exists for Carmichael numbers, cf. Exercise 2.21.)

B.3   Rings

3.5 Assume that $0 = 1$, and let $a \in R$ be any element. We have that $a \cdot 1 = a$ by axiom (vi); on the other hand, $a \cdot 0 = 0$ by Corollary 3.18. Therefore

$$a = a \cdot 1 = a \cdot 0 = 0 :$$

this shows that 0 is the only element of R.

3.6 Use distributivity and Corollary 3.18:

$$a + ((-1) \cdot a) = 1 \cdot a + (-1) \cdot a = (1 - 1) \cdot a = 0 \cdot a = 0 ;$$

and similarly $((-1) \cdot a) + a = 0$. Therefore $(-1) \cdot a$ satisfies the requirement of an additive inverse of a (axiom (iii)), and it follows that $(-1) \cdot a = -a$ by the uniqueness of inverses (Proposition 3.15).

3.7 Using the commutativity and associativity of addition, we have

$$((-a)+(-b))+a+b = ((-a)+(-b))+b+a = (-a)+((-b)+b)+a = (-a)+0+a = (-a)+a = 0$$

and similarly $a+b+((-a)+(-b)) = 0$. Therefore $(-a)+(-b)$ satisfies the requirement for an additive inverse of $a + b$, and it follows that $-(a + b) = (-a) + (-b)$ by the uniqueness of additive inverses (Proposition 3.15).

The conclusion $(-n)a = -(na)$ follows by induction. This statement is true for $n = 0$ (both sides equal 0 in this case, by Corollary 3.18). For the induction step, we have

$$-((n + 1)a) = -((na) + a) \stackrel{(1)}{=} (-(na)) + (-a) \stackrel{(2)}{=} ((-n)a) + (-a) = (-(n + 1))a :$$

(1) holds by the fact we just proved, and (2) by the induction hypothesis.

3.8 We have

$$(2(1_R)) \cdot (3(1_R)) = (1_R + 1_R) \cdot (3(1_R)) \stackrel{(1)}{=} 3(1_R) + 3(1_R) \stackrel{(2)}{=} 6(1_R)$$

where (1) holds by distributivity and (2) by definition of multiples: adding 1_R to itself 3 times and then again 3 times is the same as adding 1_R to itself $3 + 3 = 6$ times.

For the general case: First, note that $(-m)1_R = -(m1_R)$ (Exercise 3.7); so we may pull out signs if necessary and assume that m and n are nonnegative. I claim that for all

$a \in R$ and $m \geq 0$, $(m1_R) \cdot a = ma$. Indeed, this is the case for $m = 0$ by Corollary 3.18; arguing by induction, we have

$$((m + 1)1_R) \cdot a = ((m1_R + 1_R) \cdot a \overset{(1)}{=} (m1_R) \cdot a + a \overset{(2)}{=} ma + a \overset{(3)}{=} (m + 1)a$$

where (1) holds by distributivity, (2) by the induction hypothesis, and (3) by definition of multiples. Applying this fact to $a = n1_R$ gives

$$(m1_R) \cdot (n1_R) = m(n1_R)$$

and this equals $(mn)1_R$ again by definition of multiples.

3.12 To show that the characteristic of $\mathbb{Z}$ is 0, just observe that $n(1_\mathbb{Z}) = n$ equals 0 only when $n = 0$; therefore, there is no positive n for which $n(1_\mathbb{Z}) = 0$. For $\mathbb{Z}/m\mathbb{Z}$: we have $n(1_{\mathbb{Z}/m\mathbb{Z}}) = n[1]_m = [n]_m$, and this equals $[0]_m$ if and only if $m \mid n$. Therefore the characteristic of $\mathbb{Z}/m\mathbb{Z}$ is the smallest positive integer n such that $m \mid n$, and that is m itself.

If R has characteristic n, let $a \in R$ be any element. Then

$$na = n(1_R \cdot a) \overset{!}{=} (n(1_R)) \cdot a = 0 \cdot a = 0$$

by Corollary 3.18. (Equality $\overset{!}{=}$ holds by distributivity.)

3.13 Let R be an integral domain, and let n be its characteristic. If $n = 0$, then n is prime; so we may assume that $n > 0$, and it suffices to prove that n is not irreducible (Theorem 1.21).

If n is *not* irreducible, then there exist positive integers $a < n$ and $b < n$ such that $n = ab$. Then we have

$$0 = n(1_R) = (ab)(1_R) = (a1_R) \cdot (b1_R)$$

(using Exercise 3.8). Since R is an integral domain, it follows that $a1_R = 0$ or $b1_R = 0$. This contradicts the assumption that n is the *smallest* positive integer for which $n1_R = 0$. Therefore n is necessarily irreducible.

3.14 Since R is nontrivial, $0 \neq 1$ in R, and it follows that $0 \neq 1$ in $R[x]$; and since R is commutative, $R[x]$ is also (easily seen to be) commutative. We have to verify that $R[x]$ has no nonzero zero-divisors, that is, that the product of two nonzero elements of $R[x]$ is not equal to 0.

Let $f(x) = a_0 + a_1 + \cdots + a_m x^m$ and $g(x) = b_0 + b_1 + \cdots + b_n x^n$ be two nonzero elements of $R[x]$. Since they are nonzero, we may assume $a_m \neq 0$ and $b_n \neq 0$. We have

$$f(x)g(x) = (a_0 + \cdots + a_m x^m) \cdot (b_0 + \cdots + b_n x^n) = a_0 b_0 + (a_0 b_1 + a_1 b_0)x + \cdots + a_m b_n x^{m+n} :$$

since R is an integral domain and $a_m \neq 0$, $b_n \neq 0$, we have $a_m b_n \neq 0$, and this proves that $f(x)g(x) \neq 0$ as needed.

3.17 If 0_R is invertible, then there exists an element $a \in R$ such that $a \cdot 0_R = 1_R$. On the other hand, $a \cdot 0_R = 0_R$ by Corollary 3.18. Therefore we have

$$0_R = a \cdot 0_R = 1_R.$$

Since $0_R = 1_R$, R is trivial (by Exercise 3.5).

3.20 If a is nilpotent, then $a^k = 0$ for some $k \geq 1$. Let k be the smallest such exponent. Then $0 = a^k = a \cdot a^{k-1} = a^{k-1} \cdot a$, and $a^{k-1} \neq 0$ by assumption. Therefore we have found $a b \neq 0$ such that $a \cdot b = b \cdot a = 0$, and this shows that a is a zero-divisor.

A nonzero nilpotent in one of the rings $\mathbb{Z}/n\mathbb{Z}$: for example, $[2]_8 \in \mathbb{Z}/8\mathbb{Z}$. Indeed, $[2]_8 \neq 0$, and $[2]_8^3 = [2^3]_8 = [8]_8 = 0$.

Let a be nilpotent, and let $k \geq 1$ be an exponent such that $a^k = 0$. Note that

$$(1 + a) \cdot (1 - a + a^2 - \cdots + (-1)^{k-1}a^{k-1})$$
$$= 1 - a + a^2 - a^3 + \cdots + (-1)^{k-1}a^{k-1}$$
$$+ a - a^2 + a^3 - \cdots - (-1)^{k-1}a^{k-1} + (-1)^k a^k) \qquad \text{(B.1)}$$
$$= 1 + (-1)^k a^k = 1 \pm 0 = 1,$$

and similarly

$$(1 - a + a^2 - \cdots + (-1)^{k-1}a^{k-1}) \cdot (1 + a) = 1.$$

Therefore $1 + a$ is invertible. (About the hint: (B.1) is the key step in proving that the geometric series $\sum_{k \geq 0} r^k$ converges if $|r| < 1$.)

3.22 The ring R is nontrivial, and commutative (indeed, the multiplication table is symmetric with respect to the diagonal). To illustrate associativity and distributivity: for example,

$$(1 + x) + y = y + y = 0 \quad \text{vs.} \quad 1 + (x + y) = 1 + 1 = 0,$$
$$(x \cdot y) \cdot y = 1 \cdot y = y \quad \text{vs.} \quad x \cdot (y \cdot y) = x \cdot x = y,$$
$$x \cdot (x + y) = x \cdot 1 = x \quad \text{vs.} \quad (x \cdot x) + (x \cdot y) = y + 1 = x.$$

The key property defining fields is the existence of inverses for nonzero elements. Here the nonzero elements are $1, x, y$; it suffices to observe that according to the given table we have

$$1 \cdot 1 = 1, \quad x \cdot y = 1, \quad y \cdot x = 1:$$

that is, x and y are inverses of each other. Every nonzero element has a multiplicative inverse, so indeed R is a field.

B.4   The Category of Rings

4.2 The elements $(1_R, 0_S)$ and $(0_R, 1_S)$ are both nonzero, and $(1_R, 0_S) \cdot (0_R, 1_S) = (0_R, 0_S) = 0_{R \times S}$.

4.3 Let $(r, s) \in R \times S$. We have to prove that (r, s) is a unit in $R \times S$ if and only if r is a unit in R and s is a unit in S.

If r, s are units, then there exists $a \in R$, $b \in S$ such that $ar = ra = 1_R$ and $bs = sb = 1_S$. Then $(a, b) \cdot (r, s) = (r, s) \cdot (a, b) = (1_R, 1_S) = 1_{R \times S}$ and this shows that (r, s) has a multiplicative inverse, as needed.

Conversely, if (r, s) is a unit, then there exists $(a, b) \in R \times S$ such that $(a, b) \cdot (r, s) =$

$(r, s) \cdot (a, b) = 1_{R \times S}$. It is immediate to verify that then a is a multiplicative inverse of r and b is a multiplicative inverse of s, and it follows that r and s are units.

4.4 One difference: $\mathbb{Z} \times \mathbb{Z}$ has nonzero zero-divisors (cf. Exercise 4.2) while $\mathbb{Z}[i]$ has no nonzero zero-divisors. (Since $\mathbb{Z}[i]$ is a subring of $\mathbb{C}$, any zero-divisor in $\mathbb{Z}[i]$ is necessarily a zero-divisor in $\mathbb{C}$, and $\mathbb{C}$ has no nonzero zero-divisors since it is a field. Cf. Exercise 4.8.)

Another difference: In $\mathbb{Z}[i]$ there is an element, that is, i, whose square is -1; in $\mathbb{Z} \times \mathbb{Z}$ there is no element whose square is $-1 = (-1, -1)$. Indeed, for all elements (a, b) of $\mathbb{Z} \times \mathbb{Z}$, $(a, b)^2 = (a^2, b^2)$ has nonnegative entries.

4.5 Use Proposition 4.14. Since 0_R and 1_R are in S, and S is closed under $\cdot$ by the first requirement, it suffices to verify that S is closed under $+$ and contains the additive inverse of every $s \in S$.

Let $s \in S$. Since $0_R \in S$ and S is closed under subtraction, $-s = 0_R - s$ is an element of S; so S contains additive inverses as needed. To see that S is closed under $+$, let $a, b \in S$. Then $-b \in S$ as we have just verified, and it follows that $a + b = a - (-b) \in S$ as needed, since S is closed under subtraction.

4.11 Since f is a homomorphism,

$$f(b - a) + f(a) = f((b - a) + a) = f(b).$$

Adding $-f(a)$ to both sides gives $f(b - a) = f(b) - f(a)$ as needed.

4.13 In Example 4.20, $\sigma: \mathbb{Z} \to R$ was defined by setting $\sigma(n) = n 1_R$. We have to verify that every ring homomorphism $\mathbb{Z} \to R$ agrees with σ.

If $\tau: \mathbb{Z} \to R$ is any ring homomorphism, then necessarily $\tau(1_{\mathbb{Z}}) = 1_R$. It follows that for $n > 0$

$$\tau(n) = \tau(n 1_{\mathbb{Z}}) = n\tau(1_{\mathbb{Z}}) = n 1_R = \sigma(n),$$

and

$$\tau(-n) = -\tau(n) = -\sigma(n) = \sigma(-n).$$

Therefore $\tau(n) = \sigma(n)$ for all $n \in \mathbb{Z}$, and this means that $\tau = \sigma$ as functions.

4.14 An object I is 'initial' if for every object X there is a unique morphism from I to X. An object F is 'final' if for every object X there is a unique morphism from X to F.

For rings, this would be a ring T such that for every ring R there exists a unique ring homomorphism $R \to T$. Well, the trivial ring T does have this property: if t is the only element of T, the function $\tau: R \to T$ given by letting $\tau(r) = t$ for every r is a homomorphism (cf. Example 4.17), and it is in fact the only *function* one can define from R to T.

Thus, we can say that 'the trivial ring is final in the category of rings'.

In the category of sets the empty set is initial, and every set consisting of a single element (every 'singleton') is final.

4.19 Let $\ell = \text{lcm}(m, n)$. If m, n are relatively prime, then $\ell < mn$ (see, e.g., Exercise 1.23). For every $(a, b) \in \mathbb{Z}/m\mathbb{Z} \times \mathbb{Z}/n\mathbb{Z}$, we have

$$\ell(a, b) = (\ell a, \ell b) = (0, 0),$$

since ℓ is a multiple of both m, n. If $\mathbb{Z}/m\mathbb{Z} \times \mathbb{Z}/n\mathbb{Z}$ and $\mathbb{Z}/mn\mathbb{Z}$ were isomorphic, then necessarily we would have that $\ell[a]_{mn} = 0$ for all a. This is not the case: for example, $\ell 1_{\mathbb{Z}/mn\mathbb{Z}} = [\ell]_{mn} \neq 0$, since ℓ is not a multiple of mn.

B.5   Canonical Decomposition, Quotients, and Isomorphism Theorems

5.4 The set consists of matrices of the form

$$\begin{pmatrix} a & 0 \\ c & 0 \end{pmatrix},$$

that is, with vanishing second column. In order to verify that this set is not an ideal, it suffices to observe that a product

$$\begin{pmatrix} a & 0 \\ c & 0 \end{pmatrix} \cdot \begin{pmatrix} 0 & 1 \\ 0 & 0 \end{pmatrix} = \begin{pmatrix} 0 & a \\ 0 & c \end{pmatrix}$$

is *not* a matrix of this form. Thus the set does not satisfy the 'absorption property' on the right. (It does on the left; the set is a so-called 'left-ideal'.)

5.5 The set $(a_1, \ldots, a_n)$ of linear combinations of $a_1, \ldots, a_n$ is not empty. For example, a_1 is an element of $(a_1, \ldots, a_n)$ since we can view it as a linear combination: $a_1 = 1 \cdot a_1 + 0 \cdot a_2 + \cdots + 0 \cdot a_r$. The sum of two linear combinations is a linear combination:

$$(r_1 a_1 + \cdots + r_n a_n) + (s_1 a_1 + \cdots + s_n a_n) = (r_1 + s_1)a_1 + \cdots + (r_n + s_n)a_n.$$

Finally, the set satisfies the absorption property. Indeed, let b be a linear combination of the elements a_i: $b = c_1 a_1 + \cdots + c_n a_n$. We have to show that for all $r \in R$, rb and br are linear combinations. *Since R is commutative, $br = rb$*; so it suffices to show that rb is a linear combination of the elements a_i. Well, it is:

$$rb = r(c_1 a_1 + \cdots + c_n a_n) = (rc_1)a_1 + \cdots + (rc_n)a_n.$$

By Proposition 5.9, $(a_1, \ldots, a_n)$ is an ideal.

5.7 Since I is not empty, and it is closed under addition, so is $f(I)$, even if f is not surjective. By Proposition 5.9, it suffices to show that $f(I)$ satisfies the absorption property. Let then $f(a)$ be an arbitrary element of $f(I)$ (that is, let $a \in I$), and let $s \in S$ be any element of the ring S. *Since f is surjective*, there exists an element $r \in R$ such that $s = f(r)$. Then

$$sf(a) = f(r)f(a) = f(ra) \in f(I) \quad \text{and} \quad f(a)s = f(a)f(r) = f(ar) \in f(I)$$

since I satisfies absorption as it is an ideal of R. This proves that $f(I)$ satisfies absorption, as needed: so $f(I)$ is an ideal.

For an example showing that this is not the case if f is not surjective, let f be the inclusion of $R = \mathbb{Z}$ in $S = \mathbb{Q}$, and let $I = R$. Then I is an ideal, but $f(I) = \mathbb{Z} \subseteq \mathbb{Q}$ is *not* an ideal (Example 5.13).

5.9 • Since I and J are ideals, 0 is in both I and J. Therefore $0 \in I \cap J$, showing that $I \cap J \neq \emptyset$. To verify that $I \cap J$ is closed under addition, let $a, b \in I \cap J$; then a, b are in both I and J, therefore so is $a + b$ since I and J are both closed with respect to addition (as they are ideals). Therefore $a + b \in I \cap J$, as needed. Similarly, let $a \in I \cap J$ and let $r \in R$. Then $a \in I$, so $ra \in I$ and $ar \in I$ since I satisfies absorption. By the same token, $ra \in J$ and $ar \in J$. It follows that $ra \in I \cap J$ and $ar \in I \cap J$, showing that $I \cap J$ satisfies absorption. By Proposition 5.9, $I \cap J$ is an ideal of R.

• The set IJ is not empty, since it contains 0. A sum of two sums of products of elements of I by elements of J is still a sum of products of elements of I by elements of J, so IJ is closed under addition. If $r \in R$, then

$$r\left(\sum_i a_i b_i\right) = \sum_i (ra_i) b_i$$

is a sum of products of elements of I by elements of J, since $ra_i \in I$ if $a_i \in I$ (as I is an ideal). Therefore this element belongs to IJ. Similarly, $(\sum_i a_i b_i) r \in IJ$, since J is an ideal. By Proposition 5.9, IJ is an ideal of R.

To see that $IJ \subseteq I \cap J$, note that each $a_i b_i$ is an element of I, since $a_i \in I$ and I satisfies absorption; and it is an element of J since $b_i \in J$ and J satisfies absorption. Therefore each $a_i b_i$ is in $I \cap J$, and it follows that $\sum_i a_i b_i \in I \cap J$ since $I \cap J$ is closed under addition.

• $0 = 0 + 0 \in I + J$, so $I + J \neq \emptyset$. To see that $I + J$ is closed under addition, let $a_1, a_2 \in I$, $b_1, b_2 \in J$; so $a_1 + b_1$ and $a_2 + b_2$ are arbitrary elements of $I + J$. We have

$$(a_1 + b_1) + (a_2 + b_2) = (a_1 + a_2) + (b_1 + b_2)$$

since addition is commutative; and this element belongs to $I + J$ since $a_1 + a_2 \in I$ and $b_1 + b_2 \in J$. To show that $I + J$ satisfies absorption: for all $r \in R$, $a \in I$, $b \in J$

$$r(a + b) = ra + rb \in I + J$$

by distributivity and using the fact that I and J satisfy absorption since they are ideals. By Proposition 5.9, $I + J$ is an ideal.

To see that it is the *smallest* ideal containing both I and J, simply note that if K is an ideal containing I and J, then K must contain all $a \in I$ and all $b \in J$, and therefore all elements of the form $a + b$ with $a \in I$ and $b \in J$, since K is closed with respect to addition. So $K \supseteq I + J$ as needed.

5.10 Since $0 \in I_\alpha$ for all α, $0 \in \cap_\alpha I_\alpha$. Therefore $\cap_\alpha I_\alpha \neq \emptyset$.

If a and b are in $\cap_\alpha I_\alpha$, then $a, b \in I_\alpha$ for all α. It follows that $a + b \in I_\alpha$ for all α, since each I_α is an ideal, and therefore $a + b \in \cap_\alpha I_\alpha$. So this set is closed under addition.

For absorption, let $a \in \cap_\alpha I_\alpha$ and let $r \in R$. Since $a \in I_\alpha$ for all α, and each I_α is an ideal, we have $ra \in I_\alpha$ and $ar \in I_\alpha$ for all α. Therefore $ra \in \cap_\alpha I_\alpha$, $ar \in \cap_\alpha I_\alpha$, proving absorption. By Proposition 5.9, $\cap_\alpha I_\alpha$ is an ideal.

For the last part, consider the family of ideals containing the subset A; for example, R itself belongs to this family since R is an ideal of R containing A. The intersection of this family of ideals is itself an ideal, as we have just proved; it is the smallest ideal containing A in the sense that every ideal containing A must contain it.

5.12 Since I contains the ideals I_j, it is not empty. If $a, b \in I$, then $a \in I_{j_1}$ and $b \in I_{j_2}$ for two indices j_1, j_2. Letting $j = \max(j_1, j_2)$, we have that $a, b \in I_j$. Since I_j is an ideal, it is closed under addition, so $a + b \in I_j$. Since $I_j \subseteq I$, we see that $a + b \in I$, and this shows that I is closed under addition. If $r \in R$ and $a \in I$, then $a \in I_j$ for some index j. Since I_j is an ideal, $ra \in I_j$ and $ar \in I_j$. Since $I_j \subseteq I$, it follows that $ra \in I$ and $ar \in I$, proving that I satisfies the absorption property.

By Proposition 5.9, I is an ideal of R.

5.13 Since $a = bq + r$, a is a linear combination of b and r; this says that $a \in (r, b)$. At the same time $r = a - bq$: so r is a linear combination of a and b, and this says $r \in (a, b)$. It follows that $(a, b) \subseteq (r, b)$ and $(r, b) \subseteq (a, b)$, and therefore $(a, b) = (r, b)$.

5.14 If $I = \{0\}$, then $I = n\mathbb{Z}$ for $n = 0$, so the fact is true in this case. If $I \neq \{0\}$, let $a \in I$, $a \neq 0$. If $a < 0$, then $-a > 0$ and $-a \in I$: therefore I contains some positive number. By the well-ordering principle, it must contain a *smallest* positive number; let n be this number. I claim that $I = n\mathbb{Z}$.

Indeed, since $n \in I$, we have $n\mathbb{Z} \subseteq I$; we have to prove the converse inclusion $I \subseteq n\mathbb{Z}$. For this, let $a \in I$. Apply 'division with remainder' (Theorem 1.3): there exist $q, r \in \mathbb{Z}$ such that $a = qn + r$ and $0 \leq r < n$. But then $r = a - qn \in I$: and since n is the smallest positive integer in I, and $r < n$, r cannot be positive. Therefore $r = 0$, that is, $a = qn$. This proves that $a \in n\mathbb{Z}$, as needed.

5.15 If $I = (0)$, we are done. If not, we have to prove that $I = (1)$. The ideal (1) equals R, so $I \subseteq (1)$; we have to prove that $(1) \subseteq I$, or equivalently $1 \in I$. For this let $a \in I$, $a \neq 0$. In a field every nonzero element is a unit; therefore a is a unit, i.e., there exists a $b \in F$ such that $ab = 1$. Since $a \in I$ and I satisfies absorption, we see that $1 \in I$, as needed.

5.16 We have to prove that every nonzero element of R is a unit. Let then $c \in R$ and assume $c \neq 0$. Consider the ideal (c). Since $c \neq 0$, this ideal is not the ideal (0); by the hypothesis, we must have $(c) = (1)$.

This shows that $1 \in (c)$, that is, there exists an element $r \in R$ such that $cr = 1$; and since R is commutative, it follows that $rc = 1$ as well. This says precisely that c is a unit, so we are done.

5.18 Consider $\ker(\varphi)$. This is an ideal of F; since F is a field, we must have $\ker(\varphi) = (0)$ or $\ker(\varphi) = (1)$, by Exercise 5.15. If $\ker(\varphi) = (1)$, then $\varphi(1) = 0$; but $\varphi(1) = 1$ since φ is a ring homomorphism, so this would imply $0 = 1$ in R, contrary to assumption.

Therefore necessarily $\ker(\varphi) = (0)$, and it follows that φ is injective (Exercise 5.2).

5.20 Let $\varphi: \mathbb{R}[x]/(x^2 - 1) \to \mathbb{R} \times \mathbb{R}$ be the function defined in the statement, that is,

$$\varphi(a + bx + (x^2 - 1)) = (a + b, a - b).$$

For $a_1, b_1, a_2, b_2 \in \mathbb{R}$, we have

$$(a_1 + b_1 x + (x^2 - 1)) + (a_2 + b_2 x + (x^2 - 1)) = (a_1 + a_2) + (b_1 + b_2)x + (x^2 - 1)$$

and

$$(a_1 + b_1x + (x^2 - 1)) \cdot (a_2 + b_2x + (x^2 - 1))$$
$$= (a_1a_2) + (a_1b_2 + b_1a_2)x + (b_1b_2)x^2 + (x^2 - 1)$$
$$= (a_1a_2 + b_1b_2) + (a_1b_2 + b_1a_2)x + (x^2 - 1)$$

(since x^2 is congruent to 1 mod $(x^2 - 1)$). Therefore

$$\varphi((a_1 + b_1x + (x^2 - 1)) + (a_2 + b_2x + (x^2 - 1))) = (a_1 + a_2 + b_1 + b + 2, a_1 + a_2 - b_1 - b_2);$$

this agrees with $\varphi(a_1 + b_1x + (x^2 - 1)) + \varphi(a_2 + b_2x + (x^2 - 1))$, so φ preserves addition. As for multiplication, the above computation shows that

$$\varphi((a_1 + b_1x + (x^2 - 1)) \cdot (a_2 + b_2x + (x^2 - 1)))$$
$$= ((a_1a_2 + b_1b_2) + (a_1b_2 + b_1a_2), (a_1a_2 + b_1b_2) - (a_1b_2 + b_1a_2)),$$

while

$$\varphi(a_1 + b_1x + (x^2 - 1)) \cdot \varphi(a_2 + b_2x + (x^2 - 1))$$
$$= (a_1 + b_1, a_1 - b_1) \cdot (a_2 + b_2, a_2 - b_2)$$
$$= (a_1a_2 + a_1b_2 + b_1a_2 + b_1b_2, a_1a_2 - a_1b_2 - b_1a_2 + b_1b_2).$$

The two results coincide, so φ also preserves multiplication.

Further, $\varphi(1 + (x^2 - 1)) = (1, 1)$, and we can conclude that φ is a ring homomorphism. It is also clearly a bijection: an inverse is given by

$$(a, b) \mapsto \frac{a + b}{2} + \frac{a - b}{2}x + (x^2 - 1),$$

as is easy to verify. Therefore φ is a bijective ring homomorphism, i.e., an isomorphism of rings.

5.23 We have a ring homomorphism $\varphi \colon R \to S/J$, obtained by composing f with the projection $S \to S/J$. Since f is surjective, so is φ. Note that

$$\ker(\varphi) = \varphi^{-1}(0 + J) = f^{-1}(J) = I.$$

By the First Isomorphism Theorem (Theorem 5.38) we can deduce that $S/J \cong R/\ker(\varphi) = R/I$, as needed.

5.25 The natural projection $\pi \colon R \to R/I$ is a surjective ring homomorphism; therefore we have a surjective ring homomorphism $f \colon R[x] \to (R/I)[x]$, defined by applying π to the coefficients of a polynomial. Note that

$$\ker f = \{a_0 + a_1x + \cdots + a_dx^d \mid \forall i, \pi(a_i) = 0\}$$
$$= \{a_0 + a_1x + \cdots + a_dx^d \mid \forall i, a_i \in I\}$$
$$= I(x).$$

This shows that $I(x)$ is a kernel of a ring homomorphism, therefore it is an ideal (Example 5.10). Further, since f is surjective, then applying the First Isomorphism Theorem

(Theorem 5.38) gives

$$(R/I)[x] = R[x]/\ker f = R[x]/I[x],$$

as needed.

5.26 The statement that the ideal (d) is the smallest principal ideal containing both a and b says that

(i) $a, b \in (d)$, that is, d is a divisor of both a and b; and
(ii) if $a, b \in (c)$ for some other principal ideal c, i.e., if $c \mid a$ and $c \mid b$, then $(d) \subseteq (c)$, i.e., $c \mid d$.

These two conditions are precisely the conditions defining $\gcd(a, b)$ as in Definition 1.5. (Note that $d \geq 0$ by assumption.)

5.33 We have shown (Proposition 5.56) that the totient function ϕ is 'multiplicative' in the sense that $\phi(rs) = \phi(r)\phi(s)$ if r, s are relatively prime. It follows (by an immediate induction) that with notation as in the statement of the problem,

$$\phi(n) = \phi(p_1^{a_1}) \cdots \phi(p_r^{a_r}).$$

Therefore we are reduced to computing $\phi(p^a)$ for a positive irreducible number p. A number is relatively prime to p^a if and only if it is *not* a multiple of p. A moment's thought reveals that there are exactly p^{a-1} multiples of p between 1 and p^a, therefore $\phi(p^a) = p^a - p^{a-1}$. The conclusion is that

$$\phi(n) = (p_1^{a_1} - p^{a_1-1}) \cdots (p_r^{a_r} - p^{a_r-1}) = n \cdot \left(1 - \frac{1}{p_1}\right) \cdots \left(1 - \frac{1}{p_r}\right).$$

5.34 Since $a \in J$, we have $a + I \in J/I$, and hence $(a + I) \subseteq J/I$. We have to prove the converse inclusion: $J/I \subseteq (a+I)$. That is, we have to prove that if $b + I \in J/I$, then there is an $r \in R$ such that $b + I = (r + I)(a + I) = ra + I$.

Let then $b \in J$. Since $J = (a) + I$ by assumption, there exists $r \in R$ and $i \in I$ such that $b = ra + i$. It follows that $b - ra = i \in I$, and this says precisely that $b + I = ra + I$, as needed.

The extension to several generators uses the same idea. First, since each a_k is in J, then each $a_k + I$ is in J/I, and it follows that $(a_1 + I, \ldots, a_n + I) \subseteq J/I$. We have to prove the converse inclusion. For this, let $b \in J$. By assumption there exist $r_1, \ldots, r_n \in R$ and $i \in I$ such that $b = r_1 a_1 + \cdots + r_n a_n + i$. This says that $b - (r_1 a_1 + \cdots + r_n a_n) \in I$, and it follows that

$$b + I = (r_1 a_1 + \cdots + r_n a_n) + I = (r_1 + I)(a_1 + I) + \cdots + (r_n + I)(a_n + I)$$

in R/I. Therefore every element of J/I is a linear combination of $a_1 + I, \ldots, a_n + I$, and this concludes the proof.

5.36 In Example 5.41 (using the indeterminate y rather than x) we noted that for any commutative ring R, the surjective ring homomorphism $R[y] \to R$ sending a polynomial $f(y)$ to $f(r)$ has kernel $(y-r)$; and therefore it induces an isomorphism $R[y]/(y-r) \to R$.

To adapt to the situation in this problem, take $R = \mathbb{C}[x]$ and let $r = x^2$. Also recall that

$\mathbb{C}[x, y] \cong \mathbb{C}[x][y]$ (Example 4.43): given a polynomial $\alpha(x, y)$ in two indeterminates x, y and coefficients in $\mathbb{C}$, simply interpret it as a polynomial in y with coefficients in $\mathbb{C}[x]$. Evaluating this polynomial at $y = x^2$ gives $\alpha(x, x^2)$.

Example 5.41 tells us then that the function $\mathbb{C}[x, y]/(y - x^2) \to \mathbb{C}[x]$ mapping a coset $\alpha(x, y) + (y - x^2)$ to $\alpha(x, x^2) \in \mathbb{C}[x]$ is an isomorphism, as needed.

B.6　Integral Domains

6.8　By Exercise 5.35,

$$\mathbb{C}[x, y]/(x - a, y - b) \cong \mathbb{C}.$$

Since $\mathbb{C}$ is a field, this implies that $(x - a, y - b)$ is a maximal ideal in $\mathbb{C}[x, y]$.

If the ideal $(y - x^2)$ is contained in $(y - a, y - b)$, then

$$y - x^2 = f(x, y) \cdot (x - a) + g(x, y) \cdot (y - b)$$

for some polynomials $f(x, y), g(x, y)$. Then

$$b - a^2 = f(a, b) \cdot (a - a) + g(a, b) \cdot (b - b) = 0 :$$

this shows that the point (a, b) is on the parabola $y = x^2$ in this case.

Conversely, assume that (a, b) is on the parabola, i.e., $b = a^2$. Then

$$y - x^2 = (y - b + b) - (x - a + a)^2 = (y - b) + \not{b} - (x - a)^2 - 2(x - a)\,a - \not{a}^2 ,$$

showing that $y - x^2 \in (x - a, y - b)$. Therefore the ideal $(y - x^2)$ is contained in $(x - a, y - b)$ if $b = a^2$.

Summarizing, we have a one-to-one correspondence between points of the parabola $y = x^2$ and maximal ideals $(x - a, y - b)$ of $\mathbb{C}[x, y]$ containing $(y - x^2)$. By the Third Isomorphism Theorem (Theorem 5.57), these maximal ideals determine maximal ideals of the quotient $\mathbb{C}[x, y]/(y - x^2)$, as requested.

6.10　If we had $(2, x) = (f(x))$ for some $f(x) \in \mathbb{Z}[x]$, then in particular $f(x)$ would be a divisor of 2. The only divisors of 2 in $\mathbb{Z}[x]$ are $\pm 1, \pm 2$, so the only possibilities for $(f(x))$ are (1) and (2). Neither possibility works: $(2, x) \neq (1)$ (because 1 is not a linear combination of 2 and x), and $(2, x) \neq (2)$ (because x is not a multiple of 2).

Therefore $(2, x)$ cannot be generated by a single element of $\mathbb{Z}[x]$, and that means that it is not a principal ideal.

6.11　Assume that $a = bu$, with u a unit. This implies $a \in (b)$, so $(a) \subseteq (b)$. If v is the inverse of u, we have $b = av$, so the same reasoning gives $(b) \subseteq (a)$, and we can conclude that $(a) = (b)$.

For the converse implication: if $b = 0$ and $(a) = (b)$, then $(a) = 0$ and hence $a = 0 = b \cdot 1$, so the fact is true in this case. So we may assume that $b \neq 0$. The equality $(a) = (b)$ implies that $b = av$ for some $u \in R$, and $a = bu$ for some $u \in R$. With this notation,

$$b = av = buv ,$$

therefore $b(1 - uv) = 0$. Since R is an integral domain and $b \neq 0$, necessarily $1 - uv = 0$: therefore $uv = 1$, and this tells us that u is a unit, as needed.

6.12 First we show that $\ker(\varphi) = (y^2 - x^3)$. It is clear that $y^2 - x^3 \in \ker(\varphi)$, since $\varphi(y^2 - x^3) = (t^3)^2 - (t^2)^3 = t^6 - t^6 = 0$. This shows that $(y^2 - x^3) \subseteq \ker(\varphi)$. For the converse implication, we follow the hint. Let $f(x, y) \in \ker(\varphi)$. There are polynomials $g(x, y), h(x), k(x)$ such that

$$f(x, y) = g(x, y)(y^2 - x^3) + h(x)y + k(x).$$

Since φ is a ring homomorphism and $f(x, y)$ is in its kernel, we get

$$0 = \varphi(f(x, y)) = f(t^2, t^3) = g(t^2, t^3) \cdot ((t^3)^2 - (t^2)^3) + h(t^2)t^3 + k(t^2) = h(t^2)t^3 + k(t^2).$$

That is, we discover that

$$k(t^2) = -h(t^2)t^3.$$

But only even powers of t appear on the left, while the right-hand side is a multiple of t^3 and only odd powers of t appear in it. So this equality can only hold if $h(t^2) = 0$ and $k(t^2) = 0$. It follows that $h(x) = k(x) = 0$, and the conclusion is that

$$f(x, y) = g(x, y)(y^2 - x^3),$$

proving that $f(x, y) \in (y^2 - x^3)$ as needed.

By the First Isomorphism Theorem, the ring $\mathbb{C}[x, y]/(y^2 - x^3)$ is isomorphic to the image of the morphism φ, which is a subring R of $\mathbb{C}[t]$. Since $\mathbb{C}[t]$ is an integral domain, R is an integral domain (Exercise 4.8). This shows that $\mathbb{C}[x, y]/(y^2 - x^3)$ is isomorphic to an integral domain, and it follows that it is an integral domain.

Note that this subring of $\mathbb{C}[t]$ is

$$R = \{a_0 + a_2t^2 + a_3t^3 + a_4t^4 + \cdots\},$$

that is, it consists of the polynomials whose coefficient of t vanishes.

We have to show that the coset $\underline{x}$ of x is irreducible. Notice that $\underline{x}$ is identified with $\varphi(x) = t^2$ in R. Its divisors in R must in particular be divisors in $\mathbb{C}[t]$, and these are of the form $u \in \mathbb{C}$, ut, and ut^2, with u a unit (that is, $u \neq 0$ in $\mathbb{C}$). Now the key observation is that $ut \notin R$, so the only divisors of t^2 in R are of the form u or ut^2, with u a unit. Correspondingly, the only divisors of $\underline{x}$ in $\mathbb{C}[x, y]/(y^2 - x^3)$ are of the form u or $u\underline{x}$, with u a unit, and this proves that $\underline{x}$ is irreducible.

6.15 By definition, q is irreducible if whenever $q = ab$ we have that a is a unit or b is a unit. If $(ab) = (q)$, then $q = abu$ for some unit u (Exercise 6.11). Since $q = a(bu)$ and q is irreducible, then a is a unit, implying $(q) = (b)$, or bu is a unit, which implies $(q) = (a)$.

Conversely, assume that $(ab) = (q)$ implies $(a) = (q)$ or $(b) = (q)$. If $q = ab$, then in particular $(q) = (ab)$, so $(a) = (q)$ or $(b) = (q)$ by hypothesis. If $(a) = (q)$, we find that $q = au$ with u a unit, by Exercise 6.11; and then $ab = au$, so $b = u$ (R is an integral domain and $a \neq 0$ since $q \neq 0$), and hence b is a unit. If $(b) = (q)$, the same token gives us that a is a unit. This shows that $q = ab$ implies that a is a unit or b is a unit, so q is irreducible.

(A proof can also be obtained by applying Lemma 6.22.)

6.16 Assume that $q \in \mathbb{Z}$ is irreducible in the sense of Definition 1.17. Then $q \neq \pm 1$, so that q is not a unit. If $q = ab$, then a is a divisor of q, so according to Definition 1.17 $a = \pm 1$ or $a = \pm q$. In the first case a is a unit in $\mathbb{Z}$; in the second, we must have $b = \pm 1$, so b is a unit. Therefore q is irreducible in the sense of Definition 6.20.

Conversely, assume that $q \in \mathbb{Z}$ is irreducible in the sense of Definition 6.20. Then q is not a unit, so $q \neq \pm 1$. If a is a divisor of q, then $q = ab$ for some $b \in \mathbb{Z}$; according to Definition 6.20, either a is a unit, i.e., $a = \pm 1$, or b is a unit: and then $b = \pm 1$, so $a = \pm q$. Therefore q is irreducible in the sense of Definition 1.17.

6.18 Let R be any ring. If R contains a subring isomorphic to $\mathbb{Z}/n\mathbb{Z}$, then I claim that this subring is unique. Indeed, we can compose the inclusion homomorphism $\mathbb{Z}/n\mathbb{Z} \hookrightarrow R$ with the projection $\mathbb{Z} \to \mathbb{Z}/n\mathbb{Z}$, getting a ring homomorphism $\sigma \colon \mathbb{Z} \to R$:

$$\mathbb{Z} \overset{\sigma}{\underset{}{\twoheadrightarrow}} \mathbb{Z}/n\mathbb{Z} \hookrightarrow R \,.$$

However, there is *only one* ring homomorphism $\mathbb{Z} \to R$ (Exercise 4.13), so σ is unique, and $\mathbb{Z}/n\mathbb{Z}$, which must be the image of σ, is also uniquely determined by R.

This observation also tells us how to construct the subring. Let $\sigma \colon \mathbb{Z} \to R$ be the (unique) homomorphism from $\mathbb{Z}$ to R. By the First Isomorphism Theorem, the image of σ is isomorphic to $\mathbb{Z}/\ker \sigma$. Since $\mathbb{Z}$ is a PID, $\ker \sigma = (n)$ for an integer n; and n is uniquely determined if we require it to be nonnegative.

The conclusion is that R contains one and only one subring isomorphic to $\mathbb{Z}/n\mathbb{Z}$ for a nonnegative n.

To see what n must be, recall that $\sigma \colon \mathbb{Z} \to R$ is defined by $\sigma(n) = n \, 1_R$. If $\ker(\sigma) = (0)$, then $n = 0$; there is no positive n such that $n \, 1_R = 0$. Recall that we say that the 'characteristic' of R is 0 in this case (cf. Exercise 3.12). If there *is* a positive integer n such that $n \, 1_R = 0$, i.e., such that $\sigma(n) = 0$, then the smallest such integer generates $\ker(\sigma)$. This integer is the characteristic of R in this case (Exercise 3.12 again).

Therefore we see that the characteristic of R is simply the nonnegative generator of $\ker(\sigma)$; equivalently, it is the (unique) nonnegative integer n such that R contains a subring isomorphic to $\mathbb{Z}/n\mathbb{Z}$.

6.19 If F is a field, then (0) is its *only* prime ideal. Therefore the only chain of prime ideals in F consists of (0) itself: it has length 0, and hence the Krull dimension of F is 0.

If R is a PID, but not a field, then R has some non-invertible element a. By Theorem 6.35, a has an irreducible factor q. Then $(q) \neq (0)$ and (q) is a prime, and in fact maximal, ideal (Proposition 6.33). Therefore we have a length-1 chain of prime ideals: $(0) \subsetneq (q)$, and no longer chain as (q) is maximal. The maximal length of a chain of prime ideals is 1, so the Krull dimension of R is 1.

6.20 Let R be the ring $\mathbb{Z}[x_1, x_2, x_3, \dots]$. The chain of ideals

$$(0) \subsetneq (x_1) \subsetneq (x_1, x_2) \subsetneq (x_1, x_2, x_3) \subsetneq \cdots \subsetneq (x_1, x_2, \dots, x_k) \subsetneq \cdots$$

has infinite length.

6.26 Arguing by contrapositive, assume that a and b *are* relatively prime in R. Then

$(a, b) = (1)$ in R: therefore, $\exists c, d \in R$ such that $ar + bd = 1$. This implies that $(a, b) = (1)$ in S, so a and b are relatively prime in S, as needed.

6.27 Since R is a PID, $I = (a)$ for some $a \in R$. By unique factorization we have irreducible elements $q_1, \ldots, q_r$ and positive integers $m_1, \ldots, m_r$ such that

$$a = uq_1^{m_1} \cdots q_r^{m_r}$$

with u a unit, and $(q_i) \neq (q_j)$ for $i \neq j$. (Just collect together irreducible factors generating the same ideal; it may be necessary to collect a unit in the process. For example, $-4 = -2^2$ in $\mathbb{Z}$: the unit $u = -1$ cannot be omitted.) Since $q_i^{m_i}$ and $q_j^{n_j}$ have no common irreducible factor for $i \neq j$, $(q_i^{m_i}) + (q_j^{m_j}) = (1)$ for $i \neq j$ (Exercise 6.25). By Lemma 5.51, $(q_i^{m_i}) \cap (q_j^{m_j}) = (q_i^{m_i}) \cdot (q_j^{m_j})$; and an induction argument then shows that

$$(q_1^{m_1}) \cap \cdots \cap (q_r^{m_r}) = (q_1^{m_1}) \cdots (q_r^{m_r}) = (a),$$

as needed.

6.28 If F_1, F_2 both satisfy the given requirements, in particular we have injective homomorphisms $\iota_1 : R \to F_1$, $\iota_2 : R \to F_2$. Since F_1 satisfies the requirement and F_2 is a field, there must be a unique ring homomorphism $j_1 : F_1 \to F_2$ such that the diagram

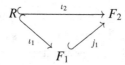

commutes. Since F_2 satisfies the requirement and F_1 is a field, there must be a unique ring homomorphism $j_2 : F_2 \to F_1$ such that the diagram

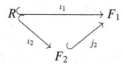

commutes. By composition, we get the commutative diagram

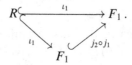

However, the requirement prescribes that the homomorphism making this diagram commute is *unique*, and since the diagram

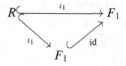

also commutes, it follows that $j_2 \circ j_1 = \mathrm{id}$. By the same argument, $j_1 \circ j_2 = \mathrm{id}$. Therefore j_1 and j_2 are inverses of each other, and in particular they are ring isomorphisms (keep in mind Corollary 4.31!). Therefore $F_1 \cong F_2$.

6.30 Associativity of addition is verified in the main text.

—*Existence of additive identity:* If $[(r, s)] \in F$, then

$$[(0, 1)] + [(r, s)] = [(0 \cdot s + 1 \cdot r, 1 \cdot s)] = [(r, s)]$$

and similarly $[(r, s)] + [(0, 1)] = [(r, s)]$.

—*Existence of additive inverses:* If $[(r, s)] \in F$, then

$$[(r, s)] + [(-r, s)] = [(r \cdot s - r \cdot s, s \cdot s)] = [(0, s^2)] = [(0, 1)],$$

since $(0, s^2) \cong (0, 1)$. Similarly $[(-r, s)] + [(r, s)] = [(0, 1)]$.

—*Commutativity for addition:* If $[(r_1, s_1)], [(r_2, s_2)] \in F$, then

$$[(r_1, s_1)] + [(r_2, s_2)] = [(r_1 s_2 + r_2 s_1, s_1 s_2)]$$

while

$$[(r_2, s_2)] + [(r_1, s_1)] = [(r_2 s_1 + r_1 s_2, s_2 s_1)].$$

These elements coincide, by the commutativity of addition and multiplication in R.

—*Associativity of multiplication:* If $[(r_1, s_1)], [(r_2, s_2)], [(r_3, s_3)] \in F$, then

$$[(r_1, s_1)] \cdot ([(r_2, s_2)] \cdot [(r_3, s_3)]) = [(r_1(r_2 r_3), s_1(s_2 s_3))],$$

while

$$([(r_1, s_1)] \cdot [(r_2, s_2)]) \cdot [(r_3, s_3)] = [((r_1 r_2) r_3, (s_1 s_2) s_3)].$$

These elements coincide by associativity of multiplication in R.

—*Existence of multiplicative identity:* If $[(r, s)] \in F$, then

$$[(1, 1)] \cdot [(r, s)] = [(1 \cdot r, 1 \cdot s)] = [(r, s)]$$

and similarly $[(r, s)] \cdot [(1, 1)] = [(r, s)]$.

—*Distributivity:* If $[(r_1, s_1)], [(r_2, s_2)], [(r_3, s_3)] \in F$, then

$$[(r_1, s_1)] \cdot ([(r_2, s_2)] + [(r_3, s_3)]) = [(r_1, s_1)] \cdot [(r_2 s_3 + r_3 s_2, s_2 s_3)]$$
$$= [(r_1(r_2 s_3 + r_3 s_2), s_1(s_2 s_3))]$$
$$= [(r_1 r_2 s_3 + r_1 r_3 s_2, s_1 s_2 s_3)]$$

while

$$[(r_1, s_1)] \cdot [(r_2, s_2)] + [(r_1, s_1)] \cdot [(r_3, s_3)] = [(r_1 r_2, s_1 s_2)] + [(r_1 r_3, s_1 s_3)]$$
$$= [(r_1 r_2 s_1 s_3 + r_1 r_3 s_1 s_2, s_1^2 s_2 s_3)].$$

These expressions coincide because the representatives are equivalent:

$$(r_1 r_2 s_3 + r_1 r_3 s_2) s_1^2 s_2 s_3 = (r_1 r_2 s_1 s_3 + r_1 r_3 s_1 s_2) s_1 s_2 s_3.$$

The verification that

$$([(r_1, s_1)] + [(r_2, s_2)]) \cdot [(r_3, s_3)] = [(r_1, s_1)] \cdot [(r_3, s_3)] + [(r_2, s_2)] \cdot [(r_3, s_3)]$$

is just as straightforward, and in fact unnecessary since multiplication is commutative in F as observed in the main text.

B.7   Polynomial Rings and Factorization

7.1 Let $\deg f = m$, $\deg g = n$, that is, $f(x) = a_0 + a_1x + \cdots + a_mx^m$, $g(x) = b_0 + b_1x + \cdots + b_nx^n$, with $a_m \neq 0$, $b_n \neq 0$. Then

$$f(x)g(x) = a_0b_a + (a_1b_0 + a_0b_1)x + \cdots + a_mb_nx^{m+n} \ .$$

Since $a_m \neq 0$ and $b_n \neq 0$, and R is an integral domain, it follows that $a_mb_n \neq 0$. This shows that $\deg(fg) = m + n = \deg f + \deg g$, as stated.

7.2 If $f(x) = a$ is a unit in R, then it has an inverse in R, and therefore it has an inverse in $R[x]$; thus, $f(x)$ is a unit in $R[x]$.

Conversely, let $f(x) \in R[x]$ be a unit; so $\exists g(x) \in R[x]$ such that $f(x)g(x) = 1$. In particular, $\deg(fg) = 0$; it then follows that $\deg f = \deg g = 0$, by Exercise 7.1. Therefore $f(x) = a \in R$, $g(x) = b \in R$; and $ab = f(x)g(x) = 1$, which implies that a is a unit in R.

If k is a field, then the units in k are the nonzero elements; this implies the last assertion.

7.11 Recall that $\mathbb{Z}/p\mathbb{Z}$ is a field if p is irreducible (Theorem 2.15, Example 3.30). According to Exercise 7.10, there are infinitely many irreducible polynomials in $(\mathbb{Z}/p\mathbb{Z})[x]$. Since $\mathbb{Z}/p\mathbb{Z}$ is finite, the number of polynomials of any given degree is finite, hence the number of polynomials of degree $\leq d$ in $(\mathbb{Z}/p\mathbb{Z})[x]$ is finite, for all $d > 0$. Since there are infinitely many irreducible polynomials, it follows that not all of them can have degree $\leq d$. This is just a reformulation of the required statement.

7.13 Let $f(x)$ have degree 1, and assume $f(x) = g(x)h(x)$. By Exercise 7.1, we have that $\deg g = 0$ or $\deg h = 0$. Then either g or h is a nonzero constant, hence a unit (Exercise 7.2). Thus whenever $f(x) = g(x)h(x)$, either $g(x)$ or $h(x)$ is a unit: this says that $f(x)$ is irreducible.

For this argument to work, we need nonzero constants to be units, and this is not necessarily true over arbitrary integral domains. And indeed, e.g., $2x \in \mathbb{Z}[x]$ has degree 1, but is *not* irreducile.

7.14 (i) Since unique factorization holds in $k[x]$ (Corollary 7.8), $f(x)$ admits a unique decomposition as a product of irreducible polynomials: $f(x) = \prod_i q_i(x)$, and $\deg f = \sum_i \deg q_i$ (Exercise 7.1). It follows that $f(x)$ has at most d irreducible factors up to units, and in particular there are at most d factors of the form $x - a$. Since $x - a \mid f(x)$ if and only if a is a root of $f(x)$ (Example 7.5), this implies that $f(x)$ has at most d roots.

(ii) The roots of $x^2 + x \in (\mathbb{Z}/6\mathbb{Z})[x]$ are $[0]_6, [2]_6, [3]_6, [5]_6$: there are four roots, even though $x^2 - x$ has degree 2. This does not contradict part (i) because $\mathbb{Z}/6\mathbb{Z}$ is not a field.

7.18 Let $f(x) = a_0 + a_1x + \cdots + a_dx^d \in \mathbb{R}[x]$; so all coefficients a_i are real numbers. If $z \in \mathbb{C}$, we have

$$f(\bar{z}) = a_0 + a_1\bar{z} + \cdots + a_d\bar{z}^d = \overline{a_0 + a_1z + \cdots + a_dz^d} = \overline{f(z)} :$$

indeed conjugation preserves addition and products, and the conjugate of a real number a is a itself (cf. Example 4.23).

7.21 We can argue as in the proof of Lemma 7.23: since q is irreducible in $\mathbb{Z}$, then $I = (q) = q\mathbb{Z}$ is a prime ideal in $\mathbb{Z}$, and it follows that $q\mathbb{Z}[x]$ is a (nonzero) prime ideal in $\mathbb{Z}[x]$; and then q is irreducible in $\mathbb{Z}[x]$ by Theorem 6.23.

(Even) more concretely: if $q \in \mathbb{Z}$ may be written as a product of polynomials in $\mathbb{Z}[x]$, $q = f(x)g(x)$, then necessarily $\deg f = \deg g = 0$ (Exercise 7.1); that is, $f(x)$ and $g(x)$ are themselves integers. Since q is irreducible as an integer, either $f(x)$ or $g(x)$ must equal ± 1; and then it is a unit in $\mathbb{Z}[x]$. This proves directly that q is irreducible as an element of $\mathbb{Z}[x]$.

7.23 We have to prove that there exist $a, b \in \mathbb{Q}$ such that $a\,g(x)$ and $b\,h(x)$ have integer coefficients and at the same time such that $ab = 1$. As in the argument given in the text, we consider all the integers ab obtained by multiplying rational numbers a, b for which both $a\,g(x)$ and $b\,h(x)$ have integer coefficients. Multiplying any such choice of a, b by sufficiently large integers, and changing signs if necessary, we see that among these integers ab there are positive integers. Every such positive integer has an irreducible factorization in $\mathbb{Z}$, and a corresponding number n of irreducible factors, counting repetitions (cf. Remark 7.24). This number is 0 precisely if $ab = 1$. Therefore, we will consider the set T of numbers of factors of positive integers ab, and it will be enough to prove that 0 is in this set.

Here is a complete definition of the set:

$$T = \{n \text{ such that } \exists a, b \in \mathbb{Q} \text{ s.t. } ag(x) \in \mathbb{Z}[x], bh(x) \in \mathbb{Z}[x], ab \in \mathbb{Z}, ab > 0,$$

$$\text{and } n \text{ is the number of positive irreducible factors of } ab\}.$$

By the well-ordering principle, T contains a smallest element m, corresponding to a specific choice of a and b. We will obtain a contradiction from the assumption that $m > 0$.

If $m \neq 0$, then consider the specific choice of a and b as in the definition of T, such that ab has m positive irreducible factors, and let q be one of these factors. Note that $\frac{ab}{q} \in \mathbb{Z}$.

Since $q \mid ab$, q divides all coefficients of $abf(x) = (ag(x))\,(bh(x))$, and by Lemma 7.23 we can conclude that q divides all the coefficients of $ag(x)$ or all the coefficients of $bh(x)$. Without loss of generality, assume that q divides all the coefficients of $ag(x)$, and let $a' = q^{-1}a$. Then

$$a'g(x) = q^{-1}(ag(x))$$

has integer coefficients. But then we have that a' and b are rational numbers such that

$$a'g(x) \in \mathbb{Z}[x], bh(x) \in \mathbb{Z}[x], a'b = \frac{ab}{q} \in \mathbb{Z}, a'b > 0 \,;$$

and the number of irreducible factors of $a'b = \frac{ab}{q}$ is $m - 1$.

This shows that $m - 1 \in T$, contradicting the choice of m as the smallest element of T. This contradiction shows that the smallest element must be $m = 0$, and we are done.

As mentioned in the text, this argument generalizes to arbitrary UFDs. (The argument given in the text does not, since it uses directly the ordering in $\mathbb{Z}$; in a general UFD we do not have an ordering as we do in $\mathbb{Z}$.)

7.24 This may be done by induction. The case $n = 0$ is the statement that k is a UFD, which is true since k is a field. Assuming that we know that $k[x_1, \ldots, x_n]$ is a UFD for a fixed $n \geq 0$, then $k[x_1, \ldots, x_n][x_{n+1}]$ is a UFD by the quoted result, and we just need to observe that $k[x_1, \ldots, x_{n+1}] \cong k[x_1, \ldots, x_n][x_{n+1}]$ (Example 4.43) to conclude the induction step.

7.26 Here is one possible formulation. Let R be an integral domain, I a prime ideal of R, and let $f(x) \in R[x]$ be a polynomial. Assume that the leading term of $f(x)$ is not in I, and that the image $\underline{f(x)}$ of $f(x)$ in R/I cannot be written as a product of positive-degree polynomials in $(R/I)[x]$. Then $f(x)$ cannot be written as a product of positive-degree polynomials in $R[x]$.

The proof of this statement matches closely the proof of Proposition 7.33. Arguing contrapositively, assume that $f(x) = g(x)h(x)$ for two polynomials $g(x), h(x) \in R[x]$ of degree > 0. The leading coefficient of $f(x)$ is the product of the leading coefficients of $g(x)$ and $h(x)$, and by hypothesis it does not belong to I. Since I is a prime ideal, the leading coefficients of $g(x)$ and $h(x)$ also do not belong to I. Therefore the degree of the image $\underline{g(x)}$, resp. $\underline{h(x)}$, of $g(x)$, resp. $h(x)$, in R/I is the same as the degree of $g(x)$, resp. $h(x)$; in particular, $\underline{g(x)}$ and $\underline{h(x)}$ are positive-degree polynomials.

This implies that $\underline{f(x)}$ may be written as a product of two polynomials of degree > 0, negating the hypothesis, as needed.

7.32 Here is one possible formulation. Let R be an integral domain, and let $f(x) \in R[x]$ be a polynomial of degree d:

$$f(x) = a_0 + a_1 x + \cdots + a_{d-1}x^{d-1} + a_d x^d.$$

Let I be a prime ideal of R, and assume that

- I does not contain the leading coefficient a_d;
- I contains all the other coefficients a_i, $i = 0, \ldots, d - 1$;
- I^2 does not contain the constant term a_0.

Then $f(x)$ cannot be written as a product of two polynomials of positive degree.

The proof of this statement follows closely the proof for $R = \mathbb{Z}$ given in the text. It relies on Lemma 7.36, which holds in this generality.

Arguing by contradiction, assume that $f(x) = g(x)h(x)$ for polynomials $g(x), h(x) \in R[x]$ with $\deg g > 0$, $\deg h > 0$. Write

$$g(x) = b_0 + b_1 x + \cdots + b_m x^m, \quad h(x) = c_0 + c_1 x + \cdots + c_n x^n,$$

with $m > 0, n > 0, m + n = d$. Note that

$$\underline{f(x)} = \underline{g(x)} \cdot \underline{h(x)},$$

where I am underlining a term to mean its image modulo I; so this equality holds in $(R/I)[x]$. By the first hypothesis, $\underline{a_d} \neq 0$; by the second,

$$\underline{f(x)} = \underline{a_d} x^d$$

is a monomial: all the other coefficients of $f(x)$ are in I, so their class modulo I is 0.

The ring R/I is an integral domain (by definition of prime ideal), so $g(x)$ and $h(x)$ are both monomials, by Lemma 7.36: necessarily $g(x) = b_m x^m$, $h(x) = c_n x^n$. As $m > 0$ and $n > 0$, it follows that $b_0 = c_0 = 0$, that is, b_0 and c_0 are both elements of I. But then $a_0 = b_0 c_0 \in I^2$, contradicting the third hypothesis. This contradiction proves the statement.

7.33 The odd primes $p < 30$ for which $x^2 + 1$ is reducible are those $p < 30$ for which $x^2 + 1$ has a root (Proposition 7.15). These can be found by inspection. For example,

$$[0]_3^2 + [1]_3 = [1]_3 \neq [0]_3, \quad [1]_3^2 + [1]_3 = [2]_3 \neq [0]_3, \quad [2]_3^2 + [1]_3 = [2]_3 \neq [0]_3.$$

Therefore, $x^2 + 1$ has *no* roots in $\mathbb{Z}/3\mathbb{Z}$. On the other hand,

$$[2]_5^2 + [1]_5 = [0]_5,$$

therefore $x^2 + 1$ *has* roots in $\mathbb{Z}/5\mathbb{Z}$. Carrying the same analysis out for all $p < 30$ reveals that the odd primes < 30 for which $x^2 + 1$ *is* reducible are

$$5, 13, 17, 29.$$

This is not much to go on, but you could notice that these numbers are precisely those primes $p < 30$ that are $\equiv 1 \bmod 4$.

This happens to be a general fact; we will deal with it in Lemma 10.33.

B.8 Modules and Abelian Groups

8.1 For every $c \in k$, multiplication by the scalar c determines a linear transformation $\mu_c \colon V \to V$ defined by $\mu_c(v) = cv$.

Indeed, this is a linear transformation because it preserves sums and multiplication by scalars: $\forall v_1, v_2 \in V$,

$$\mu_c(v_1 + v_2) = c(v_1 + v_2) = c\,v_1 + c\,v_2 = \mu_c(v_1) + \mu_c(v_2)$$

and $\forall r \in k, \forall v \in V$,

$$\mu_c(rv) = c(rv) = (cr)v = (rc)v = r(cv) = r\mu_c(v).$$

In proving the first equality I have used axiom (viii) in the definition of module (Definition 8.1), and in proving the second I have used (vi) and the fact that multiplication is commutative in k since k is a field.

Thus, we have a function $\mu \colon k \to \mathrm{End}(V)$, obtained by setting $\mu(c) = \mu_c$. I claim that μ is a ring homomorphism. Indeed, first of all $\mu(1) = 1 = \mathrm{id}_V$, because $\forall v \in V$ we have

$$\mu(1)(v) = \mu_1(v) = 1 \cdot v = v = \mathrm{id}_V(v)$$

by axiom (v); next, μ preserves addition because $\forall c_1, c_2 \in k, \forall v \in V$,

$$\mu(c_1 + c_2)(v) = \mu_{c_1+c_2}(v) = (c_1 + c_2)v = c_1 v + c_2 v = \mu_{c_1}(v) + \mu_{c_2}(v) = \mu(c_1)(v) + \mu(c_2)(v),$$

where I have used axiom (vii). Since this holds for all $v \in V$, it proves that

$$\mu(c_1 + c_2) = \mu(c_1) + \mu(c_2)$$

in $\text{End}(V)$. Finally, μ preserves multiplication: $\forall c_1, c_2 \in k$, $\forall v \in V$,

$$\mu(c_1 c_2)(v) = (c_1 c_2)(v) = c_1(c_2 v) = c_1(\mu(c_2)(v)) = \mu(c_1)(\mu(c_2)(v)) = (\mu(c_1) \circ \mu(c_2))(v),$$

by axiom (vi) again. Since this holds for all $v \in V$, it proves that

$$\mu(c_1 c_2) = \mu(c_1) \circ \mu(c_2)$$

in $\text{End}(V)$, and this concludes the proof that μ is a ring homomorphism.   □

This may look complicated—it certainly is notationally challenging. Well, that's all smoke and mirrors: if you go through it with a steady hand, you will realize that it is a trivial (!) verification. The point of this problem is to show that the four axioms (v)–(viii) are nothing but a spelled-out version of a more concise and possibly conceptually simpler statement, that is, that there is a certain naturally defined ring homomorphism. This exercise deals with vector spaces—I have chosen so because you are more likely to be familiar with vector spaces from previous encounters with linear algebra; but the same holds, with the same proof, for arbitrary modules.

8.2 Given a ring homomorphism $r: R \to S$, define 'multiplication by $r \in R$' on S by setting $\forall s \in S$

$$r \cdot s = f(r)s,$$

where the product on the right-hand side is just multiplication in S. Since S is a ring, it satisfies axioms (i)–(iv) from Definition 8.1 with respect to its addition. To verify the other axioms, with evident quantifiers:

- $1_R \cdot s = f(1_R)s = 1_S s = s$;
- $(r_1 r_2) \cdot s = f(r_1 r_2)s = (f(r_1)f(r_2))s = f(r_1)(f(r_2)s) = f(r_1)(r_2 \cdot s) = r_1 \cdot (r_2 \cdot s)$;
- $(r_1 + r_2) \cdot s = f(r_1 + r_2)s = (f(r_1) + f(r_2))s = f(r_1)s + f(r_2)s = r_1 \cdot s + r_2 \cdot s$;
- $r \cdot (s_1 + s_2) = f(r)(s_1 + s_2) = f(r)s_1 + f(r)s_2 = r \cdot s_1 + r \cdot s_2$.

These are axioms (v)–(viii), and in the verifications I have used the ring axioms (which hold in S) and the fact that f is a ring homomorphism.

8.3 The inclusion $k \subseteq K$ determines an injective ring homomorphism $k \to K$. This is all that is needed to view K as a k-module (cf. Example 8.5, Exercise 8.2), i.e., as a k-vector space.

Concretely, the multiplication of $c \in k$ on K is simply defined by the multiplication in K: for all $a \in K$, $c \cdot a$ is just the element ca in K. The verification of axioms (v)–(viii) is immediate, and in fact unnecessary since Exercise 8.2 already does this in a more general case.

8.6 Assuming that r is in the center of R, define $f: M \to M$ by $f(m) = rm$. Then $\forall m_1, m_2 \in M$ we have

$$f(m_1 + m_2) = r(m_1 + m_2) = rm_1 + rm_2 = f(m_1) + f(m_2)$$

by axiom (viii) in Definition 8.1, and $\forall a \in R, \forall m \in M$

$$f(am) = r(am) = (ra)m \stackrel{!}{=} (ar)m = a(rm) = af(m)$$

by axiom (vi) and, crucially, since r commutes with a. This concludes the verification that f is an R-module homomorphism.

If r is not in the center, the equality $\stackrel{!}{=}$ will not hold in general. To construct an example exploiting this, I can let R be the ring of 2×2 integer matrices, and choose $r = \begin{pmatrix} 1 & 1 \\ 0 & 1 \end{pmatrix}$; since

$$\begin{pmatrix} 1 & 1 \\ 0 & 1 \end{pmatrix} \cdot \begin{pmatrix} 1 & 0 \\ 1 & 1 \end{pmatrix} \neq \begin{pmatrix} 1 & 0 \\ 1 & 1 \end{pmatrix} \cdot \begin{pmatrix} 1 & 1 \\ 0 & 1 \end{pmatrix}$$

this element is not in the center (cf. Example 3.19). We can choose $M = R$ itself, so certainly M is an R-module. If we try and define f as above, i.e., if we set

$$f(A) := rA = \begin{pmatrix} 1 & 1 \\ 0 & 1 \end{pmatrix} \cdot A,$$

the resulting function preserves addition but does not preserve multiplication by a scalar. Indeed, for example, letting I denote the identity matrix, we have

$$f\left(\begin{pmatrix} 1 & 0 \\ 1 & 1 \end{pmatrix} \cdot I \right) = \begin{pmatrix} 1 & 1 \\ 0 & 1 \end{pmatrix} \cdot \begin{pmatrix} 1 & 0 \\ 1 & 1 \end{pmatrix} \cdot I = \begin{pmatrix} 2 & 1 \\ 1 & 1 \end{pmatrix}$$

while

$$\begin{pmatrix} 1 & 0 \\ 1 & 1 \end{pmatrix} \cdot f(I) = \begin{pmatrix} 1 & 0 \\ 1 & 1 \end{pmatrix} \cdot \begin{pmatrix} 1 & 1 \\ 0 & 1 \end{pmatrix} \cdot I = \begin{pmatrix} 1 & 1 \\ 1 & 2 \end{pmatrix} :$$

that is,

$$f\left(\begin{pmatrix} 1 & 0 \\ 1 & 1 \end{pmatrix} \cdot I \right) \neq \begin{pmatrix} 1 & 0 \\ 1 & 1 \end{pmatrix} \cdot f(I).$$

8.10 Let R be a PID, and let I be a nonzero ideal of R. Then $I = (a)$ for some nonzero $a \in R$ (every ideal in I is principal, by definition of PID!). We can define a function $f : R \to I$ by letting $\forall r \in R$

$$f(r) = ra \in (a) = I.$$

This function is an R-module homomorphism: indeed $\forall r_1, r_2 \in R$ we have

$$f(r_1 + r_2) = (r_1 + r_2)a = r_1 a + r_2 a = f(r_1) + f(r_2)$$

by distributivity in R, and $\forall r, s \in R$

$$f(sr) = (sr)a = s(ra) = sf(r)$$

by associativity.

This function is surjective by definition of (a): (a) consists of all multiples of a (Definition 5.17). To see that it is injective, note that

$$f(r_1) = f(r_2) \implies r_1 a = r_2 a \implies r_1 = r_2$$

since R is an integral domain, and multiplicative cancellation by nonzero elements holds in integral domains, see Proposition 3.26.

Therefore, f is a bijective R-module homomorphism, hence an isomorphism of R-modules.

In general, we need r to commute with all elements of R (to be in the *center* of R, cf. Exercise 8.6) to ensure that (r) is an ideal of R; if r is a non-zero-divisor in R, the same argument given above will then show that $R \cong (r)$ as R-modules, again by Proposition 3.26.

8.12 Since $f(0) = 0 \in N'$, $f^{-1}(N')$ is not empty. We have to verify that it is closed under addition and multiplication by scalars (Proposition 8.23).

Let $m_1, m_2 \in f^{-1}(N')$. Then $n_1 = f(m_1)$ and $n_2 = f(m_2)$ are elements of N'. Therefore

$$f(m_1 + m_2) = f(m_1) + f(m_2) = n_1 + n_2 \in N'$$

since f is an R-module homomorphism; this shows that $m_1 + m_2 \in f^{-1}(N')$. Next, let $m \in f^{-1}(N')$ and let $r \in R$. Then $n = f(m)$ is an element of N', therefore

$$f(rm) = rf(m) = rn \in N'$$

again since f is an R-module homomorphism. This shows that $rm \in f^{-1}(N')$ and concludes the verification.

8.15 There is impressively little to do here. If $f \circ g = 0$, then for all $\ell \in L$ we must have

$$f(g(\ell)) = (f \circ g)(\ell) = 0,$$

that is, $g(\ell) \in \ker f$. The needed homomorphism $g' : L \to \ker f$ is nothing but g itself, viewed as acting $L \to \ker f$ rather than $L \to M$. As usual, the fact that the diagram commutes leaves no choice as to the definition of g', so the desired uniqueness holds.

8.16 For the diagram to commute, we must have

$$(\mu \oplus \nu) \circ \iota_M = \mu \quad \text{and} \quad (\mu \oplus \nu) \circ \iota_N = \nu :$$

that is (using the definitions of ι_M and ι_N, cf. §8.3) $\forall m \in M, \forall n \in N$:

$$(\mu \oplus \nu)(m, 0) = \mu(m) \quad \text{and} \quad (\mu \oplus \nu)(0, n) = \nu(n).$$

If this is the case and $\mu \oplus \nu$ is an R-module homomorphism, then necessarily

$$(\mu \oplus \nu)((m, n)) = (\mu \oplus \nu)(m, 0) + (\mu \oplus \nu)(0, n) = \mu(m) + \nu(n).$$

What this says is that *if* a homomorphism $\mu \oplus \nu$ as required exists, then it is unique: it must be given by

$$(\mu \oplus \nu)((m, n)) = \mu(m) + \nu(n)$$

for all $(m, n) \in M \oplus N$. This formula certainly defines a *function* $M \oplus N \to P$; all we have to verify is that this function *is* a homomorphism of R-modules. For all $(m_1, n_1), (m_2, n_2) \in$

$M \oplus N$ we have

$$(\mu \oplus v)((m_1, n_1) + (m_2, n_2)) = (\mu \oplus v)((m_1 + m_2, n_1 + n_2))$$
$$= \mu(m_1 + m_2) + v(n_1 + n_2)$$
$$= \mu(m_1) + \mu(m_2) + v(n_1) + v(n_2)$$
$$\overset{!}{=} \mu(m_1) + v(n_1) + \mu(m_2) + v(n_2)$$
$$= (\mu \oplus v)((m_1, n_1)) + (\mu \oplus v)((m_2 + n_2));$$

this verifies that $\mu \oplus v$ preserves addition. For all $(m, n) \in M \oplus N$ and all $r \in R$:

$$(\mu \oplus v)(r(m, n)) = (\mu \oplus v)((rm, rn))$$
$$= \mu(rm) + v(rn)$$
$$= r\mu(m) + rv(n)$$
$$= r(\mu(m) + v(n))$$
$$= r(\mu \oplus v)((m, n));$$

this verifies that $\mu \oplus v$ preserves multiplication by scalars, and completes the proof. □

This argument may look unwieldy, but as often happens, that is just because the notation is heavy. The mathematics in it is nearly trivial: every step works by a direct application of either one of the module axioms (Definition 8.1) or the fact that μ and v are assumed to be R-module homomorphisms.

There is one amusing observation: the equality $\overset{!}{=}$ holds because the addition in P is commutative; if this were not the case, the fact we are trying to prove would not hold. This is the reason why coproducts are a more complicated business for *groups*. A few comments about this are included in §11.1.

8.17 Reversing the arrows and switching from ιs to πs (and changing the names of the other homomorphisms to avoid clashing with the other case), we obtain the following statement.

Let R be a ring and let M, N, P be R-modules. Let $\alpha: P \to M, \beta: P \to N$ be R-module homomorphisms. Then there is a unique R-module homomorphism $\alpha \times \beta: P \to M \oplus N$ making the following diagram commute:

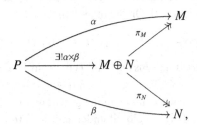

where π_M, π_N are the natural projections defined in §8.3.

I also flipped the whole diagram, since I am more used to seeing arrows point to the right.

The proof that this statement holds follows the same lines as the proof given in Exercise 8.16. First, note that if we can find $\alpha \times \beta$ as needed, then by the commutativity of the diagram we necessarily have

$$\pi_M \circ (\alpha \times \beta) = \alpha \quad \text{and} \quad \pi_N \circ (\alpha \times \beta) = \beta \, ; \tag{B.2}$$

that is, for all $p \in P$ we must have

$$(\alpha \times \beta)(p) = (\alpha(p), \beta(p)) :$$

the first identity in (B.2) tells us what the first element in the pair must be, and the second identity tells us what the second element must be.

Thus, all we have to show is that the function $\alpha \times \beta$ defined by the formula $(\alpha \times \beta)(p) = (\alpha(p), \beta(p))$ is an R-module homomorphism: the argument we just went through proves that it is then unique.

The verification that the proposed expression defines a homomorphism is of course straightforward. For p_1, p_2 in P, we have

$$\begin{aligned}
(\alpha \times \beta)(p_1 + p_2) &= (\alpha(p_1 + p_2), \beta(p_1 + p_2)) \\
&= (\alpha(p_1) + \alpha(p_2), \beta(p_1) + \beta(p_2)) \\
&= (\alpha(p_1), \beta(p_1)) + (\alpha(p_2), \beta(p_2)) \\
&= (\alpha \times \beta)(p_1) + (\alpha \times \beta)(p_2) \, ;
\end{aligned}$$

and for $p \in P, r \in R$,

$$\begin{aligned}
(\alpha \times \beta)(rp) &= (\alpha(rp), \beta(rp)) \\
&= (r\alpha(p), r\beta(p)) \\
&= r(\alpha(p), \beta(p)) \\
&= r(\alpha \times \beta)(p)
\end{aligned}$$

as needed.

8.19 Let R be an integral domain, and let T be the set of torsion elements in an R-module M. To prove that T is a submodule of M, we use Proposition 8.23: T is non-empty, since $0 \in T$, and then we have to prove that it is closed with respect to addition and to multiplication by scalars. For addition, let t_1, t_2 be torsion elements; then $\exists r_1 \neq 0, r_2 \neq 0$ in R such that $rt_1 = 0$, $rt_2 = 0$. Since r_1 and r_2 are nonzero, and R is an integral domain, then $r_1 r_2 \neq 0$. On the other hand, we have

$$(r_1 r_2)(t_1 + t_2) = (r_1 r_2)t_1 + (r_1 r_2)t_2 = r_2(r_1 t_1) + r_1(r_2 t_2) = r_2 \cdot 0 + r_1 \cdot 0 = 0 + 0 = 0 \, ;$$

here I have used axioms (vi) and (vii) from Definition 8.1 and the commutativity of multiplication in R. This shows that $t_1 + t_2$ is a torsion element—but note that we have used crucially that we are in an integral domain: this is needed to ensure that $r_1 r_2 \neq 0$.

This shows that T is closed under addition. To prove that it is closed under multiplication, let $t \in T$ and let $r \neq 0$ such that $rt = 0$. Then for all $s \in R$ we have

$$r(st) = (rs)t = (sr)t = s(rt) = s \cdot 0 = 0 :$$

this shows that $st \in T$, as needed. (We used axiom (vi) and the commutativity of multiplication in R.)

This concludes the verification that T is a submodule of M.

If R is not an integral domain, then we could have a situation in which t_1 is torsion because $r_1 t_1 = 0$ with $r_1 \neq 0$, and t_2 is torsion because $r_2 t_2 = 0$ with $r_2 \neq 0$, and yet $r_1 r_2 = 0$, so no conclusion can be drawn about $t_1 + t_2$. For a concrete example, let $R = \mathbb{Z}/6\mathbb{Z}$ and let $M = R$. Then $[3]_6$ is torsion, since $[2]_6 \neq 0$ and $[2]_6[3]_6 = [0]_6$; and $[4]_6$ is torsion, since $[3]_6 \neq 0$ and $[3]_6[4]_6 = [0]_6$. However, $[3]_6 + [4]_6 = [1]_6$ is *not* torsion.

8.22 Consider the R-module homomorphism

$$\varphi \colon M_1 \oplus M_2 \longrightarrow (M_1/L_1) \oplus (M_2/L_2)$$

defined by $\varphi(m_1, m_2) = (m_1 + L_1, m_2 + L_2)$. Since every coset is the coset of some element, φ is surjective. We have

$$\begin{aligned}
\ker(\varphi) &= \{(m_1, m_2) \mid (m_1 + L_1, m_2 + L_2) = (0, 0)\} \\
&= \{(m_1, m_2) \mid m_1 + L_1 = L_1, m_2 + L_2 = L_2\} \\
&= \{(m_1, m_2) \mid m_1 \in L_1, m_2 \in L_2\} \\
&= L_1 \oplus L_2 \,.
\end{aligned}$$

This proves that $L_1 \oplus L_2$ is a submodule of $M_1 \oplus M_2$, since kernels are submodules, and further

$$(M_1/L_1) \oplus (M_2/L_2) \cong (M_1 \oplus M_2)/\ker(\varphi) = (M_1 \oplus M_2)/(L_1 \oplus L_2)$$

by the First Isomorphism Theorem, since φ is surjective.

8.23 As usual, the commutativity of the diagram determines what $\tilde{f}$ must be, if any such homomorphism exists. Indeed, let $m + K$ be an arbitrary element of M/K. If a function $\tilde{f}$ making the diagram commute exists, we must have

$$\tilde{f}(m + K) = \tilde{f}(\pi(m)) = (\tilde{f} \circ \pi)(m) = f(m) \,.$$

Therefore, we *define* $\tilde{f}(m + K)$ to be $f(m)$, and we have to prove that this is a well-defined prescription and that the resulting function $\tilde{f}$ is an R-module homomorphism. If we are successful, the homomorphism $\tilde{f}$ is unique, since its definition was determined as above.

To prove that $\tilde{f}$ is well-defined, assume that $m + K = m' + K$; we have to verify that $f(m) = f(m')$. For this, note that $m + K = m' + K$ implies $m' - m \in K$; since by hypothesis $K \subseteq \ker f$, we have $f(m' - m) = 0$, and hence $f(m') = f(m)$ as needed.

Thus we do have a unique function $\tilde{f} \colon M/K \to N$ making the diagram commute. We have to verify that it is an R-module homomorphism, that is, that it preserves addition and multiplication by scalars. For addition: $\forall m_1 + K, m_2 + K \in M/K$ we have

$$\begin{aligned}
\tilde{f}((m_1 + K) + (m_2 + K)) &= \tilde{f}((m_1 + m_2) + K) = f(m_1 + m_2) = f(m_1) + f(m_2) \\
&= \tilde{f}(m_1 + K) + \tilde{f}(m_2 + K)
\end{aligned}$$

as needed. For multiplication by scalars: $\forall m_K \in M/K, \forall r \in R$ we have

$$\tilde{f}(r(m + K)) = \tilde{f}(rm + K) = f(rm) = rf(m) = r\tilde{f}(m + K) \,,$$

and this concludes the proof.

B.9   Modules over Integral Domains

9.2 • If A is a nontrivial free $\mathbb{Z}$-module, then $A \cong \mathbb{Z}^{\oplus r}$ for some $r > 0$. In particular, A is *infinite*. Thus, if A is nontrivial and *finite*, then it is not free.

• To see that $\mathbb{Q}$ is not free and finitely generated as a $\mathbb{Z}$-module, assume to the contrary that there is an isomorphism $\varphi \colon \mathbb{Q} \to \mathbb{Z}^{\oplus n}$ for some $n > 0$. Let $(a_1, \ldots, a_n) = \varphi(1)$. For every integer $N > 0$, we have

$$N \cdot \frac{1}{N} = 1$$

in $\mathbb{Q}$; correspondingly (as φ is a homomorphism of $\mathbb{Z}$-modules) we should have

$$(a_1, \ldots, a_n) = \varphi(1) = \varphi\left(N \cdot \frac{1}{N}\right) = N \cdot \varphi\left(\frac{1}{N}\right).$$

This would imply that every a_i is divisible by *every* positive integer N. Since the only integer divisible by all integers is 0, this says that $\varphi(1) = (a_1, \ldots, a_n) = (0, \ldots, 0) = \varphi(0)$, contradicting the assumption that φ is an isomorphism.

The same argument proves that $\mathbb{Q}$ is not free even without any further hypothesis of finite generation. The point is that $\mathbb{Q}$ is 'divisible', that is, every element can be 'divided' by every positive integer. A nontrivial free $\mathbb{Z}$-module is not divisible, because the components of its elements are integers, and $\mathbb{Z}$ is not divisible.

• If $V \neq 0$ is a free $k[t]$-module, then $V \cong k[t]^{\oplus n}$ for some $n > 0$. Since $k[t]$ is not finite-dimensional over k (the infinitely many elements $1, t, t^2, t^3, \ldots$ form a basis of $k[t]$ as a k-vector space), V is not finite-dimensional. Therefore, if $V \neq 0$ is finite-dimensional, then it is not free.

• The ideal (x, y) in $\mathbb{C}[x, y]$ is not principal, so if it were free, then a basis for it would contain at least two elements $v = f(x, y)$, $w = g(x, y)$. But then

$$g(x, y)\, v - f(x, y)\, w = g(x, y)f(x, y) - f(x, y)g(x, y) = 0$$

is a vanishing nontrivial $\mathbb{C}[x, y]$-linear combination of v and w: thus v and w cannot be part of a linearly independent set. In particular they cannot be part of a basis, contradicting the assumption.

The conclusion is that (x, y) does *not* admit a basis, and therefore it is not free (by Proposition 9.6).

9.4 The image of the first homomorphism, $0 \to M$, is 0; so by exactness the kernel of f must be 0. The statement follows, since $\ker f = 0$ if and only if f is injective (Exercise 8.13).

9.10 • Assume that every ideal of R is finitely generated, and let

$$I_1 \subseteq I_2 \subseteq I_3 \subseteq I_4 \subseteq \cdots$$

be a chain of ideals in R. Consider the subset of R defined by the *union* of all the ideals in the chain:

$$I := \bigcup_{j \geq 1} I_j = \{r \in R \mid \exists j, \, r \in I_j\}.$$

Then I is an ideal of R (Exercise 5.12). Since I is an ideal and every ideal of R is finitely generated, $I = (r_1, \ldots, r_n)$ elements $r_1, \ldots, r_n \in R$. By definition of I, there exists some index m such that I_m contains all these elements. But then we have

$$I = (r_1, \ldots, r_n) \subseteq I_m \subseteq I_{m+1} \subseteq \cdots \subseteq I$$

and it follows that $I_m = I_{m+1} = \cdots$. So the sequence stabilizes, as stated.

(*Note:* I simply copied and pasted the proof of Proposition 6.32, then minimally changed the wording to adapt it to the circumstances. That's a well-known trick.)

• Conversely, assume that every chain of ideals stabilizes. Argue by contradiction. Following the hint, let I be an ideal and let $I_0 = (0)$, $I_1 = (r_1)$ for some $r_1 \neq 0$, $I_2 = (r_1, r_2)$ for some $r_2 \in I \setminus I_1$, and so on: $I_n = I_{n-1} + (r_n)$ where r_n is any element in $I \setminus I_{n-1}$. Note that by assumption $I_{n-1} \subsetneq I_n$. Therefore, if we could do this for all $n > 0$, we would get an infinite nonstabilizing chain of ideals

$$(0) \subsetneq I_1 \subsetneq I_2 \subsetneq I_3 \subsetneq I_4 \subsetneq \cdots,$$

contradicting our assumption on R.

9.11 Let $\pi \colon R \to R/I$ be the natural projection. If

$$J_1 \subseteq J_2 \subseteq J_3 \subseteq J_4 \subseteq \cdots \tag{B.3}$$

is a chain of ideals in R/I, then

$$\pi^{-1}(J_1) \subseteq \pi^{-1}(J_2) \subseteq \pi^{-1}(J_3) \subseteq \pi^{-1}(J_4) \subseteq \cdots$$

is a chain of ideals of R; since R is Noetherian, this chain stabilizes, so that $\exists m$,

$$\pi^{-1}(J_m) = \pi^{-1}(J_{m+1}) = \pi^{-1}(J_{m+2}) = \cdots.$$

Now $J = \pi(\pi^{-1}(J))$ for all ideals J of R/I, so this implies that

$$J_m = J_{m+1} = J_{m+2} = \cdots,$$

and this proves that the chain (B.3) stabilizes.

9.12 If M is finitely generated, then (cf. Definition 9.22) there exists a surjective R-module homomorphism

$$R^{\oplus n} \twoheadrightarrow M$$

for some $n > 0$. Composing with the surjective homomorphism $M \twoheadrightarrow N$, we obtain a surjective R-module homomorphism

$$R^{\oplus n} \twoheadrightarrow N,$$

and this proves that N is finitely generated.

Of course what this argument says is that if $m_1, \ldots, m_n$ generate M, and $\varphi \colon M \to N$

is a surjective R-module homomorphism, then $\varphi(m_1), \ldots, \varphi(m_n)$ generate N. In case you feel happier with a more direct proof, here it is. Let $b \in N$; since φ is surjective, there exists an $a \in M$ such that $b = \varphi(a)$; since $m_1, \ldots, m_n$ generate M, there exist $r_1, \ldots, r_n \in R$ such that

$$a = r_1 m_1 + \cdots + r_n m_n \, ;$$

and then

$$b = \varphi(a) = r_1 \varphi(m_1) + \cdots + r_n \varphi(m_n) \, ,$$

and this should convince you that $\varphi(m_1), \ldots, \varphi(m_n)$ generate N.   ⌐

These are precisely the same argument, just dressed up differently. I have a preference for the first, because it requires less notation.

9.14 Let $\underline{a} = (0, -12, 9, -3, -6)$. We switch the first two entries to get the first entry to be nonzero:

$$(-12, 0, 9, -3, -6) \, .$$

Enter the procedure: replace each entry (except the first) by the remainder of the division by -12:

$$(-12, 0, 9, 9, 6) \, .$$

Switch the first and last entry to place in the first spot the nonzero entry with least valuation:

$$(6, 0, 9, 9, -12) \, .$$

Enter the procedure again: replace entries with the remainders of the division by 6:

$$(6, 0, 3, 3, 0) \, .$$

Switch the first and third entry to place in the first spot the nonzero entry with least valuation:

$$(3, 0, 6, 3, 0) \, .$$

Enter the procedure once more: replace entries with the remainders of the division by 3:

$$(3, 0, 0, 0, 0) \, .$$

And now the procedure exits, since the vector is in the desired form.

Of course it would have been smarter to start by switching the first and *fourth* entry, producing

$$(-3, -12, 9, 0, -6) \, ;$$

then a single run of divisions-with-remainder reduces this to

$$(-3, 0, 0, 0, 0) \, ,$$

and we can switch the sign and get the same result $(3, 0, 0, 0, 0)$. (Switching the sign amounts to multiplying by -1, which is a unit in $\mathbb{Z}$.)

9.17 We prove the two inclusions.

- First, we verify that $\text{Ann}\,(R/(d_1) \oplus \cdots \oplus R/(d_s)) \subseteq (d_s)$. If r annihilates all elements of the direct sum $R/(d_1) \oplus \cdots \oplus R/(d_s)$, then in particular it annihilates $(0, \ldots, 0, 1)$. Therefore

$$(0, \ldots, 0, r + (d_s)) = (0, \ldots, 0, 0)\,;$$

this says that $r + (d_s)$ is 0 in $R/(d_s)$, which implies $r \in (d_s)$, as needed.

- Conversely, let $r \in (d_s)$; for the other inclusion we have to prove that for all $a_i \in R$, $r(a_1 + (d_1), \ldots, a_s + (d_s)) = (0, \ldots, 0)$. Since $r \in (d_s)$, we have $r = bd_s$ for some $b \in R$. Therefore

$$r(a_1 + (d_1), \ldots, a_s + (d_s)) = b(a_1 d_s + (d_1), \ldots, a_s d_s + (d_s))\,.$$

For all i we have $d_i \mid d_s$ by hypothesis; and therefore $a_i d_s \in (d_i)$. This shows that $a_i d_s + (d_i) = 0$ for all i, and it follows that $r(a_1 + (d_1), \ldots, a_s + (d_s)) = (0, \ldots, 0)$ as needed.

9.18 By the classification theorem,

$$M \cong R^{\oplus r} \oplus (R/(d_1)) \oplus (R/(d_2)) \oplus \cdots \oplus (R/(d_s))\,, \qquad (\text{B.4})$$

where $d_1 \mid \cdots \mid d_s$ and d_1 is not a unit. We may therefore assume that M equals the right-hand side of (B.4).

Let's find what the torsion submodule of M is. An arbitrary element of M is of the form

$$(f, a_1 + (d_1), \ldots, a_s + (d_s))\,,$$

where $f \in R^{\oplus r}$ and the elements a_i are in R. This element is a torsion element if $\exists t \neq 0$ in R such that

$$t(f, a_1 + (d_1), \ldots, a_s + (d_s)) = 0$$

(see Exercise 8.19); that is,

$$(tf, ta_1 + (d_1), \ldots, ta_s + (d_s)) = 0\,.$$

If $t \neq 0$ and $f \neq 0$, then $tf \neq 0$, since R is an integral domain. Therefore a torsion element must have $f = 0$. On the other hand, if $t = d_s$, then $ta_i \in (d_i)$ for all i (since $d_i \mid d_s$), so $ta_i + (d_i) = 0$ for all i. The conclusion is that the torsion submodule is

$$T = \{(0, a_1 + (d_1), \ldots, a_s + (d_s))\} = (R/(d_1)) \oplus (R/(d_2)) \oplus \cdots \oplus (R/(d_s))\,.$$

It follows that

$$M/T \cong R^{\oplus r}\,,$$

and in particular M/T is free, as needed.

For an example of an integral domain R and a torsion-free R-module M that is not free, we can take $R = \mathbb{C}[x, y]$ and $M = (x, y)$. Since $M \subseteq R$, and R is an integral domain, the only torsion element of (x, y) is 0; so (x, y) *is* torsion free. On the other hand (x, y) is *not* free, as you checked in Exercise 9.2.

9.19 Let R be a Euclidean domain, and let N be a submodule of $R^{\oplus m}$. By Lemma 9.31,

N is finitely generated; and since N is a submodule of a free module (over an integral domain), N is torsion-free. Therefore N is free, by Exercise 9.18.

For the more precise result, use Lemma 9.38. Now that we know that $N \cong R^{\oplus n}$ is free, the inclusion $N \subseteq R^{\oplus m}$ is given by an $m \times n$ matrix A with entries in R. By Lemma 9.38, A is equivalent to a matrix of the form (9.7) for some $s \leq n$. Since A corresponds to an injective homomorphism, $s = n$. Therefore there exist invertible matrices P and Q such that $A = PA'Q$, where

$$
A' = \begin{pmatrix}
d_1 & 0 & \cdots & 0 \\
0 & d_2 & \cdots & 0 \\
\vdots & \vdots & \ddots & \vdots \\
0 & 0 & \cdots & d_n \\
0 & 0 & \cdots & 0 \\
\vdots & \vdots & \ddots & \vdots \\
0 & 0 & \cdots & 0
\end{pmatrix}
$$

and $d_1 \mid d_2 \mid \cdots \mid d_n$. If $(\underline{e}_1, \ldots, \underline{e}_m)$ is the standard basis for $R^{\oplus m}$ and $(\underline{e}'_1, \ldots, \underline{e}'_n)$ is the standard basis for $R^{\oplus n}$, we have

$$
A'\underline{e}'_i = d_i \underline{e}_i \qquad i = 1, \ldots, n. \tag{B.5}
$$

To find x_i, y_i as in the statement, let

$$
x_i := P\underline{e}_i, \qquad y_i = AQ^{-1}\underline{e}'_i.
$$

Since P is invertible, $x_1, \ldots, x_m$ form a basis for $R^{\oplus m}$ (Exercise 9.1). By the same token, the elements $Q^{-1}\underline{e}'_i$ form a basis for $R^{\oplus n}$, and $y_i = AQ^{-1}\underline{e}'_i \in R^{\oplus m}$ are the elements of $N \subseteq R^{\oplus m}$ corresponding to this basis.

Thus, we are done if we verify that $y_i = d_i x_i$ for $i = 1, \ldots, n$. And indeed, since $A = PA'Q$,

$$
y_i = AQ^{-1}\underline{e}'_i = PA'\underline{e}'_i \overset{!}{=} Pd_i\underline{e}_i = d_i P\underline{e}_i = dx_i,
$$

where I have used (B.5) at the equality $\overset{!}{=}$.

I can't blame you if this argument baffles you. *How on earth* did I come up with the definitions $x_i = P\underline{e}_i$, $y_i = AQ^{-1}\underline{e}'_i$? Truth is, once I had applied Lemma 9.38, I drew the following diagram:

$$
\begin{array}{ccc}
R^{\oplus n} & \xrightarrow{\ A\ } & R^{\oplus m} \\
{\scriptstyle Q}\big\downarrow & & \big\uparrow{\scriptstyle P} \\
R^{\oplus n} & \xrightarrow{\ A'\ } & R^{\oplus m}.
\end{array}
$$

This is just the same information as the identity $A = PA'Q$ coming from Lemma 9.38, but the visual impact of seeing it realized in a diagram helps me. All I need to do is transfer the identity $A'\underline{e}'_i = d_i\underline{e}_i$ from the bottom row to the top right-hand corner, where everything is supposed to happen. That suggests letting $x_i = P\underline{e}_i$ and $y_i = AQ^{-1}\underline{e}'_i$: this is the only reasonable way to move things around and have them all land in the free module $R^{\oplus m}$ on the top right. (Note that A and A' are not necessarily invertible, so we can't write things like A'^{-1}.) Once I have the right definitions for my elements, writing the problem up is a purely mechanical chore.

I would find such a problem much more challenging if I did not have a way to visualize the objects which I am trying to handle. But then, your brain may work differently than mine.

B.10   Abelian Groups

10.1 First, note that if f preserves addition, then $f(0) = 0$ and $f(-a) = -f(a)$ for all $a \in A$. Indeed

$$f(0) = f(0 + 0) = f(0) + f(0),$$

from which $f(0) = 0$ by cancellation, and $\forall a \in A$

$$f(-a) + f(a) = f((-a) + a) = f(0) = 0,$$

which implies $f(-a) = -f(a)$, again by cancellation.

It follows that we only need to verify that $f(n \cdot a) = n \cdot f(a)$ for all $a \in A$ and all $n \geq 0$ in $\in \mathbb{Z}$. Indeed,

$$f((-n) \cdot a) = f(-(n \cdot a)) = -f(n \cdot a) \quad \text{and} \quad (-n) \cdot f(a) = -(n \cdot f(a)).$$

Therefore, proving the statement for n is equivalent to proving it for $-n$, and we may assume $n \geq 0$.

To prove the statement for $n \geq 0$, argue by induction on n. For $n = 0$, the statement is that $f(0 \cdot a) = 0 \cdot f(a)$. This is true: both elements equal 0. Assuming that the statement is true for a fixed $n \geq 0$, we verify it for $n + 1$; and this will prove the statement for all $n \geq 0$, by induction. To verify the statement for $n + 1$, we just have to observe that

$$(n + 1) \cdot f(a) = n \cdot f(a) + f(a) \overset{1}{=} f(n \cdot a) + f(a) \overset{2}{=} f(n \cdot a + a) = f((n + 1) \cdot a),$$

where the induction hypothesis is used at $\overset{1}{=}$, and $\overset{2}{=}$ holds since we are assuming that f preserves addition.

10.6 Addition in $\text{End}(A)$ is defined by setting $f + g$ so that $\forall a \in A$

$$(f + g)(a) = f(a) + g(a);$$

we verified in §10.1 that this is indeed a homomorphism. The ring axioms, in the same order as listed in Definition 3.1, are verified as follows:

- $\forall f, g, h \in \text{End}(A), \forall a \in A,$

$$((f + g) + h)(a) = (f + g)(a) + h(a) = (f(a) + g(a)) + h(a)$$
$$= f(a) + (g(a) + h(a)) = f(a) + ((g + h)(a)) = (f + (g + h))(a);$$

since this is true $\forall a \in A$, it follows that $(f + g) + h = f + (g + h)$ in $\text{End}(A)$.

- Let $0 \in \text{End}(A)$ be the trivial homomorphism; then for all $f \in \text{End}(A)$, $\forall a \in A$,

$$(0 + f)(a) = 0(a) + f(a) = f(a) \quad \text{and} \quad (f + 0)(a) = f(a) + 0(a) = f(a).$$

It follows that $0 + f = f + 0 = f$ in $\text{End}(A)$.
- Given $f \in \text{End}(A)$, define $-f \in \text{End}(A)$ by setting $(-f)(a) = -f(a)$ for all $a \in A$. This is immediately checked to be a homomorphism, and $\forall a \in A$,

$$(f + (-f))(a) = f(a) + (-f)(a) = f(a) - f(a) = 0,$$
$$((-f) + f)(a) = (-f)(a) + f(a) = -f(a) + f(a) = 0.$$

It follows that $f + (-f) = (-f) + f = 0$ in $\text{End}(A)$.
- For all $f, g \in \text{End}(A)$ and $\forall a \in A$,

$$(f + g)(a) = f(a) + g(a) = g(a) + f(a) = (g + f)(a)$$

since addition is commutative in A. It follows that $f + g = g + f$ in $\text{End}(A)$.
- For all $f, g, h \in \text{End}(A)$ we have

$$(h \circ g) \circ f = h \circ (g \circ f)$$

since this holds for R-module homomorphisms, for all R (Proposition 8.19).
- The identity homomorphism $\text{id}_A : A \to A$ is a multiplicative identity in $\text{End}(A)$:

$$\text{id}_A \circ f = f \circ A = f$$

for all $f \in \text{End}(A)$. This is nothing but the definition of the identity morphism.
- For all $f, g, h \in \text{End}(A)$ and all $a \in A$, we have

$$(f \circ (g + h))(a) = f((g + h)(a)) = f(g(a) + h(a))$$
$$= f(g(a)) + f(h(a)) = (f \circ g)(a) + (f \circ h)(a) = (f \circ g + f \circ h)(a)$$

and

$$((f + g) \circ h)(a) = (f + g)(h(a)) = f(h(a)) + g(h(a))$$
$$= (f \circ h)(a) + (g \circ h)(a) = (f \circ h + g \circ h)(a).$$

Therefore

$$f \circ (g + h) = f \circ g + f \circ h \quad \text{and} \quad (f + g) \circ h = f \circ h + g \circ h$$

in $\text{End}(A)$.

The conclusion is that $(\text{End}(A), +, \circ)$ is indeed a ring, as stated.

10.7 If $g \in U$, then g is an element of a proper subgroup H of C_n. Then $\langle g \rangle \subseteq H \subsetneq C_n$, and in particular $\langle g \rangle \neq C_n$: so g is not a generator of C_n.

Conversely, assume that g is not a generator of C_n. Then $\langle g \rangle \neq C_n$; that is, $\langle g \rangle$ is a proper subgroup of C_n. It follows that $g \in U$, as needed.

10.11 By Proposition 10.16, the order of an element a is a power of p if and only if $p^r a = 0$ for some r. Therefore

$$O_p = \{a \in A \mid \exists r \geq 0 \text{ such that } p^r a = 0\}.$$

In particular, $0 \in O_p$, so $O_p \neq \emptyset$, and by Proposition 10.4 it suffices to verify that O_p is closed under subtraction. Let then $a, b \in O_p$. Therefore $\exists r, s$ such that $p^r a = p^s b = 0$; and hence $p^n a = p^n b = 0$ for $n = \max(r, s)$. It follows that

$$p^n(a - b) = p^n a - p^n b = 0 - 0 = 0,$$

and this proves that $a - b \in O_p$, as needed.

10.15 Let a_r be the number of exponents α_i that are equal to r, and let b_r be the number of exponents β_j equal to r. It suffices to show that $a_r = b_r$ for all r.

Let

$$e_d = a_1 + 2a_2 + \cdots + (d-1)a_{d-1} + d(a_d + a_{d+1} + \cdots),$$
$$f_d = b_1 + 2b_2 + \cdots + (d-1)b_{d-1} + d(a_d + b_{d+1} + \cdots),$$

and note that for a fixed d, (10.5) implies that

$$e_d = f_d.$$

For $d = 1$, this says

$$a_1 + a_2 + a_3 + \cdots = b_1 + b_2 + b_3 + \cdots, \tag{B.6}$$

that is, $m = n$. For $d = 2$, $e_2 = f_2$ says that

$$a_1 + 2(a_2 + a_3 + \cdots) = b_1 + 2(b_2 + b_3 + \cdots); \tag{B.7}$$

together with (B.6), this implies $a_1 = b_1$. To complete the proof, it suffices to observe that

$$a_r = 2e_r - e_{r-1} - e_{r+1} = 2f_r - f_{r-1} - f_{r+1} = b_r$$

for $r \geq 2$ (as you can verify).

10.18 By the classification theorem (Theorem 10.18), if G is an abelian group of order 1024, then there is exactly one choice of integers $1 < d_1, \ldots, d_s$ with $d_1 \mid d_2 \mid \cdots \mid d_s$ such that

$$A \cong C_{d_1} \oplus \cdots \oplus C_{d_s},$$

and $d_1 \cdots d_s = 1024 = 2^{10}$. All the numbers d_i must be powers of 2; letting $d_i = 2^{a_i}$, we see that there is one s-uple $d_1, \ldots, d_s$ as above for every nondecreasing s-uple of positive integers, $0 < a_1 \leq a_2 \leq \cdots \leq a_s$, such that $a_1 + \cdots + a_s = 10$.

These are precisely the 'partitions' of 10, and there happen to be exactly 42 of them.

(For a smaller example, the number of partitions of 5 is seven, because there are seven ways of writing 5 as a sum of nondecreasing positive integers:

$$1 + 1 + 1 + 1 + 1 = 1 + 1 + 1 + 2 = 1 + 2 + 2 = 1 + 1 + 3 = 1 + 4 = 2 + 3 = 5.$$

Therefore there are seven abelian groups of order $2^5 = 32$, up to isomorphism.)

10.22 By Theorem 10.18, there exist integers $0 < d_1, \ldots, d_s$, with $d_1 \mid \cdots \mid d_s$, such that

$$A \cong C_{d_1} \oplus \cdots \oplus C_{d_s}.$$

If $p > 0$ is a prime integer and p^k divides $|A| = d_1 \cdots d_s$, then there exist nonnegative integers $a_1, \ldots, a_s$ such that $p^{a_i} \mid d_i$ and $a_1 + \cdots + a_s = k$. We then have subgroups $C_{p^{a_i}} \subseteq C_{d_i}$ (Proposition 10.10), and therefore a subgroup

$$B := C_{p^{a_1}} \oplus \cdots \oplus C_{p^{a_s}} \subseteq C_{d_1} \oplus \cdots \oplus C_{d_s} \cong A .$$

Since $|B| = p^{a_1 + \cdots + a_s} = p^k$, this proves the statement.

10.23 By the classification theorem,

$$A \cong \left(\oplus_j \mathbb{Z}/p_1^{\alpha_{1j}}\mathbb{Z} \right) \oplus \cdots \oplus \left(\oplus_j \mathbb{Z}/p_r^{\alpha_{rj}}\mathbb{Z} \right)$$

with $\sum_j \alpha_{ij} = r_i$; therefore we may assume that A equals the direct sum shown here. Let P_i be the ith summand, so that $A = P_1 \oplus \cdots \oplus P_s$. Then P_i may be identified with a subgroup of A of order $p_i^{r_i}$, and this shows that the required subgroups exist. (This also follows from Exercise 10.22.)

To prove uniqueness, we will assume that $B \subseteq A$ is any subgroup whose order is a power of p_i, and prove that necessarily $B \subseteq P_i$. It will follow that if $|B| = p^{r_i}$, then necessarily $B = P_i$, verifying uniqueness.

Let then $a \in B$ be an element of $B \subseteq A = P_1 \oplus \cdots \oplus P_s$. Viewing a as an element of the direct sum, we can write it as $(a_1, \ldots, a_s)$ with $a_j \in P_j$. However, since $a \in B$ and $|B|$ is a power of p_i, then $|a|$ is a power of p_i, by Proposition 10.21. Therefore

$$(p_i^k a_1, \ldots, p_i^k a_s) = (0, \ldots, 0)$$

for some k. Since the order of a_j is a power of p_j (again by Proposition 10.21), it follows that for all j, a power $p_j^{k_j}$ of p_j divides p_i^k (Proposition 10.16). Since $p_j \neq p_i$ for $j \neq i$, and the p_j are all irreducible, necessarily $k_j = 0$ for $j \neq i$; that is, $a_j = 0$ for $j \neq i$. Therefore $a = (a_1, \ldots, a_s) = (0, \ldots, 0, a_i, 0, \ldots, 0) \in P_i$.

This proves that $B \subseteq P_i$, as needed.

10.24 Using the classification theorem, we have

$$A \cong C_{d_1} \oplus \cdots \oplus C_{d_s}$$

for suitable $1 < d_1 \leq \cdots \leq d_s$; if a positive integer b divides $|A| = d_1 \cdots d_s$, then there exist positive integers $e_1 \mid d_1, \ldots, e_s \mid d_s$ such that $b = e_1 \cdots e_s$. Each C_{d_i} has a subgroup C_{e_i} by Proposition 10.10, and $C_{e_1} \oplus \cdots \oplus C_{e_s}$ gives a subgroup of order b of $C_{d_1} \oplus \cdots \oplus C_{d_s} \cong A$ as required.

Here is the alternative argument sketched in the statement. Argue by induction on b. If $b = 1$, then the fact is true: every abelian group contains the trivial subgroup $\{0\}$, which has order 1. For $b > 1$, by induction we may assume that the fact is known for all positive integers $b' < b$. Let $p > 0$ be an irreducible factor of b, and let A be a group of order $|A|$, a multiple of b. Then p divides $|A|$, therefore by Cauchy's theorem (Theorem 10.22) there exists an element $g \in A$ of order p. The quotient $A/\langle g \rangle$ has order $|A|/p$, which is a multiple of $b' = b/p$; by the induction hypothesis, $A/\langle g \rangle$ contains a subgroup $\overline{B}$ of order b/p. By the Third Isomorphism Theorem (Theorem 8.47), there exists a subgroup $B \subseteq A$ containing g and such that $\overline{B} = B/\langle g \rangle$. We have $|\overline{B}| = |B|/p$, and it follows that $|B| = b$, as needed.

10.27 The question asks for the number of generators in the group $(\mathbb{Z}/87178291199\,\mathbb{Z})^*$. By Theorem 10.24, this abelian group is cyclic, isomorphic to $C_{87178291198}$. According to Proposition 10.9, this cyclic group has $\phi(87178291198)$ generators, where ϕ is Euler's totient function. According to my computer, the irreducible factorization of 87178291198 is $2 \cdot 101 \cdot 431575698$. By Proposition 5.56

$$\phi(87178291198) = \phi(2)\,\phi(101)\,\phi(431575699)$$
$$= (2-1)(101-1)(431575699-1) = 43157569800$$

as claimed. (I have also used the fact that $\phi(p) = p - 1$ if p is prime, which follows immediately from the definition of the totient function.)

10.29 Recall that $\mathbb{Z}[i] \cong \mathbb{Z}[x]/(x^2 + 1)$ (Example 5.42). Therefore

$$\mathbb{Z}[i]/(3) \cong (\mathbb{Z}[x]/(x^2 + 1))/(3) \cong \mathbb{Z}[x]/(3, x^2 + 1) \cong (\mathbb{Z}/3\mathbb{Z})[x]/(x^2 + 1),$$

invoking standard isomophisms (such as the one showing up in Exercise 5.25). Now $x^2 + 1$ is *irreducible* in $(\mathbb{Z}/3\mathbb{Z})[x]$: indeed it has no root (cf. Proposition 7.15), since

$$0^2 + 1 = 1 \not\equiv 0 \bmod 3, \quad 1^2 + 1 = 2 \not\equiv 0 \bmod 3, \quad 2^2 + 2 = 5 \not\equiv 0 \bmod 3.$$

It follows that $\mathbb{Z}[i]/(3)$ is isomorphic to an integral domain, hence it is an integral domain, hence (3) is a prime ideal in $\mathbb{Z}[i]$, hence 3 is prime in $\mathbb{Z}[i]$.

Here is a different argument. Since $\mathbb{Z}[i]$ is a Euclidean domain (Proposition 10.30), it suffices to show that 3 is irreducible. Assume to the contrary that $3 = ab$ in $\mathbb{Z}[i]$, with neither a nor b a unit. Applying the norm gives

$$N(a)N(b) = N(ab) = N(3) = 9\,;$$

since neither a nor b is a unit, we have $N(a) \neq 1$ and $N(b) \neq 1$ (Lemma 10.29), so the only possibility would be $N(a) = N(b) = 3$. However, there are *no* elements in $\mathbb{Z}[i]$ with norm equal to 3, so this is a contradiction.

B.11   Groups—Preliminaries

11.1 Assume that e_1 and e_2 satisfy the requirement for an identity in a group. Then

$$e_1 \overset{1}{=} e_1 e_2 \overset{2}{=} e_2 :$$

the equality $\overset{1}{=}$ holds because e_2 is an identity, and the equality $\overset{2}{=}$ holds because e_1 is an identity.

Next, assume that h_1 and h_2 both satisfy the requirement for the inverse of an element g in a group G. Then

$$h_1 = h_1 e_G \overset{1}{=} h_1(gh_2) \overset{2}{=} (h_1 g)h_2 \overset{3}{=} e_G h_2 = h_2 :$$

the equality $\overset{1}{=}$ holds because h_2 is an inverse of g, $\overset{2}{=}$ by associativity, and $\overset{3}{=}$ because h_1 is an inverse of g.

Note: Of course this is old fare. These proofs just reproduce the proofs of Propositions 3.14 and 3.15 in the context of groups.

11.2 Since $e_G = e_G \cdot e_G$ and φ is a group homomorphism, then

$$\varphi(e_G) = \varphi(e_G \cdot e_G) = \varphi(e_G) \cdot \varphi(e_G),$$

from which $\varphi(e_G) = e_H$ by cancellation.

Since $g^{-1} \cdot g = e_G$ and φ is a homomorphism (so that $\varphi(e_G) = e_H$ by the first part),

$$\varphi(g^{-1}) \cdot \varphi(g) = \varphi(g^{-1} \cdot g) = \varphi(e_G) = e_H,$$

and multiplying on the right by $\varphi(g)^{-1}$ we obtain

$$\varphi(g^{-1}) = \varphi(g)^{-1}$$

as needed.

Again, these proofs are just adaptations to the context of groups of arguments you
have processed already. For example, the first part reproduces the argument proving
Proposition 4.16.

11.3 The fact that the identity $\mathrm{id}_G \colon G \to G$ function is a group homomorphism is
almost a zen koan: $\forall g, h \in G$,

$$\mathrm{id}_G(gh) = gh = \mathrm{id}_G(g)\,\mathrm{id}_G(h).$$

As for compositions, assume that $\varphi \colon G_1 \to G_2$ and $\psi \colon G_2 \to G_3$ are group homomorphisms. Then $\forall g, h \in G_1$,

$$(\psi \circ \varphi)(gh) = \psi(\varphi(gh)) = \psi(\varphi(g)\varphi(h)) = \psi(\varphi(g))\psi(\varphi(h)) = (\psi \circ \varphi)(g)(\psi \circ \varphi)(h)$$

and this proves that $\psi \circ \varphi$ is a group homomorphism.

11.4 This is a rehash in the world of groups of Proposition 4.30, and I will copy the
proof almost word-for-word.

Let h_1, h_2 be elements of H, and let $g_1 = \varphi^{-1}(h_1)$, $g_2 = \varphi^{-1}(h_2)$. Equivalently, $h_1 = \varphi(g_1)$ and $h_2 = \varphi(g_2)$. Then

$$\varphi^{-1}(h_1 h_2) = \varphi^{-1}(\varphi(g_1)\,\varphi(g_2)) = \varphi^{-1}(\varphi(g_1 g_2))$$

since φ is a homomorphism

$$= (\varphi^{-1} \circ \varphi)(g_1 g_2) = g_1 g_2$$
$$= \varphi^{-1}(h_1)\,\varphi^{-1}(h_2).$$

Thus, φ^{-1} preserves the operation in the group, and this proves it is a group homomorphism.

11.7 We have injective group homomorphisms $\iota_G \colon G \to G \times H$ and $\iota_H \colon H \to G \times H$,
defined by $\iota(g) = (g, e_H)$ and $\iota(h) = (e_G, h)$. Copying the diagram from Exercise 8.16

with evident changes in the notation gives the following:

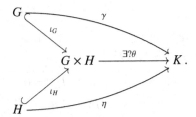

Arguing just as in the solution to Exercise 8.16, if there were a group homomorphism θ making the diagram commute, then the commutativity of the diagram would force

$$\theta \circ \iota_G = \gamma \quad \text{and} \quad \theta \circ \iota_H = \eta,$$

that is, $\forall g \in G$ and $\forall h \in H$,

$$\theta(g, e_H) = \gamma(g) \quad \text{and} \quad \theta(e_G, h) = \eta(h).$$

Necessarily we would have (as θ is supposed to be a group homomorphism)

$$\theta((g, h)) = \gamma(g)\eta(h).$$

This formula defines a set-function $G \times H \to K$, and the question is whether it defines a group homomorphism; if it does, then $G \times H$ is indeed a coproduct of G and H. In the commutative case, this turned out to be the case (as you verified in Exercise 8.16). Let's see what happens in the world of not-necessarily-commutative groups. To verify whether the function θ we just defined is a homomorphism, we see whether it preserves the group operation. Let (g_1, h_1) and (g_2, h_2) be arbitrary elements of $G \times H$. Then

$$\theta((g_1, h_1)(g_2, h_2)) = \theta((g_1 g_2, h_1 h_2)) = \gamma(g_1 g_2)\eta(h_1 h_2) = \gamma(g_1)\,\gamma(g_2)\eta(h_1)\,\eta(h_2),$$
$$\theta((g_1, h_1))\theta((g_2, h_2)) = \gamma(g_1)\,\eta(h_1)\gamma(g_2)\,\eta(h_2).$$

Problem: If K is not commutative, then in general

$$\gamma(g_2)\eta(h_1) \neq \eta(h_1)\gamma(g_2).$$

Therefore, the necessary prescription for θ is *not* a group homomorphism. We must conclude that $G \times H$ is *not* a coproduct in the category of groups, even if the analogous construction *is* a coproduct in the category of abelian groups, as we verified in Exercise 8.16.

Concerning *products,* the relevant diagram (see the solution to Exercise 8.17) is

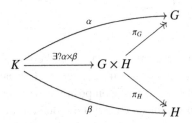

and the commutativity of the diagram forces the definition

$$(\alpha \times \beta)(k) = (\alpha(k), \beta(k))$$

for all $k \in K$. Again, the question is whether this is a group homomorphism; and in this case it *is*. Indeed, $\forall k_1, k_2 \in K$,

$$(\alpha \times \beta)(k_1 k_2) = (\alpha(k_1 k_2), \beta(k_1 k_2)) = (\alpha(k_1)\alpha(k_2), \beta(k_1)\beta(k_2))$$
$$= (\alpha(k_1), \beta(k_1))(\alpha(k_2), \beta(k_2)) = (\alpha \times \beta)(k_1)(\alpha \times \beta)(k_2).$$

The conclusion is that $G \times H$ *is* a product of G and H in the category of groups.

11.14 Let $\sigma: G \to S_A$ be a group homomorphism. We use (11.2) to define a function $\rho: G \times A \to A: \forall g \in G, \forall a \in A,$

$$\rho(g, a) = \sigma(g)(a).$$

In order to prove that this defines an action of G on A, we have to show that

- $\rho(e_G, a) = a$, and
- $\rho(gh, a) = \rho(g, \rho(h, a))$

(cf. Definition 11.19). Now using the above definition for $\rho(g, a)$, we have

$$\rho(e_G, a) = \sigma(e_G)(a) = \mathrm{id}_A(a) = a$$

since group homomorphisms send the identity to the identity; and

$$\rho(gh, a) = \sigma(gh)(a) = (\sigma(g) \circ \sigma(h))(a) = \sigma(g)(\sigma(h)(a)) = \sigma(g)(\rho(h, a)) = \rho(g, \rho(h, a))$$

since σ preserves the operation. This is exactly as needed, so we are done.

11.23 In order to perform any rigid symmetry, we have to decide to which vertex we should send a fixed vertex—and since there are four vertices, there are four such choices; and then how to rotate the tetrahedron about that vertex—and there are three choices for such a rotation. Therefore the number of symmetries is $4 \cdot 3 = 12$. All of them are even. To see this, notice that the rotations about a vertex correspond to 3-cycles, therefore to even permutations. For example, the following 120° rotation about vertex 2 corresponds to the cycle (1 4 3):

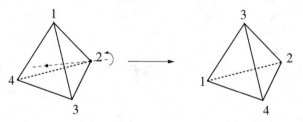

Since one can move any vertex to any other vertex by also performing a rotation about a vertex, *every* rigid symmetry of a tetrahedron can in fact be realized as a composition of rotations about vertices. Compositions of even permutations are even permutations, so all the permutations realizing rigid rotations of a tetrahedron are even.

Since $\mathcal{A}_4$ consists of the 12 even permutations in $\mathcal{S}_4$, this proves the statement. Incidentally, this argument shows that $\mathcal{A}_4$ is generated by 3-cycles. We prove in §12.6 that $\mathcal{A}_n$ is generated by 3-cyles for all n.

Even more explicitly, the 12 elements of $\mathcal{A}_4$ have the following cycle decompositions.

$$
\begin{array}{cccc}
(1) & (1\,2)(3\,4) & (1\,3)(2\,4) & (1\,4)(2\,3) \\
(1\,2\,3) & (4\,2\,1) & (2\,4\,3) & (3\,4\,1) \\
(1\,3\,2) & (4\,1\,2) & (2\,3\,4) & (3\,1\,4)
\end{array}
$$

The eight 3-cycles may be realized by rotations about a vertex, as pointed out above. A little visualization exercise shows that one can realize the three remaining nonidentity elements. For example, $(1\,2)(3\,4)$ results from a $180°$ rotation about an axis bisecting the 12 and 34 edges:

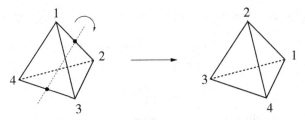

11.28 Using Exercise 11.26, we see that

$$
(1\,2\,\ldots\,n)(1\,2)(n\,\ldots\,2\,1) = (2\,3)\,;
$$

therefore $(2\,3) \in \langle (1\,2), (1\,2\,\ldots\,n) \rangle$. By the same token

$$
(1\,2\,\ldots\,n)(2\,3)(n\,\ldots\,2\,1) = (3\,4)\,;
$$

therefore $(3\,4) \in \langle (1\,2), (1\,2\,\ldots\,n) \rangle$; and so on. This shows that all transpositions of the form $(a\,a+1)$, $a = 1, \ldots, n-1$, belong to $\langle (1\,2), (1\,2\,\ldots\,n) \rangle$.

Next, using Exercise 11.27, we see that

$$
\begin{aligned}
(2\,3)(1\,2)(2\,3) &= (1\,3)\,, \\
(3\,4)(2\,3)(3\,4) &= (2\,4)\,, \\
(4\,5)(3\,4)(4\,5) &= (3\,5)\,,
\end{aligned}
$$

and so on: all transpositions $(a\,a+2)$, $a = 1, \ldots, n-2$, also belong to $\langle (1\,2), (1\,2\,\ldots\,n) \rangle$. By induction, one shows that all transpositions $(a\,a+k)$, $a = 1, \ldots, n-k$, belong to the subgroup generated by $(1\,2)$ and $(1\,2\,\ldots\,n)$.

That's *all* transpositions. Therefore the subgroup generated by $(1\,2)$ and $(1\,2\,\ldots\,n)$ contains the subgroup generated by all transpositions, and that is $\mathcal{S}_n$ by Proposition 11.47.

11.30 Notice that if $\tau = \tau_1\tau_2$, then

$$
\tau\sigma\tau^{-1} = (\tau_1\tau_2)\sigma(\tau_1\tau_2)^{-1} = (\tau_1\tau_2)\sigma(\tau_2^{-1}\tau_1^{-1}) = \tau_1(\tau_2\sigma\tau_2^{-1})\tau^{-1} \tag{B.8}
$$

(conjugation determines an *action;* this was already observed in §11.2). Since every permutation may be written as a product of transpositions (Proposition 11.47), in order to prove that the conjugate $\tau\sigma\tau^{-1}$ of σ has the same type as σ, it is enough to verify that this is the case if τ is a transposition; and this follows from the result of Exercise 11.27.

Conversely, assume that σ and σ' have the same type. Write any cycle decomposition for σ:

$$\sigma = (a_1\, a_2\, \ldots\, a_r)(b_1\, b_2\, \ldots\, b_s) \cdots (c_1\, c_2\, \ldots\, c_t);$$

then we may write a cycle decomposition for σ' of the form

$$\sigma = (\tau(a_1)\, \tau(a_2)\, \ldots\, \tau(a_r))(\tau(b_1)\, \tau(b_2)\, \ldots\, \tau(b_s)) \cdots (\tau(c_1)\, \tau(c_2)\, \ldots\, \tau(c_t)),$$

for some permutation τ. One can then verify that $\sigma' = \tau\sigma\tau^{-1}$: again, since τ may be written as a product of transpositions, and using (B.8), it is enough to verify this fact when τ is a transposition; and again this is the content of Exercise 11.27.

11.33 The elements of $\mathcal{A}_4$, in terms of their cycle decompositions, are:

(1)	(1 2)(3 4)	(1 3)(2 4)	(1 4)(2 3)
(1 2 3)	(4 2 1)	(2 4 3)	(3 4 1)
(1 3 2)	(4 1 2)	(2 3 4)	(3 1 4)

The identity (1) is its own conjugacy class.

If two permutations are conjugate in $\mathcal{A}_4$, then in particular they are conjugate in $\mathcal{S}_4$, therefore by Exercise 11.30 they have the same type. Thus the type $2 + 2$ permutations are not conjugate to the 3-cycles.

To see that the three permutations of type $2 + 2$ are conjugate in $\mathcal{A}_4$, note that they are all conjugate to (1 2)(3 4) by even permutations:

$$((1\,3)(2\,3))(1\,2)(3\,4)((2\,3)(1\,3)) = (1\,3)(1\,3)(2\,4)(1\,3) = (2\,4)(1\,3) = (1\,3)(2\,4),$$
$$((1\,4)(2\,4))(1\,2)(3\,4)((2\,4)(1\,4)) = (1\,4)(1\,4)(2\,3)(1\,4) = (2\,3)(1\,4) = (1\,4)(2\,3).$$

(Exercise 11.27 makes it quite easy to perform these computations.)

Similarly, the 3-cycles in each row are conjugate to one another, and 3-cycles in different rows are *not* conjugate to one another. Thus the conjugacy classes are the identity (size: 1), the permutations of type $2 + 2$ (size: 3), and two distinct classes each containing four 3-cycles.

This can be verified by explicit computations. For example:

$$((1\,3)(1\,4))(1\,2\,3)((1\,4)(1\,3)) = (1\,3)(4\,2\,3)(1\,3) = (4\,2\,1)$$
$$((2\,4)(1\,4))(1\,2\,3)((1\,4)(2\,4)) = (2\,4)(4\,2\,3)(2\,4) = (2\,4\,3)$$
$$((2\,4)(1\,3))(1\,2\,3)((1\,3)(2\,4)) = (2\,4)(3\,2\,1)(2\,4) = (3\,4\,1)$$

shows that the four 3-cycles in the second row are in the same conjugacy class, and a similar computation shows that the four 3-cycles in the third row are in the same conjugacy class. To verify that, e.g., (1 2 3) is not conjugate to (1 3 2), one can argue as in the solution to Exercise 11.30: $\tau(1\,2\,3)\tau^{-1} = (1\,3\,2)$ in $\mathcal{S}_4$ precisely when

$(\tau(1)\,\tau(2)\,\tau(3)) = (1\,3\,2)$ as cycles. This gives three possibilities for τ:

$$
\begin{cases} 1 \mapsto 1 \\ 2 \mapsto 3 \\ 3 \mapsto 2 \end{cases}
\qquad
\begin{cases} 1 \mapsto 3 \\ 2 \mapsto 2 \\ 3 \mapsto 1 \end{cases}
\qquad
\begin{cases} 1 \mapsto 2 \\ 2 \mapsto 1 \\ 3 \mapsto 3 \end{cases}
$$

That is, $\tau = (23)$, (13), or (12). These are all *odd* permutations, so (123) and (132) are *not* conjugate in $\mathcal{A}_4$.

In §12.2 we prove that the size of an orbit of an action of a finite group G necessarily divides the order of G. This implies that the sizes of the conjugacy classes in $\mathcal{A}_4$ must all divide 12. Since 8 does not divide 12, this shows that the 3-cycles cannot all form a single conjugacy class, without having to perform explicit computations.

11.34 We will use that K is a subgroup: therefore it contains the identity and it is close under taking inverses and under the group operation. These properties will imply reflexivity, symmetry, and transitivity (in this order) for both relations. Explicitly:

—For $\sim'_K$: $\forall a, b, c$,

- $a \sim'_K a$ since $a^{-1}a = e \in K$;
- $a \sim'_K b \implies a^{-1}b \in K \implies (a^{-1}b)^{-1} \in K \implies b^{-1}a \in K \implies b \sim'_K a$;
- $a \sim'_K b \wedge b \sim'_K c \implies a^{-1}b \in K \wedge b^{-1}c \in K \implies (a^{-1}b)(b^{-1}c) \in K$
  $\implies a^{-1}c \in K \implies a \sim'_K c$.

—For $\sim''_K$: $\forall a, b, c$,

- $a \sim''_K a$ since $aa^{-1} = e \in K$;
- $a \sim''_K b \implies ba^{-1} \in K \implies (ba^{-1})^{-1} \in K \implies ab^{-1} \in K \implies b \sim''_K a$;
- $a \sim''_K b \wedge b \sim''_K c \implies ba^{-1} \in K \wedge cb^{-1} \in K \implies (cb^{-1})(ba^{-1}) \in K$
  $\implies ca^{-1} \in K \implies a \sim''_K c$.

11.36 Take $G_1 = C_{2n}$, $G_2 = D_{2n}$, $H = C_n$. Both C_{2n} and D_{2n} contain normal subgroups isomorphic to C_n, and $C_{2n}/C_n \cong D_{2n}/C_n \cong C_2$. However, $C_{2n} \not\cong D_{2n}$. Therefore the stated fact is not true.

11.39 We can adapt the argument given for Exercise 8.23; given that we have gone through this already, I will be a little more concise.

The commutativity of the diagram determines $\tilde{\varphi}$: we must have

$$\tilde{\varphi}(gK) = \tilde{\varphi}(\pi(g)) = \varphi(g)$$

for all $g \in G$. We have to prove that this prescription is well-defined, and that $\tilde{\varphi}$ is a group homomorphism. Since the definition is forced by the commutativity of the diagram, $\tilde{\varphi}$ will automatically be unique.

To prove that $\tilde{\varphi}$ is well-defined, let $gK = g'K$, so that $g^{-1}g' \in K \subseteq \ker \varphi$. It follows that $\varphi(g^{-1}g') = e$, and hence $\varphi(g') = \varphi(g)$, as needed.

To prove that $\tilde{\varphi}$ is a group homomorphism, let $g_1, g_2 \in G$; then using the fact that φ is a group homomorphism, we get

$$\tilde{\varphi}((g_1 K)(g_2 K)) = \tilde{\varphi}(g_1 g_2 K) = \varphi(g_1 g_2) = \varphi(g_1)\varphi(g_2) = \tilde{\varphi}(g_1 K)\,\tilde{\varphi}(g_2 K).$$

Therefore $\tilde{\varphi}$ preserves the operation, and we are done.

The last statement is indeed a particular case, since the kernel of the natural projection $G \to G/H$ is H.

B.12   Basic Results on Finite Groups

12.1 We have

$$g_1 H = g_2 H \iff g_1^{-1} g_2 \in H \iff H g_1^{-1} g_2 = H \iff H g_1^{-1} = H g_2^{-1},$$

and this is the first assertion. With this understood, we can consider the function from the set of left-cosets G/H to the set of right-cosets $H \backslash G$ defined by $gH \mapsto Hg^{-1}$. This is evidently surjective, and the first assertion proves it is injective, so there is a one-to-one correspondence between G/H and $H \backslash G$. In particular, if these sets are finite, then they have the same number of elements, and this proves the second assertion.

12.2 The 'new' proof of Fermat's little theorem hinges on the fact that $|(\mathbb{Z}/p\mathbb{Z})^*| = p - 1$. This suggests that, for an arbitrary $n > 0$, we should be looking at the group $(\mathbb{Z}/n\mathbb{Z})^*$ of units in $\mathbb{Z}/n\mathbb{Z}$.

The order of this group equals Euler's totient function $\phi(n)$: this observation goes as far back as Proposition 2.16 (cf. Proposition 3.28).

Therefore, according to Proposition 12.4, if $[a]_n$ is a unit in $\mathbb{Z}/n\mathbb{Z}$, then $[a]_n^{\phi(n)} = [1]_n$. In other words, if a, n are integers, $n > 0$, and $\gcd(a, n) = 1$, then

$$a^{\phi(n)} \equiv 1 \bmod n.$$

This is the Euler–Fermat theorem. If n is prime, then $\phi(n) = n - 1$, and we recover Fermat's little theorem, Theorem 2.18. But this holds for all positive integers!

12.14 Induction on the exponent n in $|G| = p^n$. If $n = 0$, then G is trivial and there is nothing to prove. Let then $n > 0$, and let i be such that $0 \le i \le n$. We have to prove that G contains a subgroup H with $|H| = p^i$. If $i = 0$, then $H = \langle e \rangle$ works; so we may assume $i > 0$. Now let $z \in G$ be an element of order p and in $Z(G)$; such an element exists by Exercise 12.13. Since $z \in Z(G)$, the subgroup $\langle z \rangle$ is normal in $Z(G)$, and we can consider the p-group $G' := G/\langle z \rangle$. The order of G' is p^{n-1}, therefore by induction G' contains a subgroup H' with $|H'| = p^{i-1}$. By the Third Isomorphism Theorem, there is a subgroup H of G containing z, such that $H' = H/\langle z \rangle$. We have $|H| = p|H'| = p^i$, so we are done.

12.17 The group D_{12} has a subgroup of order 6, that is, the subgroup of 'rotations', a copy of C_6. On the other hand, we have verified that A_4 has *no* subgroup of order 6 (Example 12.6). Therefore D_{12} and A_4 are not isomorphic.

12.18 • By Lagrange's theorem, the possible orders for $Z(G)$ are $1, 2, 3$, or 6. However, 6 is not an option since $Z(G)$ is noncommutative; and 2 or 3 are also not an option, because in each of these cases $G/Z(G)$ would be cyclic, contradicting Exercise 12.11. Therefore $|Z(G)| = 1$, that is, $Z(G)$ is trivial.

• By Corollary 12.17, the size of a conjugacy class of G must divide 6; and it cannot be 6 since the conjugacy class of the identity is the identity itself. Therefore, nontrivial conjugacy classes must have order 2 or 3.

Since $|Z(G)| = 1$, the class equation tells us that

$$6 = 1 + \text{sum of sizes of the nontrivial conjugacy classes}.$$

The only way to get 5 with numbers from $\{2, 3\}$ is $2 + 3$, and the conclusion is that there must be two nontrivial conjugacy classes in G, of sizes 2 and 3.

- The order of the elements of G can only be 1 (for the identity), 2, or 3. (There is no element of order 6 since G is not commutative, and in particular not cyclic.) To see that not all elements can have order 2, assume to the contrary that this is the case, and let $g \neq h$ be two distinct such elements; then $gh \neq e$ (since g and h are their own inverses, since they have order 2); and $(gh)(gh) = e$, implying $gh = (gh)^{-1} = h^{-1}g^{-1} = hg$. This would hold for all choices of g and h, and it would follow that G is commutative, a contradiction.

Therefore there exists an element y of order 3. The subgroup $\langle y \rangle$ has order $3 = \frac{6}{2}$, and it follows that this subgroup is normal, cf. Example 11.59.

- Since $\langle y \rangle$ is normal, it is a union of conjugacy classes. The identity is its own conjugacy class, so the order-2 conjugacy class in G must be $\{y, y^2\}$; this says that y and y^2 are conjugate.
- Therefore, there exists an element $x \in G$ such that $y = xy^2x^{-1}$, i.e., $yx = xy^2$.
- The identity $yx = xy^2$ implies that $x \neq y$ and $x \neq y^2$; so $x \notin \langle y \rangle$. If x had order 3, the same reasoning we applied above would show that its conjugacy class would have order 2, and we know that there is only one conjugacy class of order 2, so this is not possible. Therefore x has order 2.
- Since $\langle x, y \rangle$ is strictly larger than $\langle y \rangle$, its order is > 3 and a divisor of 6 by Lagrange's theorem; so $\langle x, y \rangle = G$.
- At this point we know that $G = \langle x, y \rangle$, and x and y satisfy the relations $x^2 = e$, $y^3 = e$, $xy^2 = yx$. This implies that G may be obtained from D_6 by possibly imposing more relations (cf. Proposition 11.35). But if there were more relations, then we would have $|G| < |D_6| = 6$, contrary to assumption. Therefore necessarily $G \cong D_6 \cong S_3$, and we are done.

12.23 • $36 = 2^2 \cdot 3^2$. By Sylow I, the group has a subgroup of order $3^2 = 9$, i.e., index 4. If it were simple, Exercise 12.22 would imply that $36 \mid 4!$, which is not the case.

A 'standalone' argument: By Sylow III, a group G such that $|G| = 36$ has either one or four subgroups of order 9. If there is only one such subgroup, then it is normal and G is not simple. If there are four, they are all conjugate of one another by Sylow II. Therefore G acts (by conjugation) on this set of conjugate subgroup, that has four elements. This gives a homomorphism $G \to S_4$, and since $|G| = 36 > 24$, this homomorphism cannot be injective. It also cannot be trivial, or else conjugation would fix these subgroups, and they would be normal, a contradiction. Therefore the kernel of this homomorphism is a nontrivial proper normal subgroup of G, and it follows that G is not simple.

- Since $40 = 2^3 \cdot 5$, and the only one among $1, 2, 4, 8$ that is $\equiv 1 \bmod 5$ is 1, Sylow III tells us that the group has a unique 5-Sylow subgroup; hence it is not simple.

- Since $45 = 3^2 \cdot 5$, and the only one among $1, 3, 9$ that is $\equiv 1 \bmod 5$ is 1, Sylow III tells us that the group has a unique 5-Sylow subgroup; hence it is not simple.
- By Sylow I, a group G of order 48 has a subgroup of order 16, that is, index 3. Since $|G| = 48$ does not divide $3! = 6$, G must be simple by Exercise 12.22.

Standalone argument: By Sylow III, a group G of order 48 has one or three groups of order 16. If it has one, then that subgroup is normal and G is not simple. If it has three, then they are conjugate to each other by Sylow II; G acts by conjugation on this set of three subgroups; this gives a group homomorphism $\sigma \colon G \to S_3$. Since the action is not trivial, σ is not the trivial homomorphism. The action is not trivial, and $|G| = 48 > 6 = |S_3|$, so the kernel of σ is a nontrivial proper subgroup of G, and G is not simple.

- Let G be a group of order $56 = 2^3 \cdot 7$. By Sylow III, G has either one or eight 7-Sylow subgroups. If G has a unique 7-Sylow subgroup, then this subgroup is normal and G is not simple. If it has eight subgroups of order 7, note that these pairwise intersect at the identity; therefore, these subgroups account for $8 \cdot (7 - 1) = 48$ nonidentity elements of G. These must be the element of a 2-Sylow subgroup, whose existence is guaranteed by Sylow I. But then this subgroup is unique, hence normal, and it again follows that G is not simple.

12.28 We can argue by induction on the number r of quotients. If $r = 1$, there is nothing to prove. Assume $r > 1$, and assume that the result is known for $r - 1$; we will prove it for r quotients.

Let G be a group with an r-step normal sequence and quotients $Q_1, \ldots, Q_r$, as in the statement. Then G_1 has a normal sequence with $r - 1$ steps:

$$G_1 \supsetneq G_2 \supsetneq G_3 \supsetneq \cdots \supsetneq G_r = \{e\}$$

and quotients $Q_2, \ldots, Q_r$. By the induction hypothesis, G_1 has a normal series

$$G_1 \supsetneq \widehat{Q_{21}} \supsetneq \widehat{Q_{22}} \supsetneq \cdots \supsetneq \widehat{Q_{rs_r}} = \{e\}$$

with quotients $R_{21}, R_{22}, \ldots, R_{rs_r}$, using notation as in the statement. Then G has a normal series with quotients $Q_1, R_{21}, R_{22}, \ldots, R_{rs_r}$, and what we have to show is that this series may be refined to one with quotients $R_{11}, \ldots, R_{1s_1}, R_{21}$, etc.

Each Q_{1j} in a series for Q_1 is a subgroup of $Q_1 \cong G/G_1$. By the Third Isomorhism Theorem (Theorem 11.75) it corresponds to a subgroup $\widehat{Q_{1j}}$ of G containing G_1, and as Q_{1s_1} is the identity in G/G_1, $\widehat{Q_{1s_1}} = G_1$. Consider then the following normal series for G:

$$G = G_0 \supsetneq \widehat{Q_{11}} \supsetneq \cdots \supsetneq \widehat{Q_{1s_1}} = G_1 \supsetneq \widehat{Q_{21}} \supsetneq \widehat{Q_{22}} \supsetneq \cdots \supsetneq \widehat{Q_{rs_r}} = \{e\}.$$

By the Third Isomorphism Theorem, $\widehat{Q_{1j-1}}/\widehat{Q_{1j}} \cong Q_{1j-1}/Q_{1j} = R_{1j}$ for $j = 1, \ldots, s_1$, and this completes the proof.

12.31 Let G be a finite group.

If G admits a normal series with cyclic quotients, then these quotients are in particular abelian, and it follows that G is solvable.

Conversely, assume that G is solvable; so it has a normal series with abelian quotients. By Exercise 12.30, each of the quotients admit a *cyclic* normal series. Stringing together these normal series (cf. Exercise 12.28) gives a normal series for G with cyclic quotients, and we are done.

12.32 In the proof of Proposition 12.60 we have verified that the quotients H_i/H_{i+1} of a normal series for H are isomorphic to subgroups of the quotients G_i/G_{i+1} of a normal series for G. If each G_i/G_{i+1} is cyclic, of order m_i, then every H_i/H_{i+1} must be cyclic, of order dividing m_i, since subgroups of cyclic groups are cyclic (Proposition 10.10).

Therefore H admits a normal series as stated.

12.33 Work by induction on the order of the group. If the group is trivial, there is nothing to prove. Assume then that G is a p-group, of size $|G| = p^r$, with $r \geq 1$; and we may assume that the result is known for all groups of order $< p^r$. By Corollary 12.22, $Z(G)$ is not trivial. Note that $Z(G)$ is abelian, hence in particular solvable. Since $Z(G)$ is normal in G, we may consider the quotient $G/Z(G)$. This is a p-group of order $< p^r$, therefore by the induction hypothesis $G/Z(G)$ is solvable. But then $G \supseteq Z(G) \supseteq \{e\}$ is a normal series (with possibly repeated terms) with solvable quotients, so G is solvable by Exercise 12.29.

12.35 We have to prove that if G is solvable, then the derived series $G \supseteq G^{(1)} \supseteq G^{(2)} \supseteq \cdots$ terminates at $\{e\}$. Here $G^{(k)}$ is the *commutator subgroup* of $G^{(k)}$, generated by all commutators $ghg^{-1}h^{-1}$ as $g, h \in G^{(k)}$.

Since G is solvable, it admits a normal series $G = G_0 \supseteq G_1 \supseteq \cdots \supseteq G_r = \{e\}$ with abelian quotients. It is enough to prove that $G^{(k)} \subseteq G_k$ for all $k \geq 0$: in particular, we will have $G^{(r)} = \{e\}$, and this will show that the derived series terminates at $\{e\}$.

To prove that $G^{(k)} \subseteq G_k$ for all $k \geq 0$, work by induction. For $k = 0$ there is nothing to prove. It is then enough to show that for all $k > 0$, if the inclusion holds for $k - 1$, then it holds for k.

Thus, we may assume that $G^{(k-1)} \subseteq G_{k-1}$. Consider the projection $\pi \colon G_{k-1} \to G_{k-1}/G_k$. By assumption, this quotient is abelian. By the result of Exercise 11.37, it follows that G_k contains all commutators of G_{k-1}:

$$\forall g, h \in G_{k-1}: \quad ghg^{-1}h^{-1} \in G_k.$$

As we are assuming that $G^{(k-1)} \subseteq G_{k-1}$, this implies

$$\forall g, h \in G^{(k-1)}: \quad ghg^{-1}h^{-1} \in G_k;$$

and by definition of commutator subgroup, this in turn implies that $G^{(k)} \subseteq G_k$, and we are done.

B.13   Field Extensions

13.5 In particular, E is a subspace of the k-vector space F. Since $\dim_k F$ is finite, so is $\dim_k E$. (This fact is likely familiar from ordinary linear algebra. It is also implied by the fact that fields are Noetherian rings, so that subspaces of finitely generated vector spaces are finitely generated.) Therefore $k \subseteq E$ is a finite extension.

Since $\dim_k F$ is finite, F is generated by finitely many elements $\alpha_1, \ldots, \alpha_n$ over k. Since $k \subseteq E$, the same elements generate F over E. Therefore F is a finitely generated E-vector space, i.e., the extension $E \subseteq F$ is finite, as needed.

13.7 Let $[F : k] = p$ be a positive prime integer. Let E be an intermediate field: $k \subseteq E \subseteq F$. By Proposition 13.13,

$$[F : E][E : k] = [F : k] = p.$$

Since p is prime, then either $[F : E] = 1$ or $[E : k] = 1$; and this says precisely that $E = F$ or $E = k$. Thus the only intermediate fields are k and F.

13.8 Since $\varphi|_k = \mathrm{id}_k$, the image $E = \varphi(F)$ is an intermediate extension: $k \subseteq E \subseteq F$. Also, since φ is a homomorphism of fields, it is injective, hence (by the First Isomorphism Theorem) it is an isomorphism of F to E. Again since $\varphi|_k = \mathrm{id}_k$, it is in particular an isomorphism of k-vector spaces, so $[E : k] = [F : k]$. By Exercise 13.6, $E = F$: that is, φ is also surjective, and it follows that it is an isomorphism of F to itself.

Without the finiteness hypothesis, the conclusion does not necessarily hold. For example, take the transcendental extension $k \subseteq k(t)$. We can define a homomorphism $\varphi \colon k(t) \to k(t)$ by prescribing that $\varphi|_k = \mathrm{id}_k$ and that $\varphi(t) = t^2$. This homomorphism is not an isomorphism, since it is not surjective ($t \notin \varphi(k(t))$).

13.9 We have to prove that the set of all roots of all nonzero polynomials with rational coefficients is countable. Let P_d be the set of nonzero polynomials of degree d with rational coefficients. Note that P_d is countable: indeed, the set of polynomials of degree d is a $d + 1$-dimensional $\mathbb{Q}$-vector space, so it is a finite product of countable sets.

Now consider the subset R_d' of $P_d \times \mathbb{C}$ consisting of pairs $(f(t), z)$ where $f(t) \in P(t)$ and $z \in \mathbb{C}$ is a root of $f(t)$. Since every polynomial $f(t) \in P_d$ has at most d roots, R_d' is a countable union of finite sets of size at most d, hence R_d' is countable.

There is a projection $\pi \colon R_d' \to \mathbb{C}$, mapping a pair $(f(t), z)$ to z. The image $R_d = \pi(R_d')$ consists of all roots of all nonzero polynomials with rational coefficients. Since R_d' is countable, and $\pi \colon R_d' \to R_d$ is onto, it follows that R_d is countable.

Finally $\overline{\mathbb{Q}} = \cup_{d \geq 0} R_d$, therefore $\overline{\mathbb{Q}}$ is countable since it is a countable union of countable sets.

13.10 The polynomial $t^3 - 2 \in \mathbb{Q}[t]$ is irreducible (for example by Eisenstein's criterion); so every cube root α of 2 has degree 3 over $\mathbb{Q}$. If $\alpha \in F$, we would have $\mathbb{Q} \subseteq \mathbb{Q}(\alpha) \subseteq F$, therefore

$$2^r = [F : \mathbb{Q}] = [F : \mathbb{Q}(\alpha)][\mathbb{Q}(\alpha) : \mathbb{Q}] = [F : \mathbb{Q}(\alpha)] \cdot 3$$

by Proposition 13.13. Since $3 \nmid 2^r$, this is a contradiction.

13.15 The subfields $k(t^r)$ of $k(t)$ are all distinct, since $t^r \notin k(t^s)$ if $r < s$, and they give infinitely many intermediate fields of the transcendental extension $k \subseteq k(t)$.

13.16 We have the sequence of extensions

$$k \subseteq k(\alpha_1) \subseteq k(\alpha_1, \alpha_2) \subseteq \cdots \subseteq k(\alpha_1, \ldots, \alpha_n).$$

By (the evident generalization of) Proposition 13.13,

$$[F : k] = [k(\alpha_1, \ldots, \alpha_n) : k(\alpha_1, \ldots, \alpha_{n-1})] \cdots [k(\alpha_1, \alpha_2) : k(\alpha_1)][k(\alpha_1) : k].$$

Since all the factors on the right-hand side are finite by hypothesis, $[F : k]$ is finite.

13.19 We have to prove that every polynomial $f(t) \in E[t]$ has roots in E. Since $E \subseteq F$, we may view $f(t)$ as a polynomial in F; and since F is algebraically closed, $f(t)$ must have roots in F. Let $\alpha \in F$ be a root of $f(t)$. Since $f(t) \in E[t]$, α is algebraic over E (by Lemma 13.19). By Exercise 13.18, $\alpha \in E$ as needed.

13.20 We have seen how to construct a line perpendicular to the line through two points and containing one of them. Given two points A, B and the line through them, construct the two perpendicular lines ℓ_A, ℓ_B through A and B, respectively.

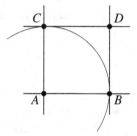

Construct the circle with center at A and through B. Let C be one of the two points of intersection of this circle with ℓ_A. Finally, construct the line through C and perpendicular to ℓ_A, and let D be the point of intersection of this line with ℓ_B. The points A, B, C, D are the vertices of a square.

Assuming we have constructed two lines through a point A, let B be any other constructed point on one of them; and let C be the point of intersection of the other line with the circle with center A and through B.

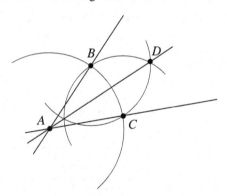

Let D be one of the points of intersection of the two circles with center at B through C and with center at C through B. Then the line through A and D bisects the given angle.

(There are many ways to perform such constructions, and these may very well not be the 'best' ones.)

13.21 First note that one can construct a line perpendicular to a given line L, through a point P not on L. Indeed (for example):

— The line L must contain another constructed point A.

— If the circle centered at P and through A is tangent to L, then the line through A and P is perpendicular to L, as needed.

— If not, this circle meets the line L at another point $B \neq A$.

— The circles through P centered at A and at B meet at another point $C \neq P$; the line through C and P is perpendicular to L.

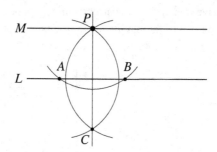

Next, we have seen how to construct a line perpendicular to a given line and through a point of the line. Applying this construction to the new line and P produces a line M parallel to a given line L and containing a point P not on L.

With this understood, let p and q be positive integers. Placing as usual the initial points O, P at $(0,0)$, $(0,1)$, we can

— Construct the line through O and perpendicular to the line through O and P; this is the y-axis.

— Construct the point $(0, 1)$ as one of the points of intersection of the y-axis with the circle centered at O and through P.

— Construct the point $A = (q, 0)$, by using the compass q times along the x-axis, as shown.

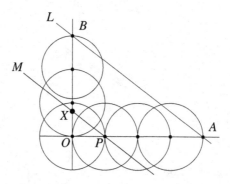

— Likewise construct the point $B = (0, p)$.

— Construct the line L through A and B.

— Construct the line M parallel to L and through P (as discussed earlier).

— And finally, let X be the point of intersection of M with the y-axis.

The point X has coordinates $\left(0, \frac{p}{q}\right)$, and this proves that $\frac{p}{q}$ is constructible.

13.24 Let $\alpha = \cos(20°) + i\sin(20°)$. Note that $\alpha = e^{\frac{2\pi i}{18}}$, and in particular $\alpha^9 = e^{\pi i} = -1$. This shows that α is a root of the polynomial $t^9 + 1$. Note that

$$t^9 + 1 = (t^3 + 1)(t^6 - t^3 + 1);$$

the roots of $t^3 + 1$ are -1 and $e^{\pm\frac{2\pi i}{3}}$, and in particular α is not one of them. It follows that α is a root of the polynomial $t^6 - t^3 + 1$. I claim that this polynomial is *irreducible* over $\mathbb{Q}$. One way to verify this is to note that setting $t = u - 1$ gives

$$(u - 1)^6 - (u - 1)^3 + 1 = u^6 - 6u^5 + 15u^4 - 21u^3 + 18u^2 - 9u + 3,$$

and this polynomial is irreducible by Eisenstein's criterion.

Therefore α has degree 6 over $\mathbb{Q}$. This proves that it is not constructible, by the result of Exercise 13.23.

If a $20°$ angle could be constructed, then both $\cos(20°)$ and $\sin(20°)$ would be constructible, therefore so would be α. So this again shows that the $60°$ angle cannot be trisected.

B.14 Normal and Separable Extensions, and Splitting Fields

14.2 If $t^2 + \alpha t + \alpha^2$ were reducible over $\mathbb{Q}(\alpha)$, then we could find its roots by using the quadratic formula—therefore its discriminant $\alpha^2 - 4\alpha^2 = -3\alpha^2$ would be a perfect square in $\mathbb{Q}(\alpha)$. It would follow that -3 is a perfect square, that is, $\sqrt{-3} \in \mathbb{Q}(\alpha)$. But then we would have extensions

$$\mathbb{Q} \subseteq \mathbb{Q}(\sqrt{-3}) \subseteq \mathbb{Q}(\alpha),$$

and by multiplicativity of degrees we would have

$$3 = [\mathbb{Q}(\alpha) : \mathbb{Q}] = [\mathbb{Q}(\alpha) : \mathbb{Q}(\sqrt{-3})] \, [\mathbb{Q}(\sqrt{-3}) : \mathbb{Q}] = [\mathbb{Q}(\alpha) : \mathbb{Q}(\sqrt{-3})] \cdot 2,$$

a contradiction (cf. Exercise 13.11).

For the last assertion: if any embedding of $\mathbb{Q}(\alpha)$ equaled another embedding, then it would contain two roots of $t^3 - 2$, hence a root of $t^2 + \alpha t + \alpha^2$, and the latter would be *reducible*, contradicting the first part.

14.5 Since $k \subseteq F$ is an algebraic extension, then $E \subseteq F$ is an algebraic extension. By Exercise 14.4, if E were algebraically closed, then we would necessarily have $E = F$. Contrapositively, E is not algebraically closed since $E \neq F$ by assumption.

14.6 Since $k \subseteq \bar{k}$ is algebraic by assumption, $E \subseteq \bar{k}$ is an algebraic extension. Therefore $\bar{k}$ is an algebraically closed field that is algebraic over E, and this makes it an algebraic closure of E by Definition 14.4.

14.12 Let $f(x) = \sum_{i \geq 0} a_i x^i$, $g(x) = \sum_{i \geq 0} b_i x^i$ (where $a_i = 0$ and $b_i = 0$ for $i \gg 0$). Then

$$f'(x) + g'(x) = \sum_{i \geq 0} a_i i x^{i-1} + \sum_{i \geq 0} b_i i x^{i-1} = \sum_{i \geq 0} (a_i + b_i) i x^{i-1} = (f + g)'(x).$$

More generally (with the same argument), $(\sum \lambda_i f_i)'(x) = \sum \lambda_i(f_i'(x))$ for all $\lambda_i \in k$ and all polynomials $f_i(x) \in k[x]$. It follows that

$$(fg)'(x) = \left(\sum_i a_i x^i g(x)\right)' = \sum_i a_i(x^i g(x))'$$

and

$$f'(x)g(x) + f(x)g'(x) = \sum_i a_i((x_i)'g(x) + x^i g'(x)).$$

Therefore, in proving the second identity we may assume that $f = x^i$. By the same token, we may assume $g = x^j$. In this case, the identity is evident:

$$(x^i x^j)' = (x^{i+j})' = (i + j)x^{i+j-1}$$

and

$$(x^i)'x^j + x^i(x^j)' = ix^{i-1}x^j + jx^i x^{j-1} = (i + j)x^{i+j-1}$$

agree.

14.13 The ring $k[u]$ is a UFD; the identity $f(u)^2 = ug(u)^2$ cannot hold because the irreducible factor u appears an even number of times on the left-hand side and an odd number of times on the right-hand side.

14.15 If $f(x)$ has a multiple root c in F, then $f(x) = (x - c)^2 g(x)$ in $F[x]$. It follows that $(f(x), f'(x)) \neq 1$ in $F[x]$, and this implies that $(f(x), f'(x)) \neq 1$ in $k[x]$ (cf. Remark 14.22). Therefore $f(x)$ is inseparable, by Proposition 14.21.

14.20 If k is finite, then *every* ring homomorphism $\varphi: k \to k$ is an isomorphism: φ is necessarily injective since all homomorphisms between fields are injective, and therefore we have $\varphi(k) \subseteq k$ and $|\varphi(k)| = |k|$, implying $\varphi(k) = k$, so φ is also surjective.

14.23 If $k \subseteq F$ is finite and separable, then it is simple by Theorem 14.24, and therefore $[F : k]_s = [F : k]$ by Corollary 14.25.

Conversely, assume that $k \subseteq F$ is finite and $[F : k]_s = [F : k]$. Given $\alpha \in F$, we have to prove that α is separable over k. By Exercise 14.22, $[F : k(\alpha)]_s \leq [F : k(\alpha)]$ and $[k(\alpha) : k]_s \leq [k(\alpha) : k]$. Under the assumption $[F : k]_s = [F : k]$, we have (using multiplicativity of both ordinary and separable degrees)

$$[F : k] = [F : k]_s = [F : k(\alpha)]_s[k(\alpha) : k]_s \leq [F : k(\alpha)][k(\alpha) : k] = [F : k].$$

It follows that the inequality in the middle must be an equality, and this implies that $[F : k(\alpha)]_s = [F : k(\alpha)]$ and $[k(\alpha) : k]_s = [k(\alpha) : k]$. The equality $[k(\alpha) : k]_s = [k(\alpha) : k]$ proves that α is separable as needed (cf. Example 14.27).

14.24 If $k \subseteq k(\alpha_1, \ldots, \alpha_r)$ is separable, then in particular each α_i is separable over k, by definition of separable extension.

For the converse, assume that each α_i is separable over k and consider the tower of extensions

$$k \subseteq k(\alpha_1) \subseteq \cdots \subseteq k(\alpha_1, \ldots, \alpha_r).$$

Each step is a simple extension

$$k(\alpha_1, \ldots, \alpha_{i-1}) \subseteq k(\alpha_1, \ldots, \alpha_{i-1})(\alpha_i) \,;$$

since each α_i is separable over k, it is separable over $k(\alpha_1, \ldots, \alpha_{i-1})$ (its minimal poly-nomial over the larger field is a factor of the minimal polynomial over k, so it does not have multiple roots in a splitting field). It follows that

$$[k(\alpha_1, \ldots, \alpha_i) : k(\alpha_1, \ldots, \alpha_{i-1})]_s = [k(\alpha_1, \ldots, \alpha_i) : k(\alpha_1, \ldots, \alpha_{i-1})]$$

(cf. Example 14.27). By multiplicativity of both ordinary and separable degrees we deduce that

$$[k(\alpha_1, \ldots, \alpha_r) : k]_s = [k(\alpha_1, \ldots, \alpha_r) : k] \,,$$

and this implies that the extension is separable by Exercise 14.23.

14.27 • Assuming that $e \mid d$, let $y = x^e$; so $x^d = y^r$ for $r = \frac{e}{d}$. The statement is then that $y - 1$ divides $y^r - 1$, which is evident.

• Letting $x = p$ in the first point gives that $p^e - 1$ divides $p^d - 1$ if $e \mid d$.

• Since now we have $(p^e - 1) \mid (p^d - 1)$, we can apply the first point again and obtain that

$$(x^{p^e-1} - 1) \quad \text{divides} \quad (x^{p^d-1} - 1) \,.$$

Multiplying through by x gives the third point.

14.28 One can proceed in the same way as in Example 14.38, and use Proposition 14.36 to count irreducible polynomials of degree d for increasing values of d. Since we are interested in degree 12, and the divisors of 12 are $1, 2, 3, 4$, and 6, we focus on these cases. Let I_d be the number of monic irreducible polynomials of degree d in $\mathbb{F}_7[x]$.

• $d = 1$: There are seven monic irreducible polynomials of degree 1, namely $x - a$ as $a \in \mathbb{F}_7$. Therefore $I_1 = 7$.

• $d = 2$: By Proposition 14.36, the factorization of $x^{7^2} - x$ in $\mathbb{F}_7[x]$ consists of all irreducible polynomials of degree 1 and 2. Therefore $I_1 + 2I_2 = 7^2$, and it follows that $I_2 = \frac{7^2 - 7}{2} = 21$.

• $d = 3$: Again applying Proposition 14.36, we obtain $7^3 = I_1 + 3I_3$, and therefore $I_3 = \frac{7^3 - 7}{3} = 112$.

• $d = 4$: The same method shows that $7^4 = I_1 + 2I_2 + 4I_4$, from which

$$I_4 = \frac{7^4 - 7 - 2 \cdot 21}{4} = 588 \,.$$

• $d = 6$: $7^6 = I_1 + 2I_2 + 3I_3 + 6I_6$ gives

$$I_6 = \frac{7^6 - 7 - 2 \cdot 21 - 3 \cdot 112}{4} = 19544 \,.$$

• $d = 12$: Finally, $7^{12} = I_1 + 2I_2 + 3I_3 + 4I_4 + 6I_6$, and therefore

$$I_{12} = \frac{7^{12} - 7 - 2 \cdot 21 - 3 \cdot 112 - 4 \cdot 588 - 6 \cdot 19544}{12} = 1153430600 \,.$$

This is actually a little more work than strictly necessary, but it's nice to get the various number I_j as subproducts of the computation. You can find the sequence $I_1, I_2, I_3, \ldots$:

$$7, 21, 112, 588, 3360, 19544, 117648, 720300, 4483696, 28245840,$$
$$179756976, 1153430600, 7453000800, 48444446376, 316504099520, \ldots$$

in the *Online Encyclopedia of Integer Sequences,* sequence A001693. I have underlined the terms we have computed; isn't it nice that you could compute the others as well? And if you look the sequence up, you will learn that it may be described as the *number of aperiodic necklaces with n beads of 7 colors.* Isn't this marvelous?

B.15   Galois Theory

15.1 Let $g(x)$ be the polynomial obtained from the same factors in an irreducible decomposition of $f(x)$, each taken to the first power. Different irreducible factors are relatively prime over k, hence over the splitting field of $f(x)$ (Exercise 6.26), and in particular they do not have a common factor $x - \alpha$ over the splitting field; that is, they do not share roots in the splitting field. Each irreducible factor is separable by the hypothesis on the characteristic (Proposition 14.23), therefore $g(x)$ is a separable polynomial with the same roots, and hence the same splitting field, as $f(x)$.

15.2 Since 0 is fixed by every automorphism, F^G is not empty. We have to verify that F^G is closed under subtraction and multiplication, and that every nonzero element is a unit in F^G. If $a, b \in F^G$, then for all $\gamma \in G$, $\gamma a = a$ and $\gamma b = b$. It follows that

$$\gamma(a - b) = \gamma(a) - \gamma(b) = a - b \quad \text{and} \quad \gamma(a \cdot b) = \gamma(a) \cdot \gamma(b) = a \cdot b$$

since γ is a ring homomorphism. This ensures that $a - b \in F^G$ and $a \cdot b \in F^G$. Therefore F^G is a subring of F. If $c \neq 0$ is in F^G, then c^{-1} exists in F as F is a field; and for all $\gamma \in G$ we have

$$\gamma(c^{-1}) = \gamma(c)^{-1} = c^{-1}$$

again because γ is a homomorphism. Therefore $c^{-1} \in F^G$, and we can conclude that F^G is a subfield of F. Since every element of $\mathrm{Gal}_k(F)$ restricts to the identity on k, we have $k \subseteq F^G$, confirming that F^G is an intermediate field of the extension $k \subseteq F$.

15.3 Note that $(x - 1)f(x) = x^p - 1$: thus, the roots of $f(x)$ consist of all pth roots of 1 except 1 itself. Letting $\zeta_p = e^{2\pi i/p}$, the roots are $\zeta_p, \zeta_p^2, \ldots, \zeta_p^{p-1}$. It follows that the splitting field F of $f(x)$ is $\mathbb{Q}(\zeta_p)$, and $|\mathrm{Gal}_\mathbb{Q}(F)| = [\mathbb{Q}(\zeta_p) : \mathbb{Q}] = p - 1$.

By Proposition 15.5, elements $\gamma \in \mathrm{Gal}_\mathbb{Q}(F)$ are determined by $\gamma(\zeta_p)$, and by Proposition 15.6, $\gamma(\zeta_p)$ must be another root of $f(x)$. By Proposition 14.1, every choice of a root will determine an element of the Galois group. Therefore $\mathrm{Gal}_\mathbb{Q}(F)$ consists of the $p - 1$ automorphisms γ_i determined by $\zeta_p \mapsto \zeta_p^i$, $i = 1, \ldots, p - 1$. With this notation, γ_1 is the identity.

Define a function $\varphi: (\mathbb{Z}/p\mathbb{Z})^* \to \mathrm{Gal}_{\mathbb{Q}}(F)$ by mapping $[i]_p$, $i = 1, \ldots, p - 1$, to γ_i. Then φ is a group homomorphism: indeed, if $ij \equiv k \bmod p$, with $1 \leq i, j, k \leq p - 1$, then

$$(\gamma_i \circ \gamma_j)(\zeta_p) = (\zeta_p^j)^i = \zeta_p^{ij} \overset{!}{=} \zeta_p^k = \gamma_k(\zeta_p),$$

where $\overset{!}{=}$ holds since $\zeta_p^p = 1$. The kernel of φ is $[1]_p$, therefore φ is injective. Both groups have $p - 1$ elements, so φ is an isomorphism, as required.

15.4 This is really a straightforward modification of the argument given for Exercise 15.3. Let $\zeta_n = e^{2\pi i/n}$, so that ζ_n is an nth root of 1 and the roots of $f(x) = x^n - 1$ are $\zeta_n^0 = 1, \zeta_n, \zeta_n^2, \ldots, \zeta_n^{n-1}$; these form a cyclic subgroup $C \cong \mathbb{Z}/n\mathbb{Z}$ of $\mathbb{C}^*$, and ζ_n is a generator of this subgroup (cf. Example 15.43). The splitting field F of $f(x)$ is $\mathbb{Q}(\zeta_n)$. The size of $\mathrm{Gal}_{\mathbb{Q}}(F)$ equals $[\mathbb{Q}(\zeta_n) : \mathbb{Q}]$, that is, the degree of the nth cyclotomic polynomial.

Every element γ of $\mathrm{Gal}_{\mathbb{Q}}(F)$ is determined by where it maps ζ_n; and $\gamma(\zeta_n)$ must be another root of $x^n - 1$, therefore $\gamma(\zeta_n) = \zeta_n^i$ for some i. The image $\gamma(\zeta_n)$ must be another generator for F over $\mathbb{Q}$; that is, ζ_n^i must be another generator for the cyclic group C. Therefore, $[i]_n$ must be a generator of $\mathbb{Z}/n\mathbb{Z}$; equivalently, $[i]_n \in (\mathbb{Z}/n\mathbb{Z})^*$ (Proposition 10.9).

This shows that $\mathrm{Gal}_{\mathbb{Q}}(F)$ consists of the automorphisms γ_i determined by $\zeta \mapsto \zeta^i$, where i is such that $[i]_n \in (\mathbb{Z}/n\mathbb{Z})^*$, i.e., $0 < i < n$ and $\gcd(i, n) = 1$.

Define then a function $\varphi: (\mathbb{Z}/n\mathbb{Z})^* \to \mathrm{Gal}_{\mathbb{Q}}(F)$ by mapping $[i]_n$ to γ_i. The same argument given at the end of the solution to Exercise 15.3 proves that φ is an isomorphism, and we are done.

15.6 By definition of Galois conjugates (Definition 15.22), the conjugates of α form its orbit under the action of the Galois group.

Therefore, by Theorem 12.10, the number of conjugates equals the index of the stabilizer of α, that is, of the subgroup fixing $k(\alpha)$. This subgroup is precisely $\mathrm{Gal}_{k(\alpha)}(F)$, so indeed the number of conjugates of α equals $[\mathrm{Gal}_k(F) : \mathrm{Gal}_{k(\alpha)}(F)]$ as stated.

As for the last assertion, $F = k(\alpha)$ if and only if $\mathrm{Gal}_{k(\alpha)}(F)$ is trivial, so this happens precisely when the number of conjugages equals $[\mathrm{Gal}_k(F) : \{e\}] = |\mathrm{Gal}_k(F)|$, which equals $[F : k]$ since the extension is Galois.

15.7 If $k \subseteq F$ is Galois, then it is a splitting field of a separable polynomial f(x). By Remark 14.12, the image of any homomorphism $F \to \bar{k}$ extending the inclusion $k \subseteq K$ must be the subfield of $\bar{k}$ generated by the roots of $f(x)$, so it is independent of the homomorphism.

Conversely, let $k \subseteq F$ be a finite separable extension, and assume that it is *not* Galois. By the primitive element theorem, $F = k(\alpha)$ for some α; let $f(x)$ be the minimal polynomial of α. Then the extension is not Galois precisely if $f(x)$ does not split completely in F. In this case, if $\varphi: F \to \bar{k}$ is any homomorphism extending the inclusion $k \subseteq \bar{k}$, then some root β of $f(x)$ in $\bar{k}$ is not contained in $\varphi(F)$. We can then define a different homomorphism $\psi: F \to \bar{k}$ extending $k \subseteq \bar{k}$ by mapping α to β, and by construction $\psi(F) \neq \varphi(F)$.

Contrapositively, if all homomorphisms of the extension $k \subseteq F$ to the extension $k \subseteq \bar{k}$ have the same image, then the extension is Galois, and this completes the proof.

15.8 The two functions are σ, defined by $\sigma(E) = \mathrm{Gal}_E(F)$ (viewed as a subgroup of $\mathrm{Gal}_k(F)$), and τ, defined by $\tau(G) = F^G$.

Let $E_1 \subseteq E_2$ be intermediate fields, and let $\gamma \in \mathrm{Gal}_{E_2}(F)$. Then γ is an automorphism of F restricting to the identity on E_2. Since $E_1 \subseteq E_2$, γ restricts to the identity on E_1; this shows that $\gamma \in \mathrm{Gal}_{E_1}(F)$. Thus $\sigma(E_2) \subseteq \sigma(E_1)$.

Let $G_1 \subseteq G_2$ be subgroups of $\mathrm{Gal}_k(F)$, and let $c \in F^{G_2}$. Then for all $\gamma \in G_2$, $\gamma(c) = c$. Since $G_1 \subseteq G_2$, it follows that for all $\gamma \in G_1$, $\gamma(c) = c$; this shows that $c \in F^{G_1}$. Thus $F^{G_2} \subseteq F^{G_1}$, i.e., $\tau(G_2) \subseteq \tau(G_1)$, as required.

15.19 By the result of Exercise 10.25, we know that if G is a finite abelian group, then there exists a positive integer n and a subgroup K of $(\mathbb{Z}/n\mathbb{Z})^*$ such that G is isomorphic to the quotient $(\mathbb{Z}/n\mathbb{Z})^*/K$.

Now let F be the nth cyclotomic extension, i.e., the splitting field of $x^n - 1$ over $\mathbb{Q}$. By Exercise 15.4, $\mathrm{Gal}_{\mathbb{Q}}(F) \cong (\mathbb{Z}/n\mathbb{Z})^*$. Let $E = F^K$ be the fixed field of K. By the second part of the Fundamental Theorem of Galois Theory (Theorem 15.35), the extension $\mathbb{Q} \subseteq E$ is Galois (since $(\mathbb{Z}/n\mathbb{Z})^*$ is abelian, so K is automatically normal) and

$$\mathrm{Gal}_{\mathbb{Q}}(E) = \mathrm{Gal}_{\mathbb{Q}}(F)/\mathrm{Gal}_E(F) \cong (\mathbb{Z}/n\mathbb{Z})^*/K \cong G$$

as needed.

15.22 By Proposition 14.1, if α and β are any two roots of $f(x)$, then there exists an isomorphism $k(\alpha) \to k(\beta)$ extending the identity on k. By Corollary 14.11, this isomorphism extends to an automorphism φ of the splitting field F of $f(x)$, and by construction $\varphi(\alpha) = \beta$.

Thus any two roots of $f(x)$ are in the same orbit under the Galois action, and this is the statement.

Note: This result is also encapsulated within Remark 15.17. Taking $G = \mathrm{Gal}_k(F)$ in that remark, we see that in fact $f(x)$ splits in $F[x]$ as a constant times the product of factors $x - \gamma(\alpha)$, where α is any root and γ ranges over $\mathrm{Gal}_k(F)$. This implies that the action is transitive. The roots of $f(x)$ are precisely the Galois conjugates of any one root.

15.27 The action of the Galois group G of $f(x)$ on the roots of $f(x)$ identifies G with a subgroup of S_p. Since $f(x)$ has rational coefficients, it is invariant under complex conjugation; so conjugation acts on its roots, and it may be viewed as an element ι of G. By hypothesis, ι swaps exactly two roots, i.e., it acts as a transposition in S_p. Therefore G contains a transposition.

On the other hand, G acts transitively on $\{1, \ldots, p\}$. By Exercise 12.7, G contains a p-cycle. Therefore G contains a p-cycle and a transposition, and by Exercise 11.29 we can conclude that $G = S_p$.

The polynomial $x^5 - 8x + 2 \in \mathbb{Q}[x]$ is irreducible by Eisenstein, and verifying that it has exactly three real roots is a calculus exercise. By the first part, the Galois group of this polynomial is S_5.

15.28 • The polynomial $x^m - D$ has at most m roots. The m elements $u, \zeta u, \ldots, \zeta^{m-1}u$ are roots of $x^m - D$ and are distinct, since ζ is a *primitive* mth root of 1, so they must be all of the roots.

• It follows that $x^m - D$ is separable and it splits completely in $E(u)$; since $E(u)$ is generated over E by the roots of this polynomial (in fact, it is already generated by the single root u), $E(u)$ is a splitting field for the separable polynomial $x^m - D \in E[x]$. Therefore the extension $E \subseteq E(u)$ is Galois.

- Since u is a root of $x^m - D$, $x^m - D$ must be a multiple of the minimal polynomial of u. By hypothesis, $[E(u) : E] = m$; it follows that the minimal polynomial of u has degree m, and therefore it must equal $x^m - D$. In particular, $x^m - D$ is irreducible.

- Since $x^m - D$ is irreducible, we have a homomorphism $E(u) \to E(u)$ mapping u to any other root of $x^m - D$, by Proposition 14.1. Thus there is an element $\sigma_i \in \mathrm{Gal}_E(E(u))$ mapping u to $\zeta^i u$, for every $i = 0, \dots, m - 1$.

- Define a function $\varphi \colon \mathbb{Z}/m\mathbb{Z} \to \mathrm{Gal}_E(E(u))$ by mapping $[i]_m$, $i = 0, \dots, m - 1$ to the automorphism σ_i.

This function is a group homomorphism: for i, j, k such that $i + j \equiv k \bmod m$ and $0 \le i, j, k \le m - 1$,

$$(\sigma_i \circ \sigma_j)(u) = \zeta^i(\zeta^j u) = \zeta^{i+j} u \overset{!}{=} \zeta^k u = \sigma_k(u),$$

where $\overset{!}{=}$ uses the fact that $\zeta^m = 1$. The kernel of φ is $[0]_m$, so φ is injective. Both $\mathbb{Z}/m\mathbb{Z}$ and $\mathrm{Gal}_E(E(u))$ consist of m elements, so φ must be an isomorphism.

15.30 By Theorem 15.56, the regular n-gon can be constructed by straightedge and compass if and only if the totient function $\phi(n)$ of n is a power of 2. We have to verify that this is the case if and only if n is a power of 2 times a product of distinct Fermat primes. Recall that a 'Fermat prime' is a prime of the form $2^{(2^r)} + 1$.

By Proposition 5.56, Euler's totient function is multiplicative in the sense that if a and b are relatively prime, then $\phi(ab) = \phi(a)\phi(b)$. It follows that if the prime factorization of n is $p_1^{a_1} \cdots p_m^{a_m}$, then $\phi(n)$ is a power of 2 if and only if each $\phi(p_i^{a_i})$ is a power of 2. Therefore, we are reduced to proving that if $p > 0$ is prime, and $a > 0$ is an integer, then $\phi(p^a)$ is a power of 2 if and only if $p = 2$ or $a = 1$ and p is a Fermat prime.

Recall that $\phi(p^a) = p^{a-1}(p-1)$: this was verified in Example 10.11, and you probably had figured it our already when you solved Exercise 5.33. It follows that $\phi(p^a)$ is a power of 2 if and only if (i) p^{a-1} is a power of 2 and (ii) $p - 1$ is a power of 2.

If $p = 2$, both (i) and (ii) are clearly satisfied.

If $p = 2^{(2^r)} + 1$ is a Fermat prime and $a = 1$, then necessarily (i) is satisfied as $p^{a-1} = p^0 = 1$, and so is (ii) since $p - 1 = 2^{(2^r)}$ is a power of 2.

This establishes one implication.

For the other implication, assume that (i) and (ii) are satisfied. Then by (i) either $p = 2$ or $a = 1$. If $p = 2$, we are done. If $p > 2$, so that $a = 1$ by (i), then (ii) implies that $p = 1 + 2^s$ for some integer $s > 0$.

Therefore, we are reduced to showing that if $1 + 2^s$ is prime, with $s > 0$, then necessarily s is a power of 2. Let then $s = 2^r \cdot c$, with c odd. Since

$$2^s + 1 = 2^{2^r c} + 1 = (2^{2^r} + 1)(2^{2^r(c-1)} - 2^{2^r(c-2)} + \cdots + 1),$$

we see that if $c > 1$, so that $(2^{2^r(c-1)} - 2^{2^r(c-2)}) + \cdots + 1 > 1$, then this number is not irreducible.

Therefore $c = 1$, $s = 2^r$ is a power of 2, and this completes the argument.

Index of Definitions

Index of Theorems

Subject Index

Printed in the United States
by Baker & Taylor Publisher Services